D0860359

BASIC ALGEBRA

Richard G. Brown
Geraldine D. Smith
Mary P. Dolciani

Editorial Advisers: Andrew M. Gleason, Robert H. Sorgenfrey

Houghton Mifflin Company • Boston
Atlanta Dallas Geneva, Illinois
Lawrenceville, New Jersey Palo Alto Toronto

The Authors

Richard G. Brown, Mathematics teacher at The Phillips Exeter Academy, Exeter, New Hampshire. Mr. Brown has taught a wide range of mathematics courses for both students and teachers at several schools and universities, including Newton (Massachusetts) High School, the University of New Hampshire, Arizona State University, and the North Carolina School for Science and Mathematics.

Geraldine D. Smith, well known both in publishing and in mathematics education for many years. Miss Smith has done extensive work in developing and directing a wide variety of major elementary and secondary mathematics programs. She also serves as a special consultant for many other educational materials.

Mary P. Dolciani, formerly Professor of Mathematical Sciences, Hunter College of the City University of New York. Dr. Dolciani wrote many mathematics texts, both elementary and secondary, and developed programs of basic mathematics review for beginning college students.

The Editorial Advisers

Andrew M. Gleason, Hollis Professor of Mathematics and Natural Philosophy, Harvard University, Cambridge, Massachusetts. Professor Gleason is a well-known research mathematician. His many affiliations include membership in the National Academy of Sciences.

Robert H. Sorgenfrey, Professor of Mathematics, University of California, Los Angeles. Dr. Sorgenfrey has won the Distinguished Teaching Award at U.C.L.A. and has been Chairman of the Committee on Teaching there.

Printed in U.S.A.

ISBN: 0-395-41188-2

ABCDEFGHIJ–RM–9543210/8987

CONTENTS

UNIT A

UNIT B

UNIT C

CHAPTER 5 Working with Polynomials

CHAPTER 6 Factoring

UNIT D

CHAPTER 7 Graphs

CHAPTER 8 Equations with Two Variables

UNIT E

CHAPTER 9 Working with Fractions

CHAPTER 10 Decimals and Percents

UNIT F

CHAPTER 11 Squares and Square Roots

CHAPTER 12 Quadratic Equations

Diagnostic Tests in Arithmetic

1 Whole Numbers—Addition

1. $\begin{array}{r}6\\+9\\\hline\end{array}$ **2.** $\begin{array}{r}2\\2\\+3\\\hline\end{array}$ **3.** $\begin{array}{r}4\\6\\+7\\\hline\end{array}$ **4.** $\begin{array}{r}55\\+25\\\hline\end{array}$ **5.** $\begin{array}{r}68\\+15\\\hline\end{array}$

6. $\begin{array}{r}575\\+371\\\hline\end{array}$ **7.** $\begin{array}{r}778\\+177\\\hline\end{array}$ **8.** $\begin{array}{r}814\\857\\+311\\\hline\end{array}$ **9.** $\begin{array}{r}7456\\3701\\6101\\+5592\\\hline\end{array}$

2 Whole Numbers—Subtraction

1. $\begin{array}{r}7\\-2\\\hline\end{array}$ **2.** $\begin{array}{r}11\\-9\\\hline\end{array}$ **3.** $\begin{array}{r}97\\-54\\\hline\end{array}$ **4.** $\begin{array}{r}769\\-608\\\hline\end{array}$ **5.** $\begin{array}{r}684\\-308\\\hline\end{array}$

6. $\begin{array}{r}736\\-365\\\hline\end{array}$ **7.** $\begin{array}{r}6508\\-4395\\\hline\end{array}$ **8.** $\begin{array}{r}643\\-177\\\hline\end{array}$ **9.** $\begin{array}{r}4220\\-3686\\\hline\end{array}$ **10.** $\begin{array}{r}40{,}704\\-27{,}978\\\hline\end{array}$

3 Whole Numbers—Multiplication

1. $\begin{array}{r}6\\\times 6\\\hline\end{array}$ **2.** $\begin{array}{r}21\\\times 4\\\hline\end{array}$ **3.** $\begin{array}{r}28\\\times 3\\\hline\end{array}$ **4.** $\begin{array}{r}66\\\times 8\\\hline\end{array}$ **5.** $\begin{array}{r}375\\\times 3\\\hline\end{array}$ **6.** $\begin{array}{r}69\\\times 20\\\hline\end{array}$

7. $\begin{array}{r}125\\\times 200\\\hline\end{array}$ **8.** $\begin{array}{r}1291\\\times 2000\\\hline\end{array}$ **9.** $\begin{array}{r}96\\\times 65\\\hline\end{array}$ **10.** $\begin{array}{r}734\\\times 29\\\hline\end{array}$ **11.** $\begin{array}{r}825\\\times 955\\\hline\end{array}$ **12.** $\begin{array}{r}801\\\times 203\\\hline\end{array}$

4 Whole Numbers—Division

1. $9\overline{)54}$ **2.** $2\overline{)42}$ **3.** $8\overline{)448}$ **4.** $5\overline{)2285}$ **5.** $13\overline{)455}$ **6.** $16\overline{)216}$

7. $568\overline{)416090}$ **8.** Express the quotient as a mixed numeral: $\begin{array}{r}592\text{ R41}\\482\overline{)285385}\end{array}$

9. Express the quotient to the nearest hundredth: $6\overline{)38}$

5 Fractions—Basic Skills

1. Identify by letter the figure that is divided into thirds.

A

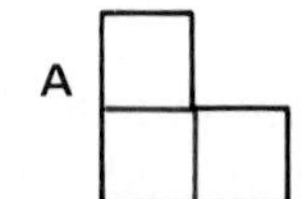

B

C

2. What is the denominator in $\frac{1}{3}$?

3. The area shaded in figure D is represented by which fraction in row E?

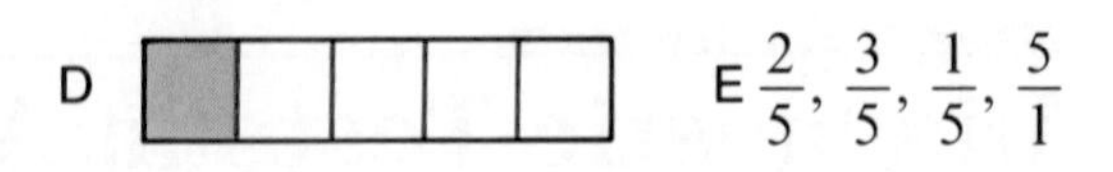

4. Which of the figures below represents the fraction $\frac{1}{3}$?

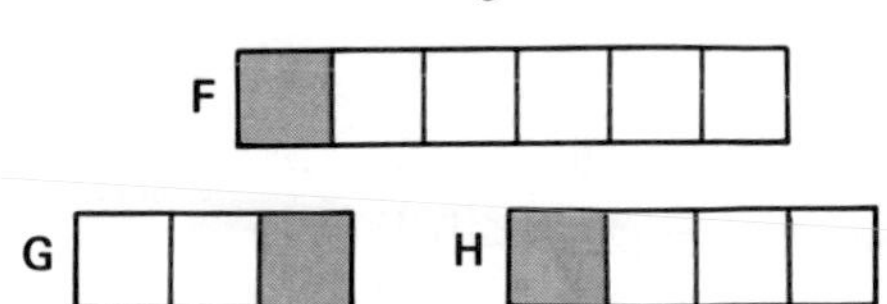

5. Write the fraction represented by the set diagram shown below.

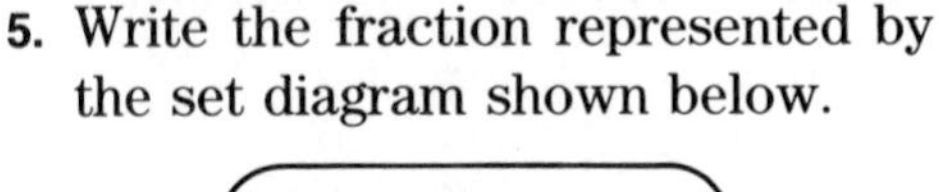

6. Identify by letter the set diagram at the right that represents the fraction $\frac{2}{6}$.

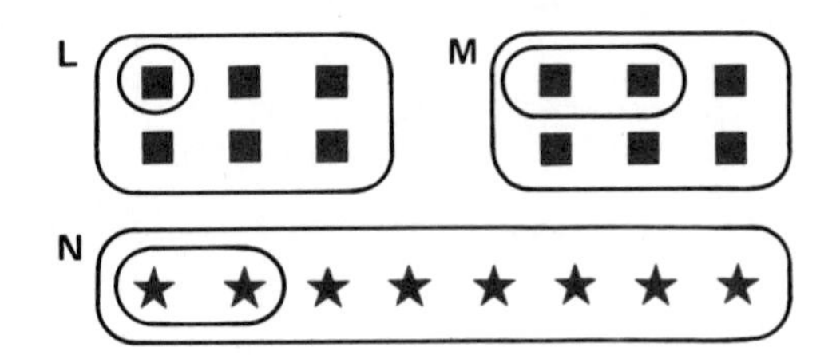

7. Which fraction represents the number 1? $\frac{4}{1}, \frac{4}{4}, \frac{4}{8}, \frac{4}{4}$

8. Which fraction represents C on the number line? $\frac{5}{3}, \frac{3}{2}, \frac{3}{3}, \frac{5}{2}$

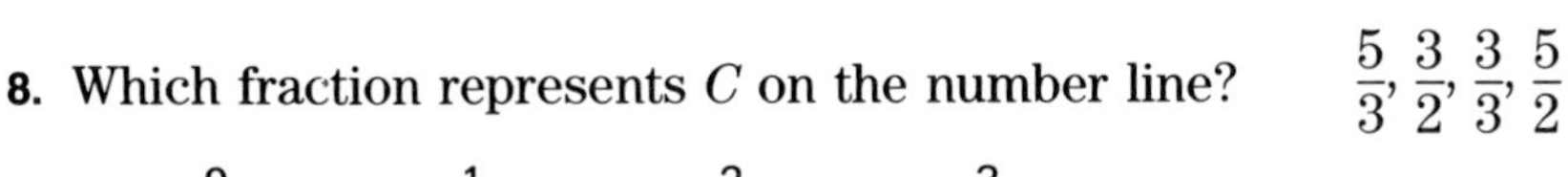

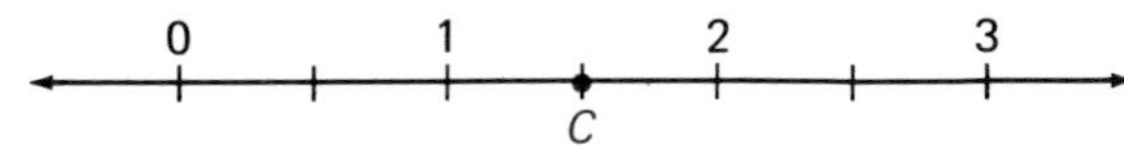

9. Write the consecutive multiples of 6: 6, __?__, __?__, __?__

10. Find the least common multiple of 3, 4, and 6.

11. Write 36 as the product of prime factors.

12. Find the greatest common factor of 12 and 28.

13. Find the fraction in row Y that is equal to each fraction or mixed numeral in row X.

X: (a) $\frac{4}{6}$ (b) $\frac{3}{5}$ (c) $2\frac{3}{4}$

Y: $\frac{2}{2}, \frac{6}{4}, \frac{11}{4}, \frac{3}{4}, \frac{9}{15}, \frac{6}{9}$

14. Find the least common denominator for the fractions $\frac{2}{4}, \frac{1}{6}$, and $\frac{7}{8}$.

15. Find the fraction or mixed numeral in row Y that is equal to each fraction or mixed numeral in row X.

X: (a) $\frac{5}{4}$ (b) $2\frac{3}{5}$

Y: $\frac{1}{4}, 1\frac{1}{4}, \frac{13}{5}, \frac{5}{13}$

6 Fractions—Addition

1. $\frac{1}{5} + \frac{2}{5} = \underline{\ ?\ }$ 2. $3\frac{1}{8} + 4\frac{1}{8} = \underline{\ ?\ }$ 3. $\begin{array}{r} 3\frac{7}{12} \\ +4\frac{11}{12} \\ \hline \end{array}$ 4. $\frac{1}{5} + \frac{1}{4} = \underline{\ ?\ }$ 5. $\begin{array}{r} 4\frac{2}{3} \\ +3\frac{4}{7} \\ \hline \end{array}$

7 Fractions—Subtraction

1. $\begin{array}{r} \frac{2}{4} \\ -\frac{1}{4} \\ \hline \end{array}$ 2. $\begin{array}{r} 3\frac{3}{8} \\ -1\frac{1}{8} \\ \hline \end{array}$ 3. $\begin{array}{r} 7 \\ -5\frac{1}{6} \\ \hline \end{array}$ 4. $\begin{array}{r} 7\frac{1}{9} \\ -1\frac{7}{9} \\ \hline \end{array}$ 5. $\begin{array}{r} \frac{2}{3} \\ -\frac{1}{7} \\ \hline \end{array}$ 6. $\begin{array}{r} 14\frac{4}{6} \\ -3\frac{4}{5} \\ \hline \end{array}$

8 Fractions—Multiplication

1. $\frac{1}{3} \times \frac{1}{9} = \underline{\ ?\ }$ 2. $\frac{7}{8} \times \frac{6}{7} = \underline{\ ?\ }$ 3. $\frac{1}{2} \times 4 = \underline{\ ?\ }$

4. $4\frac{5}{6} \times \frac{3}{5} = \underline{\ ?\ }$ 5. $1\frac{3}{5} \times 2\frac{1}{2} = \underline{\ ?\ }$

9 Fractions—Division

1. $5 \div \frac{1}{9} = \underline{\ ?\ }$ 2. $6 \div \frac{3}{4} = \underline{\ ?\ }$ 3. $\frac{1}{6} \div \frac{1}{2} = \underline{\ ?\ }$ 4. $5\frac{5}{9} \div \frac{2}{7} = \underline{\ ?\ }$

5. $5\frac{1}{8} \div 4 = \underline{\ ?\ }$ 6. $5\frac{5}{8} \div 5\frac{5}{6} = \underline{\ ?\ }$

10 Decimals

1. Write the decimal which represents "two and six thousandths."

2. $\frac{5}{1000} = \underline{\ ?\ }$ (Decimal) 3. $0.206 = \underline{\ ?\ }$ (Fraction) 4. $3\frac{4}{5} = \underline{\ ?\ }$ (Decimal)

5. Round 40.5656 to the nearest tenth. 6. $1.033 + 0.1 + 10.066 = \underline{\ ?\ }$

7. $856.175 - 20.05 = \underline{\ ?\ }$ 8. $\begin{array}{r} 71.60 \\ 500.42 \\ +53.96 \\ \hline \end{array}$ 9. $\begin{array}{r} 735.02 \\ -7.709 \\ \hline \end{array}$ 10. $\begin{array}{r} 6639 \\ \times 0.0031 \\ \hline \end{array}$ 11. $0.07\overline{)21.707}$

11 Percents

1. $\frac{7}{4} = \frac{?}{100}$ 2. $0.69 = \underline{\ ?\ }\%$ 3. $\frac{1}{3} = \underline{\ ?\ }\%$ 4. $6\frac{1}{2} = \underline{\ ?\ }\%$

5. $2.405 = \underline{\ ?\ }\%$ 6. $38\frac{1}{2}\% = \underline{\ ?\ }$ (Decimal)

7. 25% of $23 = \underline{\ ?\ }$ 8. $\underline{\ ?\ }\%$ of $20 = 8$ 9. 20% of $\underline{\ ?\ } = 340$

SYMBOLS

Symbol	Meaning	Page
$=$	equals, is equal to	5
$\cdot$	$\times$ (times)	5
( )	parentheses—a grouping symbol	8
$<$	is less than	27
$>$	is greater than	27
$\neq$	is not equal to	27
$+$	positive	78
$-$	negative	78
$\circ$	degree	78
-8	opposite of 8	85
$\|6\|$	absolute value of 6	99
π	pi, a Greek letter standing for a value about 3.14 or $\frac{22}{7}$	190
$(3, -2)$	ordered pair describing a point	237
$\approx$	is approximately equal to	350
$\sqrt{\ }$	positive square root	388
$\pm$	read: positive or negative	388

METRIC SYSTEM SYMBOLS

Symbol	Meaning
mm	millimeter
cm	centimeter
m	meter
km	kilometer
km/h	kilometers per hour
cm^2	square centimeter
km^2	square kilometer
cm^3	cubic centimeter
g	gram
kg	kilogram
L	liter
°C	degrees Celsius
s	second
min	minute
h	hour

READING ALGEBRA

HOW TO READ YOUR ALGEBRA TEXTBOOK

To read a textbook with understanding you need some special skills that you might not use in reading a story. You should not hurry through the lessons or do any skipping. Read slowly and think about each sentence. Make notes and draw diagrams if necessary. This book has been planned to help you become a good reader of mathematics.

VOCABULARY

You will find many new words in this book. Some of these words, such as *coefficient* and *monomial,* belong to the special vocabulary of algebra. Others, such as *power* and *variable,* are everyday words that have a special meaning in mathematics. Important words, whose meanings you will learn, are printed in heavy type when they first appear (see *variable* and *variable expressions* on page 5, for example). If you do not recall the meaning of a certain word, you can look it up in the Glossary or the Index at the back of the book. The Glossary gives definitions, and the pages listed in the Index usually provide more information.

SYMBOLS

In algebra you will use the familiar symbols of arithmetic and you will learn some new symbols. By using symbols you can often say a great deal in a small space. Look at page 27, for example. The symbols $<$ and $>$ are great time-savers. You would soon get tired of writing out "is less than" and "is greater than" in inequality statements; the symbols replace these phrases. If you need to look up the meaning of any symbol, you can use the list on the facing page.

DIAGRAMS

Diagrams are used in this book to help you understand important ideas that are being discussed. For example, look at the drawings of scales on page 38. These diagrams help you see the relationship between the two sides of an equation and the need for keeping the sides in balance as you solve the equation. Be sure to study the diagrams along with the text.

DISPLAYED MATERIAL

Throughout the book important information is displayed in red boxes. This information includes properties, rules, formulas, and summaries of methods. Be sure to read all displayed material carefully and to return to it when you are reviewing.

READING AIDS

If you turn to pages 4 and 5, you will see that many items have notes in red boxes. Notes of this kind are used throughout to call your attention to important parts of sentences, expressions, and diagrams, and to provide brief explanations. Sometimes the red boxes contain suggestions for solving an example or directions to guide you through a solution step by step. Use the boxes as "highway signs" that show you the right direction.

Additional help in reading is provided by sections called Reading Algebra throughout the book. Be sure to study these sections; they will help you with important topics such as solving word problems.

EXERCISES, TESTS, AND REVIEWS

Most of the lessons in this book have two sets of exercises: Classroom Practice and Written Exercises. At the end of each chapter, the Chapter Review and Chapter Test provide review exercises for each lesson. There is a Cumulative Review at the end of each unit, and five more Cumulative Reviews appear at the back of the book. Each odd-numbered chapter contains a Mixed Review and every chapter contains a Skills Review. You can use the reviews and tests to check your progress and to discover your strengths and weaknesses. The page references in the Chapter Reviews and the section references in the Chapter Tests will help you locate sections that you may need to reread.

Another good way to keep track of your progress is to use the Self-Tests that appear throughout the chapters. Answers to the Self-Tests are given at the back of the book. If you find that you need more practice on any topic, turn to the Extra Practice Exercises at the back of the book. You will find exercises for every chapter, with page references to the lessons.

UNIT

A

Here's what you'll learn in this chapter:

To find the values of expressions with variables.

To simplify expressions.

To find the value of expressions with exponents.

To check whether a given number is a solution of an equation or inequality.

Hoover Dam, 726 feet high, holds back the waters of the Colorado River to form Lake Mead. Algebra and many other branches of mathematics are used in the design and construction of a large-scale engineering project such as this one.

Chapter 1

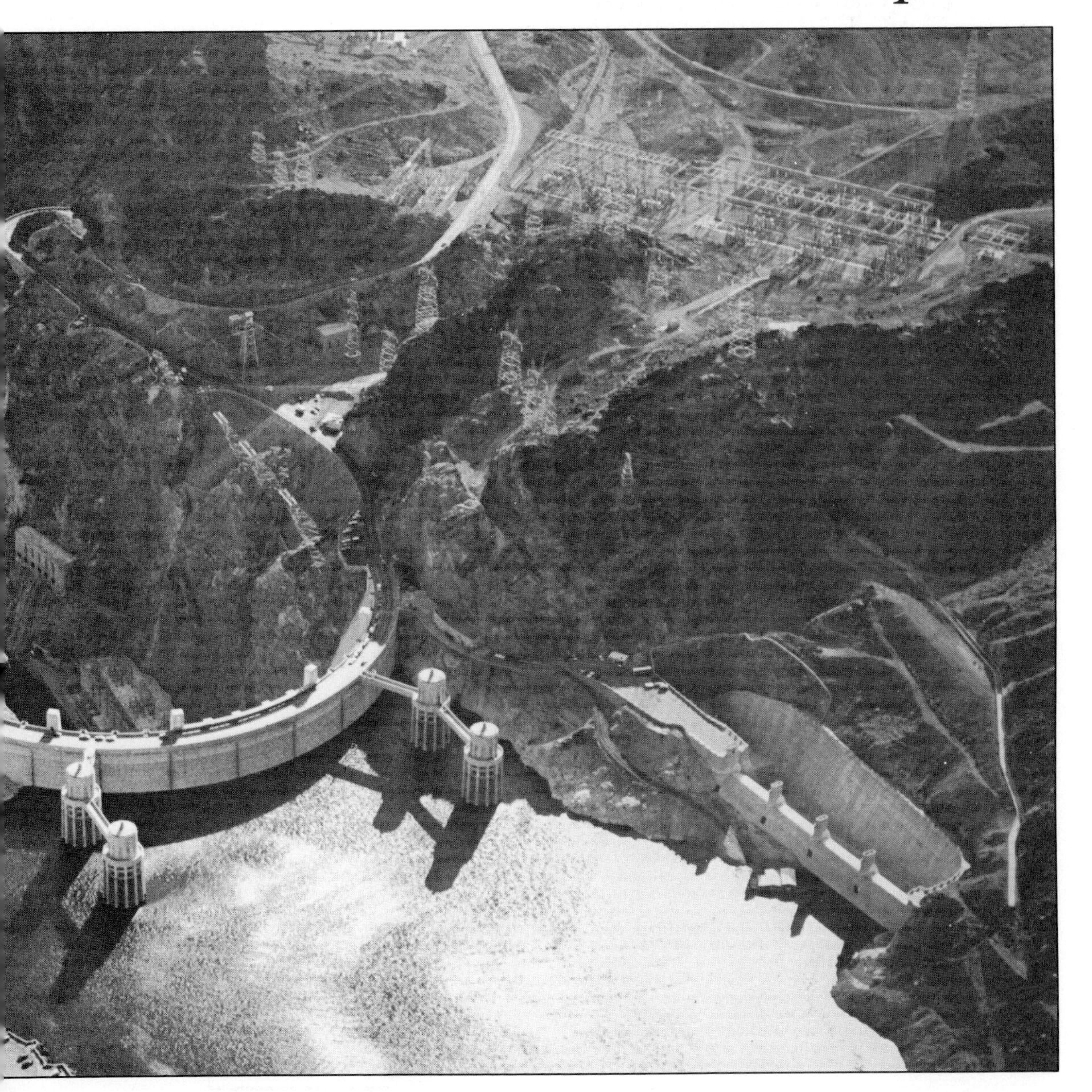

Working with Variables

1 Using Letters for Numbers

Imagine a calculating machine that adds 5 to any number you put into it. It might look like this.

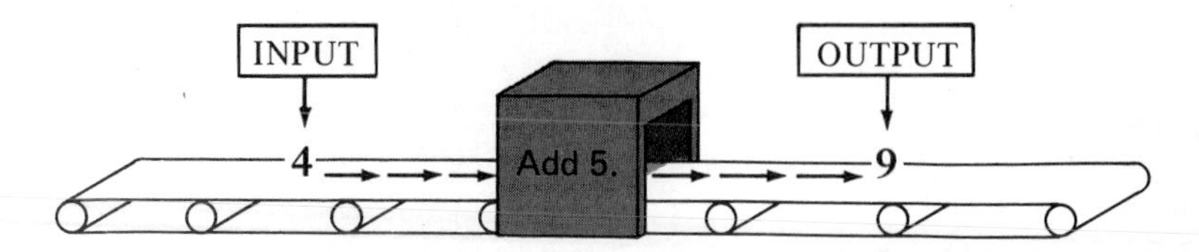

Now suppose you put 7 into the machine.

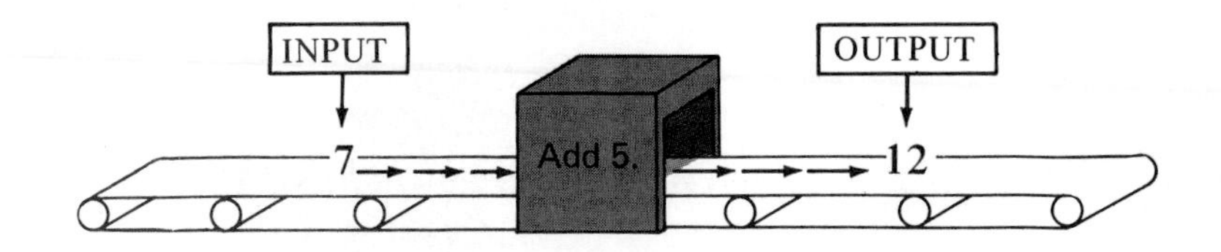

You could go on drawing pictures like this all day. However, you could draw just one picture to represent all possibilities. You can use a letter to represent the INPUT.

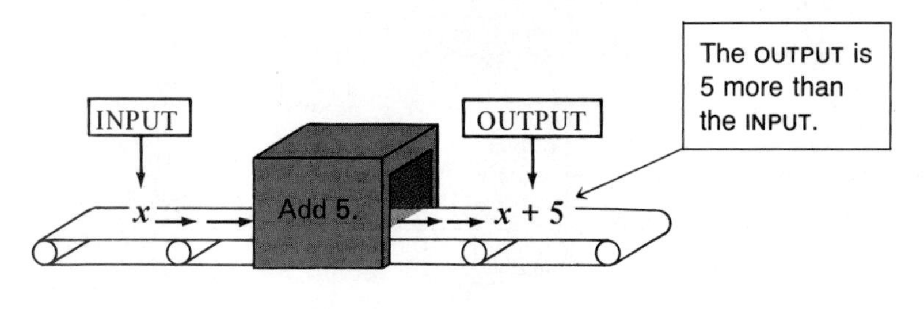

INPUT	OUTPUT
4	9
7	12
20	25
x	$x + 5$

Here is a machine that subtracts 3. Let's use the letter y to represent the INPUT this time.

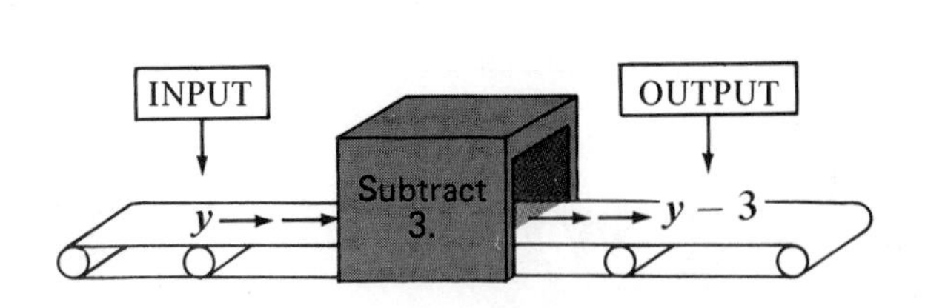

INPUT y	OUTPUT $y - 3$
3	0
8	5
9	6
20	17

When we use a letter like y to represent a number we call the letter a **variable.** We call $x + 5$ and $y - 3$ **variable expressions.**

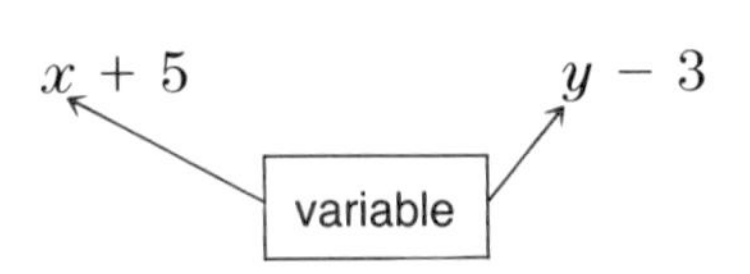

Here's a hint to help you remember the word *variable:* A variable may stand for one of *various* numbers.

We find the value of an expression in the following way.

EXAMPLE 1

If $n = 10$, find the value of $n + 2$.

$$\begin{aligned} n + 2 &= 10 + 2 \\ &= 12 \end{aligned}$$

EXAMPLE 2

If $x = 14$ and $y = 3$, find the value of $x - y$.

$$\begin{aligned} x - y &= 14 - 3 \\ &= 11 \end{aligned}$$

Suppose you had a machine that multiplied by 7.

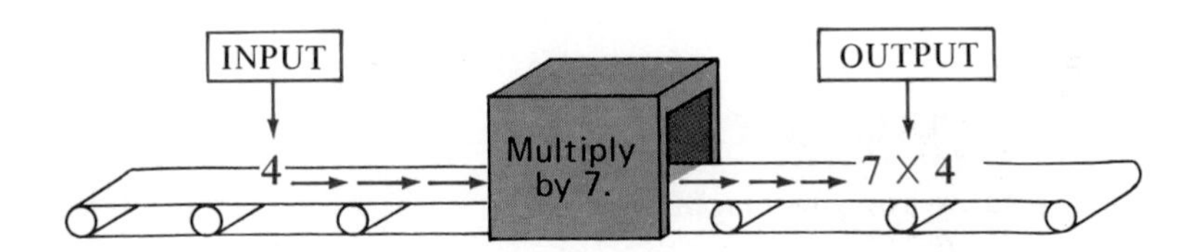

There is another way to show multiplication, too. It is with a dot. You'll see this used very often.

$7 \times 4 \longleftarrow$ means the same as $\longrightarrow 7 \cdot 4$

In algebra, the times sign and the dot are usually left out when a variable is used.

$2n \longleftarrow$ means the same as $\longrightarrow 2 \cdot n$

$ax \longleftarrow$ means the same as $\longrightarrow a \cdot x$

EXAMPLE 3

If $n = 5$, find the value of $7n$.

$$\begin{aligned} 7n &= 7 \cdot 5 \\ &= 35 \end{aligned}$$

Classroom Practice

Complete the table for the machine shown.

	INPUT x	OUTPUT $7x$
Sample	3	21
1.	4	?
2.	8	?
3.	5	?

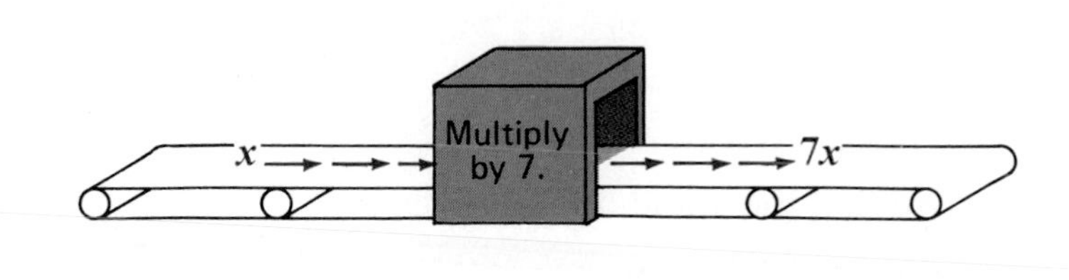

Find the value of the expression.

4. $6 \cdot 7$ **5.** $8 \cdot 3$ **6.** $3 \cdot 5$ **7.** $4 \cdot 8$

If $x = 10$, find the value of the expression.

8. $x + 5$ **9.** $x - 5$ **10.** $4x$ **11.** $x \div 2$

If $s = 12$ and $t = 4$, find the value of the expression.

12. $s + t$ **13.** $s - t$ **14.** st **15.** $s \div t$

Written Exercises

Complete the table for the machine shown.

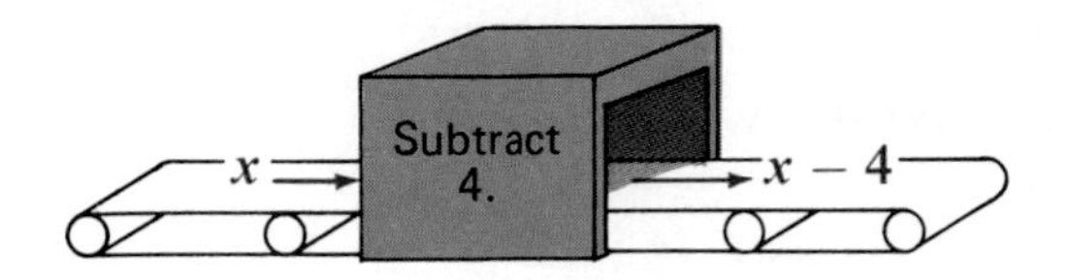

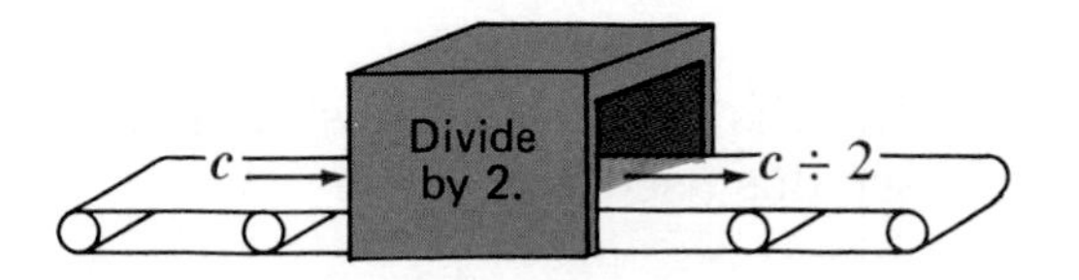

A

	INPUT x	OUTPUT $x - 4$
1.	5	?
2.	7	?
3.	14	?
4.	27	?

	INPUT c	OUTPUT $c \div 2$
5.	6	?
6.	16	?
7.	20	?
8.	30	?

If $s = 4$, find the value of the expression.

9. $4s$
10. $4 + s$
11. $4 - s$
12. $4 \div s$

If $n = 12$, find the value of the expression.

13. $n - 3$
14. $n + 7$
15. $2n$
16. $n \div 4$

If $k = 8$, find the value of the expression.

17. $3k$
18. $4k$
19. $k \div 2$
20. $k \div 4$

If $r = 15$, find the value of the expression.

21. $r + 2$
22. $r \div 3$
23. $r - 5$
24. $2r$

If $y = 20$, find the value of the expression.

25. $y - 10$
26. $5y$
27. $y \div 2$
28. $30 - y$

If $p = 12$ and $q = 8$, find the value of the expression.

29. $p + q + 6$
30. $p - q + 6$
31. pq
32. $p + 4 - q$

Tell what is missing in the table.

	INPUT	Operation	OUTPUT
Sample	x	Add 2.	$x + 2$
33.	x	Add 7.	?
34.	x	Subtract 4.	?
35.	n	Multiply by 2.	?
36.	n	Divide by 2.	?
37.	a	Multiply by b.	?
38.	a	Subtract 7.	?
39.	x	?	$x - 4$
40.	y	?	$3y$
41.	z	?	$z \div 8$
42.	?	Add 3.	$x + 3$
43.	?	Multiply by s.	st

2 Parentheses, Order of Operations

When you have several numbers to add, subtract, multiply, or divide, it helps to show which operation to do first. Parentheses are used.

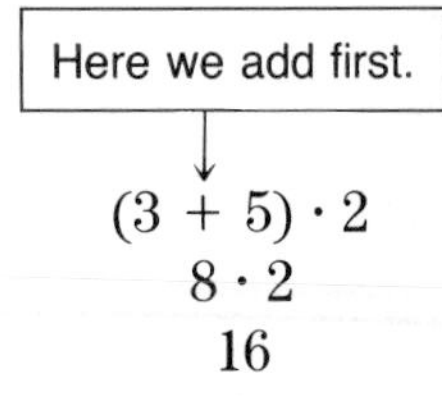

Multiply first.

$7 - (2 \cdot 3)$
$7 - 6$
1

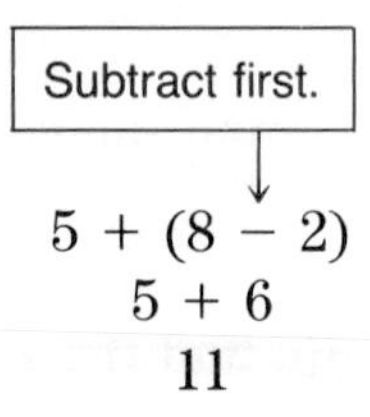

When there are no parentheses, there are only two steps to follow.

Step 1: Do all multiplications and divisions in order from left to right.
Step 2: Do all additions and subtractions in order from left to right.

$3 + 5 \cdot 2$
$3 + 10$
13

$5 + 8 - 2 \cdot 3$
$5 + 8 - 6$
$13 - 6$
7

Now let's see how to use these ideas when there are variables.

EXAMPLE If $x = 4$, find the value of $2(x + 5)$ and $2x + 5$.

This means that you multiply here.

$$2(x + 5) = 2(4 + 5) = 2 \cdot 9 = 18$$

$$2x + 5 = 2 \cdot 4 + 5 = 8 + 5 = 13$$

Classroom Practice

Find the value of the expression.

1. $(7 - 3) \cdot 2$
2. $2 + 3 \cdot 4$
3. $3 + (4 - 1)$
4. $7 - (3 \cdot 2)$

If $x = 3$, find the value of the expression.

5. $2(x + 5)$
6. $2x + 5$
7. $3(x - 1)$
8. $3x - 1$

Written Exercises

Find the value of the expression.

A

1. $(16 - 4) \cdot 2$
2. $16 - (4 \cdot 2)$
3. $3 \cdot (5 + 7)$
4. $(16 - 4) \div (3 + 1)$
5. $(8 + 4) \div (4 + 2)$
6. $9 \div 1 + 3$
7. $3 \cdot 4 + 5$
8. $6 + 8 \cdot 4$
9. $(8 + 6) \div (7 - 5)$
10. $2 \cdot 2 - 1$
11. $3 + 7 \cdot 4$
12. $8 \cdot 2 + 5 \cdot 4$
13. $8 + (2 \cdot 3)$
14. $(2 + 3) \cdot (4 + 1)$
15. $4 \cdot 4 + 4 \div 2$
16. $4 + 8 \div 2$
17. $6 + 2 \div 2 + 2$
18. $(3 - 2) \cdot (9 - 3)$
19. If $x = 4$, find the value of $2(x + 6)$ and $2x + 6$.
20. If $x = 7$, find the value of $5(x + 1)$ and $5x + 1$.
21. If $b = 9$, find the value of $8(b - 4)$ and $8b - 4$.
22. If $y = 8$, find the value of $3(y + 2)$ and $3y + 2$.
23. If $a = 5$, find the value of $4(a - 2)$ and $4a - 2$.
24. If $s = 3$, find the value of $6(s + 4)$ and $6s + 4$.

Tell what is missing in the table.

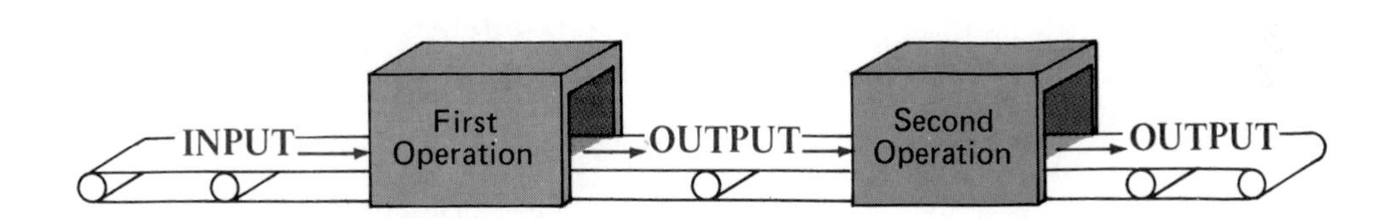

	INPUT →	First Operation →	OUTPUT →	Second Operation →	OUTPUT
25.	3	Multiply by 2.	?	Add 7.	?
26.	4	Multiply by 2.	?	Add 7.	?
27.	x	Multiply by 2.	?	Add 7.	?
28.	3	Add 7.	?	Multiply by 2.	?
29.	4	Add 7.	?	Multiply by 2.	?
30.	x	Add 7.	?	Multiply by 2.	?

3 Coefficients and Terms

In arithmetic you know that $5 + 5 = 2 \cdot 5$, that $5 + 5 + 5 = 3 \cdot 5$, and so on. Working with variables is very similar.

$$x = \mathbf{1}x$$
$$x + x = \mathbf{2}x$$
$$x + x + x = \mathbf{3}x$$
$$x + x + x + x = \mathbf{4}x$$

Note that x and $1x$ mean the same.

The numbers 1, 2, 3, and 4 are called **coefficients** of x.

Now you can rewrite expressions.

$$\begin{aligned} x + x + y + y + y &= (x + x) + (y + y + y) \\ &= 2x + 3y \end{aligned}$$

In the expression $2x + 3y$, we call $2x$ and $3y$ the **terms.**

$2x + 3y$ — terms of the expression

$3a + 4b - 8c$ — terms of the expression

Sometimes the terms of an expression can be combined. We can only combine *like* terms. We cannot combine *unlike* terms. Here are some examples of like and unlike terms.

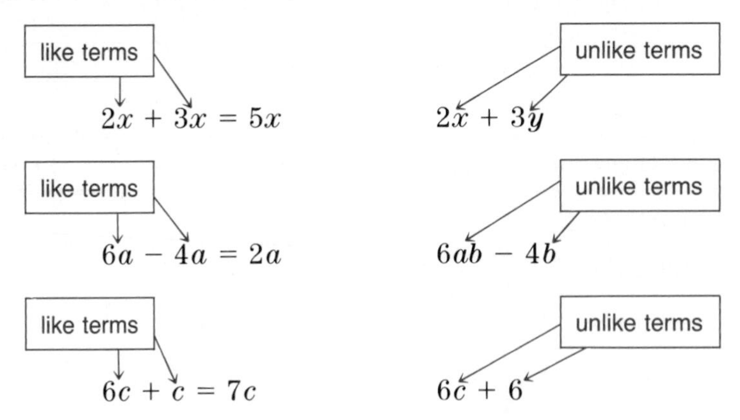

When you combine like terms you **simplify** the expression. Look at the examples at the top of the following page.

EXAMPLES

1. Simplify $2a + 6a + 3$.
   $2a + 6a + 3 = 8a + 3$

2. Simplify $3ab + 4ab - 9$.
   $3ab + 4ab - 9 = 7ab - 9$

3. Simplify $4x + 5x + 7 - 2$.
   $4x + 5x + 7 - 2 = 9x + 5$

4. Simplify $5a + 3xy - xy$.
   $5a + 3xy - xy = 5a + 2xy$

Classroom Practice

Complete the statement.

1. In the expression $5x$ the coefficient of x is _?_.
2. In the expression $6x + 7y$ the coefficient of x is _?_.
3. In the expression $4a + b$ the coefficient of b is not written. It is understood to be _?_.

Simplify.

4. $x + x$
5. $y + y + y$
6. $2ab + 3ab$
7. $3a + a + 3 + 2$

Written Exercises

Simplify.

A

1. $2x + 4x$
2. $3x + 5x$
3. $3a + 7a$
4. $5b - 2b$
5. $9y - 6y$
6. $6xy - 2xy$
7. $2a + 5a + 6$
8. $5y + 3y + 4y$
9. $3a + 5a + 4a$
10. $6b + 3b - b$
11. $4 + 5y + 6y$
12. $2x + 4x + 7$
13. $8s - 2s - 2$
14. $3a + a - 2$
15. $3 + 4y + 5y$
16. $3a + 3a + 3$
17. $3a + a - 4$
18. $2a + 5a + 3a$
19. $6 - 2 + 4x$
20. $y + 2y - 7$
21. $6xy - xy + 7$
22. $9c - 2c + 4c$
23. $5d + 6d - 3d$
24. $8y + 3y - 4y$
25. $7y - 2y + 9x$
26. $2x + x + 4y$
27. $7ab - 2ab + 5b$
28. $cd + 4cd - 2a$
29. $3xy - xy + 2x$
30. $2m + 5n + 6 - 4$

Simplify if possible. If you cannot simplify, write *not possible*.

31. $3n + 3$
32. $4s - 4s$
33. $3ab + 3b$
34. $2a + 5a$
35. $ar + 4ar$
36. $r + 4s$
37. $8n + 2$
38. $9a - 3$

4 Properties of Addition

Addition has some properties that are probably familiar to you. Here are two of them.

THE COMMUTATIVE PROPERTY	
In addition, the order of the terms doesn't matter.	
Arithmetic Example	***Algebra Example***
$1 + 2 = 2 + 1$	$a + b = b + a$
THE ASSOCIATIVE PROPERTY	
In addition, the grouping of the terms doesn't matter.	
Arithmetic Example	***Algebra Example***
$(1 + 2) + 3 = 1 + (2 + 3)$	$(a + b) + c = a + (b + c)$

You don't need to know the names of the properties. You should, however, be able to use them to simplify your work.

EXAMPLE 1 Find the value of 97 + 54 + 3.

$$\begin{aligned} 97 + 54 + 3 &= 97 + 3 + 54 \\ &= (97 + 3) + 54 \\ &= 100 + 54 \\ &= 154 \end{aligned}$$

EXAMPLE 2 Simplify $3n + 5 + 6n - 2$.

$$\begin{aligned} 3n + 5 + 6n - 2 &= (3n + 6n) + (5 - 2) \\ &= 9n + 3 \end{aligned}$$

Classroom Practice

Simplify.

1. $2x + 3 + 4x + 5$ **2.** $3a + 4 + a - 2$ **3.** $8 + 5xy - 2xy + 3$

4. $9r + 5s + 3s + 7r$ **5.** $5mn + 3 + 4mn - 2$ **6.** $8c + 5d - 4c - 3d$

7. $5m + 6 - 3m + 9$ **8.** $10 + 2x - x + 6$ **9.** $3a + 2b - a + 4b$

10. $2z + 12 - 6 - z$ **11.** $5 + 4pq - 3 - 2pq$ **12.** $8f + 2g + 7f - g$

Written Exercises

Find the value of the expression. Show how you use the properties.

A

1. $26 + 5 + 15$
2. $7 + 63 + 3$
3. $49 + 93 + 1$
4. $7 + 25 + 5 + 3$
5. $998 + 571 + 2$
6. $187 + 38 - 87$

Simplify.

7. $3n + 6 + 5n$
8. $2n + 3 + 4n$
9. $5c + 7 - 3 + 2c$
10. $4x + 3 - 2 + 4x$
11. $30 + 5a + a + 1$
12. $9 + 2ac + 3ac + 7$
13. $8m + 5 + 6 + 2m$
14. $4xy + 5 - xy + 2$
15. $7ab + 7 - 6ab + 3$
16. $12m + 3 - 5m + 8m$
17. $6s + 5 + 2s - 4s$
18. $12a - 3a + 5 - 2a$
19. $6k + 9n - 5n - 2k$
20. $4p - 2p + 9a + 3p$
21. $4n - 9b + 3n - 2n$
22. $3c + 7d + 5c + 5d$
23. $8a + 9b - 3b - a$
24. $3p + 4q + 7p - 2q$
25. $8a + 2b + 4 + 3b$
26. $7x + 3y - 2y + 5$
27. $4d - 7a - 2d + 5d$
28. $5m + 3m - 3 + 4n$
29. $6a - 2a + 4b - a$
30. $cd + 4a + 5cd - 2$

SELF-TEST

Vocabulary

variable (p. 5)
variable expression (p. 5)
coefficient (p. 10)
term (p. 10)
simplify (p. 10)
commutative property of addition (p. 12)
associative property of addition (p. 12)

If $x = 14$, find the value of the expression.

Chapter 1, Section 1

1. $x + 6$
2. $x - 6$
3. $3x$
4. $x \div 7$ *(1-1)*

Find the value of the expression.

5. $3 \cdot 2 + 5$
6. $(8 - 4) \cdot 2$
7. $8 - 4 \cdot 2$
8. $4 + 6 \div 2$ *(1-2)*

Simplify if possible. If you cannot simplify, write *not possible*.

9. $4xy - 2xy$
10. $5s - 4$ *(1-3)*
11. $3y + 4y + y$
12. $7w + 2w - 5$
13. $3a - 5 - a$
14. $4m + 2m - 3 - m$ *(1-4)*
15. $4x + 3y - y - x$
16. $9k + j - 3k + 4j$

5 Exponents and Factors

In both addition and multiplication we can simplify the expressions.

Adding	Multiplying
$x = \mathbf{1}x$ $x + x = \mathbf{2}x$ $x + x + x = \mathbf{3}x$	$x = x^{\mathbf{1}}$ $x \cdot x = x^{\mathbf{2}}$ $x \cdot x \cdot x = x^{\mathbf{3}}$
Recall that the numbers 1, 2, and 3 in these expressions are called **coefficients.**	Here, the numbers 1, 2, and 3 are called **exponents.**

x^1, x^2, and x^3 are called **powers** of x.

x to the first power ⟶ $x^{\mathbf{1}}$ ⟵ Usually written as just x.

x to the second power ⟶ $x^{\mathbf{2}}$

x to the third power ⟶ $x^{\mathbf{3}}$

Often the second power of x is called x **squared.** To remember this, think of a square.

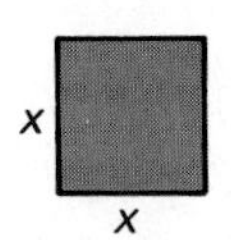

Area $= x^2$ or *x squared*

The third power of x is often called x **cubed.** Think of a cube like this one.

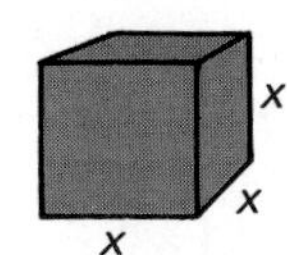

Volume $= x^3$ or *x cubed*

EXAMPLE 1 Find the value of 2^5.

$$2^5 = 2 \cdot 2 \cdot 2 \cdot 2 \cdot 2 = 32$$

EXAMPLE 2 Simplify $x \cdot x \cdot y \cdot y \cdot y$.

$$x \cdot x \cdot y \cdot y \cdot y = (x \cdot x) \cdot (y \cdot y \cdot y) = x^2y^3$$

Each term in $3x - 5y + 4xy$ expresses multiplication. Each part of the term is called a **factor** of the term.

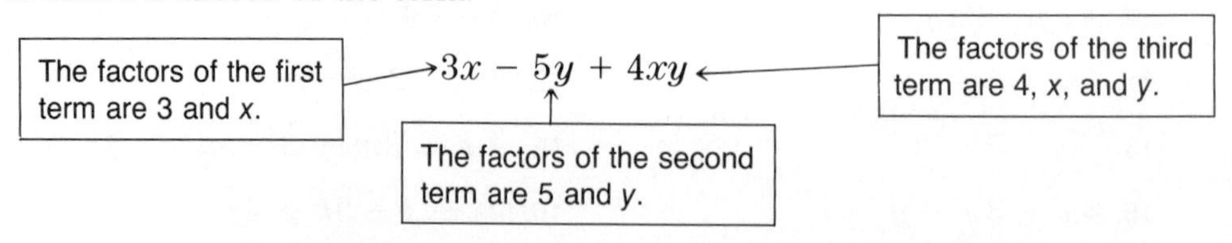

Classroom Practice

Find the value of the expression.

1. 2^2
2. 2^3
3. 2^4
4. 2^1
5. 6^2

Simplify.

6. $x \cdot x \cdot x$
7. $a \cdot a \cdot a \cdot a$
8. $x \cdot x \cdot y \cdot y$
9. $6 \cdot c \cdot c \cdot c$
10. What is the third power of 10?
11. What is the third term of the expression $3a + 4b + 7ab$?
12. Name the factors in the third term of $3a + 4b + 7ab$.

Written Exercises

Find the value of the expression.

1. 3^2
2. 3^3
3. 4^1
4. 5^2
5. 10^2

Simplify.

6. $x \cdot x \cdot x$
7. $y \cdot y \cdot y \cdot y$
8. $a \cdot a \cdot a \cdot a$
9. $x \cdot x \cdot y \cdot y \cdot y$
10. $p \cdot q \cdot q$
11. $r \cdot s \cdot s \cdot s$
12. $c \cdot c \cdot c \cdot c \cdot c$
13. $t \cdot t \cdot m \cdot m$
14. $n \cdot r \cdot r \cdot r$
15. $f \cdot f \cdot f \cdot g \cdot g \cdot h$
16. $a \cdot a \cdot b \cdot b \cdot b \cdot c$
17. $d \cdot e \cdot e \cdot f \cdot f$
18. $5 \cdot n \cdot n$
19. $7 \cdot m \cdot m \cdot m$
20. $6 \cdot a \cdot b \cdot b$
21. $2 \cdot a \cdot a \cdot m \cdot m$
22. $c \cdot c \cdot c \cdot m \cdot m$
23. $5 \cdot n \cdot n \cdot n \cdot p$
24. $7 \cdot s \cdot s \cdot t \cdot t \cdot t$
25. $3 \cdot 2 \cdot a \cdot a \cdot b$
26. $9 \cdot r \cdot r \cdot s \cdot s$

PUZZLE ◆ PROBLEMS

Copy these nine dots on a piece of paper.

Can you draw 4 straight lines through the dots so that each dot is touched?

You must not lift your pencil from the paper. And be sure you do not retrace a line or go through any dot more than once.

6 Properties of Multiplication

As shown in the box below, multiplication has some properties that are like the properties of addition.

THE COMMUTATIVE PROPERTY	
In multiplication, the order of the factors doesn't matter.	
Arithmetic Example	***Algebra Example***
$2 \cdot 3 = 3 \cdot 2$	$ab = ba$
THE ASSOCIATIVE PROPERTY	
In multiplication, the grouping of the factors doesn't matter.	
Arithmetic Example	***Algebra Example***
$(2 \cdot 3) \cdot 4 = 2 \cdot (3 \cdot 4)$	$(ab)c = a(bc)$

As with the addition properties, you do not need to memorize the names. Just be sure you can use the properties. They can make your work easier.

EXAMPLE 1 Evaluate $5 \cdot 82 \cdot 2$.

$$\begin{aligned} 5 \cdot 82 \cdot 2 &= 5 \cdot 2 \cdot 82 \\ &= (5 \cdot 2) \cdot 82 \\ &= 10 \cdot 82 \\ &= 820 \end{aligned}$$

EXAMPLE 2 Simplify $2 \cdot (5x)$.

$$\begin{aligned} 2 \cdot (5x) &= (2 \cdot 5) \cdot x \\ &= 10x \end{aligned}$$

EXAMPLE 3 Simplify $(2r)(4r)$.

$$\begin{aligned} (2r)(4r) &= (2 \cdot 4)(r \cdot r) \\ &= 8r^2 \end{aligned}$$

EXAMPLE 4 Simplify $6 \cdot a \cdot a \cdot 5 \cdot a$.

$$\begin{aligned} 6 \cdot a \cdot a \cdot 5 \cdot a &= (6 \cdot 5)(a \cdot a \cdot a) \\ &= 30a^3 \end{aligned}$$

Classroom Practice

Find the value of the expression. Show how you use the properties.

1. $17 \cdot 25 \cdot 4$
2. $(2 \cdot 43) \cdot 5$
3. $2 \cdot 2 \cdot 5 \cdot 5$

Simplify.

4. $3 \cdot (4x)$
5. $5 \cdot 6a$
6. $3n \cdot 2$
7. $a \cdot (4a)$
8. $(3x) \cdot x$
9. $(2x)(3x)$
10. $(3a)(4b)$
11. $4 \cdot x \cdot x$

Written Exercises

Find the value of the expression.

A

1. $4 \cdot 29 \cdot 25$
2. $92 \cdot 5 \cdot 2$
3. $(33 \cdot 25) \cdot 4$
4. $(50 \cdot 83) \cdot 2$
5. $2 \cdot (73 \cdot 5)$
6. $10 \cdot 3 \cdot 10 \cdot 2 \cdot 10$

Simplify.

7. $2 \cdot (3x)$
8. $3 \cdot (4x)$
9. $5 \cdot (3a)$
10. $4 \cdot (4a)$
11. $b \cdot 2b$
12. $c \cdot 3c$
13. $(3y) \cdot 6$
14. $(5k) \cdot 2$
15. $(4n) \cdot 5$
16. $2s \cdot s$
17. $8 \cdot 9f$
18. $m \cdot 6n$
19. $(2a)(2a)$
20. $(3x)(5y)$
21. $(2a)(4b)$
22. $(6s)(3s)$
23. $8 \cdot x \cdot x \cdot 2$
24. $3 \cdot b \cdot b \cdot 2$
25. $5 \cdot a \cdot 7 \cdot a \cdot a$
26. $4 \cdot y \cdot y \cdot y \cdot 2$
27. $2 \cdot b \cdot 5 \cdot b \cdot b$
28. $3 \cdot a \cdot a \cdot 4 \cdot a$

Tell what is missing in the table.

	INPUT ⟶	⟶ Operation ⟶	⟶ OUTPUT
29.	$3x$	Multiply by 2.	?
30.	$5a$	Multiply by 3.	?
31.	$2x$	Multiply by x.	?
32.	?	Multiply by 5.	$15c$
33.	$5x$	?	$20x$
34.	$3b$	?	$18b$

7 The Distributive Property

Jan is buying two felt-tip pens at $1 each and two note pads at $2 each. The total bill can be figured in two ways.

2($1 + $2)	2 · $1 + 2 · $2
2 · $3	$2 + $4
$6	$6

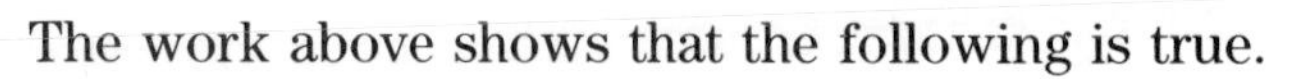

The work above shows that the following is true.

$$2(1 + 2) = 2 \cdot 1 + 2 \cdot 2$$

Let's check a similar statement to see if it is also true.

Is $3(4 + 6) = 3 \cdot 4 + 3 \cdot 6$ true?

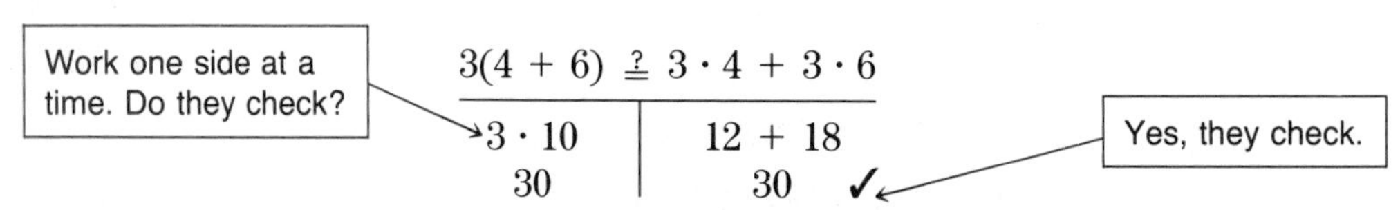

$$3(4 + 6) \stackrel{?}{=} 3 \cdot 4 + 3 \cdot 6$$

$3 \cdot 10$	$12 + 18$
30	30 ✓

This example shows the **distributive property.**

THE DISTRIBUTIVE PROPERTY	
Arithmetic Examples	***Algebra Examples***
$2(1 + 3) = 2 \cdot 1 + 2 \cdot 3$	$a(b + c) = ab + ac$
$3(4 - 1) = 3 \cdot 4 - 3 \cdot 1$	$a(b - c) = ab - ac$

You can use the distributive property to change an expression with parentheses to one without them.

EXAMPLE 1 $\quad 3(x + 6) = 3x + 3 \cdot 6$
$\qquad\qquad\qquad\qquad = 3x + 18$

EXAMPLE 2 $\quad 2(a - 3b) = 2a - 2 \cdot 3b$
$\qquad\qquad\qquad\qquad = 2a - 6b$

EXAMPLE 3 $\quad m(m - n) = m \cdot m - m \cdot n$
$\qquad\qquad\qquad\qquad = m^2 - mn$

Classroom Practice

1. Check to see if $3(2 + 1) = (3 \cdot 2) + (3 \cdot 1)$ is true.
2. Is $2(4 \cdot 3) = (2 \cdot 4) \cdot (2 \cdot 3)$ true?
3. Once you've checked to see if the statement in Exercise 2 is true, you should be able to tell if $2(ab) = (2a) \cdot (2b)$ is true. Is it?

True or false?

4. $3(a + b) = 3a + 3b$
5. $3(cd) = (3c) \cdot (3d)$

State the expression without parentheses.

6. $2(x + 5)$
7. $3(x + 2)$
8. $4(a + 1)$
9. $5(7 + a)$
10. $3(x - 4)$
11. $a(a - 3)$
12. $b(b - 3)$
13. $y(2 - y)$

Written Exercises

Check to see if the statement is true.

A

1. $3(4 + 5) = (3 \cdot 4) + (3 \cdot 5)$
2. $4(8 + 3) = (4 \cdot 8) + (4 \cdot 3)$
3. $5(7 - 3) = (5 \cdot 7) - (5 \cdot 3)$
4. $6(21 - 11) = (6 \cdot 21) - (6 \cdot 11)$

State the expression without parentheses.

5. $2(x + 3)$
6. $3(x + 5)$
7. $4(x + 2)$
8. $5(a + 4)$
9. $7(a + 3)$
10. $8(b + 1)$
11. $9(x + 4)$
12. $6(y + 2)$
13. $3(x - 2)$
14. $4(y - 7)$
15. $c(c - 3)$
16. $d(d - 4)$
17. $7(2x + 1)$
18. $4(3x + 2)$
19. $5(3x - 2)$
20. $3(7x - 3)$
21. $2(3x + 4y)$
22. $b(4a + 2b)$
23. $x(3x - y)$
24. $a(2a - 3b)$

Use the distributive property. Then combine like terms.

25. $2(x + 5) + 3$
26. $3(x - 1) + 5x$
27. $5(x + 3) + 2x$
28. $3(2x - 1) + 4x$
29. $4(9 + 2y) - 3y$
30. $7(1 + 2c) - 8c$

B

31. $5(3x + 5) - 7x - 2$
32. $2(1 + 3x) + 7x - 1$
33. $5(2 + 7x) - 3x - 2$
34. $4(a + 3) + 3(a - 2)$
35. $8(2b + 3) + 4(b - 5)$
36. $6(c + 3) + 4(1 - c)$
37. $5(m + 5) + 2(m - 8)$
38. $3(4p + 3) + 4(p - 1)$
39. $x(x + 2y) - xy + 2$

8 Properties of 0

Any number times 0 is 0.

$$7 \cdot 0 = 0 \qquad a \cdot 0 = 0$$

No number can be divided by 0.

IMPOSSIBLE $\frac{a}{0}$

Zero divided by any number (except 0) is 0.

$$\frac{0}{6} = 0 \qquad \frac{0}{n} = 0$$ ← Divisions are often written with a bar.

Adding 0 to a number gives that number.

$$6 + 0 = 6 \qquad n + 0 = n$$

Adding and subtracting the same number is like adding 0.

$$n + 5 - 5 = n + 0 = n$$

Let's put these ideas to work.

EXAMPLE 1

Let $x = 4$. Find the value of $9(x - 4)$.

$$\begin{aligned} 9(x - 4) &= 9(4 - 4) \\ &= 9 \cdot 0 \\ &= 0 \end{aligned}$$

EXAMPLE 2

Simplify $2 + 3x - 2$.

$$\begin{aligned} 2 + 3x - 2 &= 3x + 2 - 2 \\ &= 3x + 0 \\ &= 3x \end{aligned}$$

EXAMPLE 3

Simplify $3r + 4 + 6r - 4$.

$$\begin{aligned} 3r + 4 + 6r - 4 &= 3r + 6r + 4 - 4 \\ &= 9r + 0 \\ &= 9r \end{aligned}$$

Classroom Practice

Find the value if possible. If not, say *impossible*.

1. $5 \cdot 0$
2. $\frac{0}{7}$
3. $\frac{7}{0}$
4. $\frac{8 \cdot 0}{9}$

Let $x = 5$. Find the value of the expression.

5. $\frac{0}{x}$
6. $x + 0$
7. $\frac{x - 5}{11}$
8. $35(x - 5)$

Written Exercises

Find the value if possible. If not, write *impossible*.

A

1. $12 \cdot 0$
2. $0 \cdot 31$
3. $18 + 35 - 18$
4. $16 + 0$
5. $\frac{0}{16}$
6. $\frac{31}{0}$
7. $\frac{5}{5 \cdot 0}$
8. $16 \cdot 8 \cdot 0$
9. Let $x = 1$. Find the value of $5(x - 1)$.
10. Let $x = 2$. Find the value of $(x - 2) \div 4$.
11. Let $x = 3$. Find the value of $(2x - 6) \div 4$.
12. Let $x = 0$. Find the value of $x \div 3$.

Simplify.

13. $3x + 6 + 5x - 6$
14. $5a + 3 + 8a - 3$
15. $6 + 4c + 2c - 6$
16. $4(x + 2) - 8$
17. $3(2 + 5a) - 6$
18. $2(c + 5) - 10$

Think of an INPUT-OUTPUT machine. Tell what is missing in the table.

	Sample	19.	20.	21.	22.	23.	24.
INPUT	$x - 4$	$x - 5$	$x - 9$	$x - 1$	$x - 3$	$x + 2$	$x + 6$
↓ Operation	Add 4.	Add 5.	Add 9.	Add 1.	Add 3.	Subtract 2.	Subtract 6.
↓ OUTPUT	x	?	?	?	?	?	?

	25.	26.	27.	28.	29.	30.	31.
INPUT	$y - 7$	$x - 13$	$y + 9$	$y + 4$	$x - 16$	$14 + n$	$n - 11$
↓ Operation	?	?	?	?	?	?	?
↓ OUTPUT	y	x	y	y	x	n	n

9 Properties of 1

Whether you add 0 to a number or multiply the number by 1, you get the same result—the original number.

original number + 0 = original number
original number × 1 = original number

The second sentence above states a property of 1.

Multiplying a number by 1 gives that number.

$$6 \cdot 1 = 6 \qquad n \cdot 1 = n$$

Any number (except 0) divided by itself is 1.

$$\frac{4}{4} = 1 \qquad \frac{n}{n} = 1$$

Remember, you can write a division with a bar.

Any number divided by 1 is that number.

$$\frac{8}{1} = 8 \qquad \frac{n}{1} = n$$

Here are a couple of examples to show these properties.

EXAMPLE 1 Simplify $8 \cdot \frac{x}{8}$.

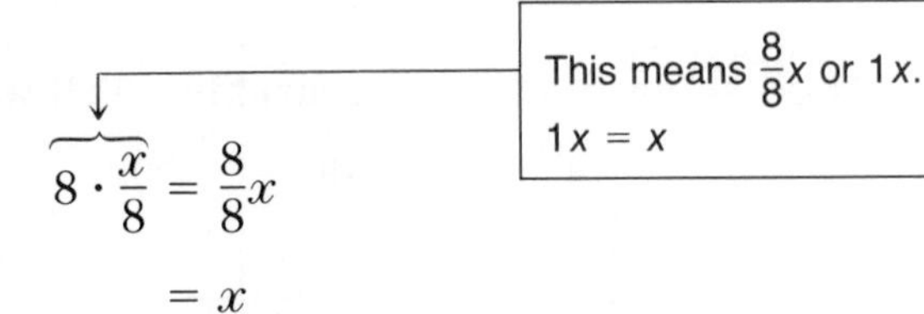

$$8 \cdot \frac{x}{8} = \frac{8}{8}x$$
$$= x$$

EXAMPLE 2 Simplify $4x \div 4$.

This gives you $\frac{4}{4}x$, or x.

$$4x \div 4 = \frac{4x}{4}$$
$$= \frac{4}{4}x$$
$$= x$$

Classroom Practice

Find the value.

1. $14 \cdot 1$
2. $\frac{16}{1}$
3. $\frac{4 \cdot 3}{12}$
4. $7(5 - 4)$

Let $z = 2$. Find the value of the expression.

5. $z \div 2$
6. $4z \div 4$
7. $3 \cdot \frac{z}{3}$
8. $\frac{10z}{z}$

Written Exercises

Let $x = 4$. Find the value of the expression.

A

1. $\frac{x}{4}$
2. $4x$
3. $(4 \div x) \cdot 1$
4. $\frac{x + 2}{6}$

Simplify.

5. $b \div b$
6. $5x \div 5$
7. $3x \div x$
8. $10x \div 10$
9. $\frac{5c}{5}$
10. $\frac{3}{3}x$
11. $\frac{4n}{n}$
12. $\frac{6}{6}y$
13. $5 \cdot \frac{x}{5}$
14. $\frac{x}{7} \cdot 7$
15. $\frac{2a}{2}$
16. $\frac{c}{9} \cdot 9$

Think of an INPUT-OUTPUT machine. Tell what is missing in the table.

	17.	18.	19.	20.	21.
INPUT	$4x$	$5x$	$3x$	$7x$	$6x$
↓ Operation	Divide by 4.	Divide by 5.	Divide by 3.	?	?
↓ OUTPUT	?	?	?	x	x

	22.	23.	24.	25.	26.
INPUT	$\frac{x}{8}$	$\frac{x}{4}$	$\frac{x}{9}$	$\frac{x}{7}$	$\frac{x}{10}$
↓ Operation	Multiply by 8.	Multiply by 4.	Multiply by 9.	?	?
↓ OUTPUT	?	?	?	x	x

27. Let $x = 3$. Find the value of $4(x - 2)$.
28. Let $y = 5$. Find the value of $(2y - 10) \div 1$.

SELF-TEST

Vocabulary

exponent (p. 14)
power (p. 14)
squared (p. 14)
cubed (p. 14)
factor (p. 14)
commutative property of multiplication (p. 16)
associative property of multiplication (p. 16)
distributive property (p. 18)

Simplify.

1. $r \cdot r \cdot r \cdot s \cdot s$ 2. $6 \cdot m \cdot m \cdot n \cdot p$ 3. $5 \cdot 4 \cdot b \cdot c \cdot c$ *(1-5)*

4. $5 \cdot (6x)$ 5. $(4a)(7a)$ 6. $2 \cdot e \cdot e \cdot 4 \cdot e$ *(1-6)*

State the expression without parentheses.

7. $3(x - 2)$ 8. $y(3y + 1)$ *(1-7)*

9. $2(4a - 6b)$ 10. $6(3d + 5)$

Simplify.

11. $2(a + 4) - 8$ 12. $5 + 3m - 2m - 5$ *(1-8)*

13. $13m \div 13$ 14. $\frac{a}{12} \cdot 12$ *(1-9)*

CALCULATOR ACTIVITIES

Check to see whether your calculator follows the order of operations discussed on page 8. Evaluate the expression $2 + 5 \cdot 6$ by entering

$$2 + 5 \times 6 =$$

If your calculator displays 32, then it performed the operations of addition and multiplication in the order used in algebra.

If your calculator displays 42, then it performed the operations in the order they were entered. For this type of calculator, you should enter

$$5 \times 6 + 2 =$$

Use a calculator to find the value of each expression.

1. $10 - 4 \cdot 2$ 2. $7 - 6 \div 3$ 3. $4 \cdot 16 \div 2$

4. $10 \div 5 \cdot 2 - 1$ 5. $8 \cdot 2 + 5$ 6. $8(2 + 5)$

10 Introduction to Equations

Consider whether the following statements are true or false.

Babe Ruth was a baseball hero . . . . true
George Washington was a baseball hero . . . . false
He was a baseball hero . . . . maybe true, maybe false
(It depends on who *he* was.)

The sentences below are like the ones above.

$6 + 4 = 10$ . . . . true
$5 + 1 = 8$ . . . . false
$x + 2 = 7$ . . . . maybe true, maybe false
(It depends on what x is.)

These number sentences are called **equations.** Equations always have an = sign, but are not always true.

Is $x + 2 = 7$ true if $x = 6$?

$x + 2 = 7$		
$6 + 2$	7	
8		No!

Is $x + 2 = 7$ true if $x = 5$?

$x + 2 = 7$		
$5 + 2$	7	
7		✓ Yes!

We call 5 a **solution,** or **root,** of the equation $x + 2 = 7$ because it is a value for x that makes the equation true.

EXAMPLE Is 4 a solution of $9 - x = 3$? Is 6 a solution?

Check 4:

$9 - x = 3$		
$9 - 4$	3	
5		No

Check 6:

$9 - x = 3$		
$9 - 6$	3	
3		✓ Yes

Classroom Practice

Tell which of the numbers shown in color is a solution.

1. $x + 3 = 9$ 7 or 6?

2. $y - 7 = 11$ 18 or 4?

3. $4 + x = 10$ 9 or 6?

4. $3n = 12$ 4 or 9?

5. $3x - 1 = 8$ 2 or 3?

6. $4x - 12 = 20$ 7 or 8?

7. $m - 9 = 25$ 16 or 34?

8. $5a = 35$ 30 or 7?

Written Exercises

Tell which of the numbers shown in color is a solution.

A

1. $x + 9 = 12$ 3 or 21?
2. $y - 7 = 15$ 8 or 22?
3. $3n = 36$ 12 or 33?
4. $4x = 24$ 6 or 20?
5. $3a + 7 = 13$ 2 or 4?
6. $2a - 6 = 4$ 4 or 5?
7. $5(r + 2) = 30$ 3 or 4?
8. $7(s - 3) = 21$ 5 or 6?
9. $x + 9 = 4x$ 2, 3, or 4?
10. $2b + 5 = b + 7$ 1, 2, or 3?
11. $5a - 1 = 3a + 9$ 5, 6, or 7?
12. $4(a - 5) = a + 1$ 6, 7, or 8?
13. $n^2 = 25$ 0, 5, or 10?
14. $a^3 = 1$ 0, 1, or 2?

B

15. Write an equation that has 9 as a solution.
16. Write an equation that has 11 as a solution.
17. Write an equation that has 0 as a solution.
18. Write an equation that has no number as a solution.

Find at least one solution. You may need to try many numbers for some exercises.

19. $x + 5 = 13$
20. $y - 17 = 20$
21. $5n = 30$
22. $6x = 36$
23. $b \div 4 = 2$
24. $b \div 5 = 5$
25. $n \cdot 0 = 0$
26. $x + 5 = 11$
27. $y + 1 = 41$
28. $x + 5 = 25$
29. $x + 3 = x + 3$
30. $m - 9 = 27$

C

31. $3a \cdot 7 = 42$
32. $2m + 9 = 21$
33. $2b + 4 = 4$
34. $x^2 = x$

PUZZLE ◆ PROBLEMS

Amos, Christopher, and Geraldine work in the circus. They are the ringmaster, lion tamer, and elephant trainer, not necessarily in that order.

1. Geraldine has red hair.
2. Amos has curly hair.
3. The ringmaster is taller than Amos.
4. The lion tamer is bald.

Who is the elephant trainer?

11 Introduction to Inequalities

Not all number sentences are equations. Some number sentences are statements that two numbers are not equal. We call a number sentence like this an **inequality.** Here are some examples of inequalities.

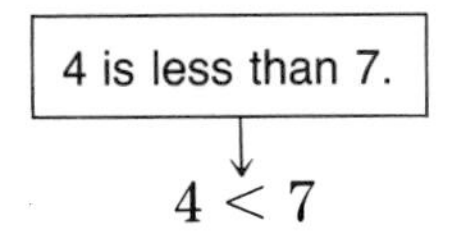

$4 < 7$

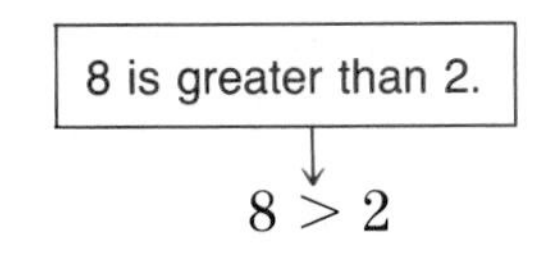

$8 > 2$

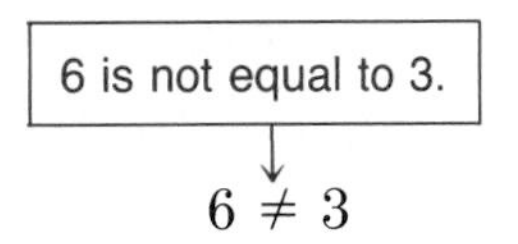

$6 \neq 3$

Some inequalities have variables. A **solution** of the inequality is a number that makes the statement true.

EXAMPLE 1 Is 4 a solution of $x < 6$? Is 9 a solution?

Check 4: $x < 6$

$4 \overset{?}{<} 6$ ✓ Yes

Check 9: $x < 6$

$9 \overset{?}{<} 6$ No

EXAMPLE 2 Is 2 a solution of $y > 6$? Is 4 a solution?

Check 2: $y > 6$

$2 \overset{?}{>} 6$ No

Check 4: $y > 6$

$4 \overset{?}{>} 6$ No

Classroom Practice

Read the statement aloud.

1. $17 \neq 7$ **2.** $6 > 3$ **3.** $7 < 8$ **4.** $5 > 2$

5. $9 < 10$ **6.** $4 \neq 9$ **7.** $2x > 7$ **8.** $2x + 1 < 5$

Tell which of the numbers shown in color are solutions.

9. $x < 3$ 5, 2, 1, 0

10. $a > 6$ 4, 9, 10, 12

11. $y > 12$ 3, 5, 7, 15

12. $y < 21$ 12, 15, 25, 35

13. $s > 13$ 9, 11, 13, 15

14. $m < 2$ 0, 1, 2, 3

15. For Exercises 9–14, are there more solutions that are not given? If yes, can you name a few?

Written Exercises

Compare. Write < or >.

A

1. 6 __?__ 7
2. 0 __?__ 4
3. 3 __?__ 1
4. 4 __?__ 9
5. 7 __?__ 4
6. 5 __?__ 11
7. 6 __?__ 3
8. 5 __?__ 0

Tell which of the numbers shown in color are solutions.

9. $x < 6$ 4, 6, 7, 10
10. $c > 2$ 0, 3, 5, 7
11. $x > 0$ 0, 1, 3, 7
12. $y > 9$ 4, 6, 10, 14
13. $x > 14$ 7, 8, 10, 20
14. $y < 9$ 6, 2, 10, 4
15. $a < 7$ 7, 6, 2, 12
16. $c > 10$ 4, 8, 12, 35
17. $x < 9$ 4, 6, 9, 10
18. $y < 12$ 6, 12, 7, 11
19. $x + 7 > 16$ 8, 9, 10, 11
20. $3y - 1 < 9$ 4, 3, 6, 2
21. $3x + 3 < 24$ 2, 5, 7, 9
22. $4y - 7 > 13$ 2, 4, 5, 7
23. $2y - 8 < 6$ 4, 5, 6, 7
24. $4a > a + 6$ 1, 0, 2, 5

Compare. Write < or > or =.

25. $3(8 + 4)$ __?__ $7 \cdot 6 - 5$
26. $2 \cdot 3 + 7$ __?__ $10 + 6 \div 2$
27. $83 \cdot 5 \cdot 0 \div 9$ __?__ 0
28. $791 \cdot (17 - 17)$ __?__ 2
29. $(24 \div 8) \cdot 8$ __?__ 3^3
30. $(5 \cdot 4 - 2) \div 6$ __?__ 4^2

SELF-TEST

Vocabulary

equation (p. 25)
solution of an equation (p. 25)
root of an equation (p. 25)
inequality (p. 27)
solution of an inequality (p. 27)

Tell which of the numbers shown in color are solutions.

1. $x - 4 = 9$ 13 or 15?
2. $5m = 35$ 7 or 30? *(1-10)*
3. $2a + 4 = 18$ 7 or 8?
4. $9(s - 3) = 36$ 4 or 7?
5. $b \neq 13$ 5, 3, 13, 21
6. $y + 6 < 12$ 4, 6, 5, 7 *(1-11)*

Compare. Write < or >.

7. 5 __?__ 6
8. 0 __?__ 2
9. 9 __?__ 7
10. 4 __?__ 8

READING ALGEBRA

SYMBOLS

Symbols are very useful in algebra as well as in arithmetic. They make it easy to express mathematical ideas without using a great number of words. Some familiar symbols can be read in several different ways. The examples below will remind you of some of these ways.

$6 + 4$

six plus four
add four to six
the sum of six and four
four more than six

$10 - 2$

ten minus two
subtract two from ten
the difference of ten and two
two less than ten

6×4, or $6 \cdot 4$

six times four
multiply four by six
the product of six and four

$10 \div 2$

ten divided by two
divide ten by two
the quotient of ten and two

9^2

nine times nine
nine squared
nine to the second power

$\frac{7}{2}$

seven divided by two
the quotient of seven and two
seven halves

EXERCISES

Read each expression in at least two ways.

1. $15 - 8$ **2.** $5 \cdot 12$ **3.** $17 + 4$ **4.** $18 \div 9$

Write each word expression in symbols. Then write the common numeral for each expression.

SAMPLE add two to ten $10 + 2 = 12$

5. five less than thirty

6. the product of nine and zero

7. three to the second power

8. thirty-two fourths

9. eighty plus eight

10. divide one hundred by ten

SKILLS REVIEW

WHOLE NUMBER ARITHMETIC

Take time out to work some arithmetic problems. See if you can solve them quickly, but *correctly.* Maybe you can time yourself. Ten minutes is very good!

Add.

1. $\begin{array}{r} 24 \\ +32 \\ \hline \end{array}$

2. $\begin{array}{r} 146 \\ +253 \\ \hline \end{array}$

3. $\begin{array}{r} 566 \\ +427 \\ \hline \end{array}$

4. $824 + 56$

5. $137 + 92$

6. $270 + 125$

7. $\begin{array}{r} 452 \\ 76 \\ +184 \\ \hline \end{array}$

8. $\begin{array}{r} 760 \\ 135 \\ +85 \\ \hline \end{array}$

9. $\begin{array}{r} 621 \\ 6 \\ 58 \\ +137 \\ \hline \end{array}$

10. $\begin{array}{r} 584 \\ 861 \\ 76 \\ +35 \\ \hline \end{array}$

Subtract.

11. $\begin{array}{r} 58 \\ -24 \\ \hline \end{array}$

12. $\begin{array}{r} 67 \\ -35 \\ \hline \end{array}$

13. $\begin{array}{r} 126 \\ -49 \\ \hline \end{array}$

14. $\begin{array}{r} 376 \\ -298 \\ \hline \end{array}$

15. $\begin{array}{r} 620 \\ -85 \\ \hline \end{array}$

16. $\begin{array}{r} 704 \\ -367 \\ \hline \end{array}$

17. $\begin{array}{r} 491 \\ -137 \\ \hline \end{array}$

18. $852 - 68$

19. $266 - 49$

20. $306 - 87$

Multiply.

21. $\begin{array}{r} 36 \\ \times 8 \\ \hline \end{array}$

22. $\begin{array}{r} 59 \\ \times 4 \\ \hline \end{array}$

23. $\begin{array}{r} 73 \\ \times 25 \\ \hline \end{array}$

24. $\begin{array}{r} 87 \\ \times 42 \\ \hline \end{array}$

25. $\begin{array}{r} 120 \\ \times 17 \\ \hline \end{array}$

26. $\begin{array}{r} 311 \\ \times 54 \\ \hline \end{array}$

27. $\begin{array}{r} 672 \\ \times 65 \\ \hline \end{array}$

28. $\begin{array}{r} 247 \\ \times 132 \\ \hline \end{array}$

29. $\begin{array}{r} 721 \\ \times 105 \\ \hline \end{array}$

30. $\begin{array}{r} 630 \\ \times 116 \\ \hline \end{array}$

Divide.

31. $6\overline{)102}$

32. $4\overline{)632}$

33. $7\overline{)478}$

34. $9\overline{)367}$

35. $2\overline{)149}$

36. $7\overline{)492}$

37. $6\overline{)1427}$

38. $4\overline{)8562}$

39. $57\overline{)5941}$

40. $24\overline{)675}$

41. $38\overline{)8412}$

42. $72\overline{)5629}$

CHAPTER REVIEW

CHAPTER SUMMARY

1. A variable is a letter used to represent a number.
2. To find the value of an expression, do operations inside parentheses first. Next do multiplications and divisions in order from left to right. Finally, do additions and subtractions in order from left to right.
3. To simplify an algebraic expression, combine like terms.
4. An expression of the form x^5 is a power of x. The second power of x is x squared, or x^2. The third power of x is x cubed, or x^3.
5. The properties of addition and multiplication are illustrated below.

$a + b = b + a$ | $(a + b) + c = a + (b + c)$

$ab = ba$ | $(ab)c = a(bc)$

$a(b + c) = ab + ac$ | $a(b - c) = ab - ac$

$a \cdot 0 = 0$ and $a + 0 = a$ | $a \cdot 1 = a$ and $\frac{a}{1} = a$

$\frac{0}{a} = 0$ if a is not 0 | $\frac{a}{a} = 1$ if a is not 0

$\frac{a}{0}$ Impossible | $a + b - b = a + 0 = a$

6. An equation states that two numbers are equal. An inequality states that two numbers are not equal. A solution of an equation or inequality is a value for the variable that makes the statement true.

REVIEW EXERCISES

If $x = 8$, find the value of the expression. ***(See pp. 4–9.)***

1. $x + 3$
2. $2x$
3. $x \div 4$
4. $2x + 5$
5. $2(x + 5)$
6. $3x - 3$
7. $3(x - 3)$
8. $5x - 2$

Find the value of the expression. ***(See pp. 8–9.)***

9. $3 \cdot 4 + 2$
10. $5 + 8 \cdot 2$
11. $(3 + 4)(5 - 2)$
12. $(8 + 4) \div 4 - 2$
13. $8 + 4 \div 4 - 2$
14. $8 + 4 \div (4 - 2)$

Simplify. ***(See pp. 10–13.)***

15. $2x + 5x$ **16.** $8a - 6a$ **17.** $3b + 4b + 2$

18. $7c - c + 3c$ **19.** $7c - c + 3$ **20.** $8y - 2y + 7 + 5$

21. $5x + 7 + 3x - 4$ **22.** $4a - 4 + 6a - a$ **23.** $3a + 2b + 7a - b$

Find the value of the expression. ***(See pp. 14–15.)***

24. 2^4 **25.** 3^5 **26.** 5^3 **27.** 3 to the third power

28. 5^2 **29.** 4^3 **30.** 7^2 **31.** 1 to the first power

Simplify. ***(See pp. 16–17.)***

32. $2 \cdot (5x)$ **33.** $3 \cdot 4a$ **34.** $8 \cdot 2b$

35. $4 \cdot x \cdot x \cdot x$ **36.** $3 \cdot b \cdot b \cdot b \cdot 2$ **37.** $a \cdot b \cdot b \cdot a \cdot a$

38. $(2x)(3x)$ **39.** $(4y)(3y)$ **40.** $n(2n)(3n)$

State the expression without parentheses. ***(See pp. 18–19.)***

41. $2(x + 5)$ **42.** $3(2a - 7)$ **43.** $5(4x - 2y)$ **44.** $5(3y - 4)$

45. $4(2c - 3d)$ **46.** $2(9x + 7y)$ **47.** $3(a - 6b)$ **48.** $6(4x + 3)$

Let $x = 4$. Find the value if possible. If not, write *impossible*.
(See pp. 20–23.)

49. $6(x - 4)$ **50.** $\frac{x - 4}{x + 4}$ **51.** $\frac{x - 4}{x - 4}$ **52.** $\frac{x - 4}{x}$

53. $\frac{x}{x - 4}$ **54.** $\frac{x}{4}$ **55.** $0 \div x$ **56.** $x \div 0$

Simplify. ***(See pp. 20–21.)***

57. $3x + 7 + 5x - 7$ **58.** $9 + 4a - 4a - 3$ **59.** $4a + 7b - 3a - 7b$

60. $2b + 5c - 5c - 2b$ **61.** $3(2x + 3) - 9$ **62.** $3(x + 4) + 2(x - 6)$

63. $4s + 9 + 6s - 9$ **64.** $8t + 6 - 8t - 5$ **65.** $5(m + n) - 5m$

Tell which of the numbers shown in color is a solution. ***(See pp. 25–28.)***

66. $5x - 7 = 13$ 3, 4, or 5? **67.** $3x + 9 = 36$ 5, 7, or 9?

68. $2(b + 6) = 18$ 3, 6, or 9? **69.** $x + 3 = 4x$ 0, 1, or 3?

70. $x + 8 > 15$ 6, 7, or 8? **71.** $3y - 4 < 6$ 3, 4, or 5?

CHAPTER TEST

If $a = 9$ and $b = 6$, find the value of the expression.

1. $a - 2$ 2. $b \div 2$ 3. ab 4. $a + b - 8$ *(1-1)*

5. $7b + 2$ 6. $7(b + 2)$ *(1-2)*

Find the value of the expression.

7. $15 - 9 \div 3$ 8. $9 \cdot (8 + 2)$

Simplify if possible. If you cannot simplify, write *not possible*.

9. $7k - k - 3k$ 10. $s + 4t + 5st$ *(1-3)*

11. $9 + 3x - 3x$ 12. $6yz - 2yz + 4y$

13. $8r + 12 + r$ 14. $5h + 7t - 6t + 6h$ *(1-4)*

15. $3b + 4c - 2d - c$ 16. $v + 4v + 7y - 3v$

17. $4 \cdot a \cdot a \cdot a \cdot b$ 18. $d \cdot e \cdot e \cdot e \cdot e$ *(1-5)*

19. $(5z) \cdot 7$ 20. $9 \cdot b \cdot 2 \cdot b$ 21. $(4x)(3y)$ *(1-6)*

22. Find the value of $50 \cdot (7 \cdot 2)$.

State the expression without parentheses.

23. $6(r + 3)$ 24. $2(9x - 8)$ 25. $t(7t + 2q)$ *(1-7)*

26. Use the distributive property and simplify: $4(2y + 1) + 6 - 7y$

Find the value if possible. If not, write *impossible*.

27. $\frac{4 + 4}{4 - 4}$ 28. $\frac{0 \cdot 1}{0 + 1}$ 29. $\frac{2 \cdot 4 - 8}{5}$ *(1-8)*

Simplify.

30. $5x \div x$ 31. $4t \div 1$ 32. $\frac{r}{8} \cdot 8$ *(1-9)*

Tell which of the numbers shown in color are solutions.

33. $5j - 2 = 2j + 7$ 3 or 4? 34. $x^3 = 8$ 1, 2, or 3? *(1-10)*

Tell which of the numbers 0, 5, 8, and 10 are solutions.

35. $x > 8$ 36. $2x < 18$ 37. $x + 1 < 2x$ *(1-11)*

38. Compare. Write $<$ or $>$ or $=$: $21 \ \underline{\ ?\ } \ 2 + 5 \cdot 3$

MIXED REVIEW

ARITHMETIC REVIEW

Add, subtract, multiply, or divide.

1. $774 \div 18$
2. 521×36
3. $723 + 417$
4. $8\overline{)312}$
5. $\begin{array}{r} 686 \\ -592 \\ \hline \end{array}$
6. $\frac{1073}{29}$
7. $\begin{array}{r} 37 \\ \times 9 \\ \hline \end{array}$
8. $\begin{array}{r} 104 \\ -48 \\ \hline \end{array}$
9. $\begin{array}{r} 14 \\ \times 35 \\ \hline \end{array}$
10. $\begin{array}{r} 273 \\ 45 \\ +190 \\ \hline \end{array}$
11. $8251 - 762$
12. $8 + 29 + 746$
13. $7\overline{)531}$ (to the nearest tenth)

Match each expression with its value in the right-hand column.

14. $\frac{2}{9} + \frac{4}{9}$ — a. $\frac{1}{3}$
15. $\frac{7}{6} - \frac{5}{6}$ — b. $\frac{41}{30}$
16. $\frac{7}{8} \cdot \frac{4}{3}$ — c. $\frac{4}{3}$
17. $1\frac{7}{8} \div 1\frac{1}{4}$ — d. $\frac{7}{6}$
18. $6 \div \frac{2}{3}$ — e. 9
19. $4\frac{1}{6} - 2\frac{4}{5}$ — f. $\frac{2}{3}$
20. $\frac{7}{12} + \frac{3}{4}$ — g. $\frac{3}{2}$

21. Find the least common denominator of $\frac{7}{12}$, $\frac{5}{9}$, and $\frac{11}{18}$.
22. Find the greatest common factor of 42 and 60.
23. Write $\frac{129}{10{,}000}$ as a decimal.
24. Write 3.175 as a mixed number in simplest form.

Add, subtract, multiply, or divide.

25. $2.93 + 0.8 + 4.279$
26. $25.1 - 7.263$
27. 4.8×0.0571
28. $(7.5 - 3.5)(28 + 3.71)$
29. $5.13 \div 0.012$
30. $\frac{1.6 + 0.48}{6.5}$

CHAPTER 1 REVIEW

Simplify.

1. $7(3 + 5x) - 21$
2. $5ab - 2a - 4ab$
3. $2 \cdot 5 \cdot x \cdot x \cdot x \cdot y$
4. $\frac{c}{6} \cdot 6$
5. $y \cdot 3 \cdot y \cdot 8 \cdot x \cdot y$
6. $9k + 8 - 3k + 2kt$

Think of an INPUT-OUTPUT machine. Tell what is missing in the table.

	7.	8.	9.	10.	11.	12.
INPUT ↓	$x - 8$	n	y	$4b$	$2 + x$	$5x$
Operation ↓	Add 8.	Divide by 5.	?	?	Subtract 2.	?
OUTPUT	?	?	$y - 1$	$4b^2$	?	x

If $a = 6$, find the value of the expression.

13. $7(a - 4)$
14. $7a - 4$
15. $18 \div a$
16. a^1
17. $\frac{6 - a}{6 + a}$
18. $\frac{a}{a} \cdot 3$
19. $9a$
20. $a \cdot a$

Find the value if possible. If not, write *impossible*.

21. $9 + 6 \div 3$
22. $8 - 4 \div 2 + 1$
23. $496 + 13 + 4$
24. 4^3
25. $\frac{3 \cdot 3}{3 - 3}$
26. $(20 \cdot 18) \cdot 5$
27. Compare by writing <, >, or =: $(7 \cdot 4 + 2) \div 5$ __?__ 2^3
28. If $x = 7$ and $y = 10$, find the value of $x + 4 - y$.
29. Is it possible to simplify $2xy + 3y$?
30. Find at least one solution: $a \div 2 > 8$
31. Use the distributive property and combine like terms: $6(2x + 1) + 5(2 - x)$
32. Write an equation which has 1 as a solution.

Tell which of the numbers shown in color is a solution of the equation or inequality.

33. $4x + 7 = 2x + 11$ 1, 2, or 3?
34. $x + 4 > 2$ 0, 1, 2, or 3?
35. $5a < a + 9$ 0, 2, 4, or 6?
36. $5(b - 8) = b - 4$ 8, 9, or 11?

Here's what you'll learn in this chapter:

To solve equations with one variable.

To write algebraic expressions for everyday situations.

To use equations to solve word problems.

Circus performers—animals as well as people—need a good sense of balance. You will discover in this chapter that balance is also important in working with equations.

Chapter 2

Solving Equations

1 Solving Equations by Addition

The figure at the left shows a balanced scale.

If you add the same weight to both sides, the scale will remain balanced.

You can use this idea with equations too. For example, the equation $x - 4 = 19$ is like a balanced scale.

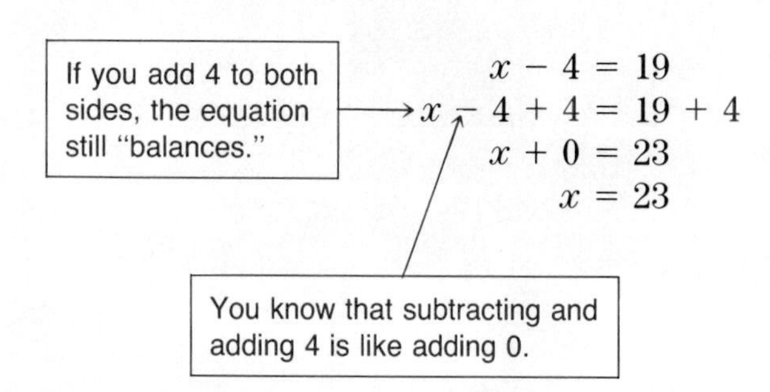

$$x - 4 = 19$$
$$x - 4 + 4 = 19 + 4$$
$$x + 0 = 23$$
$$x = 23$$

The basic plan in solving any equation is to get the variable by itself on one side of the equation. For the equations in this section, you do this by adding the same number to both sides of the equation.

EXAMPLE 1

$$x - 5 = 12$$
$$x - 5 + 5 = 12 + 5 \leftarrow \text{Add 5 to both sides.}$$
$$x = 17$$

EXAMPLE 2

$$50 = n - 8$$
$$50 + 8 = n - 8 + 8 \leftarrow \text{Add 8 to both sides.}$$
$$58 = n$$

Classroom Practice

You want to solve for x. Tell what must be added to both sides of the equation. Solve.

1. $x - 5 = 7$	2. $x - 11 = 19$	3. $x - 3 = 21$	4. $x - 27 = 4$
5. $x - 2 = 19$	6. $x - 4 = 11$	7. $9 = x - 6$	8. $25 = x - 7$
9. $10 = x - 8$	10. $x - 12 = 2$	11. $x - 1 = 99$	12. $48 = x - 2$
13. $0 = x - 10$	14. $x - 15 = 30$	15. $13 = x - 27$	16. $x - 9 = 81$

Written Exercises

Solve.

A

1. $x - 5 = 11$	2. $x - 7 = 5$	3. $x - 8 = 2$	4. $y - 3 = 17$
5. $x - 3 = 9$	6. $y - 4 = 7$	7. $a - 7 = 6$	8. $y - 6 = 2$
9. $b - 9 = 15$	10. $a - 12 = 12$	11. $x - 9 = 10$	12. $x - 3 = 12$
13. $y - 2 = 13$	14. $x - 5 = 2$	15. $y - 7 = 2$	16. $a - 9 = 4$
17. $n - 1 = 8$	18. $x - 8 = 20$	19. $x - 9 = 21$	20. $x - 7 = 19$
21. $x - 15 = 10$	22. $a - 9 = 25$	23. $6 = x - 2$	24. $9 = y - 4$
25. $10 = x - 3$	26. $6 = x - 5$	27. $0 = y - 5$	28. $15 = x - 5$
29. $21 = b - 14$	30. $43 = x - 21$	31. $36 = y - 15$	32. $x - 52 = 100$

Solve. Do a step in your head, if you can.

SAMPLE

All steps

$$x - 5 = 8$$
$$x - 5 + 5 = 8 + 5$$
$$x = 13$$

Can you do this step in your head?

A step left out

$$x - 5 = 8$$
$$x = 13$$

B

33. $x - 3 = 7$	34. $x - 4 = 13$	35. $x - 2 = 4$	36. $y - 5 = 3$
37. $y - 2 = 9$	38. $n - 6 = 4$	39. $n - 10 = 11$	40. $x - 3 = 0$
41. $a - 5 = 10$	42. $x - 7 = 20$	43. $7 = y - 2$	44. $10 = x - 7$
45. $12 = x - 3$	46. $20 = n - 5$	47. $22 = x - 3$	48. $14 = y - 9$

2 Solving Equations by Subtraction

Suppose you have an equation like $x + 4 = 18$ to solve. You want to get the variable alone on one side of the equation. How do you do it? *Subtract* 4 from both sides.

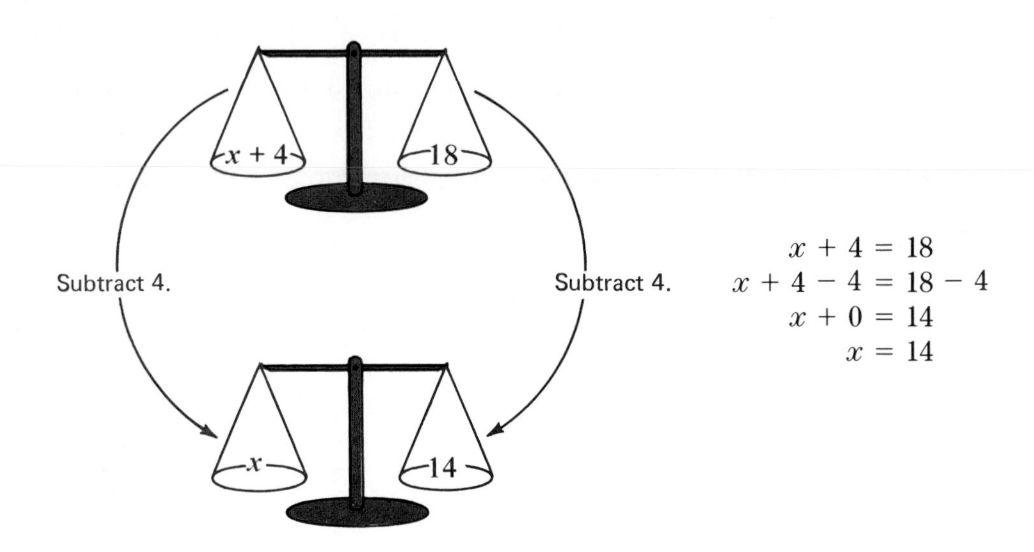

$$x + 4 = 18$$
$$x + 4 - 4 = 18 - 4$$
$$x + 0 = 14$$
$$x = 14$$

Here are some other examples.

EXAMPLE 1

$$x + 5 = 28$$
$$x + 5 - 5 = 28 - 5 \leftarrow \text{Subtract 5 from both sides.}$$
$$x = 23$$

EXAMPLE 2

$$19 = a + 7$$
$$19 - 7 = a + 7 - 7 \leftarrow \text{Subtract 7 from both sides.}$$
$$12 = a$$

Classroom Practice

You want to solve for *x*. Tell what to subtract from each side of the equation. Solve.

1. $x + 9 = 14$ **2.** $x + 3 = 7$ **3.** $x + 5 = 9$ **4.** $x + 2 = 12$

5. $x + 7 = 9$ **6.** $x + 4 = 34$ **7.** $20 = x + 11$ **8.** $60 = x + 45$

Written Exercises

Solve.

A

1. $x + 5 = 15$
2. $x + 8 = 20$
3. $x + 2 = 10$
4. $a + 7 = 25$
5. $y + 7 = 11$
6. $n + 6 = 14$
7. $n + 4 = 16$
8. $y + 6 = 21$
9. $x + 20 = 34$
10. $8 + y = 15$
11. $5 + x = 20$
12. $28 + y = 42$
13. $24 = x + 7$
14. $30 = y + 8$
15. $60 = b + 48$
16. $88 = x + 7$

Solve. Do a step in your head, if you wish.

SAMPLE

$$x + 7 = 15$$
$$x + 7 - 7 = 15 - 7$$
$$x = 8$$

You might do this step in your head.

B

17. $x + 4 = 9$
18. $x + 7 = 14$
19. $y + 3 = 20$
20. $n + 8 = 12$
21. $x + 3 = 15$
22. $y + 6 = 10$
23. $n + 4 = 17$
24. $y + 9 = 25$
25. $6 + x = 16$
26. $9 + y = 10$
27. $8 + x = 21$
28. $4 + x = 27$
29. $18 = x + 3$
30. $20 = x + 4$
31. $14 = x + 6$
32. $24 = x + 6$
33. $15 = 8 + x$
34. $x + 12 = 16$
35. $20 + x = 24$
36. $50 = x + 10$
37. $19 + y = 19$
38. $x + 14 = 37$
39. $z + 52 = 76$
40. $34 = m + 9$

Mixed Practice Exercises

Solve.

1. $x - 2 = 7$
2. $x - 5 = 4$
3. $x - 9 = 9$
4. $y - 7 = 3$
5. $y - 6 = 5$
6. $x - 4 = 6$
7. $y - 7 = 11$
8. $x - 2 = 14$
9. $x + 5 = 11$
10. $x + 6 = 13$
11. $y + 7 = 12$
12. $n + 2 = 17$
13. $n + 8 = 14$
14. $x + 3 = 22$
15. $n + 9 = 26$
16. $x + 12 = 19$
17. $y - 2 = 17$
18. $x - 12 = 42$
19. $y + 6 = 35$
20. $n + 17 = 31$
21. $x - 6 = 27$
22. $y + 7 = 40$
23. $x - 18 = 12$
24. $y + 9 = 37$
25. $20 = 16 + x$
26. $12 = x - 7$
27. $15 = x - 2$
28. $25 = y + 6$
29. $36 = y + 9$
30. $40 = x - 16$
31. $29 = x - 17$
32. $48 = n + 26$
33. $45 = z - 15$
34. $28 = x + 12$
35. $x + 18 = 36$
36. $x - 16 = 54$

3 Solving More Equations

Solving a multiplication equation like $2x = 16$ should be no problem. All you do is *divide* both sides by 2.

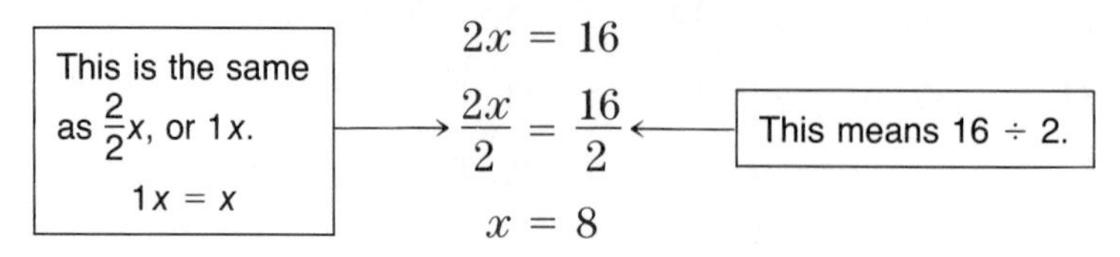

$$2x = 16$$
$$\frac{2x}{2} = \frac{16}{2}$$
$$x = 8$$

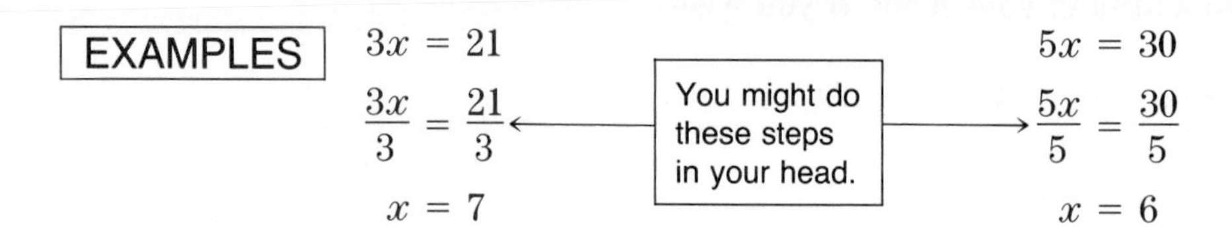

$$3x = 21 \qquad 5x = 30$$
$$\frac{3x}{3} = \frac{21}{3} \qquad \frac{5x}{5} = \frac{30}{5}$$
$$x = 7 \qquad x = 6$$

When you see a division equation like $\frac{x}{3} = 12$, you can solve it by *multiplying* each side by 3.

This gives you $\frac{3}{3}x$, or x.

$$\frac{x}{3} = 12$$
$$3 \cdot \frac{x}{3} = 3 \cdot 12$$
$$x = 36$$

EXAMPLES

$$\frac{x}{2} = 5 \qquad \frac{x}{4} = 6$$
$$2 \cdot \frac{x}{2} = 2 \cdot 5 \qquad 4 \cdot \frac{x}{4} = 4 \cdot 6$$
$$x = 10 \qquad x = 24$$

Classroom Practice

You want to solve for *x*. By what number would you divide each side of the equation? Solve.

1. $2x = 10$ **2.** $4x = 24$ **3.** $3x = 18$ **4.** $5x = 40$

5. $7x = 21$ **6.** $8x = 88$ **7.** $9x = 36$ **8.** $12x = 36$

You want to solve for y. By what number would you multiply each side of the equation? Solve.

9. $\frac{y}{2} = 8$
10. $\frac{y}{5} = 9$
11. $\frac{y}{4} = 3$
12. $\frac{y}{6} = 5$
13. $\frac{y}{3} = 6$
14. $\frac{y}{8} = 7$
15. $\frac{y}{10} = 4$
16. $\frac{y}{7} = 12$
17. $\frac{y}{15} = 3$
18. $\frac{y}{9} = 8$
19. $\frac{y}{30} = 6$
20. $\frac{y}{8} = 20$

Written Exercises

Solve. Do one step in your head, if you wish.

SAMPLE

$$2x = 18$$
$$\frac{2x}{2} = \frac{18}{2}$$
$$x = 9$$

You might do this step in your head.

A

1. $3x = 12$
2. $5x = 15$
3. $6x = 42$
4. $4a = 32$
5. $6b = 72$
6. $4y = 16$
7. $7x = 35$
8. $5x = 25$
9. $2x = 20$
10. $3x = 15$
11. $6y = 18$
12. $7y = 28$
13. $8m = 24$
14. $6x = 48$
15. $7n = 49$
16. $4a = 36$
17. $9a = 81$
18. $8x = 72$
19. $9x = 108$
20. $10a = 100$
21. $7x = 91$
22. $12a = 84$
23. $8x = 112$
24. $9b = 513$
25. $3b = 540$
26. $11n = 231$
27. $7y = 175$
28. $4m = 244$

Solve. Do one step in your head, if you wish.

29. $\frac{x}{2} = 9$
30. $\frac{x}{3} = 8$
31. $\frac{x}{4} = 11$
32. $\frac{x}{5} = 7$
33. $\frac{a}{6} = 9$
34. $\frac{a}{3} = 7$
35. $\frac{n}{2} = 6$
36. $\frac{x}{6} = 4$
37. $\frac{n}{5} = 3$
38. $\frac{x}{2} = 10$
39. $\frac{n}{7} = 8$
40. $\frac{x}{3} = 11$
41. $\frac{x}{6} = 8$
42. $\frac{a}{9} = 9$
43. $\frac{a}{9} = 8$
44. $\frac{x}{10} = 13$

Mixed Practice Exercises

Think of an INPUT-OUTPUT machine. Tell what is missing in the table.

	Sample	1.	2.	3.	4.
INPUT	$x + 4 = 11$	$x + 2 = 13$	$7x = 14$	$9x = 45$	$x - 6 = 7$
↓ Operation ↓	Subtract 4 from both sides.	Subtract 2 from both sides.	Divide both sides by 7.	Divide both sides by 9.	Add 6 to both sides.
OUTPUT	$x = 7$	?	?	?	?

	5.	6.	7.	8.	9.
INPUT	$x + 6 = 10$	$y - 4 = 8$	$5x = 35$	$8y = 32$	$x - 7 = 13$
↓ Operation ↓	Subtract 6 from both sides.	Add 4 to both sides.	Divide both sides by 5.	Divide both sides by 8.	Add 7 to both sides.
OUTPUT	?	?	?	?	?

	10.	11.	12.	13.	14.
INPUT	$y - 3 = 11$	$x + 6 = 12$	$7x = 84$	$x - 8 = 15$	$n + 7 = 22$
↓ Operation ↓	Add 3 to both sides.	Subtract 6 from both sides.	Divide both sides by 7.	Add 8 to both sides.	Subtract 7 from both sides.
OUTPUT	?	?	?	?	?

Solve.

15. $x - 4 = 10$
16. $y + 2 = 7$
17. $x + 4 = 10$
18. $y - 6 = 3$
19. $n - 7 = 4$
20. $n + 8 = 17$
21. $x - 7 = 14$
22. $y + 9 = 11$
23. $x - 9 = 12$
24. $4x = 20$
25. $8y = 16$
26. $n + 6 = 32$
27. $\frac{a}{7} = 3$
28. $\frac{x}{4} = 9$
29. $\frac{n}{5} = 2$
30. $\frac{n}{8} = 9$
31. $8 + y = 13$
32. $a - 6 = 18$
33. $x + 12 = 26$
34. $y + 17 = 30$
35. $a - 15 = 12$
36. $12x = 48$
37. $m - 14 = 37$
38. $n + 36 = 52$
39. $\frac{x}{6} = 3$
40. $\frac{n}{8} = 1$
41. $\frac{x}{5} = 6$
42. $\frac{a}{7} = 9$

43. $y + 10 = 15$

44. $x - 3 = 20$

45. $9x = 27$

46. $7n = 63$

47. $3y = 63$

48. $7x = 42$

49. $5a = 100$

50. $9x = 90$

51. $7 + x = 20$

52. $a - 12 = 14$

53. $x + 9 = 28$

54. $x - 15 = 22$

55. $x - 3 = 18$

56. $y + 7 = 26$

57. $3x = 57$

58. $2n = 104$

59. $35 = x + 7$

60. $29 = y - 6$

61. $54 = 9x$

62. $7 = 7x$

SELF-TEST

Solve.

1. $x - 3 = 19$

2. $a - 14 = 6$ *(2-1)*

3. $5 = n - 5$

4. $13 = b - 10$

5. $m + 2 = 15$

6. $n + 7 = 18$ *(2-2)*

7. $x + 9 = 21$

8. $17 = s + 7$

9. $5b = 45$

10. $9x = 63$ *(2-3)*

11. $7a = 56$

12. $4x = 12$

13. $8n = 96$

14. $12x = 132$

15. $\frac{n}{3} = 5$

16. $\frac{a}{6} = 7$

17. $\frac{b}{4} = 9$

18. $\frac{x}{10} = 3$

PUZZLE ◆ PROBLEMS

There are 3 red hats and 2 black hats in a drawer. Melba, Fran, and Brent line up in a single file, and a hat is placed on each one's head. They are asked to figure out what color hat they are wearing.

Brent, who could look ahead and see Fran and Melba, says, "I don't know."

Fran, who could see only Melba, says, "I don't know."

But Melba, who could see nobody, says, "I know."

What color is Melba's hat and how does she know?

4 Equations with More Steps

In the last sections you studied equations that needed just one operation to solve. In this section you'll learn to solve equations that need more steps to solve.

$$2x - 5 = 3$$

You'll want to get x alone on one side of the equation. First get rid of the 5. Then get rid of the 2.

$$2x - 5 = 3$$
$$2x - 5 + 5 = 3 + 5$$ ← Add 5 to both sides.
$$2x = 8$$
$$\frac{2x}{2} = \frac{8}{2}$$ ← Next divide both sides by 2.
$$x = 4$$

It's easy if you just take one step at a time.

EXAMPLE 1

$$3x + 7 = 19$$
$$3x + 7 - 7 = 19 - 7$$ ← First subtract 7 from both sides.
$$3x = 12$$
$$x = 4$$ ← Divide both sides by 3 to get $x = 4$.

EXAMPLE 2

$$6 + 2x = 10$$
$$6 - 6 + 2x = 10 - 6$$ ← Subtract 6 from both sides.
$$2x = 4$$
$$x = 2$$ ← Divide both sides by 2 to get $x = 2$.

Classroom Practice

Tell what steps you would take to solve the equation.

SAMPLE $2x - 5 = 9$ *What you say:* Add 5 to both sides. Then divide both sides by 2.

1. $2x - 4 = 16$ **2.** $5x - 2 = 8$ **3.** $2x - 1 = 9$ **4.** $7x - 2 = 12$

5. $2a + 4 = 12$ **6.** $3b + 5 = 5$ **7.** $4s + 7 = 11$ **8.** $8p + 3 = 11$

Solve.

9. $2a - 3 = 3$ **10.** $2m + 1 = 9$ **11.** $3x - 4 = 11$ **12.** $5b - 4 = 6$

13. $3r + 2 = 11$ **14.** $4y + 2 = 14$ **15.** $5a + 1 = 36$ **16.** $2x - 9 = 3$

17. $4b - 3 = 13$ **18.** $6x - 1 = 23$ **19.** $8m + 2 = 10$ **20.** $6y + 3 = 33$

Written Exercises

Think of two INPUT-OUTPUT machines. Tell what is missing in the table.

A

	Sample	**1.**	**2.**	**3.**	**4.**
INPUT	$2x - 1 = 5$	$5x - 3 = 7$	$2a + 1 = 3$	$3b + 5 = 8$	$4y + 1 = 9$
↓ First Operation ↓	Add 1 to both sides.	Add 3 to both sides.	Subtract 1 from both sides.	Subtract 5 from both sides.	Subtract 1 from both sides.
OUTPUT	$2x = 6$	?	?	?	?
↓ Second Operation ↓	Divide both sides by 2.	Divide both sides by 5.	Divide both sides by 2.	Divide both sides by 3.	Divide both sides by 4.
OUTPUT		?	?	?	?

	5.	**6.**	**7.**	**8.**	**9.**
INPUT	$2a - 5 = 13$	$5b - 9 = 31$	$4c + 3 = 15$	$8a + 7 = 55$	$6n - 3 = 15$
↓ First Operation ↓	Add 5 to both sides.	Add 9 to both sides.	Subtract 3 from both sides.	Subtract 7 from both sides.	Add 3 to both sides.
OUTPUT	?	?	?	?	?
↓ Second Operation ↓	Divide both sides by 2.	Divide both sides by 5.	Divide both sides by 4.	Divide both sides by 8.	Divide both sides by 6.
OUTPUT	?	?	?	?	?

Solve.

10. $2x - 4 = 0$ **11.** $5x - 8 = 7$ **12.** $2x + 1 = 5$

13. $5x - 1 = 14$ **14.** $3x - 2 = 10$ **15.** $2x - 1 = 7$

Solve.

16. $3x - 5 = 4$
17. $4x - 7 = 1$
18. $2x - 4 = 8$
19. $4x - 8 = 12$
20. $2x - 5 = 1$
21. $4x - 5 = 15$
22. $2x + 5 = 11$
23. $3x + 4 = 22$
24. $4x + 1 = 1$
25. $4y + 7 = 27$
26. $3y + 2 = 8$
27. $2a + 5 = 25$
28. $5n - 1 = 9$
29. $4n - 7 = 13$
30. $2y + 6 = 22$
31. $4y + 7 = 15$
32. $6y + 4 = 34$
33. $8 + 3r = 23$
34. $2x - 3 = 5$
35. $8y + 9 = 33$
36. $2a - 5 = 11$
37. $6s - 1 = 17$
38. $3c - 7 = 14$
39. $3t - 8 = 13$
40. $7y + 2 = 23$
41. $5b - 9 = 21$
42. $6r - 3 = 21$
43. $4c + 1 = 13$
44. $8a + 5 = 53$
45. $8x - 5 = 35$
46. $3y + 11 = 41$
47. $4x - 5 = 23$
48. $8a - 5 = 75$
49. $4b + 6 = 30$
50. $3x - 5 = 28$
51. $3b - 12 = 45$
52. $7s - 9 = 68$
53. $8y + 5 = 37$
54. $8t - 14 = 26$

Write each sentence as an equation. Then solve.

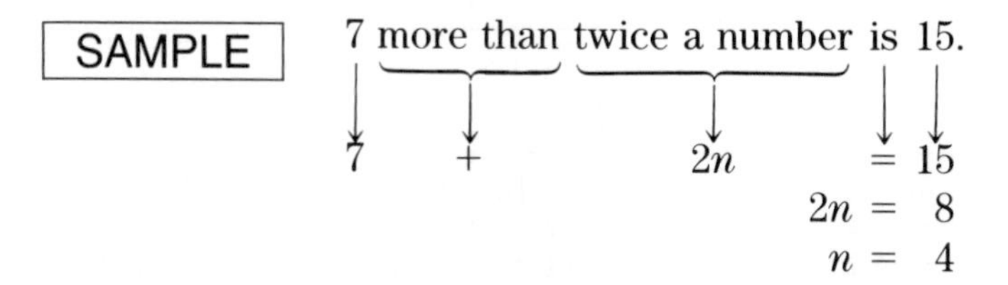

B 55. 8 more than twice a number is 32.

56. 7 more than 3 times a number is 25.

57. Twice a number increased by 4 is 16.

58. Five times a number decreased by 17 is 53.

59. When 5 is added to 4 times a number, the result is 17.

60. When 6 is subtracted from 3 times a number, the result is 21.

61. When 15 is subtracted from twice a number, the result is 25.

62. When 11 is added to 5 times a number, the result is 66.

C **Can you solve this cross-number puzzle?**

Across

1. $x + 7 = 22$

3. $m - 6 = 296$

5. $w - 34 = 4295$

8. $a + 15 = 49$

11. $\frac{x}{2} = 86$

12. $52 = n - 9$

13. $b + 9 = 874$

14. $3x = 39$

16. $5x = 105$

18. $\frac{w}{4} = 51$

20. $2m - 19 = 39$

22. $3a + 11 = 341$

24. $5y + 9 = 69$

25. $n - 50 = 1992$

27. $3x - 21 = 1479$

28. $9b + 24 = 150$

1	2			3		4	
	5	6	7			8	9
10		11				12	
13				14	15		
		16	17		18		19
20	21		22	23			
24			25			26	
	27					28	

Down

2. $b + 5 = 59$

3. $x - 12 = 3909$

4. $m + 6 = 242$

6. $\frac{a}{8} = 394$

7. $n + 5 = 32$

9. $3x = 123$

10. $2y = 36$

15. $b - 56 = 3148$

17. $2a + 9 = 2249$

19. $4x - 9 = 175$

20. $3m + 4 = 67$

21. $\frac{n}{25} = 37$

23. $9y - 47 = 43$

26. $6m + 11 = 137$

SELF-TEST

Solve. *(2-4)*

1. $3x - 6 = 0$

2. $4n - 5 = 7$

3. $2a + 3 = 11$

4. $5r + 4 = 34$

5. $3t - 4 = 17$

6. $6x - 1 = 35$

7. $7s + 3 = 66$

8. $5a + 4 = 59$

9. $9n - 4 = 32$

10. $4t - 15 = 13$

11. $2r - 9 = 31$

12. $6x + 5 = 35$

5 A Mind-Reading Trick

Here's a trick to play on a friend. You will ask your friend to think of a number, then give some steps to compute. Then you'll ask for the result. Like magic you'll be able to find your friend's original number!

	This is how your friend's arithmetic might go.	This is your algebra.
Step 1: Think of a whole number.	4	n
Step 2: Add 8.	12	$n + 8$
Step 3: Multiply by 5.	60	$5(n + 8)$, or $5n + 40$
Step 4: Subtract 37.	23	$5n + 3$
Step 5: Multiply by 20.	460	$20(5n + 3)$, or $100n + 60$
Step 6: Tell me your answer.	460	$100n + 60$

To find your friend's original number, solve.

$$460 = 100n + 60$$
$$400 = 100n$$
$$4 = n$$

This is it! → $4 = n$

COMPARING ALGEBRA AND ARITHMETIC

1. Algebra is more powerful than arithmetic.

Arithmetic	Algebra
4	n
12	$n + 8$
60	$5n + 40$
23	$5n + 3$
460	$100n + 60$

2. Algebra is often easier than arithmetic.

Arithmetic	Algebra
Try doubling 789 in your head. . . . not easy	Try doubling n in your head. . . . easy . . . $2n$

Written Exercises

Copy and complete.

A

1.

Steps	Arithmetic	Algebra
Think of a number.	8	n
Multiply it by 5.	?	?
Add 4.	?	?
Multiply by 20.	?	?
Subtract 80.	?	?

2.

Steps	Arithmetic	Algebra
Think of a number.	6	n
Add 9.	?	?
Multiply by 4.	?	?
Subtract 30.	?	?
Multiply by 25.	?	?

Using algebra, find the original number.

3.

Steps
Think of a number.
Add 8.
Multiply by 2.
Subtract 6.
Divide by 2.
Subtract 5.

Result: 9
Original number: ?

4.

Steps
Think of a number.
Add 5.
Multiply by 2.
Subtract 6.
Multiply by 5.
Subtract 4.

Result: 56
Original number: ?

5.

Steps
Think of a number.
Multiply by 5.
Add 6.
Multiply by 4.
Add 9.
Multiply by 5.

Result: 1365
Original number: ?

6. Make up a mind-reading trick of your own.

B

7. Try this mind-reading trick on a friend. The steps are given at the right. After all the steps are computed, tell your friend that the final answer is 20. It will always be 20. Why?

Steps
Think of a number.
Add 10.
Multiply by 2.
Add 5.
Multiply by 4.
Add 60.
Divide by 8.
Subtract the original number.

6 Combining Terms

Sometimes you have to combine like terms to solve an equation.

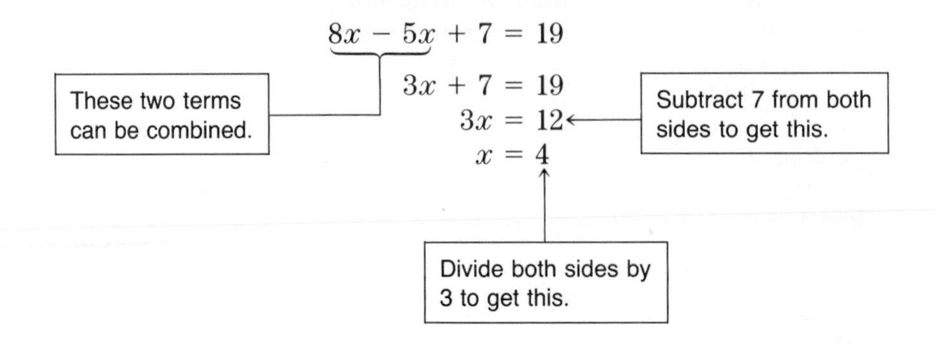

It is a good idea to check your work. Substitute your solution, 4, for x in the equation.

Check to see that both sides are equal.

$$8x - 5x + 7 = 19$$

$(8 \cdot 4) - (5 \cdot 4) + 7$	19
$32 - 20 + 7$	
$12 + 7$	
19	

✓ It checks.

Written Exercises

Solve. Check.

A

1. $4a + 9a - 6a = 42$ **2.** $6y - 4y + y = 18$ **3.** $9c + 4c - c = 48$

4. $9 + 6a - 2a = 21$ **5.** $8a + 7 - 5a = 31$ **6.** $9a - 4 + 3a = 32$

7. $13 + 5b - 2b = 37$ **8.** $8b + 4b - b = 55$ **9.** $12c - 8 - 5c = 27$

10. $8x + 4 - 3x = 24$ **11.** $7n + 2 + 3n = 12$ **12.** $13x + 5 - 4x = 23$

13. $4x + 5x - 3 = 15$ **14.** $4x - x + 1 = 22$ **15.** $7x + 3x - 5 = 15$

16. $12 = 3c - 12 + 5c$ **17.** $9 = 13x - 7 - 5x$ **18.** $35 = 19m - 4m + 5$

19. $18 = 8y - 7 - 3y$ **20.** $21 = 4a - 3a + 6a$ **21.** $16 = 3x + 2x - 9$

22. $21 = 4n - 3 - n$ **23.** $32 = 8x + 4 - 4x$ **24.** $7 = 3a + 5a - a + 7$

Write an equation. Solve.

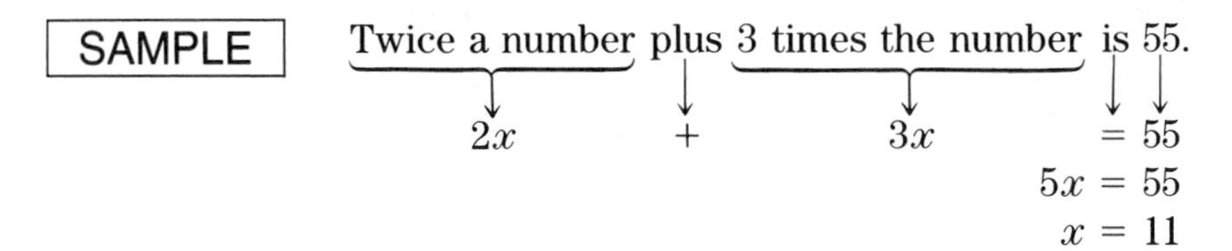

25. Four times a number plus 3 times the same number is 63. What is the number?

26. The difference between $7y$ and $3y$ is 20. Find the value of y.

27. A number is doubled, then 5 is added. The result is 73. Find the number.

28. A number is multiplied by 3 and then 7 is subtracted. The result is 83. Find the number.

B 29. I am thinking of a number. When I double it and add 9, I get 25. What is the number?

30. I am thinking of a number. When I multiply it by 3 and subtract 11, I get 37. What is the number?

SELF-TEST

Using algebra, find the original number.

(2-5)

1.

Steps
Think of a number. Add 4. Multiply by 3. Subtract 3. Divide by 3. Subtract 1.

Result: 5
Original number: ___?___

2.

Steps
Think of a number. Multiply by 5. Add 4. Multiply by 2. Subtract 4. Multiply by 3.

Result: 42
Original number: ___?___

Solve. Check.

(2-6)

3. $x + 3x - 2x = 12$

4. $4a - 2a + a = 42$

5. Three times a number plus five times the same number is 40. What is the number?

7 Writing Algebraic Expressions

Before you can use algebra to solve real-life problems you will need practice in using variables to translate sentences into equations.

The soccer team won 7 more games than they lost.
Let x be the number of games lost.
Write an expression to show the number won.

The number won is 7 more than the number lost.

The number won $= 7 \quad + \quad x$

So, the number won is $7 + x$.

Here are some more examples.

EXAMPLE 1

Laura has twice as much homework as Ann.
Let x be Ann's amount of homework.
Then ___?___ is Laura's amount.

Laura's amount is twice Ann's amount.

Laura's amount $= \quad 2 \quad x$

So, Laura's amount $= 2x$.

EXAMPLE 2

Let x be a whole number.
Then the next two whole numbers are ___?___ and ___?___.

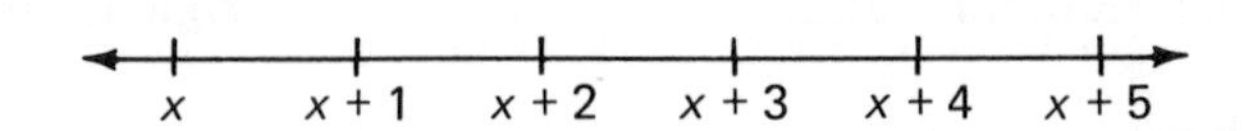

The next two whole numbers are $x + 1$ and $x + 2$.

Classroom Practice

Complete.

1. The Packers won 5 more games than they lost.
   Let x = number of games lost.
   Then __?__ = number of games won.

2. The Tigers had twice as many hits as the Yanks.
   Let x = number of hits by the Yanks.
   Then __?__ = number of hits by the Tigers.

3. The length of a rectangle is 3 units more than the width.
   Let x = the width.
   Then __?__ = the length.

4. Jo is 5 cm taller than Kathy.
   Let k = Kathy's height.
   Then __?__ = Jo's height.

5. Jo is 5 cm taller than Kathy.
   Let j = Jo's height.
   Then __?__ = Kathy's height.

6. Mike is 3 years older than Jim.
   Let j = Jim's age.
   Then __?__ = Mike's age.

7. Mike is 3 years older than Jim.
   Let m = Mike's age.
   Then __?__ = Jim's age.

8. Howard strikes out 5 times as often as Lucy.
   And Lucy strikes out twice as often as Raymond.
   Let K = number of times Raymond strikes out.
   Then __?__ = number of times Lucy strikes out,
   and __?__ = number of times Howard strikes out.

Written Exercises

Complete.

A

1. Jan is 5 years older than Nan.
   Let n = Nan's age.
   Then __?__ = Jan's age.

2. Jan is 5 years older than Nan.
   Let j = Jan's age.
   Then __?__ = Nan's age.

3. Rose is 2 cm taller than Joan.
   Let j = Joan's height.
   Then __?__ = Rose's height.

4. Rose is 2 cm taller than Joan.
   Let r = Rose's height.
   Then __?__ = Joan's height.

Complete.

5. There are 11 more girls here than boys.
   Let b = number of boys.
   Then __?__ = number of girls.

6. There are 11 more girls here than boys.
   Let g = number of girls.
   Then __?__ = number of boys.

7. Pete has three times as much money as Mike.
   Let x = Mike's amount of money.
   Then __?__ = Pete's amount of money.

8. There are 180 more students than teachers.
   Let x = number of teachers.
   Then __?__ = number of students.

9. The length of a rectangle is four times the width.
   Let x = the width.
   Then __?__ = the length.

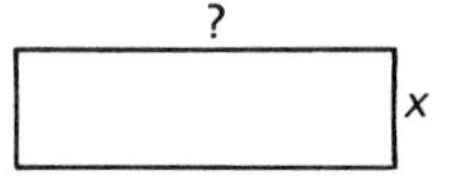

10. The temperature today is 12° higher than yesterday.
    Let t = yesterday's temperature.
    Then __?__ = today's temperature.

11. The Dragons scored 3 times as many touchdowns as the Warriors.
    Let x = number of touchdowns scored by the Warriors.
    Then __?__ = number of touchdowns scored by the Dragons.

12. The Bucs have won 6 more games than they have lost.
    Let n = number of games lost.
    Then __?__ = number of games won.

13. Mac is x years old.
How old will he be next year?

14. Judy is x years old.
How old will she be in 10 years?

15. Roger is r years old.
How old was he last year?

16. Trish is t years old.
How old was she 7 years ago?

17. Karen will be m years old next year.
How old is she this year?

18. Jeff is 15 years old.
How old will he be x years from now?

B

19. Pete worked 4 hours more than Quinn.
Quinn worked 2 hours more than Rob.
Let x = the number of hours Rob worked.
Then __?__ = the number of hours Quinn worked,
and __?__ = the number of hours Pete worked.

20. Manuel walked 4 km farther than Nick.
Ollie walked 3 km less than Nick.
Let y = Nick's distance in km.
Then __?__ = Manuel's distance,
and __?__ = Ollie's distance.

21. Karl is 3 years younger than Sherry.
Let k = Karl's age.
Then __?__ = Sherry's age.
A year ago, Karl's age was __?__,
and Sherry's age was __?__.

22. Let n = an odd number.
Then the next two odd numbers are __?__ and __?__.

23. Let x = an even number.
The next two even numbers are __?__ and __?__.

24. Frances won one game more than Keith.
Will won twice as many as Frances.
Let k = the number of games Keith won.
Then __?__ = the number Frances won,
and __?__ = the number Will won.

8 Applied Problems and Puzzles

Now you can use what you have just learned to solve problems.

A basketball team played 25 games. They won 7 more games than they lost. How many games did they win and lose?

Let x = number of games lost.
Then $x + 7$ = number of games won.

$$\underbrace{\text{number lost}}_{x} + \underbrace{\text{number won}}_{x+7} = 25$$

$$\begin{aligned} x + x + 7 &= 25 \\ 2x + 7 &= 25 \\ 2x &= 18 \\ x &= 9 \end{aligned}$$

Answer: $x = 9$ → games lost
$x + 7 = 16$ → games won

Check: games lost = 9
games won = 16
total games = 25 ✓

A GUIDE FOR PROBLEM SOLVING
1. *Read the problem. Read it more than once if you wish.*
2. *Let a variable stand for one of the unknowns. Show the other unknowns in terms of that variable.*
3. *Write an equation and solve it.*
4. *Answer the question.*
5. *Check your answer.*

EXAMPLE

1. *Read the problem.*
   Ed has twice as much money as Jake.
   Together they have \$54.
   How much money does each have?

2. Let x = amount Jake has.
   Then $2x$ = amount Ed has.

3. $\underbrace{\text{Ed's amount}} + \underbrace{\text{Jake's amount}} = \$54$

$$2x \quad + \quad x \quad = 54$$
$$3x = 54$$
$$x = 18$$

4. *Answer*: $x = 18 \longrightarrow$ Jake has \$18.
   $2x = 36 \longrightarrow$ Ed has \$36.

5. *Check*: $18 + 36 = 54$ ✓

Classroom Practice

Solve.

1. There are 15 more girls than boys in a summer camp.
   There are 135 campers in all.
   How many boys and how many girls are there?
2. Janet has three times as much money as Peter.
   Together they have \$40.
   How much money does each have?

Written Exercises

Solve.

A

1. There are 80 computers in Central School.
   There are 32 more computers with color monitors than with black and white monitors.
   How many of the computers have black and white monitors?
2. Carmen's golf score is 6 strokes less than Linda's.
   Their two scores total 156.
   What is each girl's score?

Solve.

3. South High School has 85 students more than North High School.
The total number for both schools is 1495.
How many students are at South High?

4. The Chargers played 64 games.
They won 22 more games than they lost.
How many games did they win and lose?

5. A car dealer sold 37 vehicles.
There were 5 more cars sold than trucks.
How many cars were sold?

6. Luis has $48 more than Ed.
Together they have $200.
How much money does each man have?

7. A hog weighs twice as much as a sheep.
Together they total 285 kg.
How much does each weigh?

8. An apple has twice as many calories as a peach.
The two fruits together have 105 calories.
How many calories are in each fruit?

9. The Bees played 153 ball games.
They won 17 more than they lost.
How many games did they win?

10. The Lions scored a total of 114 points in their first two games.
They scored twice as many points in the second game as in the first.
How many points did they score in their first game?

Consecutive numbers are numbers that follow each other in order.
Solve these problems about consecutive numbers.

11. Find two consecutive whole numbers that total 75.

12. Find three consecutive whole numbers that total 75.

B

13. Route 2 is twice as long as Route 1.
Route 3 is 30 km longer than Route 2.
Together the three roads are 280 km long.
How long is Route 3?

14. There were 82 people at a family reunion.
There were 4 more men than women.
There were 10 more children than adults.
How many children were there?

C **15.** "We made \$450 profit," said Liz. "How do we split it up?"
"Well," said Sue, "you worked twice as long as I did,
and Wendy worked three times as long as I."
How much should each girl get?

SELF-TEST

Complete.

1. Jan is 3 cm shorter than Tom.
   Let m = Jan's height.
   Then __?__ = Tom's height.

2. Vera is 3 years older than Len. *(2-7)*
   Let x = Len's age.
   Then __?__ = Vera's age.

Solve.

3. Peg has \$37 more than Hal.
   Together they have \$175.
   How much does Hal have?

4. The Reds played 162 games. *(2-8)*
   They won twice as many as
   they lost.
   How many games did they win?

CAREER NOTES

TV STUDIOS

Not all jobs in a TV studio are in front of the camera. There are opportunities in advertising, writing, producing, directing, and in many technical fields.

After a television program or news feature has been filmed, film editors prepare the film for broadcast. They trim the film segments to specified lengths, and cut and splice film to insert commercials.

During the airing of a television program, broadcast technicians operate equipment that controls the sounds and pictures being shown.

Before a television news program can be produced and aired, researchers and reporters must gather information and write concise, accurate news items. Before giving a weather report on a news program, meteorologists also gather information by studying data from weather satellites and from observers all over the world.

9 Variable on Both Sides of the Equation

In this section we shall consider equations with the variable on both sides. You'll use several steps to get the variable alone on one side of the equation.

$$\begin{aligned} 8x &= 10 + 3x \\ 8x - 3x &= 10 + 3x - 3x \\ 5x &= 10 \\ x &= 2 \end{aligned}$$

Check:

$8x$	$= 10 + 3x$
$8 \cdot 2$	$10 + 3 \cdot 2$
16	$10 + 6$
	16 ✓

Sometimes it's easier to get the variable alone on the right side of the equation.

$$\begin{aligned} 15 - 4x &= x \\ 15 - 4x + 4x &= x + 4x \\ 15 &= 5x \\ 3 &= x \end{aligned}$$

Check:

$15 - 4x$	$= x$
$15 - 4 \cdot 3$	3
$15 - 12$	
3	✓

Sometimes you may want to skip some steps when solving an equation. Remember to skip *only* those steps that are easy for you to do in your head.

All steps given

$$\begin{aligned} 6x - 7 &= 4x + 13 \\ 6x - 7 + 7 &= 4x + 13 + 7 \\ 6x &= 4x + 20 \\ 6x - 4x &= 4x + 20 - 4x \\ 2x &= 20 \\ x &= 10 \end{aligned}$$

Some steps skipped

$$\begin{aligned} 6x - 7 &= 4x + 13 \\ 6x &= 4x + 20 \\ 2x &= 20 \\ x &= 10 \end{aligned}$$

Classroom Practice

Tell what steps you would take to solve the equation.

SAMPLE $6x = 8 + 2x$ *What you say:* Subtract $2x$ from both sides, then divide both sides by 4.

1. $5x = 8 + 3x$ **2.** $8x = 2x + 6$ **3.** $4y = 2y + 10$ **4.** $5 - 3y = 2y$

5. $12 - a = 5a$ **6.** $6 + 3b = 5b$ **7.** $2a + 4 = 4a$ **8.** $4c + 4 = 5c$

Written Exercises

Solve.

A

1. $2x = 14 + x$
2. $7x = 2x + 5$
3. $9x = 5x + 16$
4. $6y = 18 - 3y$
5. $4n = 21 - 3n$
6. $6x = 20 - 4x$
7. $8x = 25 + 3x$
8. $9x = 26 - 4x$
9. $12x = 10 + 7x$
10. $6 - 2x = x$
11. $16 - 3x = x$
12. $20 - 6a = 4a$
13. $7 - 3x = 4x$
14. $10 + x = 6x$
15. $12 + 3y = 9y$
16. $16 + 5a = 13a$
17. $14 - 6y = 8y$
18. $25 - 2n = 3n$
19. $6y - 3 = 3y$
20. $4x - 2 = 3x$
21. $8n - 6 = 5n$
22. $5x - 9 = 2x$
23. $12n - 20 = 2n$
24. $15x - 12 = 12x$
25. $6x - 5 = 4x + 9$
26. $13x - 7 = 5x + 1$
27. $8a + 6 = 3a + 21$
28. $4x + 4 = 2x + 6$
29. $7a - 3 = 3a + 5$
30. $10n - 8 = 3n - 1$
31. $5b + 8 = 3b + 26$
32. $18 + 3x = x + 30$
33. $7 - 4x = 5x - 2$
34. $5y + 9 = 8y - 15$
35. $7c + 9 = 11c - 31$
36. $8a - 5 = 3a + 25$

Write an equation. Solve for the variable.

SAMPLE

$\underbrace{\text{5 times a number}}$ is 12 $\underbrace{\text{more than}}$ $\underbrace{\text{2 times the number}}$.

$$\begin{aligned} 5n &= 12 \quad + \quad 2n \\ 5n - 2n &= 12 + 2n - 2n \\ 3n &= 12 \\ n &= 4 \end{aligned}$$

B

37. 7 times a number is 12 more than 3 times the number.
38. 4 times a number is 18 more than the number.
39. If you double a number and add 24, the result is 8 times the number.
40. If you subtract a certain number from 30, the result is 4 times the number.
41. If you add 10 to a certain number, the result is 6 times the number.
42. If you triple a number and add 2, the result is 8 more than the number.

10 Equations with Parentheses

You can get rid of parentheses in an equation by using the distributive property. The equation will then be one you already know how to solve.

EXAMPLE 1

$$4(x + 2) = 20$$

$$\begin{aligned} 4x + 8 &= 20 \\ 4x + 8 - 8 &= 20 - 8 \\ 4x &= 12 \\ x &= 3 \end{aligned}$$

Check: $4(x + 2) = 20$

$4(3 + 2)$	20
$4 \cdot 5$	
20	✓

EXAMPLE 2

$$3(x - 1) = 2x$$

$$\begin{aligned} 3x - 3 &= 2x \\ 3x - 3 + 3 &= 2x + 3 \\ 3x &= 2x + 3 \\ 3x - 2x &= 2x + 3 - 2x \\ x &= 3 \end{aligned}$$

Check: $3(x - 1) = 2x$

$3(3 - 1)$	$2 \cdot 3$
$3 \cdot 2$	6
6	✓

Of course you may want to skip some steps after you have solved a few equations like this, but do that *only* if you are ready.

Classroom Practice

Use the distributive property to simplify.

1. $3(x - 2)$ **2.** $6(x - 1)$ **3.** $5(x + 4)$ **4.** $2(3x - 7)$

5. $4(2x + 3)$ **6.** $2(a + 4) + a$ **7.** $5(a - 1) + 5$ **8.** $4(1 + x) + 5x$

Tell the first step you would take to solve.

9. $2(x - 3) = x$ **10.** $2(x - 2) = 4$ **11.** $2(x + 3) = 3(x - 4)$

Solve.

12. $3(x - 1) = 6$ **13.** $5(x + 2) = 15$ **14.** $4(a - 2) = 3(a + 1)$

15. $4(y + 4) = 12y$ **16.** $3(a + 2) = a + 8$ **17.** $2(c + 3) = 5(c - 3)$

Written Exercises

Use the distributive property to simplify.

A

1. $3(2x - 5)$
2. $5(6 - 2a)$
3. $8(2x - 5)$
4. $7(3 - 5a)$
5. $6(8y + 4)$
6. $7(4x + 12)$
7. $8(3 + 6b)$
8. $9(2 + 5y)$

Solve.

9. $2(x + 3) = 10$
10. $3(2x - 1) = 9$
11. $2(6 + 2a) = 24$
12. $5(x - 2) = 0$
13. $8(y - 4) = 0$
14. $6(x + 1) = 36$
15. $2(x - 5) = x$
16. $3(x - 2) = 2x$
17. $4(x - 1) = 2x$
18. $3(y + 4) = 5y$
19. $6(y - 2) = 3y$
20. $2(3 - y) = 4y$
21. $4(x - 3) = 0$
22. $2(2x + 2) = 6x$
23. $7(2 + y) = 14$
24. $5(x - 3) = 2x + 3$
25. $5(x - 1) = 7 + x$
26. $7(x + 1) = 9 + 5x$
27. $4(y - 1) = 2y + 6$
28. $3(x - 3) = x + 1$
29. $5(y + 4) = 2y + 20$
30. $2(z + 2) = 3z + 2$
31. $4(m + 3) = 6m + 8$
32. $3(x - 4) = 2x + 5$

Solve and check.

B

33. $6(x - 1) = 3(x + 1)$
34. $2(x + 3) = 3(x - 3)$
35. $8(x - 1) = 4(x + 4)$
36. $7(2a - 4) = 2(a + 4)$
37. $3(a - 2) + a = 2(a + 1)$
38. $5(x + 2) = x + 6(x - 3)$
39. $4(m + 3) - 2m = 3(m - 3)$
40. $2(a + 4) = 2(a - 4) + 4a$
41. $7(b + 2) - 4b = 2(b + 10)$
42. $3(x + 2) = 3(x - 2) + 3x$

For each equation tell which of the following is true.

a. No number is a solution.
b. One number is a solution.
c. All numbers are solutions.

C

43. $2(x + 5) = 3x + 4 - x$
44. $6(2x + 3) = 2(9 + 6x)$
45. $7(x - 5) = 2(x + 5)$
46. $4(x + 1) = 4x + 1$
47. $6(x + 3) = 2(3x + 1)$
48. $2(8x + 6) = 4(4x + 3)$
49. $8(x + 1) = 4(x + 5)$
50. $5(2x + 3) = 10(x + 1)$
51. $2(3x + 9) = 3(2x + 6)$
52. $4(x + 2) = 5x + 3$

11 Puzzles

At a certain store, a pair of pants costs \$8 more than a shirt. Three pairs of pants cost the same as 5 shirts. How much does each item cost?

Let x = cost of a shirt.
Then $x + 8$ = cost of a pair of pants.

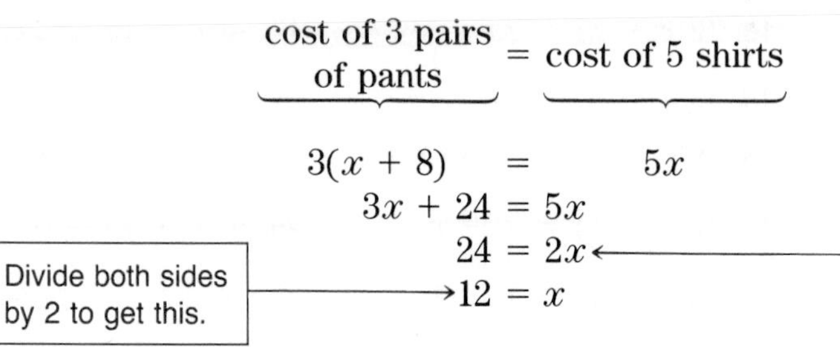

Subtract $3x$ from both sides to get this.

Answer: $x = 12$ ⟶ A shirt costs \$12.
$x + 8 = 20$ ⟶ A pair of pants costs \$20.

Check: cost of 3 pairs of pants = $3 \cdot \$20 = \$60$
cost of 5 shirts = $5 \cdot \$12 = \$60$ ✓

Written Exercises

Solve.

A

1. A whole melon has 20 more calories than a pear.
   Six pears have as many calories as five melons.
   How many calories are in each?

2. Sonya works 7 hours a week more than Marco.
   Sonya works as much in 3 weeks as Marco works in 4 weeks.
   How many hours does each work per week?

3. One number is 7 more than another.
   Twice the larger is 22 less than 4 times the smaller.
   Find the numbers.

4. One number is 5 less than another.
   Five times the smaller number is 1 less than 3 times the larger.
   Find the numbers.

5. Tran has twice as much money as Paul.
Paul has $8 more than Mona.
Together they have $236.
How much does each have?

6. Rich is 3 years older than Carla.
Ruth is twice as old as Rich.
Their ages total 33 years.
How old is each person?

B 7. Kent is 14 years older than Harlin.
Next year Kent will be 3 times as old as Harlin.
How old is Kent?

8. Beth is 14 years older than Peggy.
Last year Beth was three times as old as Peggy.
How old is Peggy?

C 9. A man has a daughter and a son.
The son is three years older than the daughter.
In one year the man will be six times as old as the daughter is now.
In ten years he will be fourteen years older than the combined ages of his children. What is the man's present age?

SELF-TEST

Solve.

(2-9)

1. $3x = 48 - x$
2. $2x - 4 = 5x - 7$
3. $24 + 4y = 10y$
4. $3z + 2 = 6z - 13$

(2-10)

5. $6(a - 3) = 12$
6. $4(m - 3) = m + 15$
7. $5(2x + 2) = 12x$
8. $2(b - 2) = b + 3$

(2-11)

9. Lou has twice as much money as Jo.
Jo has $11 more than Sherry.
Together they have $89.
How much does each have?

10. Al has twice as much money as Vic.
Vic has $5 less than Connie.
Together they have $125.
How much does each have?

PROBLEM SOLVING STRATEGIES

PARTS AND TOTALS

Word problems describe relationships between numerical quantities such as distances, amounts of money, or ages. Sometimes you can solve a problem by combining quantities to find a total. Sometimes you need to separate a quantity into its parts to solve a problem.

When you read a problem, ask yourself the following questions.

1. Do I need to separate a quantity?
2. Do I need to combine quantities?
3. Does the problem ask for a total or for some part of the total?

The following example involves parts and totals.

EXAMPLE

It takes Maria a total of one hour to walk to school and return home by bus.
If she were to take the bus both ways it would take her a total of 50 minutes.
How long would it take her to walk both ways?

Let b = the number of minutes needed for a one-way bus ride.
Use b to write an equation separating the 50-minute trip into its parts.

$$b + b = 50$$

Now find the time required for a one-way bus trip.

$$2b = 50$$
$$b = 25$$

It takes Maria 25 minutes to ride the bus one way.

Answer the following questions to solve the problem.

During her trip by foot and by bus, how many minutes did Maria spend riding the bus? 25 minutes
How many minutes did she spend walking? $60 - 25 = 35$; 35 minutes
How long will it take Maria to walk to school? 35 minutes
How long will it take her to walk to and from school?
$35 + 35 = 70$; 70 minutes

Answer: 70 minutes

EXERCISES

Solve. Check your answer.

1. It usually takes Frank a total of 40 minutes to ride his bicycle to the beach and return home.
   Yesterday he rode his bicycle to the beach but had to get a ride home by car because his bicycle chain broke.
   That round trip took a total of 25 minutes.
   How long would a trip to the beach and back by car take?
   - **a.** Separate the 40-minute bike trip into its parts. How many minutes would a one-way bike trip take?
   - **b.** In his trip by bike and by car, how many minutes did Frank spend riding his bike? How many minutes did he spend in the car?
   - **c.** How long would a car ride to the beach take? How long would it take Frank to ride by car to and from the beach?

2. Jessica earned \$44 by selling 20 plants and 12 jars of honey. The price for each jar of honey is twice the price of each plant. What is the price of each jar of honey?

   - **a.** If each plant costs p dollars, then write an expression in terms of p for the following:
     - I. the price of each jar of honey
     - II. the price of 20 plants
     - III. the price of 12 jars of honey
   - **b.** Write an equation to represent the total cost of 20 plants and 12 jars of honey.
   - **c.** Solve the equation. Answer the question.

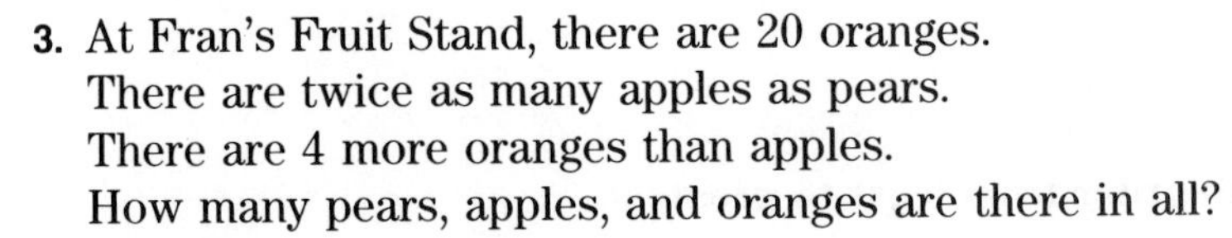

3. At Fran's Fruit Stand, there are 20 oranges.
   There are twice as many apples as pears.
   There are 4 more oranges than apples.
   How many pears, apples, and oranges are there in all?
   - **a.** Suppose there are p pears. Complete the following.
     Give your answers in terms of p.
     Then __?__ = number of apples.
     Also, __?__ = number of oranges.
   - **b.** What information has not been used in part (a)? Use this information to write an equation.
   - **c.** How many pears are there? How many apples?
   - **d.** Answer the question in the problem.

SKILLS REVIEW

ORDER AND THE NUMBER LINE

The number line helps you compare numbers.

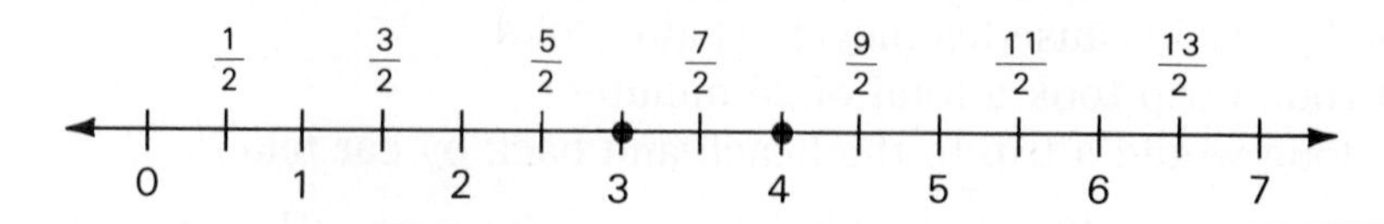

3 is to the *left* of 4. 3 is *less* than 4.

Compare the numbers. Write > or <.

1. 2 __?__ 7 **2.** 7 __?__ 5 **3.** 4 __?__ 3 **4.** 3 __?__ 7

5. 11 __?__ 3 **6.** 21 __?__ 12 **7.** 5 __?__ 4 **8.** 101 __?__ 110

9. $\frac{3}{2}$ __?__ $\frac{5}{2}$ **10.** 7 __?__ $\frac{7}{2}$ **11.** $\frac{1}{3}$ __?__ $\frac{4}{3}$ **12.** $\frac{5}{3}$ __?__ $\frac{2}{3}$

Copy the number line. Graph the numbers.

SAMPLE $3\frac{3}{4}$

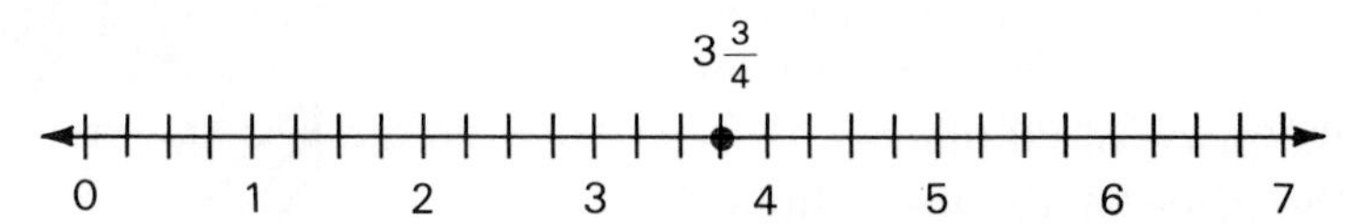

13. 2 **14.** 5 **15.** 1 **16.** 6

17. $\frac{3}{4}$ **18.** $2\frac{1}{4}$ **19.** $6\frac{3}{4}$ **20.** $1\frac{1}{2}$

Write the numbers in order from least to greatest.

21. $\frac{1}{2}$, 0, 2 **22.** 23, 32, 18 **23.** 7, $\frac{1}{7}$, 17

24. 15, 51, 5 **25.** $\frac{2}{5}$, $\frac{6}{5}$, 1 **26.** 107, 100, 170

27. $\frac{1}{2}$, $\frac{3}{4}$, $\frac{1}{4}$ **28.** 45, 54, 44 **29.** $\frac{5}{3}$, $\frac{5}{1}$, $\frac{5}{6}$

30. $\frac{7}{5}$, 1, $\frac{9}{10}$ **31.** $2\frac{1}{3}$, $2\frac{1}{5}$, $2\frac{1}{4}$ **32.** 2^4, 3^3, 5^2

CHAPTER REVIEW

CHAPTER SUMMARY

1. To solve an equation, get the variable by itself on one side. You can use one or more of the following methods.

   Add the same number to both sides of the equation.

   Subtract the same number from both sides.

   Divide both sides by the same number.

   Multiply both sides by the same number.

   You may also need to do the following.

   Combine like terms.

   Use the distributive property to get rid of parentheses.

2. To solve a word problem, follow the steps outlined in the Guide for Problem Solving on page 58.

REVIEW EXERCISES

Solve. *(See pp. 38–49.)*

1. $a - 5 = 13$

2. $x + 9 = 24$

3. $3b = 24$

4. $5m = 35$

5. $2x - 7 = 9$

6. $3y + 4 = 19$

7. $\frac{b}{2} = 7$

8. $\frac{x}{3} = 13$

9. $\frac{m}{4} = 7$

10. $5a + 9 = 29$

11. $3x - 11 = 52$

12. $8y - 15 = 41$

13. $12a - 7 = 17$

14. $19 = x + 7$

15. $25 = b - 9$

16. $3n + 4 = 40$

17. $6x - 9 = 21$

18. $4s - 11 = 41$

Write an equation. Solve for the variable. *(See pp. 46–49.)*

19. 7 more than twice a number is 25.

20. The difference between $9y$ and $3y$ is 42.

Solve. *(See pp. 52–53.)*

21. $4x - 2x + 5x = 14$

22. $15y - 12 + 3y = 24$

23. $20 = 5z - 8 - z$

Complete. *(See pp. 54–57.)*

24. Lea is 5 in. taller than José.
Let x = José's height in inches.
Then ___?___ = Lea's height.

25. Amy is 3 years older than Becky.
Let a = Amy's age.
Then ___?___ = Becky's age.

Solve. *(See pp. 58–61, 66–67.)*

26. Marta has 3 times as much money as Phil.
Together they have $44.
How much does each have?

27. The Tigers played 48 games.
They won twice as many as they lost.
How many did they win?

28. There are 900 students in Wayne High School.
There are 20 more girls than boys.
How many girls are there?

29. A board 400 cm long is cut into 2 pieces.
One piece is 66 cm longer than the other.
Find the length of the shorter piece.

30. Mona has $8 more than Cindy.
Cindy has twice as much money as Bert.
Together they have $63.
How much does each have?

31. I'm thinking of a number.
If I add 7 to it, then multiply by 3, I get 36.
What is my number?

32. I'm thinking of a number.
If I subtract 8 from it, then multiply by 6, I get 24.
What is my number?

Solve and check. *(See pp. 62–65.)*

33. $8x = 25 + 3x$

34. $9t = 40 - t$

35. $7x - 10 = 2x$

36. $3y + 12 = 7y$

37. $11a - 6 = 5a + 24$

38. $3(a - 4) = a$

39. $4(m - 4) = 2m - 2$

40. $5m + 7 = 3(m + 5)$

41. $x - 3 = 4(3 - x)$

42. $6(y + 2) = 5y + 18$

CHAPTER TEST

Solve.

1. $n - 5 = 5$ 2. $0 = x - 12$ 3. $y - 7 = 2$ *(2-1)*

4. $n + 10 = 18$ 5. $3 + a = 9$ 6. $20 = x + 8$ *(2-2)*

7. $6m = 42$ 8. $5x = 85$ 9. $70b = 210$ *(2-3)*

10. $\frac{x}{2} = 8$ 11. $\frac{m}{12} = 3$ 12. $\frac{y}{5} = 5$

13. $4x - 1 = 19$ 14. $7c + 8 = 36$ 15. $9t + 2 = 11$ *(2-4)*

16. Copy and complete the table at the right. *(2-5)*

Steps	Arithmetic	Algebra
Think of a number.	10	n
Multiply by 2.	?	?
Add 12.	?	?
Divide by 2.	?	?
Subtract 5.	?	?

Solve. Check.

17. $13 = 3a + 7 - a$ 18. $50 = 5t - 2t + 7t$ *(2-6)*

19. The cost of a notebook is four times the cost of a pencil. Let p = the cost of a pencil. Then _?_ = the cost of a notebook. *(2-7)*

20. Masami is 4 years younger than Darlene. Let m = Masami's age. Then _?_ = Darlene's age now and _?_ = her age next year.

21. A carrot has 25 more calories than a stalk of celery. Together they have 35 calories. How many calories are in each vegetable? *(2-8)*

Solve.

22. $12 - x = 2x$ 23. $9n - 7 = 5n + 17$ 24. $40 + a = 8a + 5$ *(2-9)*

25. $4(x - 3) = 3x$ 26. $6(y + 5) = 30$ 27. $8(x + 1) = x + 15$ *(2-10)*

28. Pedro has $4 more than Rita. Rita has three times as much money as Jim. Together they have $46. How much does each have? *(2-11)*

UNIT A CUMULATIVE REVIEW

If $n = 2$, find the value of the expression.

1. $5n$
2. $\frac{n - 2}{7}$
3. $6 \div n$
4. $\frac{n + 5}{7}$

Find the value of the expression.

5. $2 + 8 \div 2$
6. $(6 + 4) \div 5$
7. 5^3
8. $(25 \cdot 73) \cdot 4$

Simplify.

9. $5t + 12t$
10. $7n + 2 - 4n$
11. $4x + 8y + 2x - 3y$
12. $6 \cdot a \cdot a \cdot b \cdot b \cdot b$
13. $r \cdot 3 \cdot r \cdot 2$
14. $(3t)(8y)$
15. $y(3x - 4y)$
16. $8(a - 2) + 4a$
17. $5 + 3y - 5 - y$

Tell which of the numbers shown in color is a solution.

18. $t - 5 = 11$ 6 or 16?
19. $3(x - 4) = x + 2$ 5, 6, or 7?
20. $y - 3 < 8$ 10, 11, or 12?
21. $2n + 5 > 8$ 0, 1, or 2?

Solve.

22. $y - 3 = 8$
23. $n + 5 = 24$
24. $3x = 81$
25. $\frac{b}{7} = 21$
26. $5y + 3 = 18$
27. $7t - 2 = 19$
28. $5n + 8n - 2 = 50$
29. $7m - 4 = 3m$
30. $3(z + 2) = 8z + 1$

Write an equation and solve.

31. Seventeen less than 4 times a number is 11.
32. Five more than 8 times a number is 101.

Solve.

33. Miyoshi has $18 more than Jon.
    Together they have $56.
    How much money does each have?
34. Lea is 8 years older than Lester.
    In 4 years she will be twice his age.
    How old is each now?

UNIT B

Here's what you'll learn in this chapter:

- To compare integers.
- To draw the graph of the solution of a simple equation or inequality.
- To add, subtract, multiply, and divide positive and negative numbers.
- To simplify expressions.
- To find powers of positive and negative numbers.
- To solve equations involving negative numbers.

Colonies of penguins, such as the one shown in the photograph, are able to live in the Antarctic. The low temperatures of the world's cold regions are expressed as negative numbers.

Chapter 3

Positive and Negative Numbers

1 Positive and Negative Numbers

The number line is often used to picture numbers.

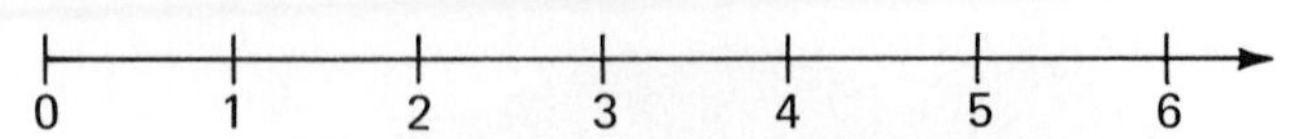

In algebra we use numbers to the left of zero too.

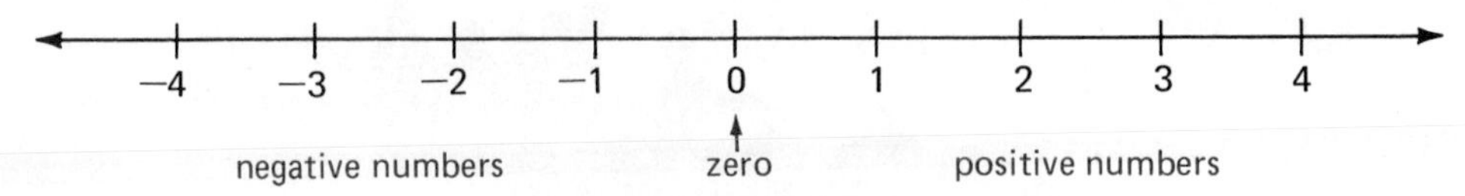

The numbers shown on the number line above form the set of **integers.**

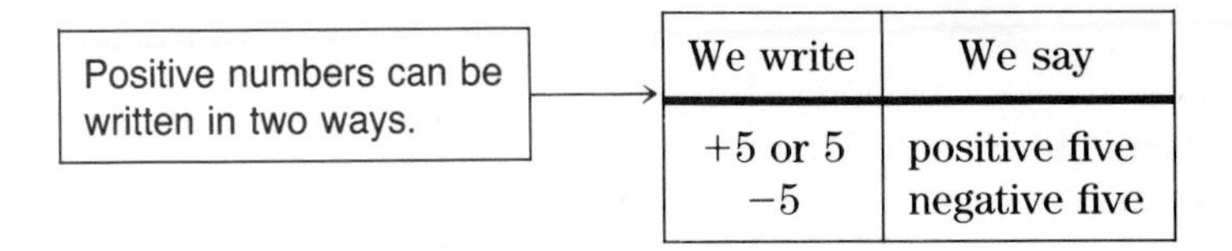

We write	We say
+5 or 5	positive five
−5	negative five

Positive and negative numbers are used in talking about temperatures. You can see that a thermometer is really like a number line.

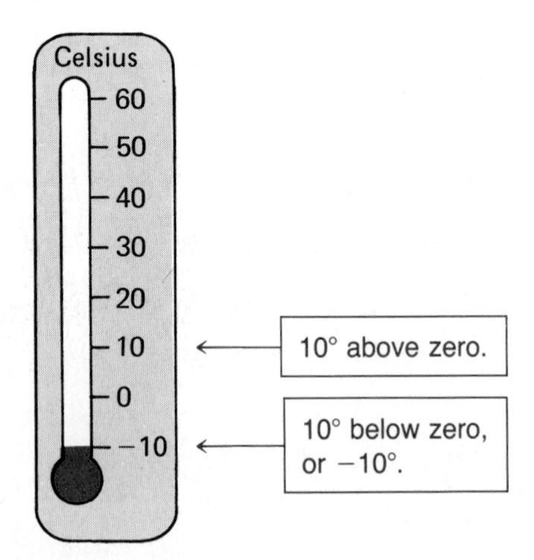

A number line helps to show relationships between numbers.

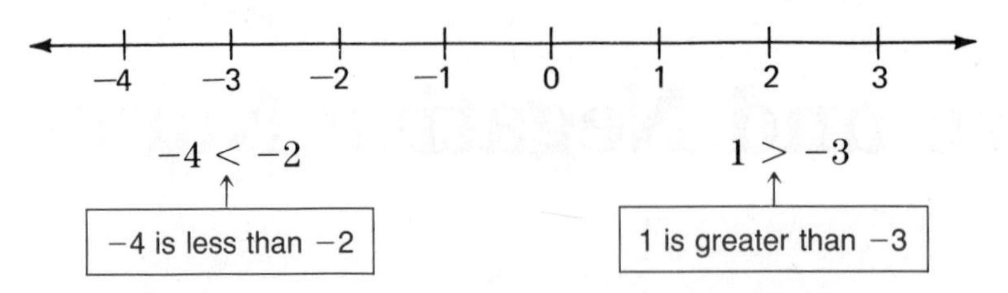

EXAMPLES

Compare −3 and −2.

$-3 < -2$

Compare 3 and −4.

$3 > -4$

Classroom Practice

1. +30 is read ___?___
2. −4 is read ___?___
3. +30° temperature means ___?___
4. −16 is read ___?___
5. If +20 means a gain, −20 means ___?___
6. If −5 means a fall, +5 means ___?___

Read aloud.

7. $4 > 3$
8. $1 < 5$
9. $-4 < -2$
10. $-2 > -4$

Written Exercises

Write a positive or negative number.

A

1. a temperature 25° above zero
2. a gain of $20
3. a loss of $10
4. a gain of 5 points
5. an ocean depth of 200 m
6. a loss of 7 points

Compare the numbers. Write > or <.

7. −3 ___?___ 0
8. −2 ___?___ 3
9. 5 ___?___ 0
10. 3 ___?___ −3
11. −1 ___?___ −4
12. −2 ___?___ 5
13. −4 ___?___ 4
14. −2 ___?___ −3
15. −6 ___?___ 3
16. −2 ___?___ −1
17. 7 ___?___ −8
18. −5 ___?___ −2

19. A diving bell is 274 meters below the surface of the water. Write this depth as a positive or negative number.

B

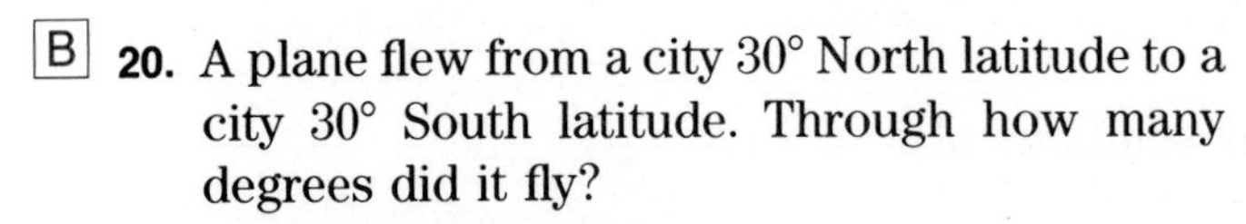

20. A plane flew from a city 30° North latitude to a city 30° South latitude. Through how many degrees did it fly?

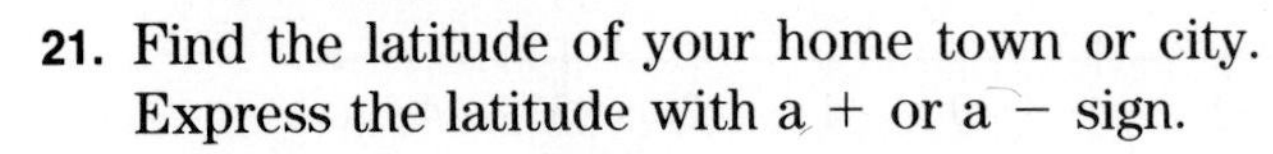

21. Find the latitude of your home town or city. Express the latitude with a + or a − sign.

2 Inequalities and Graphs

The number line is useful for picturing a solution of an equation.

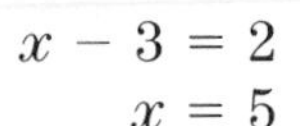

$$x - 3 = 2$$
$$x = 5$$

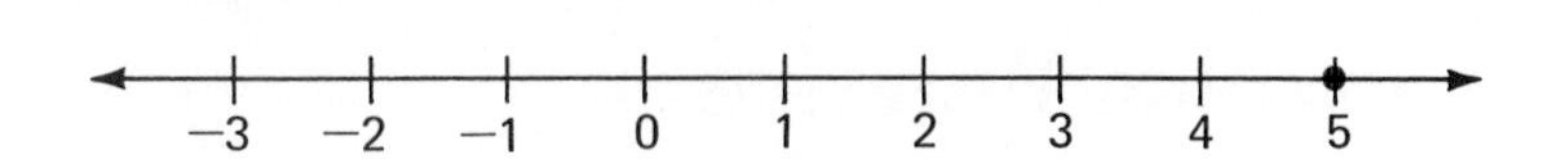

Think about inequalities. Sometimes there are an unlimited number of solutions. Take the inequality $x < 8$.

Some solutions: 7, 6, 5, 4, 3, 2, 1, 0, −1, −2, −3, . . .

means "and so on"

More solutions: all numbers *between* 8 and 7, *between* 7 and 6, . . .

In order to show *all* the solutions, we make a graph like this.

$x < 8$

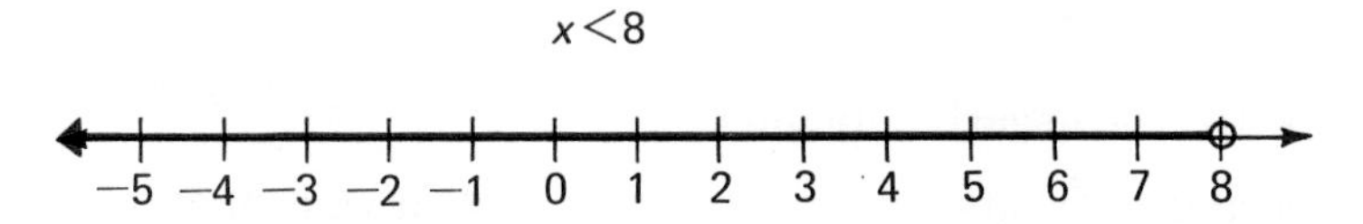

The open dot shows that 8 is not a solution.

Study these examples.

EXAMPLE 1 Graph the solution of $x + 3 = 5$.

$$x + 3 = 5$$
$$x = 2$$

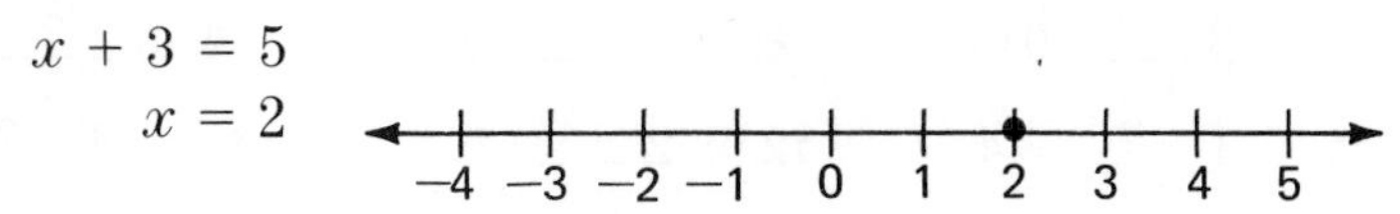

EXAMPLE 2 Graph the solutions of $x > -2$.

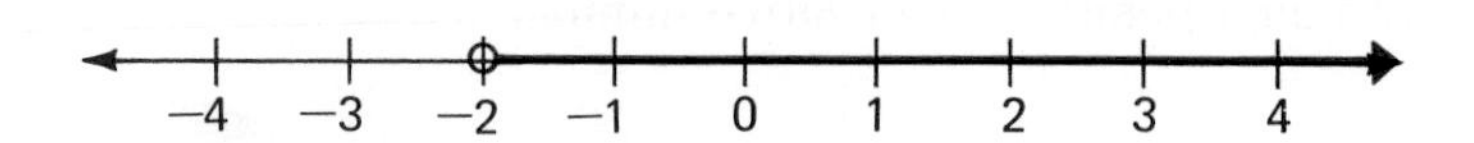

EXAMPLE 3 Graph the solutions of $x < -1$.

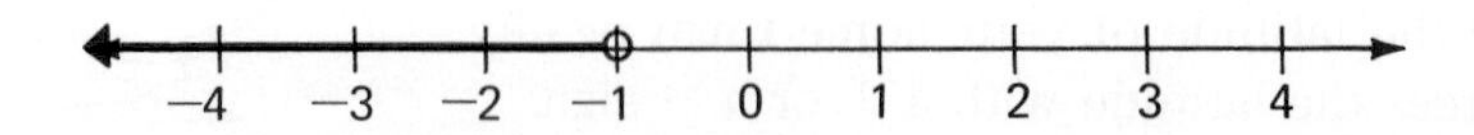

Sometimes we write an inequality sign and an equals sign together. We use just one line for the equals sign.

$x \geq 2$ means "x is greater than 2 or x is equal to 2."

$y \leq 0$ means "$y < 0$ or $y = 0$."

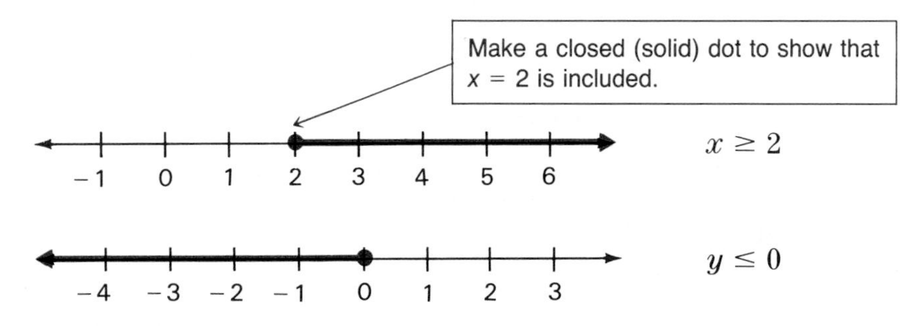

Classroom Practice

State an inequality comparing the given numbers.

1. 1, 0 **2.** 5, 2 **3.** −3, 6 **4.** 1, −1 **5.** 7, 4

6. −2, 2 **7.** 4, 3 **8.** −4, −5 **9.** −10, −6 **10.** −3, −12

Written Exercises

Graph the given numbers. Write an inequality to compare them.

−3, 1

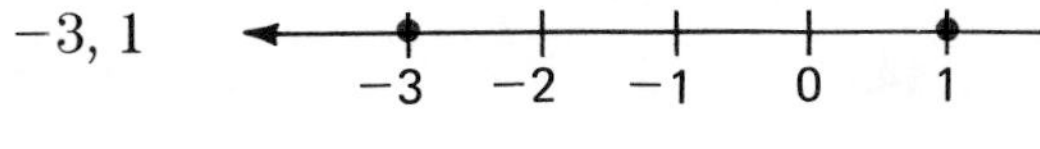

$-3 < 1$

A

1. −2, 3 **2.** 0, −5 **3.** −1, −6 **4.** 7, −2 **5.** 5, −4

6. −7, −1 **7.** −1, −7 **8.** 0, −4 **9.** 8, −8 **10.** −5, −2

Graph the solution of the equation on the number line.

11. $x + 1 = 7$ **12.** $y - 2 = 5$ **13.** $x + 6 = 9$ **14.** $5 + x = 10$

15. $y - 6 = 2$ **16.** $a + 3 = 12$ **17.** $x - 6 = 4$ **18.** $y - 2 = 7$

Graph the solutions of the inequality on the number line.

19. $x > 2$ **20.** $x > 0$ **21.** $x < -2$ **22.** $x < 3$

23. $x < 0$ **24.** $x > -3$ **25.** $x < 2$ **26.** $x > -2$

27. $y > -1$ **28.** $x < 4$ **29.** $y < -1$ **30.** $x < 7$

Write an inequality to compare the three numbers.

SAMPLE	Compare -3, 1, and -4.	$-4 < -3 < 1$

B **31.** 0, -4, and 4 **32.** -1, 2, and -6 **33.** -3, -5, and 0

34. 3, -5, and -6 **35.** -4, 6, and -2 **36.** -2, -1, and -6

Graph the solutions on the number line.

C **37.** $x \geq 3$ **38.** $y \leq 3$ **39.** $n \geq 0$ **40.** $y \leq -1$

41. $x \geq -4$ **42.** $7 \leq n$ **43.** $x \geq 6$ **44.** $x \leq -6$

SELF-TEST

Vocabulary

integers (p. 78)

Compare the numbers. Write > or <.

1. 6 __?__ -7 **2.** 3 __?__ 5 **3.** -6 __?__ 9 *(3-1)*

4. -4 __?__ 0 **5.** -2 __?__ 4 **6.** 1 __?__ -8

7. -9 __?__ -5 **8.** -7 __?__ -10 **9.** 4 __?__ -1

10. 0 __?__ 6 **11.** -3 __?__ 2 **12.** -1 __?__ -12

Graph the solutions of the inequality.

13. $x > 5$ **14.** $x > -6$ **15.** $x < -4$ *(3-2)*

16. $x < 4$ **17.** $x > 1$ **18.** $x < -3$

19. $x < -6$ **20.** $x < 5$ **21.** $x > -5$

PUZZLE ♦ PROBLEMS

This is a street map of a certain town.

Annie lives at A, Bren lives at B, Curt lives at C, and Don lives at D.

How can each person get home without running into another person's path? (They must walk along the streets. No shortcuts!)

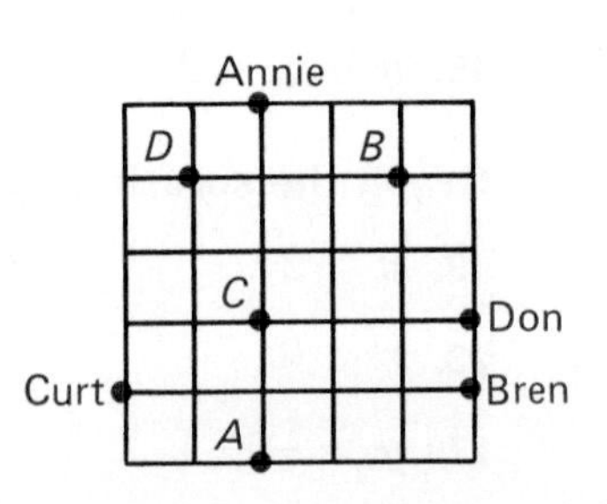

3 Addition

Addition can easily be shown on the number line.

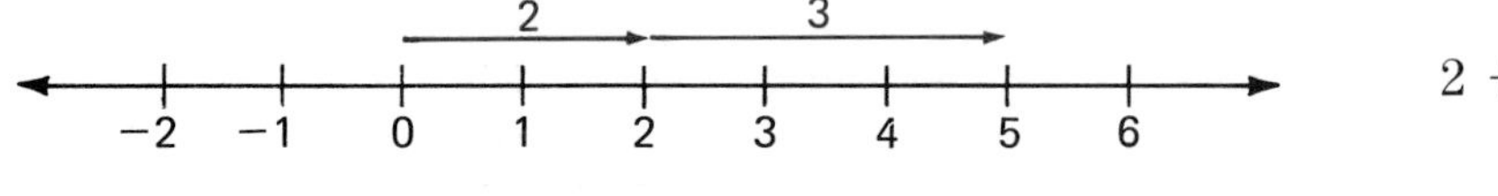

$2 + 3 = 5$

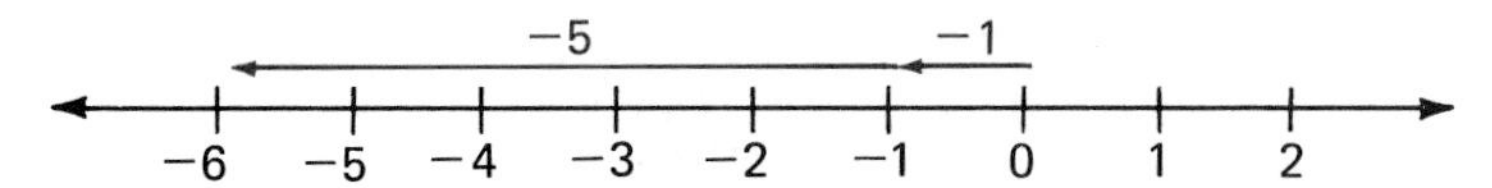

$(-1) + (-5) = -6$

The sum of two positive numbers is positive.
The sum of two negative numbers is negative.

Suppose in a football game you gain 5 yards, then lose 5 yards.

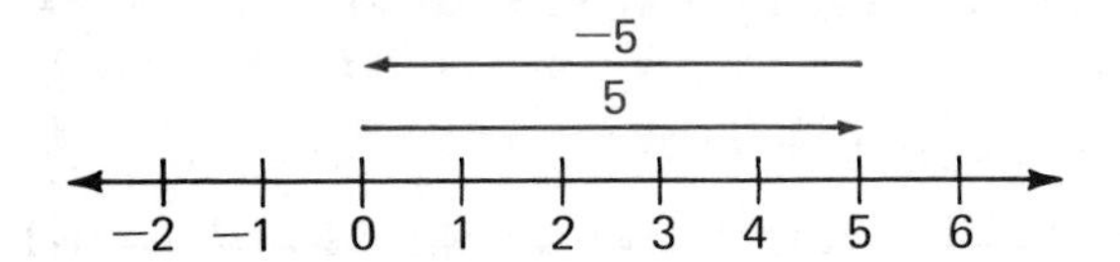

$5 + (-5) = 0$

Suppose you gain 8 yards, then lose 2 yards.

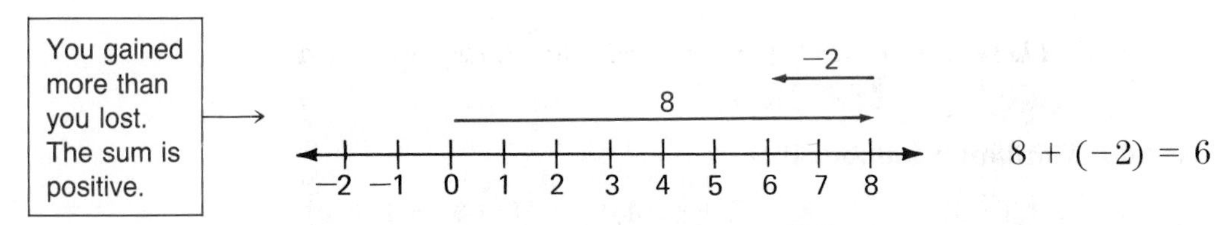

Now imagine that you gained 5 yards but lost 9 yards on the next play.

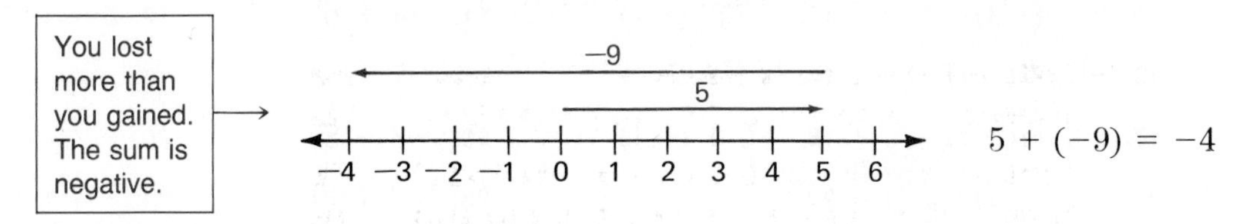

The sum of a positive number and a negative number may be zero, positive, or negative.

Consider the examples of addition on the next page.

EXAMPLES

$(-6) + 4 = -2$ $\quad$ $8 + (-3) = 5$

$(-3) + (-2) = -5$ $\quad$ $(-5) + 5 = 0$

Classroom Practice

Read the sum from the number-line drawing.

1.

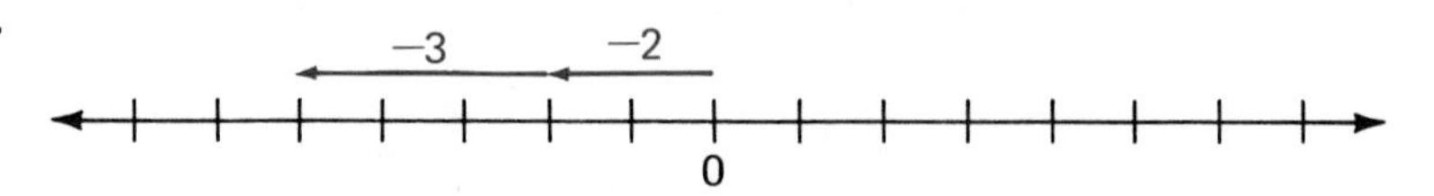

2.

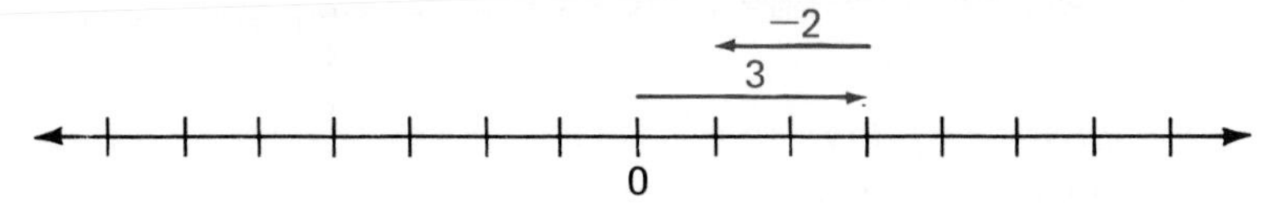

3.

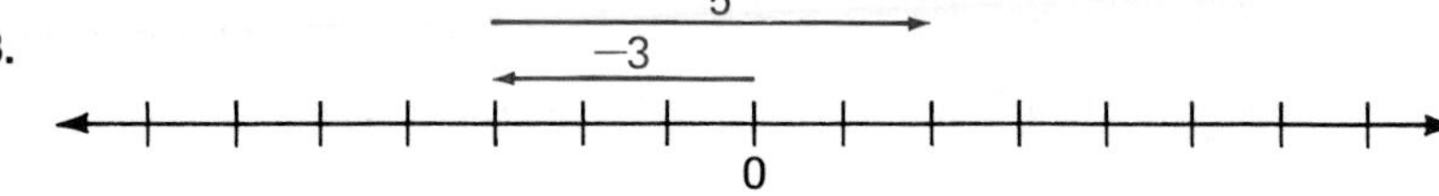

4.

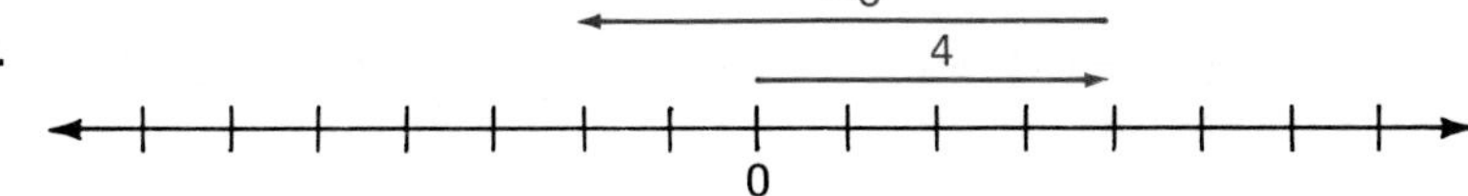

Written Exercises

Add. Think of a number line.

A

1. $-2 + (-3)$	**2.** $-7 + (-4)$	**3.** $-4 + 3$	**4.** $2 + (-4)$
5. $-6 + (-2)$	**6.** $2 + (-2)$	**7.** $-8 + 0$	**8.** $-1 + (-2)$
9. $6 + (-3)$	**10.** $-5 + (-4)$	**11.** $-4 + 6$	**12.** $3 + (-5)$
13. $-2 + (-4)$	**14.** $0 + (-9)$	**15.** $-7 + 4$	**16.** $-3 + 7$
17. $-5 + (-1)$	**18.** $-6 + (-4)$	**19.** $-8 + 8$	**20.** $-2 + 2$
21. $4 + (-4)$	**22.** $6 + (-6)$	**23.** $3 + (-9)$	**24.** $-5 + (-8)$
25. $-7 + (-2)$	**26.** $-5 + (-12)$	**27.** $6 + (-11)$	**28.** $12 + (-6)$

Add.

29. $5 + (-2) + (-6)$	**30.** $-10 + 2 + (-5)$	**31.** $1 + (-5) + 3$
32. $-6 + 6 + (-2)$	**33.** $9 + (-6) + 7$	**34.** $12 + (-12) + 0$

4 Subtraction

Every number has an opposite.

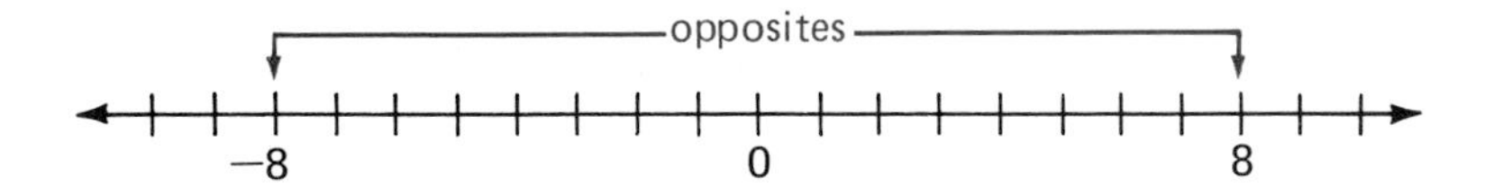

In Words	In Symbols
The opposite of 8 is −8. The opposite of −8 is 8.	$-(8) = -8$ $-(-8) = 8$

You'll use this idea of opposites in subtraction.

If you have 3 and subtract 2, you can show this on the number line.

$$3 - 2 = 1$$

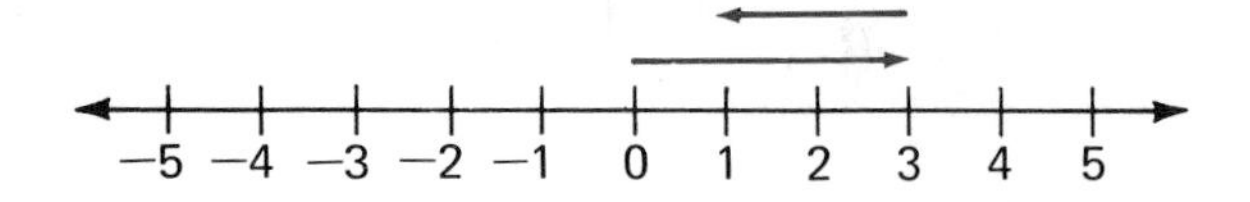

You know that the sum of 3 and −2 is also 1.

$$3 + (-2) = 1$$

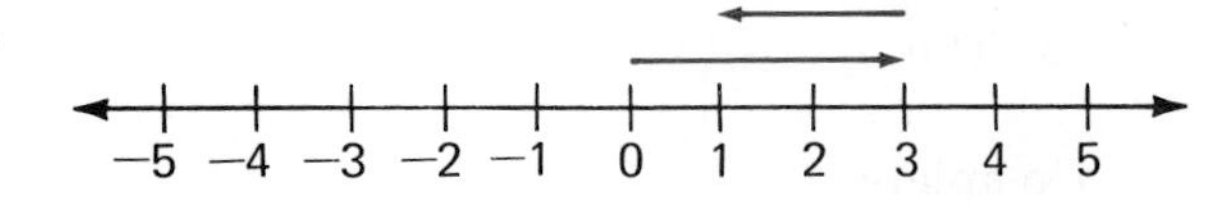

You can then write the following equation.

$$3 - 2 = 3 + (-2)$$

Let's consider another subtraction.

If you have 4 and subtract 7, you can show this on the number line.

$$4 - 7 = -3$$

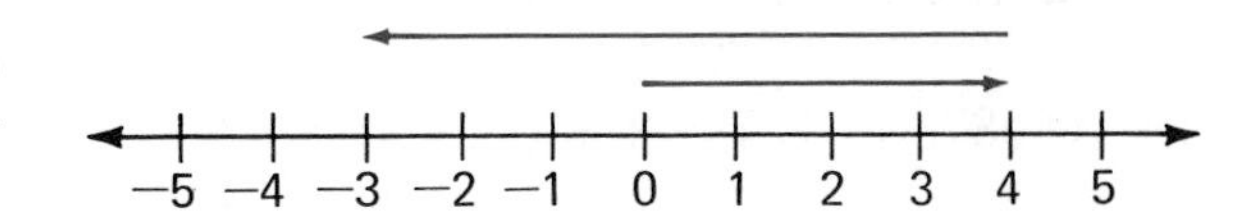

The sum of 4 and −7 is also −3.

$$4 + (-7) = -3$$

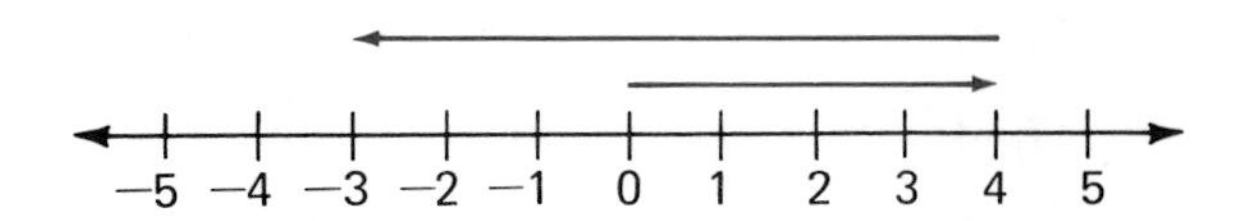

$$4 - 7 = 4 + (-7)$$

We can write a rule for any subtraction.

> *To subtract any number, we add its opposite.*

EXAMPLE 1 $5 - 2 = 5 + (-2)$
$= 3$

EXAMPLE 2 $-3 - 5 = -3 + (-5)$
$= -8$

EXAMPLE 3 $6 - (-3) = 6 + (3)$
$= 9$

EXAMPLE 4 $-4 - (-6) = -4 + (6)$
$= 2$

Classroom Practice

1. The opposite of 2 is __?__.

2. The opposite of -2 is __?__.

3. The opposite of __?__ is 6.

4. The opposite of __?__ is -6.

Complete.

5. $2 - 10 = 2 + (\underline{\ ?\ })$
$= \underline{\ ?\ }$

6. $-2 - 10 = -2 + (\underline{\ ?\ })$
$= \underline{\ ?\ }$

7. $2 - (-10) = 2 + (\underline{\ ?\ })$
$= \underline{\ ?\ }$

8. $-2 - (-10) = -2 + (\underline{\ ?\ })$
$= \underline{\ ?\ }$

9. $8 - 17 = 8 + (\underline{\ ?\ })$
$= \underline{\ ?\ }$

10. $-8 - 17 = -8 + (\underline{\ ?\ })$
$= \underline{\ ?\ }$

11. $8 - (-17) = 8 + (\underline{\ ?\ })$
$= \underline{\ ?\ }$

12. $-8 - (-17) = -8 + (\underline{\ ?\ })$
$= \underline{\ ?\ }$

Subtract.

13. $6 - 7$ **14.** $-6 - (-7)$ **15.** $4 - 9$ **16.** $7 - (-2)$

17. $3 - (-5)$ **18.** $-2 - 8$ **19.** $-7 - (-1)$ **20.** $-2 - (-10)$

21. $10 - 5$ **22.** $10 - (-5)$ **23.** $5 - 10$ **24.** $-5 - 10$

25. $-4 - 3$ **26.** $-4 - (-3)$ **27.** $20 - 25$ **28.** $8 - (-12)$

Written Exercises

Complete.

A

1. $15 - 10 = 15 + (\underline{\ ?\ })$
   $= \underline{\ ?\ }$

2. $10 - 15 = 10 + (\underline{\ ?\ })$
   $= \underline{\ ?\ }$

3. $-16 - 7 = -16 + (\underline{\ ?\ })$
   $= \underline{\ ?\ }$

4. $-3 - 7 = -3 + (\underline{\ ?\ })$
   $= \underline{\ ?\ }$

5. $4 - (-2) = 4 + (\underline{\ ?\ })$
   $= \underline{\ ?\ }$

6. $6 - (-1) = 6 + (\underline{\ ?\ })$
   $= \underline{\ ?\ }$

Subtract.

7. $2 - 3$
8. $1 - 5$
9. $6 - 9$
10. $8 - 10$
11. $-3 - 4$
12. $-5 - 2$
13. $-7 - 6$
14. $-3 - 7$
15. $-3 - 9$
16. $-1 - 6$
17. $6 - (-2)$
18. $7 - (-1)$
19. $3 - (-5)$
20. $8 - (-3)$
21. $5 - (-8)$
22. $9 - (-4)$
23. $-2 - (-1)$
24. $-7 - (-3)$
25. $-10 - (-4)$
26. $-3 - (-2)$
27. $-6 - (-5)$
28. $-2 - (-5)$
29. $-8 - (-6)$
30. $-10 - (-1)$
31. $5 - (-3)$
32. $8 - (-4)$
33. $7 - 10$
34. $-4 - 6$
35. $-2 - 5$
36. $3 - 9$
37. $-9 - (-4)$
38. $6 - (-5)$

Solve.

39. Temperatures in New York City range from about 4°C below zero in the winter to 30°C above zero in the summer. What is the difference in temperatures?

40. Temperatures in one place in the Arctic range from 34°C below zero in the winter to 16°C above zero in the summer. What is the difference between these temperatures?

41. Add 7 and 3. From the sum subtract 16.

42. Add -4 and 9. From the sum subtract 10.

43. From the sum of -7 and -17 subtract -77.

44. From the sum of 35 and -50 subtract -3.

45. A number is 12 less than -18. What is the number?

46. A number is 100 less than 10. What is the number?

Mixed Practice Exercises

Add or subtract.

1. $2 + (-3)$	**2.** $-5 + (-4)$	**3.** $2 - 7$	**4.** $-3 - (-2)$
5. $8 - (-1)$	**6.** $-7 + (-2)$	**7.** $-3 - 4$	**8.** $6 - (-4)$
9. $8 + (-4)$	**10.** $-2 + (-5)$	**11.** $6 - (-2)$	**12.** $-5 + (-5)$
13. $-2 + 2$	**14.** $-8 - (-6)$	**15.** $4 - 10$	**16.** $6 + (-2)$
17. $-5 + 4$	**18.** $-7 + 0$	**19.** $-1 + 1$	**20.** $2 - (-3)$
21. $8 - (-7)$	**22.** $-10 - (2)$	**23.** $-7 + (-2)$	**24.** $-3 - (-4)$
25. $5 - (-6)$	**26.** $-12 - (-3)$	**27.** $8 - 12$	**28.** $7 - (-1)$
29. $9 - (-2)$	**30.** $8 + (-6)$	**31.** $4 - (-6)$	**32.** $-12 - (-9)$

CAREER NOTES

COMPUTER OPERATORS AND SERVICE TECHNICIANS

Many computer-related jobs require a knowledge of computer programming. In these jobs, a programmer develops a set of instructions for the computer to follow. Other jobs, however, such as computer operator and computer service technician, do not need a specialized knowledge of programming.

A computer operator uses the instructions developed by the programmer. The operator runs the computer and inputs any data asked for in the program. The operator then gives the results, called output, to the programmer or other personnel. Computer operators work in all industries, from universities and hospitals to banks and insurance companies.

Computer equipment must be kept in good working order. A computer service technician maintains the terminals, printers, and so on. If a computer system breaks down, service technicians must quickly discover and correct the problem. The technician must often be familiar with the repair of many different models. A computer service technician may work for the bank or university but more often works for the computer company that leased or sold the equipment.

5 Simplifying Expressions

Do you remember when you studied the distributive property in Chapter 1? It allowed you to do the following.

$$2(x + y) = 2x + 2y$$

If you turn the equation around, you'll still have a true statement.

$$2x + 2y = 2(x + y)$$

The following equation is also true.

$$3a + 5a = (3 + 5)a$$
$$= 8a$$

The distributive property helps to explain why you can combine like terms. Of course you can combine like terms whether the coefficients are positive or negative numbers.

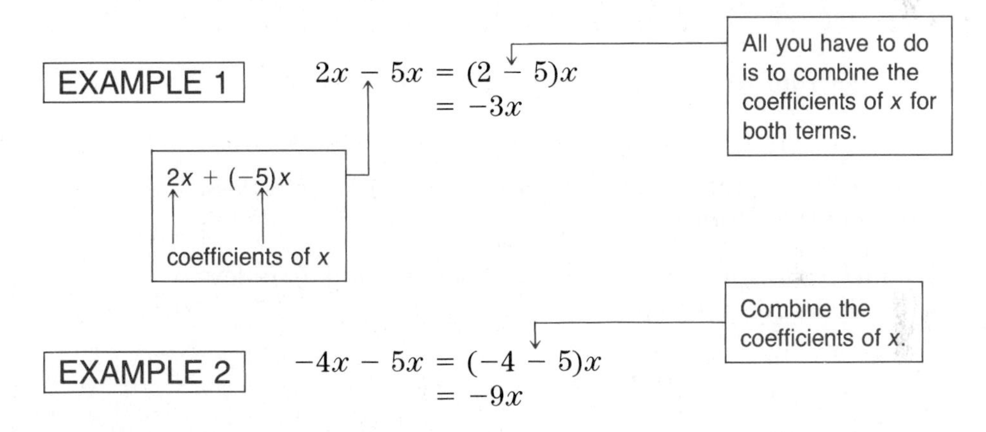

Once you understand how like terms are combined, it is easy to do one step in your head, if you wish.

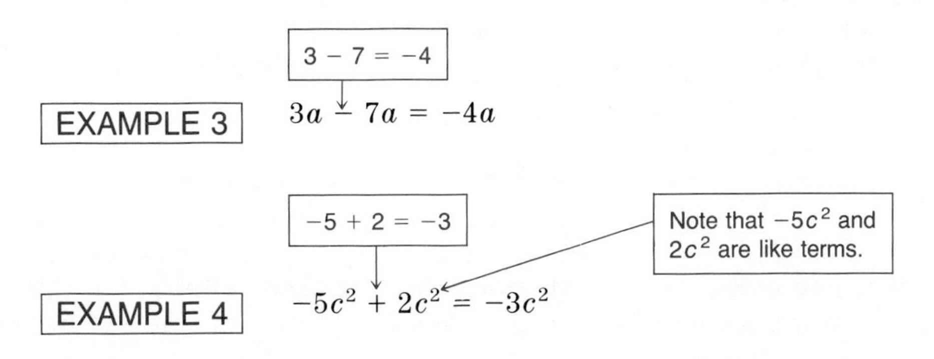

Written Exercises

Simplify.

A

1. $x + 3x$
2. $6x + 7x$
3. $3x + 7x$
4. $x - 3x$
5. $6y - 7y$
6. $a - 4a$
7. $-x^2 + 3x^2$
8. $-b + 5b$
9. $-2c^2 + 8c^2$
10. $-6y + 3y$
11. $a - 2a$
12. $7c^2 - 9c^2$
13. $-2x + 8x$
14. $-5b - 6b$
15. $-8a - 2a$
16. $-x + 5x$
17. $2n^2 - 5n^2$
18. $-10x + 4x$
19. $y - 10y + 10$
20. $x + 3 - 3x$
21. $-3y + 3y + 1$
22. $-7x - 8x - 2$
23. $-6x - (-7x)$
24. $10x - (-7x)$
25. $-4x - (-5x)$
26. $2a - (-a) + 4$
27. $x^2 - 3x^2 - (-y)$
28. $y - 2x^2 - (-3y)$
29. $4x - (-2x) + x$
30. $y - 5y - (-x^2)$

Solve.

B

31. Subtract $2r$ from $10 - r$. Then decrease the result by $4r$.
32. Subtract $5x$ from $12 + 2x$. Then decrease the result by $3x$.
33. Anita rode $(6n - 2)$ kilometers. Cindy rode n kilometers less. How far did Cindy ride?
34. John's bike cost $(2y - 8)$ dollars. Al's bike cost y dollars less. How much did Al's bike cost?

SELF-TEST

Add.

1. $(-3) + 9$ 2. $(-4) + (-2)$ 3. $6 + (-7)$ 4. $(-5) + 3$ *(3-3)*

Subtract.

5. $5 - 3$ 6. $-8 - 3$ 7. $7 - (-5)$ 8. $-4 - (-6)$ *(3-4)*

Simplify.

9. $5x - 4x$ 10. $-6n + 7n$ 11. $-3c - 8c$ 12. $2m^2 - 4m^2$ *(3-5)*

6 Multiplication

You know that multiplication and addition are related.

$$4 \times 3 = 3 + 3 + 3 + 3$$
$$= 12$$

In the same way, you can show multiplication and addition with a negative number.

$$4(-3) = (-3) + (-3) + (-3) + (-3)$$
$$= -12$$

Notice: When the sign of *one* factor is changed, the sign of the product is changed.

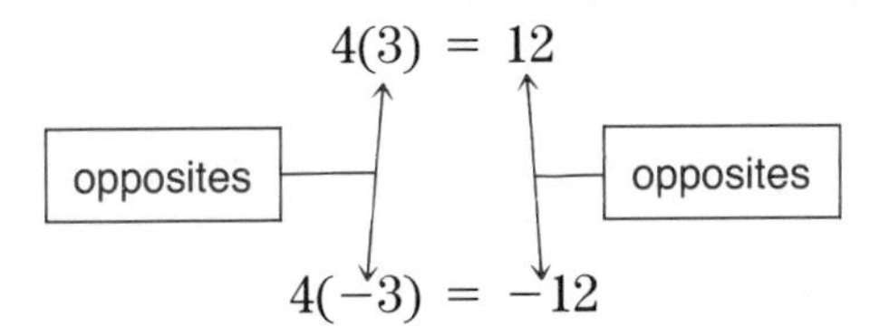

We use this pattern to multiply two negative numbers.

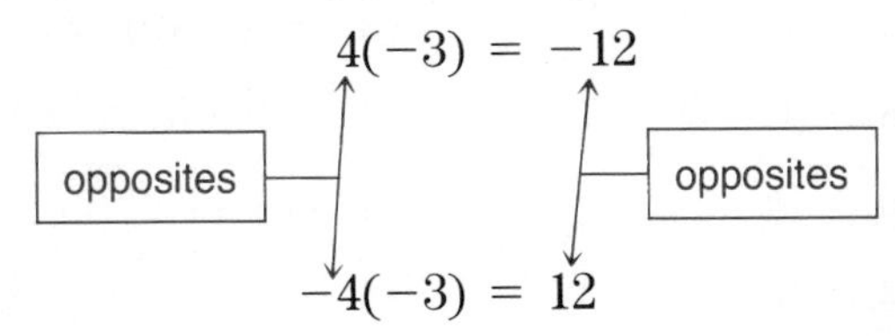

These examples lead to some rules about signs in multiplication.

When two numbers have like *signs, their product is a* positive *number.*

positive × positive = positive ⟶ $6 \cdot 2 = 12$
negative × negative = positive ⟶ $-6(-2) = 12$

When two numbers have unlike *signs, their product is a* negative *number.*

positive × negative = negative ⟶ $6(-2) = -12$
negative × positive = negative ⟶ $-6(2) = -12$

Classroom Practice

Multiply.

1. $6 \cdot 3$
2. $-6 \cdot 3$
3. $-3 \cdot 6$
4. $-3(-6)$
5. $3 \cdot 0$
6. $12 \cdot 1$
7. $(-3) \cdot 0$
8. $-1 \cdot 14$
9. $-3 \cdot -9$
10. $-4 \cdot -6$
11. $-1(1)$
12. $-1(-2)$

Simplify.

SAMPLES

$(-3)^2 = (-3)(-3) = 9$

$-(3)^2 = -(3)(3) = -9$

13. $(-1)^2$
14. $(-5)^2$
15. $(-4)^2$
16. $-(2)^2$
17. $-(-2)^2$

Written Exercises

Multiply.

A

1. $4 \cdot 8$
2. $-2(100)$
3. $-3(-40)$
4. $6 \cdot 8$
5. $-2 \cdot 7$
6. $-16 \cdot 0$
7. $-4(-5)$
8. $5(-2)$
9. $-6(4)$
10. $-9(-1)$
11. $6(-1)$
12. $-7(-3)$
13. $5 \cdot 6$
14. $-8(3)$
15. $-1(7)$
16. $-2(4)$
17. $3(-4)$
18. $(0)(-2)$
19. $4(-1)$
20. $8(-6)$
21. $-1(-4)$
22. $2(-10)$
23. $7(-4)$
24. $-3 \cdot 9$
25. $8(-2)$
26. $-7(-7)$
27. $5(-5)$
28. $9(-2)$

Simplify.

29. $(-5)^2$
30. 6^2
31. $(-6)^2$
32. 4^2
33. $(-4)^2$
34. $-(-1)^2$
35. $-(10)^2$
36. $-(7)^2$
37. $-(-5)^2$
38. $-(-3)^2$
39. $(-10)^2$
40. $(-8)^2$
41. $-(6)^2$
42. $-(-8)^2$
43. $-(-10)^2$

B

44. $3(-10) + 3(6)$
45. $-1(6) + (-1)(-6)$
46. $-7(-1) + (-4)(-1)$
47. $8(-7) - 4(-2)$
48. $3(-2) - 7(-4)$
49. $(4)(-2)(-3) + (-1)(-1)$
50. $(-3)^3$
51. $(-2)^4$
52. $(-1)^5$

7 Simplifying Expressions

The rules for the signs in multiplication also are true when you're working with variables.

EXAMPLE 1

$$4(6a) = 24a$$

EXAMPLE 2

$$-6(4a) = -24a$$

EXAMPLE 3

$$\begin{aligned} -6a(-4b) &= (-6)(-4)(a)(b) \\ &= 24ab \end{aligned}$$

You know that we can rewrite $2(x + 5)$ in the following way.

$$\begin{aligned} 2(x + 5) &= (2 \cdot x) + (2 \cdot 5) \\ &= 2x + 10 \end{aligned}$$

You can work in this way when some numbers are negative, too.

EXAMPLE 4

$$\begin{aligned} -2(x + 5) &= (-2 \cdot x) + (-2 \cdot 5) \\ &= -2x - 10 \end{aligned}$$

EXAMPLE 5

$$\begin{aligned} -3(a - 4) &= (-3 \cdot a) + (-3 \cdot -4) \\ &= -3a + 12 \end{aligned}$$

Recall from Chapter 1 that x means $1x$. You can then see that some other statements are also true.

$x = 1x$ —— and
- $-x = -1x$
- $-(x - 2) = -1(x - 2)$

These are useful facts in simplifying some expressions.

EXAMPLE 6

$$\begin{aligned} -(a - 4) &= -1(a - 4) \\ &= (-1 \cdot a) + (-1 \cdot -4) \\ &= -a + 4 \end{aligned}$$

EXAMPLE 7

$$\begin{aligned} -(-x + 3) &= -1(-x + 3) \\ &= (-1 \cdot -x) + (-1 \cdot 3) \\ &= x - 3 \end{aligned}$$

Classroom Practice

Multiply.

1. $2(-2a)$
2. $-2(-2a)$
3. $6a \cdot 0$
4. $-1 \cdot x$
5. $8(-9x)$
6. $(-8x)(8)$
7. $-3b \cdot 0$
8. $(2a)(-1)$
9. $a(-a + b)$
10. $4(-x - 2)$
11. $-c(c - 5)$
12. $-x(x - 2)$

Written Exercises

Multiply.

A

1. $-1(-2x)$
2. $-4(-3x)$
3. $2(-3x)$
4. $-1(7y)$
5. $-6a \cdot 0$
6. $-4a(2)$
7. $-4(3x)$
8. $-5(2x)$
9. $3a(-2b)$
10. $-a(4b)$
11. $-3x(-2y)$
12. $-7y(-6z)$
13. $4x(-2y)$
14. $-5x(-3y)$
15. $7a(-2b)$
16. $-4r(-5s)$
17. $4(9 - x)$
18. $-4(9 - x)$
19. $2(a - b)$
20. $-2(a - b)$
21. $-x(6 - x)$
22. $-y(y - 8)$
23. $(a + b)(-3)$
24. $(c + d)(-2)$

Simplify.

25. $-(a + 4)$
26. $-(x + 6)$
27. $-(y - 1)$
28. $-(x - 3)$
29. $-(-x + 5)$
30. $-(4 - y)$
31. $-(7 - a)$
32. $-(x - y)$
33. $x - (x + 3)$
34. $y - (y - 1)$
35. $2x - (x + 4)$
36. $3a - (a - 5)$
37. $8a - (a - 4)$
38. $2a - (a + 4)$
39. $6x - (3 - x)$
40. $5x - (7 - 2x)$

Solve for the variable.

B

41. $x + 11 = 2(x - 1)$
42. $3(n - 1) = 2n + 8$
43. $-(n - 1) = -2(n + 3)$
44. $p - (p - 2) = p$
45. $-3(5 - y) = 4(1 - 4y)$
46. $6a - (a - 2) = 3a - 8$

There is a mistake in this work. Can you correct it?

C

47. $4(7x - 1) - 3(8x - 2) = 4 - (-3x + 2)$

$$28x - 4 - 24x - 6 = 4 + 3x - 2$$
$$4x - 10 = 3x + 2$$
$$x = 12$$

8 Division

Once you know the rules for signs in multiplication, it's easy to learn the rules for division.

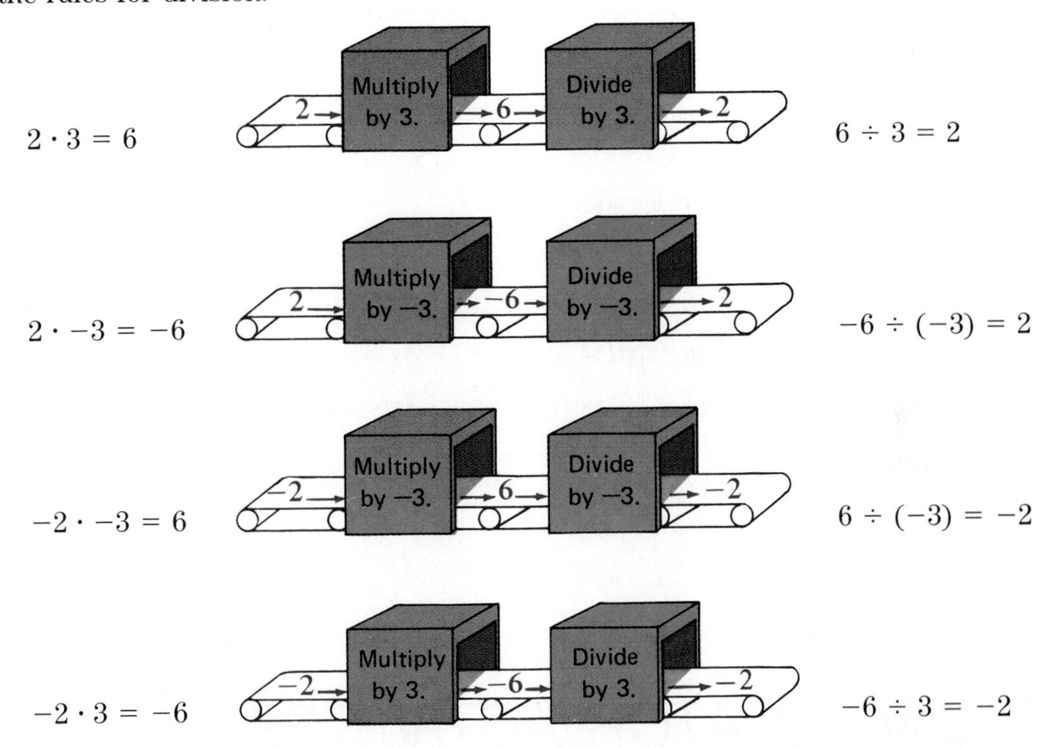

If you study the division equations closely, you'll notice that the rules for the signs in division are similar to those in multiplication.

When two numbers have like *signs, their quotient is a* positive *number.*

positive ÷ *positive* = *positive* ⟶ $14 \div 2 = 7$
negative ÷ *negative* = *positive* ⟶ $-14 \div -2 = 7$

When two numbers have unlike *signs, their quotient is a* negative *number.*

positive ÷ *negative* = *negative* ⟶ $14 \div -2 = -7$
negative ÷ *positive* = *negative* ⟶ $-14 \div 2 = -7$

Classroom Practice

Divide.

1. $8 \div -8$
2. $-10 \div -2$
3. $-4 \div 2$
4. $-12 \div 6$
5. $-15 \div -5$
6. $\frac{16}{-2}$
7. $\frac{-45}{9}$
8. $\frac{22}{2}$
9. $\frac{-49}{-7}$
10. $\frac{28}{-7}$

Written Exercises

Divide.

A

1. $3 \div -3$
2. $-3 \div -3$
3. $-56 \div 7$
4. $56 \div -7$
5. $4 \div -2$
6. $-6 \div 2$
7. $-10 \div -5$
8. $-15 \div -3$
9. $42 \div -6$
10. $-16 \div 4$
11. $-20 \div -4$
12. $25 \div -5$
13. $49 \div -7$
14. $-18 \div 6$
15. $-21 \div -7$
16. $24 \div -4$
17. $\frac{12}{-2}$
18. $\frac{14}{7}$
19. $\frac{-10}{5}$
20. $\frac{-15}{-3}$
21. $\frac{-21}{-3}$
22. $\frac{18}{-2}$
23. $\frac{-9}{-3}$
24. $\frac{-20}{5}$
25. $\frac{35}{7}$
26. $\frac{56}{-8}$

CALCULATOR ACTIVITIES

How do you indicate a negative number on your calculator?

Your calculator may have a key that can change a positive number to a negative number. You can enter 5 and press the [+/−] key to display −5.

If your calculator does not have this key, you may need to enter 0 − 5 to display −5, 0 − 13 to display −13, and so on.

Note that you do not always need to display negative numbers on your calculator. For example, to use your calculator to find the product of −17 and 20, use only positive numbers and enter 17 × 20 =. The calculator will display 340. Since negative × positive = negative, you know that your final result is −340.

Use a calculator to find the value of each expression.

1. $-2 + 5$
2. $-6 - (-3)$
3. $-4 - 8$
4. $-15 - (-20)$
5. $17(-8)$
6. $(-12)(-18)$
7. $-56 \div 14$
8. $-144 \div -9$

9 Solving Equations

Now that you can work with both negative and positive numbers, you can put the skills to work in solving equations.

EXAMPLE 1

$$n - 7 = -16$$
$$n - 7 + 7 = -16 + 7$$
$$n = -9$$

Check:

$$\begin{array}{r|l} n - 7 & = -16 \\ \hline -9 - 7 & -16 \\ -16 & \end{array} \checkmark$$

EXAMPLE 2

$$3a = -12$$
$$\frac{3a}{3} = \frac{-12}{3}$$
$$a = -4$$

Check:

$$\begin{array}{r|l} 3a & = -12 \\ \hline 3 \cdot -4 & -12 \\ -12 & \end{array} \checkmark$$

EXAMPLE 3

$$4c - 5c = 8$$
$$-c = 8$$
$$-1 \cdot -c = -1 \cdot 8$$
$$c = -8$$

If you know $-c$, multiply both sides by -1 to find c.

Check:

$$\begin{array}{r|l} 4c - 5c & = 8 \\ \hline (4 \cdot -8) - (5 \cdot -8) & 8 \\ -32 - (-40) & \\ -32 + 40 & \\ 8 & \end{array} \checkmark$$

EXAMPLE 4

$$r - 8 = -3r + 4$$
$$r - 8 + 3r = -3r + 4 + 3r$$
$$4r - 8 = 4$$
$$4r - 8 + 8 = 4 + 8$$
$$4r = 12$$
$$r = 3$$

Check:

$$\begin{array}{r|l} r - 8 & = -3r + 4 \\ \hline 3 - 8 & -3(3) + 4 \\ -5 & -9 + 4 \\ & -5 \end{array} \checkmark$$

Classroom Practice

Solve for the variable. Check.

1. $x + 1 = -3$ **2.** $y + 6 = 4$ **3.** $x + 2 = 0$ **4.** $-x + 4x = 9$

5. $-2x + 3x = 7$ **6.** $3x - 2x = -11$ **7.** $4a = 3a - 3$ **8.** $4x = -20$

9. $-5x = -40$ **10.** $-3y = 24$ **11.** $2x = 4 - 6$ **12.** $2y - 5y = -9$

13. $-x = 11$ **14.** $-a = -6$ **15.** $5n - 7n = 10$ **16.** $2y - 7y = -15$

Written Exercises

Solve for the variable. Check.

A

1. $x + 2 = -4$
2. $x + 6 = 3$
3. $y + 12 = -12$
4. $y - 12 = 12$
5. $a - 3 = -2$
6. $3x = -15$
7. $-4x = 12$
8. $-6x = 18$
9. $5y - 3y = -10$
10. $5x - 2x = -18$
11. $15x = -45 + 30$
12. $3x - 2x = -2$

Solve for the variable.

13. $-x + 2x = 9$
14. $-n + 4n = 15$
15. $-2n + 5n = 12$
16. $-x + 5x = 16$
17. $-2a + 6a = 12$
18. $-2n + 3n = -12$
19. $4c - 3c = -6$
20. $4x + 7 = 3x$
21. $2n + 5 = n$
22. $3y + 4 = 2y$
23. $7y - 2 - 5y = 0$
24. $-6m + 5 = -7m$
25. $-3x - 8 = -5x$
26. $3x + 1 = 2x$
27. $6y + 4 = 4y$
28. $-14 = -6x - 2$
29. $2n + 16 = n + 2$
30. $-3x + 2 = -x + 4$
31. $2x - (3x + 2) = -7$
32. $5y - (2y - 1) = -2$
33. $-x - (5x - 7) = -5$
34. $7 - (2 - x) = -4$
35. $6 - (4 - x) = 3x$
36. $2x - (3 - x) = x - 7$

SELF-TEST

Multiply.

1. $6(-5)$
2. $-2(-7)$
3. $(-7)^2$ *(3-6)*
4. $-9(3x)$
5. $-3a(-2b)$
6. $-4(y - 5)$ *(3-7)*

Simplify.

7. $-(x + 6)$
8. $-(a + 5)$
9. $y - (-y - 7)$
10. $2x - 3(x - 1)$

Divide.

11. $-15 \div -5$
12. $21 \div -7$
13. $-6 \div -2$ *(3-8)*

Solve for x.

14. $3x - 4x = 7$
15. $9x - 3x = -12$ *(3-9)*
16. $5x + 3 = 4x$
17. $-6x + 4 = -8x$

10 Absolute Value (Optional)

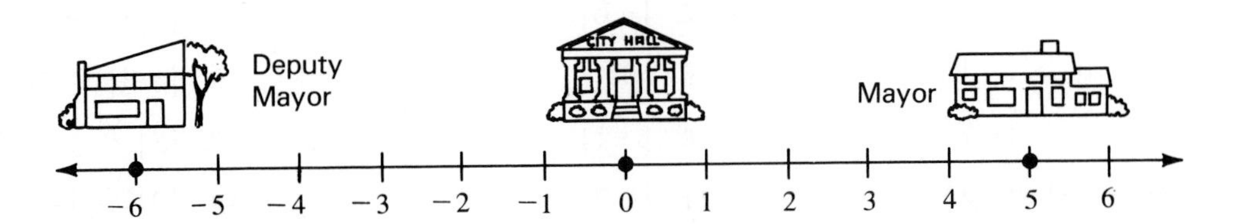

The Deputy Mayor lives 6 blocks to the left of City Hall.

The Mayor lives 5 blocks to the right of City Hall.

Who lives farther from City Hall? Of course it's the Deputy Mayor. Sometimes we care only about the number of units and not whether the number is positive or negative. In cases like this we use the following notation.

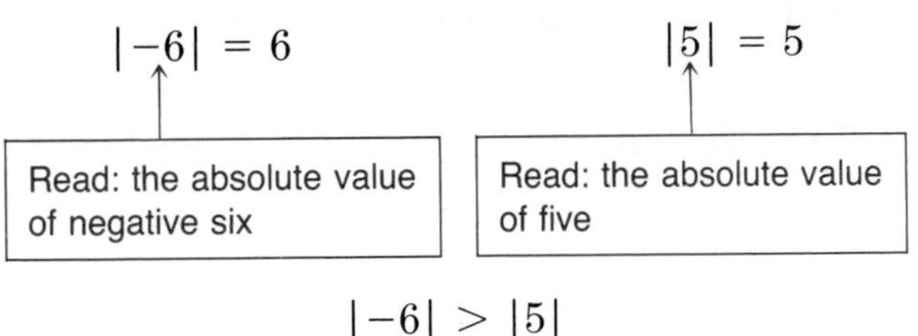

$$|-6| > |5|$$

EXAMPLE

Compare -5 and -4.

$$-5 < -4$$

Compare $|-5|$ and $|-4|$.

$$|-5| > |-4|$$

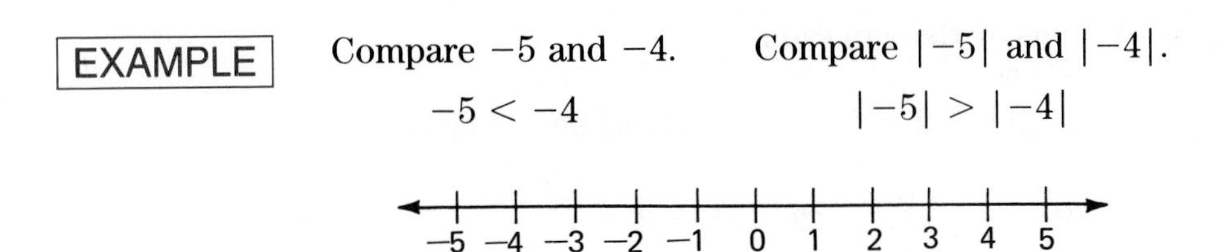

Can you guess what $|0|$ is? (Yes, it is 0.)

Classroom Practice

Find the value.

1. $|4|$

2. $|-4|$

3. $|-12|$

4. $|12|$

5. $|-100|$

6. $|-16|$

7. $\left|\frac{1}{2}\right|$

8. $\left|-\frac{2}{3}\right|$

9. $\left|-1\frac{1}{4}\right|$

10. $|2^2|$

11. $-|5|$

12. $-|-5|$

Written Exercises

True or false?

A

1. $|-3| < |-2|$
2. $|-6| > |2|$
3. $|-4| > |-3|$
4. $|5| > |-1|$
5. $|0| < |6|$
6. $|-3| < |0|$
7. $|1| > |-7|$
8. $|-1| > |-7|$
9. $|2| < |-2|$
10. $|-6| < |4|$
11. $|-7| > |7|$
12. $|3| > |-2|$

Compare. Write sentences with >, <, or =.

SAMPLE Compare $|-4|$ and $|2|$. $|-4| > |2|$

13. $|5|$ _?_ $|6|$
14. $|2|$ _?_ $|-3|$
15. $|-7|$ _?_ $|-3|$
16. $|0|$ _?_ $|-4|$
17. $|9|$ _?_ $|-10|$
18. $|-6|$ _?_ $|-2|$
19. $|-11|$ _?_ $|7|$
20. $|-1|$ _?_ $|1|$
21. $|-5|$ _?_ $|2|$

Find the value of the expression.

22. $-|3|$
23. $-|-2|$
24. $-|3| + |-2|$
25. $7 + |-4|$
26. $-|5| + |-5|$
27. $|-2| - |3|$

Find two solutions of the equation.

SAMPLE $|x| + 2 = 5$ x could be -3 or 3.

B

28. $|x| + 4 = 9$
29. $|y| + 3 = 7$
30. $|x| - 1 = 9$
31. $|a| + 5 = 11$
32. $|n| - 2 = 5$
33. $|x| - 7 = 7$
34. $|n| - 3 = 12$
35. $|c| + 1 = 4$
36. $|r| - 9 = 18$
37. $|z| + 6 = 13$
38. $|a| - 2 = 22$
39. $|x| + 5 = 35$

PUZZLE ◆ PROBLEMS

A liter is a little more than a quart. I have 2 pitchers which hold exactly 3 liters and 5 liters. Using just these two pitchers, how can I give you exactly 4 liters of water?

READING ALGEBRA

INDEPENDENT STUDY

This book has been planned to help you when you are studying on your own. Notice that your goals for each chapter in the book are listed on one of the opening pages of the chapter. This list shows you what you will learn in the lessons that follow. One good way to tackle a new lesson is to look quickly through the lesson before you begin to read it slowly. In this way you get the main idea and discover items that will need special attention as you study. The lesson title tells you the topic of the lesson. Look through the lesson for new words and phrases, which are printed in heavy type. You will need to understand these words and phrases in order to understand the lesson. Notice also any statements that are emphasized by a box.

After taking this quick look through the pages, read the lesson slowly and think as you read. If you come to any words that you do not understand, look them up in a dictionary or in the Glossary at the back of the book. Be sure to read the notes shown in red boxes. The notes are special helpers that give you important information in a few words. You may want to read some parts of the lesson more than once. If there is some idea that you do not understand, make a note of it so you can discuss it with your teacher. The presentation of the lesson almost always includes some examples. Try working these examples on your own before looking at the answers. Then test your understanding by trying some of the exercises. Try the Self-Tests, Chapter Reviews, Chapter Tests, Mixed Reviews, and Cumulative Reviews as well.

EXERCISES

Look quickly at pages 78–79 and then answer these questions.

1. What is the topic of the lesson?
2. What important new word do you see on page 78?
3. On a number line, where would you look for negative numbers?
4. What numbers make up the set of integers?
5. How do you read "-3" in words?
6. Do the statements in red boxes at the bottom of page 78 give you *new* information? What is their purpose?

COMPUTER ACTIVITIES

EQUIVALENT EXPRESSIONS

Two expressions are equivalent if they represent the same number for all values of the variable. Knowing whether two expressions are equivalent can be helpful in solving equations and simplifying expressions.

The basic number properties can show that some pairs of expressions are equivalent. For example, you know that $2(x + 3)$ and $2x + 6$ are equivalent expressions because of the distributive property.

The following computer program can be used to find equivalent expressions. The program is written in the language BASIC. In BASIC the symbols corresponding to the arithmetic signs $+$, $-$, $\times$, and $\div$ are $+$, $-$, $*$, and /. Statements in a program that begin with "REM" are ignored by the computer; their purpose is to help a user read the program.

The program below enables you to find the value of two expressions for any value of the variable you choose. After using a variety of values, including positive and negative numbers and decimals that are not whole numbers, you can probably make a good guess whether the expressions are equivalent.

To use the program, type it exactly as it appears here, then type "RUN." When you have tested enough values of x to decide whether the expressions are equivalent, you can input "999" for x to end the program.

```
10 REM**TO COMPARE EXPRESSIONS
20 REM**GIVEN A VALUE OF X
30 PRINT "WHAT IS THE VALUE OF X"
40 PRINT "(USE 999 IF NO MORE VALUES)";
50 INPUT X
60 IF X = 999 THEN 300
70 REM**FIRST EXPRESSION: A
80 LET A = -X
90 REM**SECOND EXPRESSION: B
100 LET B = (-1)*X
110 PRINT "EXPRESSION A:";A
120 PRINT "EXPRESSION B:";B
130 PRINT
140 GOTO 20
300 END
```

When you want to put in new expressions, just type in new lines 80 and 100, before typing RUN. This actually changes these two lines of the program. For example, if you wanted to use the expression "$x \div (-1)$" instead of "$-x$," you would type in a new line 80, reading

80 LET A = X/(−1).

When entering new expressions involving multiplication, remember to type 3x as 3 * X, −4x as −4 * X, and so on.

EXERCISES

1. RUN the program. Are $-x$ and $(-1)x$ equivalent expressions?

2. For each pair of expressions, change lines 80 and 100. RUN the program for several values of x. Are the expressions equivalent?

   a. $-1x$ and $0 - x$
   b. $-(x - 5)$ and $-x - 5$
   c. $-(x - 5)$ and $-x + 5$
   d. $10 - x$ and $x - 10$
   e. $(-x)(-x)$ and $-(x \cdot x)$
   f. x and $-(-x)$
   g. $x - (-x)$ and 0
   h. $(7 - x) - 4$ and $(4 - x) - 7$
   i. $x - x$ and $x + (-x)$
   j. $-1(6 - x)$ and $x - 6$

3. Of the three expressions given, one is not equivalent to the other two. RUN the program to find the expression that is not equivalent.

   a. $-x$
   $1 - x$
   $x \div (-1)$

   b. $(-2)x$
   $2(-x)$
   $(-2)(-x)$

   c. $-8x - 5$
   $-8(x - 5)$
   $-8x + 40$

   d. $2x - (x - 3)$
   $x + 3$
   $x - 3$

4. Some expressions are not equivalent. However, there may be some values of the variable for which the expressions have the same value. RUN the program for the expressions given. Find all values of x for which the two expressions have the same value.

   a. $x \cdot x$ and $x + x$
   b. $6x$ and $x + 10$
   c. $2 - x$ and $x - 2$
   d. $x \div 5$ and $5 \div x$
   e. x^2 and $-4x$ (Use $x \cdot x$ for x^2.)
   f. $|x|$ and $-x$ (In BASIC, $|x|$ is ABS(X).)

SKILLS REVIEW

ESTIMATING AND FINDING AVERAGES

Use your estimating skills. Estimate in your head.

SAMPLE $39 + 142 \longrightarrow 40 + 140 = 180$ *Answer:* about 180

1. $22 + 59$	**2.** $47 + 11$	**3.** $86 + 21$	**4.** $79 + 32$
5. $41 + 51$	**6.** $99 + 16$	**7.** $81 + 39$	**8.** $64 + 89$
9. $76 + 120$	**10.** $51 + 135$	**11.** $106 + 84$	**12.** $92 + 188$

Estimate in your head.

SAMPLE $3 \times 47 \longrightarrow 3 \times 50 = 150$ *Answer:* about 150

13. 6×18	**14.** 7×62	**15.** 8×12	**16.** 3×98
17. $4 \times \$.39$	**18.** $5 \times \$.95$	**19.** $2 \times \$.65$	**20.** $5 \times \$.75$
21. $4 \times \$.89$	**22.** $6 \times \$.35$	**23.** $3 \times \$.85$	**24.** $7 \times \$.29$

Suppose you have a calculator. Do these answers look reasonable, or have you touched a wrong button by mistake? (Don't calculate. Just check to see if the answer is *reasonable*.)

25. $16 + 47 = 99$	**26.** $86 + 92 = 178$	**27.** $153 + 89 = 532$
28. $235 + 41 = 286$	**29.** $723 + 19 = 714$	**30.** $811 + 49 = 890$
31. $4 \times 48 = 216$	**32.** $5 \times 39 = 1950$	**33.** $66 \times 23 = 158$
34. $70 \times 84 = 2048$	**35.** $49 \times 73 = 3577$	**36.** $58 \times 91 = 3048$

Find the average.

SAMPLE 10, 15, 26 (3 items)

$10 + 15 + 26 = 51$ ← sum of the items

$51 \div 3 = 17$ ← This is the average.

37. 12, 8, 19	**38.** 3, 14, 16	**39.** 98, 20, 32	**40.** 10, 12, 20
41. 2, 4, 10, 20	**42.** 16, 5, 7, 12	**43.** 5, 10, 20, 21	**44.** 10, 12, 16, 20, 27

CHAPTER REVIEW

CHAPTER SUMMARY

1. On a number line, numbers to the right of zero are positive. Numbers to the left of zero are negative.
2. A number line can be used to show solutions of equations or inequalities.
3. The sum of two positive numbers is positive.
   The sum of two negative numbers is negative.
   The sum of a positive number and a negative number may be zero, positive, or negative.
4. To subtract any number, add its opposite.
5. When two numbers have like signs, their product is positive.
   When two numbers have unlike signs, their product is negative.
6. When two numbers have like signs, their quotient is positive.
   When two numbers have unlike signs, their quotient is negative.

REVIEW EXERCISES

Compare the numbers. Write $>$ or $<$. ***(See pp. 78–79.)***

1. $10 \underline{\ \ ?\ \ } 12$ **2.** $15 \underline{\ \ ?\ \ } 4$ **3.** $-10 \underline{\ \ ?\ \ } -12$ **4.** $-6 \underline{\ \ ?\ \ } -20$

5. $2 \underline{\ \ ?\ \ } -2$ **6.** $-6 \underline{\ \ ?\ \ } 10$ **7.** $4 \underline{\ \ ?\ \ } -9$ **8.** $0 \underline{\ \ ?\ \ } 6$

Graph the solution of the equation on the number line.
(See pp. 80–81.)

9. $x + 2 = 4$ **10.** $x + 6 = 7$ **11.** $x + 9 = 3$

Graph the solutions of the inequality on the number line.
(See pp. 80–81.)

12. $x > 0$ **13.** $x < 1$ **14.** $x < -3$

Add. ***(See pp. 83–84.)***

15. $(-6) + (-7)$ **16.** $(-3) + (-8)$ **17.** $(-4) + 1$ **18.** $6 + (-6)$

19. $3 + (-2)$ **20.** $(-8) + 5$ **21.** $(-5) + (-6)$ **22.** $(-9) + 5$

23. $4 + (-10)$ **24.** $(-5) + 9$ **25.** $(-7) + (-4)$ **26.** $7 + (-11)$

Subtract. ***(See pp. 85–87.)***

27. $8 - (-10)$
28. $-6 - (-2)$
29. $3 - (-1)$
30. $6 - 7$
31. $-8 - 5$
32. $13 - (-10)$
33. $-9 - 3$
34. $8 - (-5)$
35. $6 - 11$
36. $-9 - (-5)$
37. $-5 - (-6)$
38. $3 - (-12)$

Simplify. ***(See pp. 89–90.)***

39. $x - 4x$
40. $x - 10 - 4x$
41. $5a - 15 - 5a$
42. $h - 2h + 3$
43. $-x - y + 3x$
44. $6b - 9b + 3$
45. $-b^2 + 9b^2$
46. $-a - 10 + 8a$
47. $-4x + 3y - 2x$

Multiply. ***(See pp. 91–94.)***

48. $-8(3)$
49. $6(-9)$
50. $(-3)(-8)$
51. $(-9)(-3)$
52. $-9(8)$
53. $18(-2)$
54. $(y - 5)y$
55. $x(x - 4)$
56. $-2(x + 3)$
57. $(x - 4)(-5)$
58. $-3(a - b)$
59. $6(-m + n)$

Simplify. ***(See pp. 91–94.)***

60. $-(x)^2$
61. $-(-b)^2$
62. $-(y - 7)$
63. $-a - (2a + b)$
64. $-(-x - 9)$
65. $-x - (-3 + y)$

Divide. ***(See pp. 95–96.)***

66. $-81 \div 3$
67. $-72 \div 8$
68. $-12 \div -12$
69. $-16 \div 4$
70. $27 \div 9$
71. $30 \div -3$
72. $-16 \div -4$
73. $24 \div -6$
74. $\frac{48}{-6}$
75. $\frac{-25}{5}$
76. $\frac{-24}{-2}$
77. $\frac{18}{-9}$

Solve for *x*. ***(See pp. 97–98.)***

78. $6x = -30$
79. $8x = -64$
80. $5x - 6x = -8$
81. $5x + 9 = 6x$
82. $-9x = 36$
83. $-6 + x = -9$
84. $8x - 9x = -2$
85. $4x + 10 = 6x$
86. $9x - 4x = -15$

(Optional) Find the value of the expression. ***(See pp. 99–100.)***

87. $-|-5|$
88. $-|2| + |-2|$
89. $|-4| - |-1|$

CHAPTER TEST

1. Write a positive or negative number: a temperature 5° below zero. *(3-1)*

2. Compare the numbers. Write $>$ or $<$.
   a. $-3 \ \underline{\ ?\ }\ -5$ b. $-7 \ \underline{\ ?\ }\ 1$

3. Graph the solution of $x + 7 = 11$ on the number line. *(3-2)*

4. Graph the solutions of $y < -3$ on the number line.

5. Write an inequality to compare the numbers 0, -5, and 2.

Add or subtract.

6. $-9 + 2$ 7. $-5 + (-5)$ 8. $7 + (-5) + (-3)$ *(3-3)*

9. $1 - (-6)$ 10. $8 - 12$ 11. $-2 - 4$ *(3-4)*

12. A number is 15 less than -5. What is the number?

Simplify.

13. $3x - 5 - 7x$ 14. $-7y - (-y)$ 15. $-2a - 8b + 9a$ *(3-5)*

Multiply.

16. $4(-2)$ 17. $(-8)(-3)$ 18. $-10(5)$ 19. $0(-40)$ *(3-6)*

20. Simplify. a. $-2(-3) + (-2)(7)$ b. $(-2)^5$

21. Multiply. a. $-3a(4b)$ b. $-x(x - 1)$ *(3-7)*

22. Simplify $5y - (4 - y)$.

23. Solve for x: $-(8 - x) = 2(x + 1)$

Divide.

24. $9 \div -3$ 25. $-2 \div -1$ 26. $\frac{-40}{5}$ 27. $\frac{-36}{-9}$ *(3-8)*

Solve for the variable. Check.

28. $2x - 5x = -18$ 29. $x + 1 = -3$ *(3-9)*

30. $-5y + 7 = -9y - 5$ 31. $7a - (a - 8) = a + 3$

(Optional) True or false?

32. $|-9| > |-1|$ 33. $|-4| < |3|$ 34. $|-5| < |5|$ *(3-10)*

MIXED REVIEW

Simplify.

1. $(-4)^2$
2. $y \cdot 7 \cdot x \cdot 4 \cdot y \cdot y$
3. $-3x - (-x) + 4x$
4. $2(x - 5) - 3x$
5. $2 - (4 - a)$
6. $2r + 6s - 5s - 2r$

Solve.

7. $\frac{x}{2} = 8$
8. $y - 13 = 14$
9. $3(2a - 1) = 2(a + 3) + a$
10. $9x - 8 = 37$
11. $2n + 5 = 9n - 2$
12. $7 - 3r = 6r - 11$

13. Eight less than 5 times a number is equal to the number. What is the number?

14. Write an equation which has -2 as a solution.

15. Tell which of the numbers shown in color are solutions:

$$5x - 3 > 14 \qquad 1, 2, 3, 4$$

16. Meg has \$4 more than Loc.
Pat has twice as much as Meg and Loc combined.
Let d = Loc's number of dollars.
Then __?__ = Meg's number of dollars and __?__ = Pat's number of dollars.

17. Compare. Write $>$ or $<$. $\quad -7$ __?__ -9

18. Graph the solutions of $x < -4$ on the number line.

If $x = 14$, find the value of the expression.

19. $2x + 9$
20. $2(x + 9)$
21. $(x - 4)^2$
22. $(x - 2) \div 4$

23. Write $a(7a - 5b)$ without parentheses.

24. Find the value if possible. If not, write *impossible.*

a. $\frac{x}{x - x}$ b. $\frac{2x}{x}$ c. $\frac{x - x}{4}$

Solve.

25. $9 + x = 2$
26. $-7y = 56$
27. $6(y + 1) = 22 + 2y$

28. Amos is twice as old as his brother. Last year their ages totaled 7. How old is Amos?

29. The sum of $4n$ and $6n$ is 50. Find the value of n.

30. Write an inequality to compare the numbers -2, 1, and -5.

Add, subtract, multiply, or divide.

31. $5 - 8$ **32.** $-24 \div (-4)$ **33.** $(-2)(-9)$ **34.** $-7 + (-4)$

35. $63 \div (-9)$ **36.** $3 - (-9)$ **37.** $-19 + 7$ **38.** $-(-2)(-2)$

39. A cup of whole milk has 30 more calories than a cup of orange juice.
Together they have 270 calories.
How many calories are in each?

40. Use the distributive property and combine like terms:

$$7(3a - 1) + 5(2 - 9a)$$

41. Copy and complete the table. Then find the original number.

Steps	Algebra
Think of a number.	n
Multiply by 3.	?
Subtract 6.	?
Multiply by 4.	?
Add 8.	?
Divide by 2.	?

Result: 34
Original number: __?__

42. Compare. Write < or > or =. 2^5 __?__ $(3 + 1 \cdot 5)4$

43. The length of a rectangle is 8 cm more than the width.
Let w = the width. Then __?__ = the length.

44. One number is 6 less than another.
Three times the smaller number is 2 more than the larger.
Find the numbers.

45. Write a positive or negative number: an increase of \$9

46. Graph the solution of the equation $x - 2 = 3$ on the number line.

47. Add 7 and -8. From the sum subtract 4.

48. Simplify $3(-5)(2) + (-9)(-8)$.

Here's what you'll learn in this chapter:

To use formulas to find perimeters, areas, and volumes.

To make your own formulas for problems.

To solve motion and cost problems.

To solve a formula for a given variable.

The distance around a bicycle wheel or any other circular object can be found by means of a formula. Also, the speeds of the gears on a bicycle are related by a formula that you will find in this chapter.

Chapter 4

Formulas

1 Perimeter Formulas

One lap around the inside lane of an Olympic track is 400 meters. We call the distance around a figure its **perimeter.**

Many figures have special **formulas** for their perimeters.

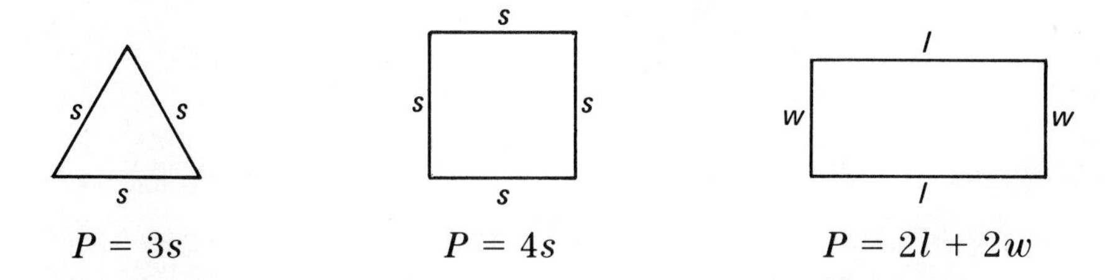

$P = 3s$ $P = 4s$ $P = 2l + 2w$

The perimeter of a circle has a special name, **circumference.**

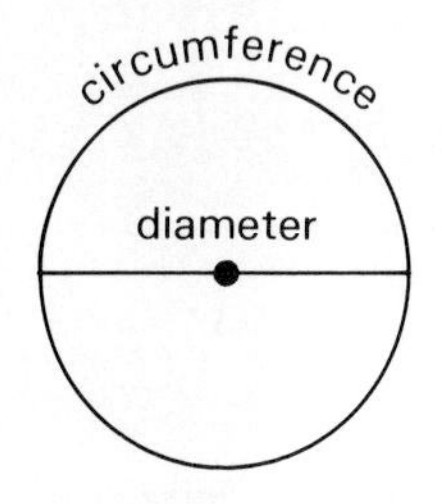

The circumference is about 3.14 times the diameter of the circle.

$$C = 3.14d$$

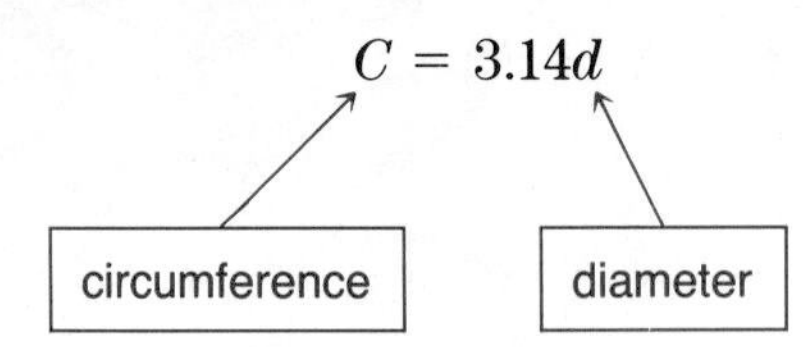

Knowing some formulas, you can solve many problems involving perimeters.

EXAMPLE 1

The length of a rectangle is 8 cm longer than the width. The perimeter is 44 cm.
Find the length and width.

Let x = width.
Then $x + 8$ = length.

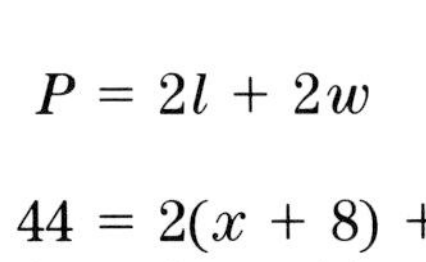

$$P = 2l + 2w$$

$$\begin{aligned} 44 &= 2(x + 8) + 2x \\ 44 &= 2x + 16 + 2x \\ 44 &= 4x + 16 \\ 44 - 16 &= 4x + 16 - 16 \\ 28 &= 4x \\ 7 &= x \end{aligned}$$

Answer: width $= x = 7 \longrightarrow 7$ cm
length $= x + 8 = 15 \longrightarrow 15$ cm

Check: $15 - 7 = 8$
$2 \cdot 15 + 2 \cdot 7 = 30 + 14$
$= 44$ ✓

A mirror has a diameter of 6 cm.
What is the circumference of the mirror?

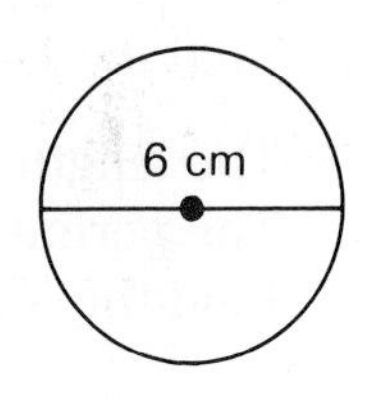

$$\begin{aligned} C &= 3.14d \\ &= 3.14 \times 6 \\ &= 18.84 \longrightarrow 18.84 \text{ cm} \end{aligned}$$

For some figures you can make up your own perimeter formula.

EXAMPLE 3

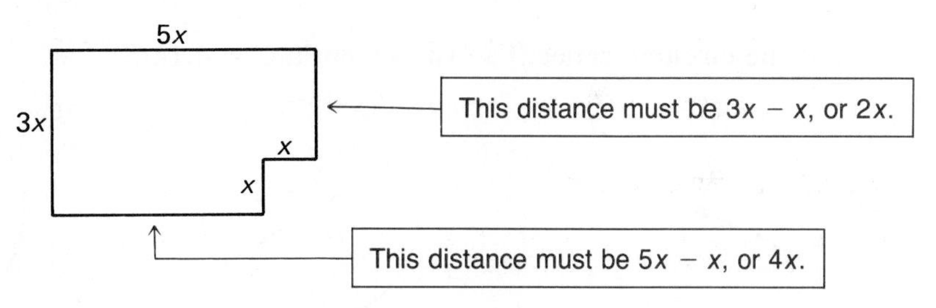

$$P = 3x + 5x + 2x + x + x + 4x$$
$$P = 16x$$

Written Exercises

Find the perimeter.

1.

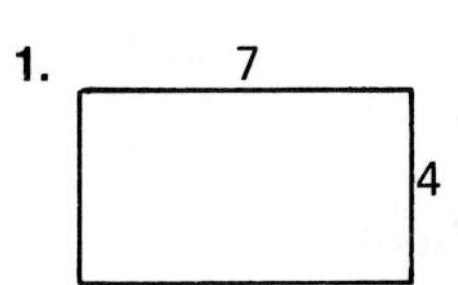

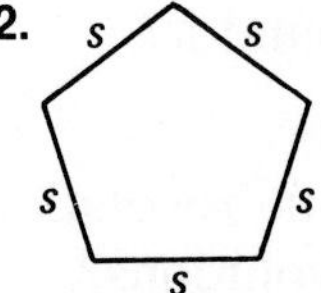

3.
2x
3x
x
3x

Find the perimeter.

4.

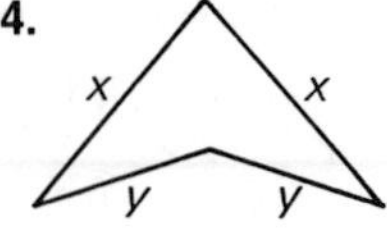

5.

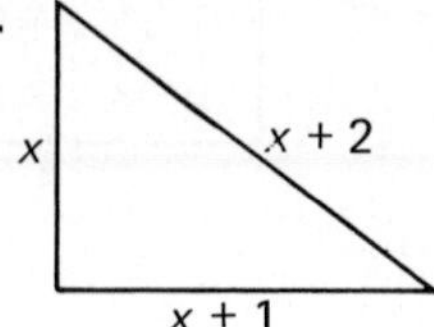

6.

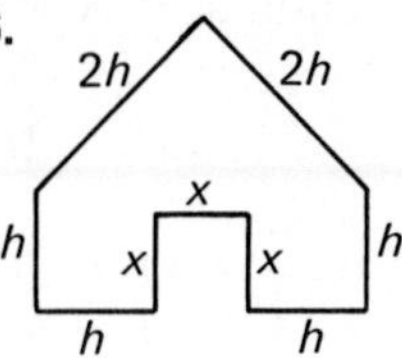

7.

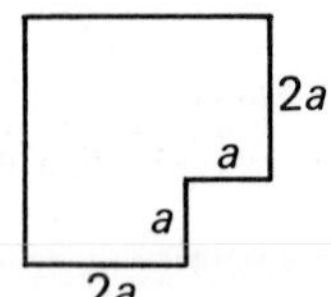

8.

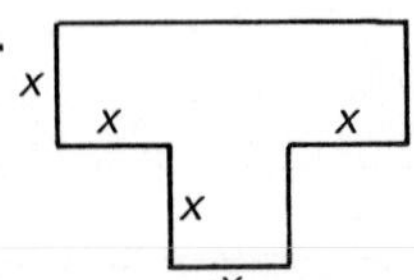

9.

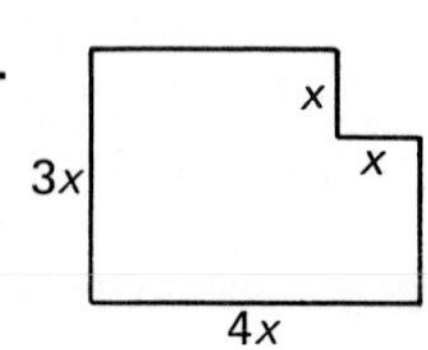

10. The length of a rectangle is 2 cm more than the width.
The perimeter is 20 cm.
Find the length and width.

11. The length of a rectangle is 5 cm more than the width.
The perimeter is 50 cm.
Find the length and width.

12. The perimeter of the triangle shown is 48.
Find the value of x.

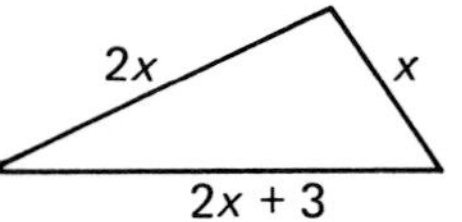

Find the circumference. Use the formula $C = 3.14d$.

B 13.

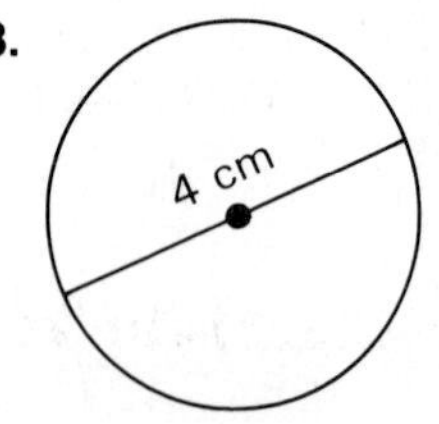

14.

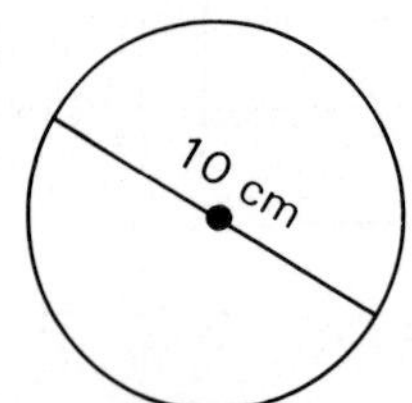

15.

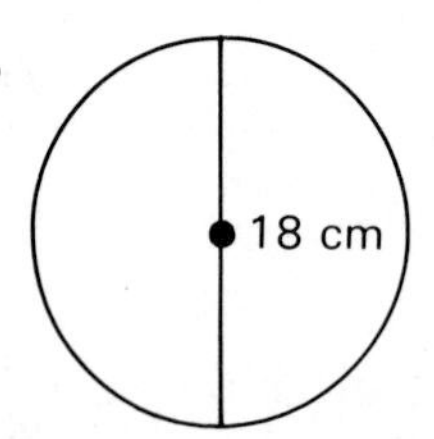

16. A turntable has a diameter of 40 cm.
What is its circumference?

17. A circular garden has a diameter of 3 m.
What length fencing is needed to go around its edge?

18. A clock has a diameter of 16 cm.
How long is the metal band around it?

C 19. A car wheel is 56 cm in diameter.
 a. How far does the car go in one revolution of the wheel?
 b. How far does it go in 100 revolutions?

2 Area Formulas

How much paint do I need to paint the walls of this room?

How much carpeting is needed to cover the floor?

Questions like those above involve area. The **area** of a figure is the amount of its surface. Small areas are often measured in square centimeters.

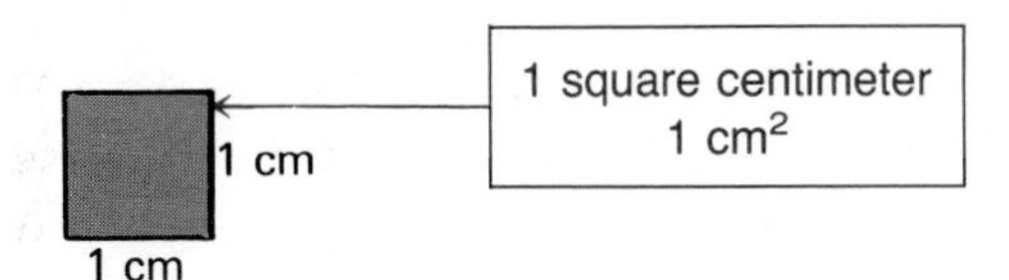

Of course larger regions are measured in square meters or square kilometers.

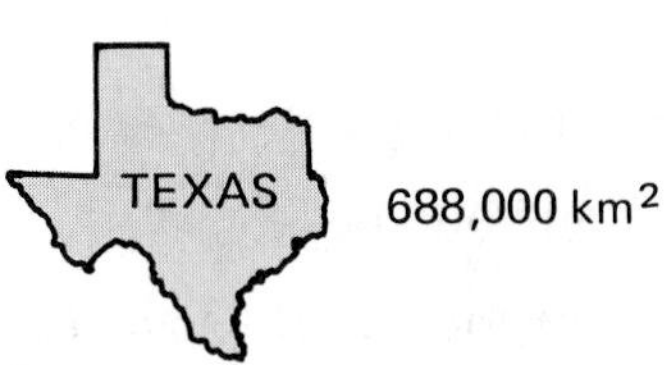

The areas of some figures can be given by formulas.

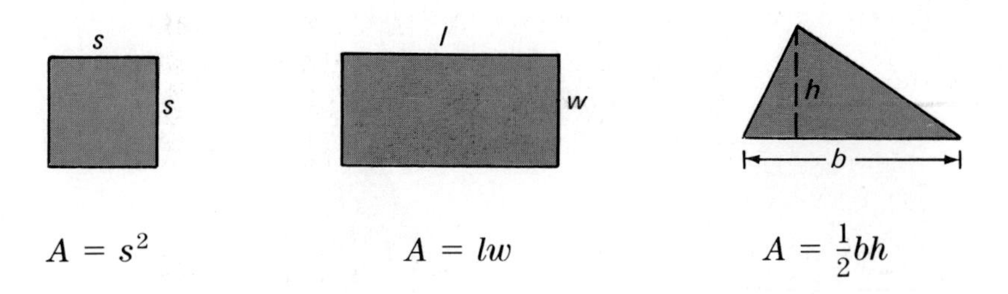

$A = s^2$ $A = lw$ $A = \frac{1}{2}bh$

EXAMPLE 1

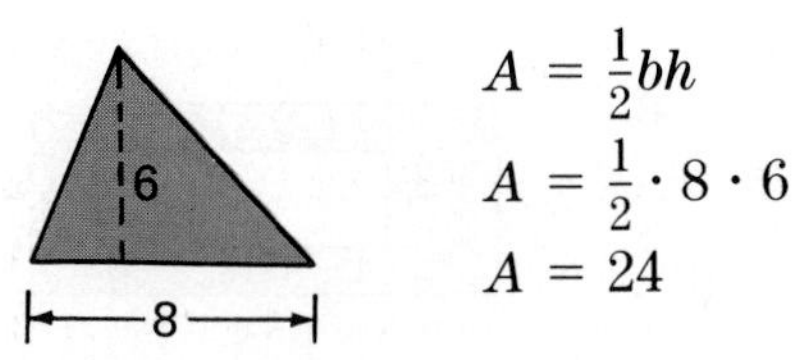

$A = \frac{1}{2}bh$

$A = \frac{1}{2} \cdot 8 \cdot 6$

$A = 24$

EXAMPLE 2

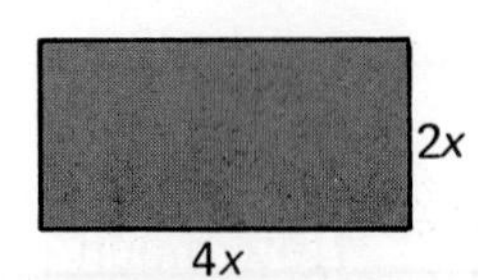

$A = lw$
$A = (4x)(2x)$
$A = 8x^2$

EXAMPLE 3

Find the area of the shaded region.

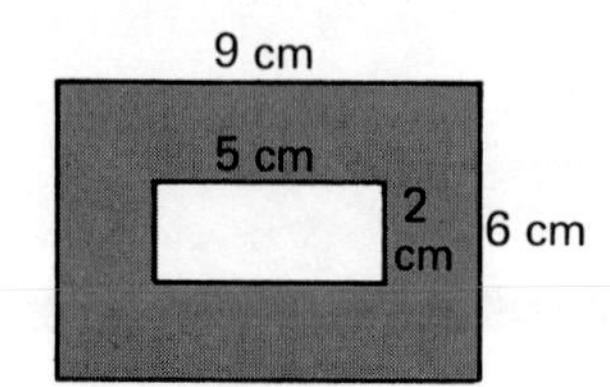

$$\text{Area} = \underbrace{\text{Area large rectangle}} - \underbrace{\text{Area small rectangle}}$$

$A = 9 \cdot 6 - 5 \cdot 2$
$A = 54 - 10$
$A = 44$
Answer: 44 cm^2

Classroom Practice

The formula for the area of a triangle is $A = \frac{1}{2}bh$.

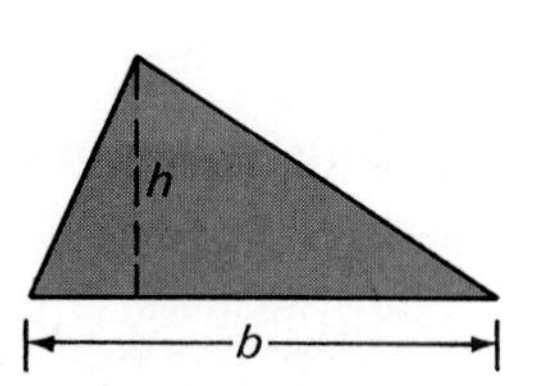

1. Can you guess what the letter b stands for?
2. Can you guess what the letter h stands for?
3. In the formula $A = lw$, what do the letters l and w stand for?

Find the area of the shaded figure.

4. 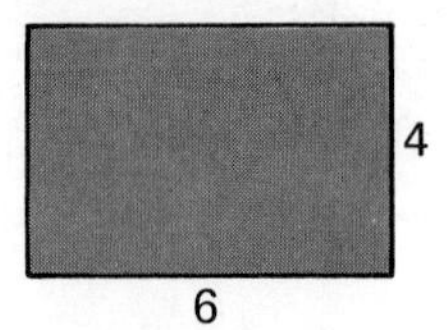

5. 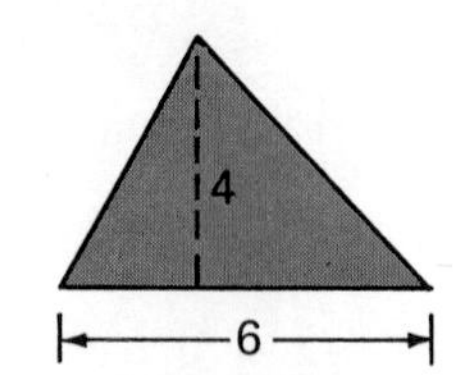

6. 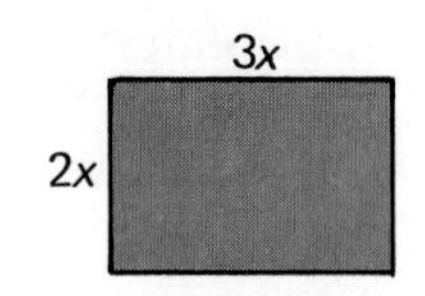

7. 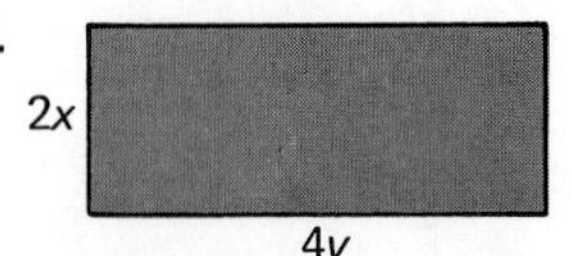

8. 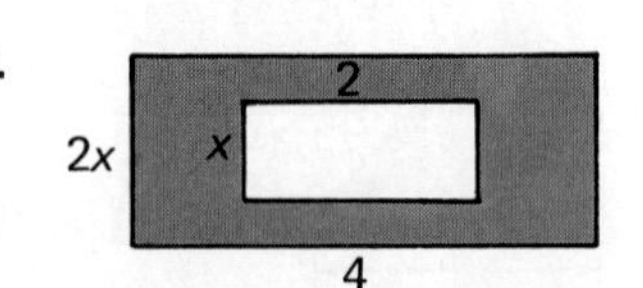

9. 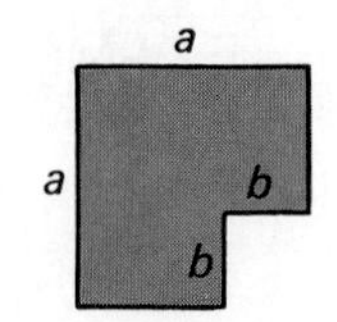

Written Exercises

Find the area.

A

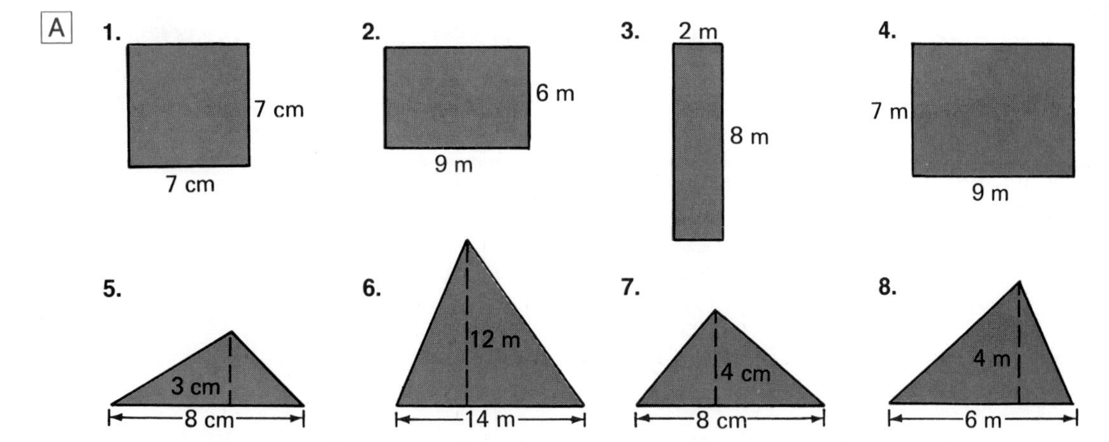

Find the area of the shaded figure. Give your answer in terms of the variables.

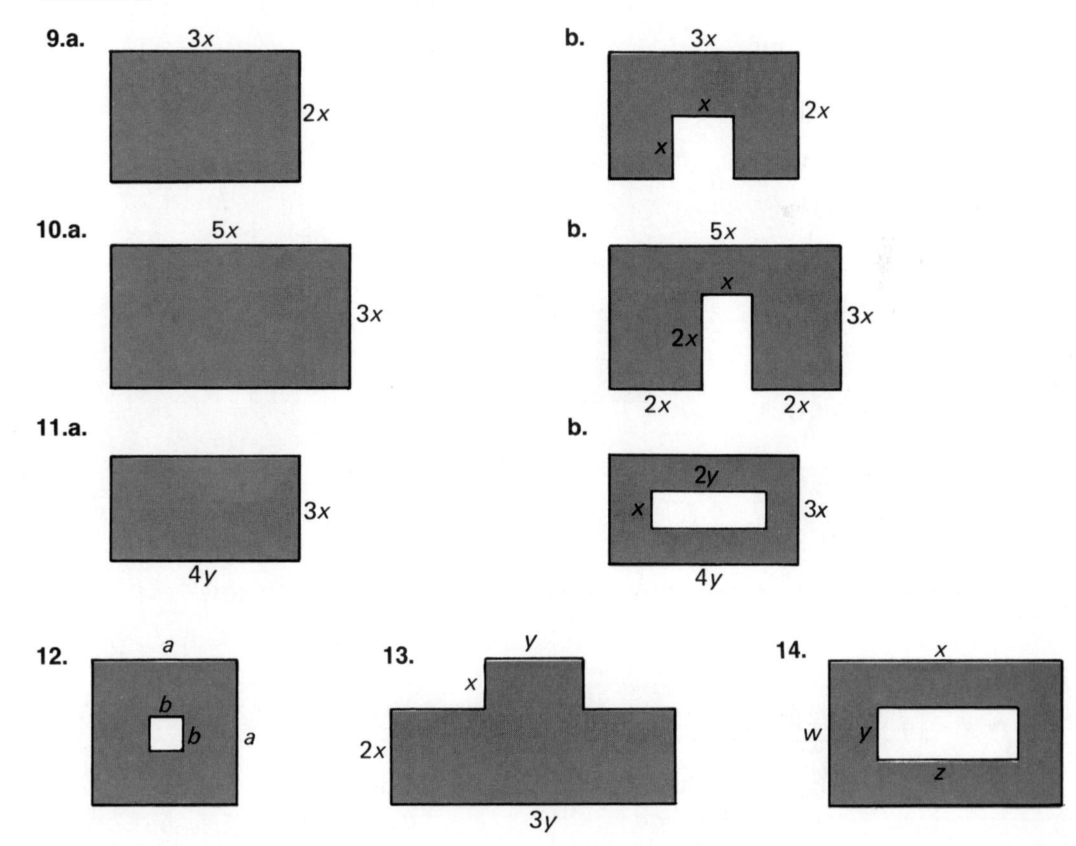

Find the area of the shaded figure. Give your answer in terms of the variables.

15.

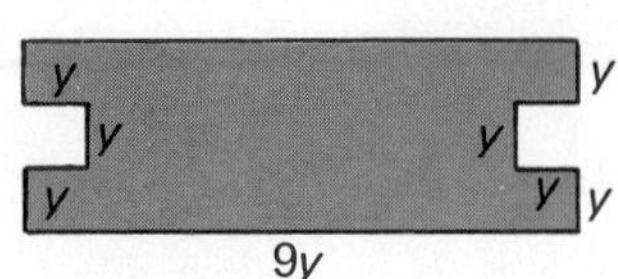

16.

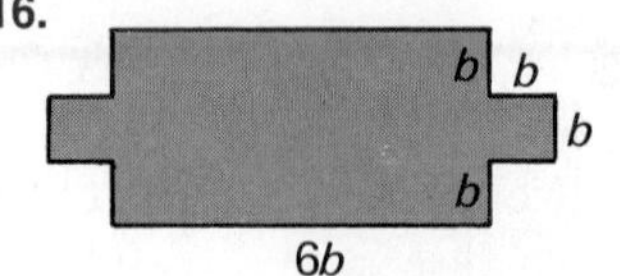

17.

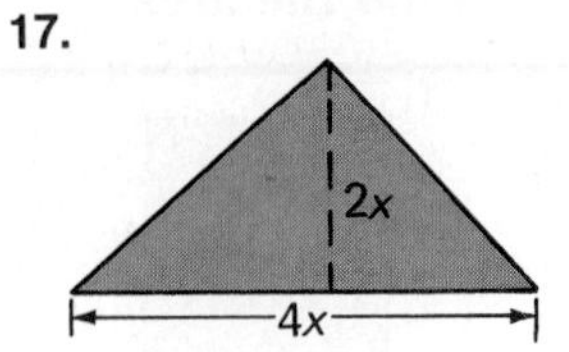

18. Would the area of a rectangle be given in *centimeters* or in *square centimeters?*

19. Would the perimeter of a rectangle be given in *centimeters* or in *square centimeters?*

20. A nylon rug is priced at \$18 per square meter. How much would it cost to buy a rug 5 meters long and 3 meters wide?

21. It costs \$24 per square meter to pave a driveway. How much would it cost to pave a driveway 20 meters long and 4 meters wide?

22. You know that 1 cm = 10 mm.

The following is then true.

$1 \text{ cm}^2 = \underline{\ ?\ } \text{ mm}^2$

1 cm

1 cm 1 mm²

Find the area.

B **23.**

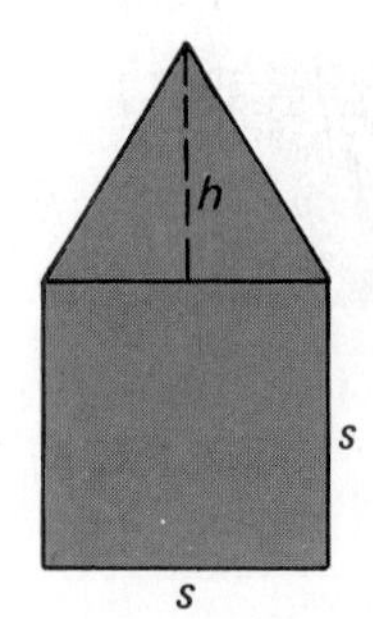

24.

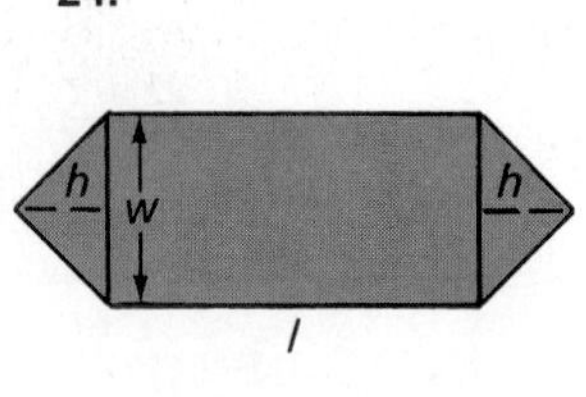

25.

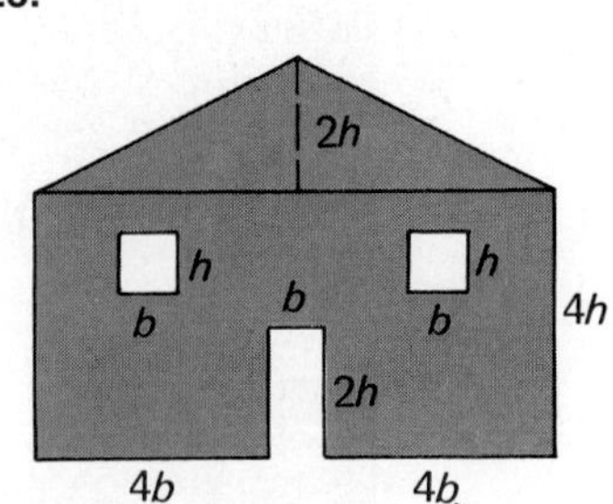

26. What property is shown by the diagram and equations below?

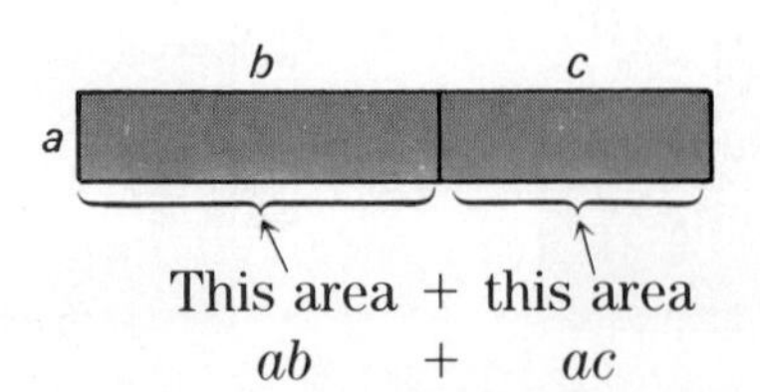

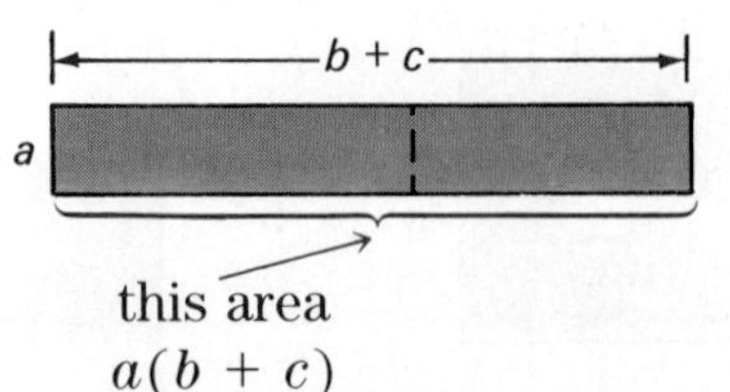

This area + this area = this area

$ab + ac = a(b + c)$

3 Volume Formulas

The **volume** of a telephone booth is the amount of space inside it. Volume can be measured in many units. One common unit is the cubic centimeter.

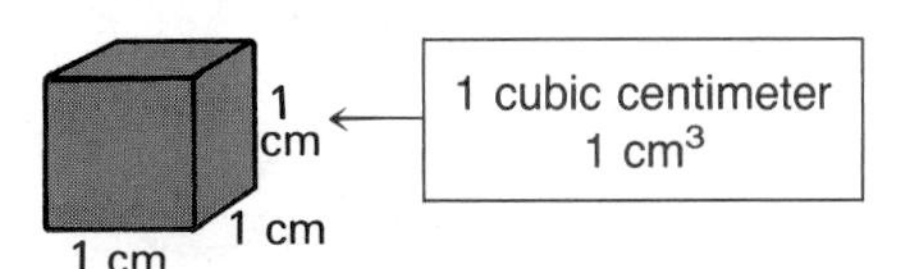

Suppose you want to find the volume of a box like this.

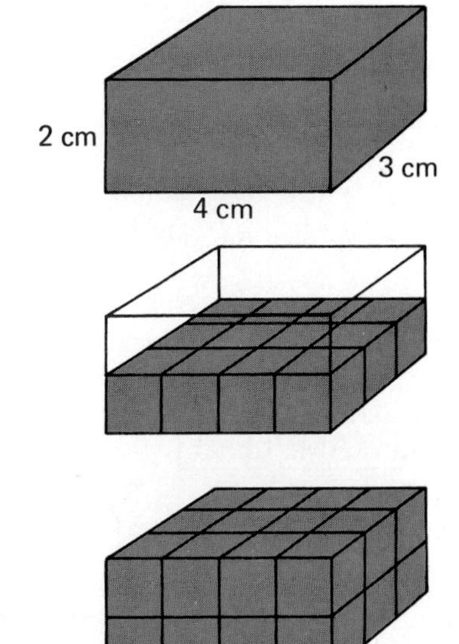

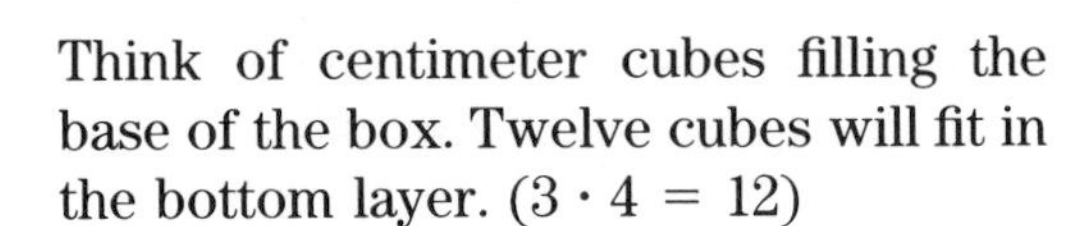

Think of centimeter cubes filling the base of the box. Twelve cubes will fit in the bottom layer. $(3 \cdot 4 = 12)$

Two layers of cubes will fit in the box. This means $2 \cdot 12$, or 24 cubes will fill the box. The volume is 24 cubic centimeters.

The box above is called a **prism.** In a prism, the bases are parallel and are the same size and shape.

Volume of a Prism

Volume = Base Area × height

$V = Bh$

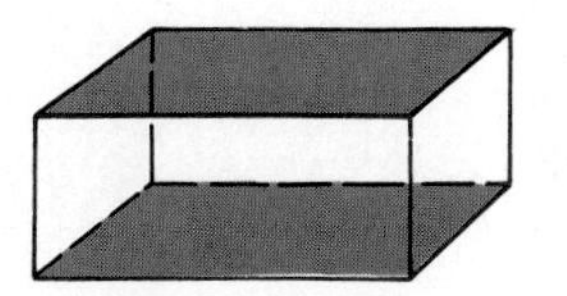

$V = Bh$

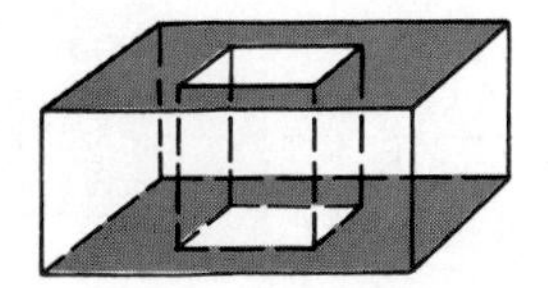

$V = Bh$

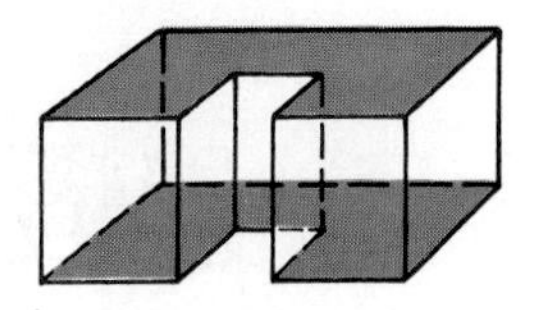

$V = Bh$

EXAMPLE 1

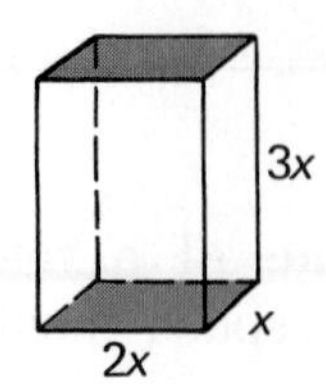

The bases are rectangles.

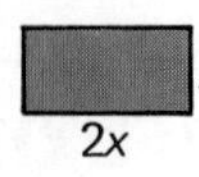

Base Area:
$(2x)(x)$

$V = Bh$

$V = (2x)(x)(3x)$
$V = 6x^3$

EXAMPLE 2

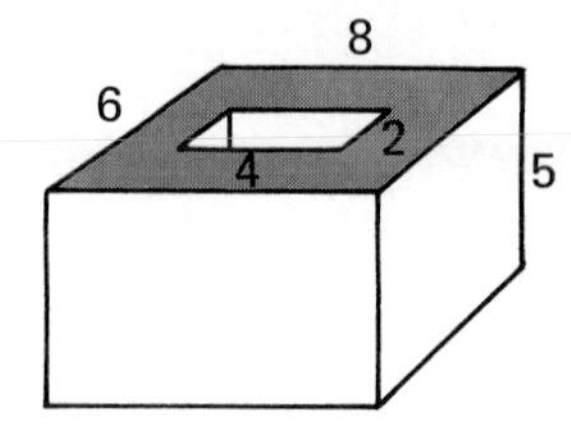

The two bases look like this.

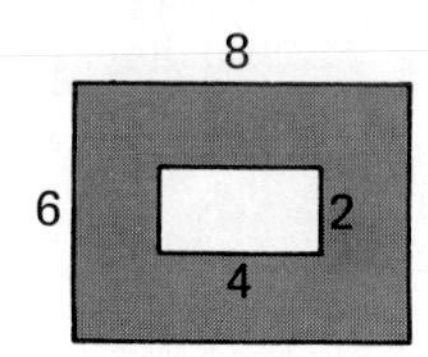

Base Area:

$(6 \cdot 8) - (4 \cdot 2) = 48 - 8$
$= 40$

$V = Bh$

$V = 40 \cdot 5$
$V = 200$

EXAMPLE 3

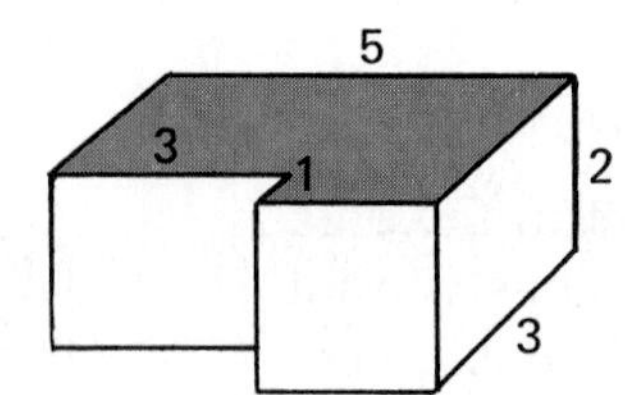

Base Area: $(5 \cdot 3) - (3 \cdot 1) = 15 - 3$
$= 12$

$V = Bh$

$V = 12 \cdot 2$
$V = 24$

Classroom Practice

Find the volume.

1.

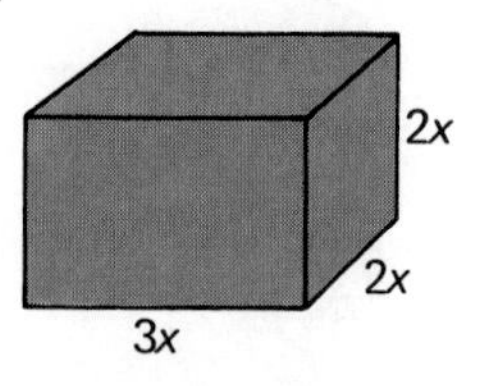

2.

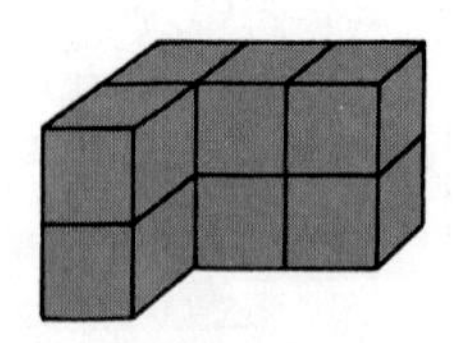

3.

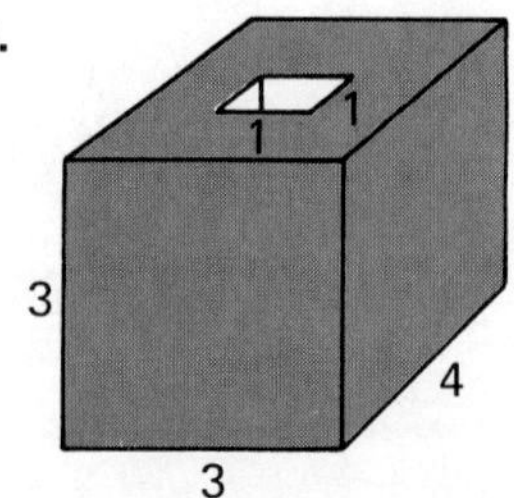

Another unit of volume often used is the liter. (The symbol for a liter is the letter L.) A liter is a little more than a quart.

1 liter = 1000 cubic centimeters

Complete.

4. 3 L = ___?___ cm^3

5. $\frac{1}{2}$ L = ___?___ cm^3

6. ___?___ L = 4000 cm^3

7. ___?___ L = 4500 cm^3

Written Exercises

Draw a sketch to help you.

A

1. A box has a rectangular base that measures 9 cm by 5 cm. The height of the box is 4 cm. Find the volume.

2. Find the volume of a cube 4 cm on a side.

3. A box measures 20 cm by 15 cm by 10 cm. Its volume is ___?___ cm^3, or ___?___ liters.

4. A box measures 40 cm by 30 cm by 50 cm. Its volume is ___?___ cm^3, or ___?___ liters.

Find the volume.

5.

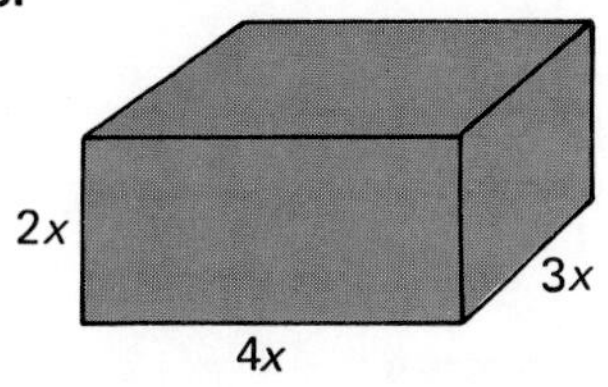

6.

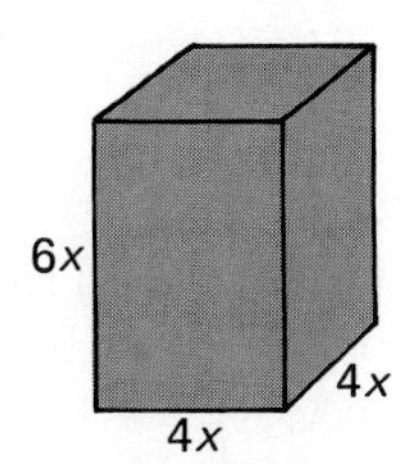

7.

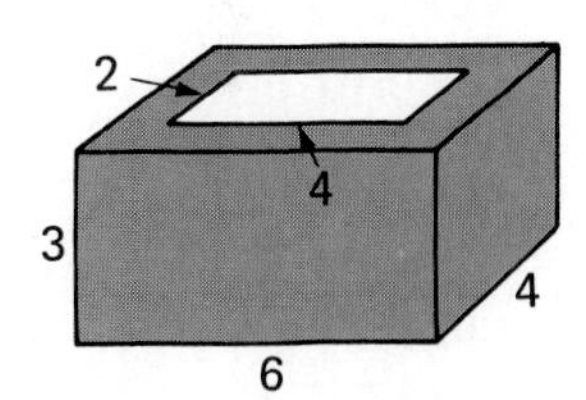

8.

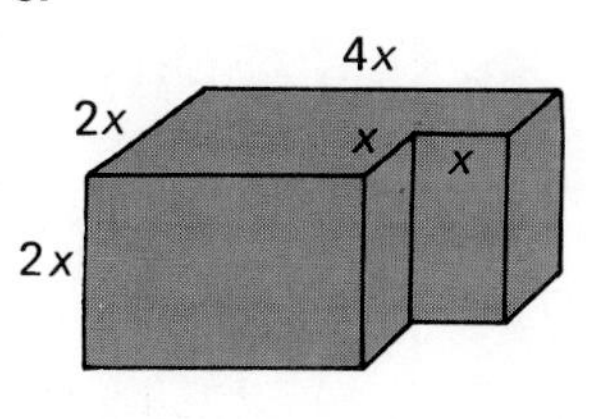

9.

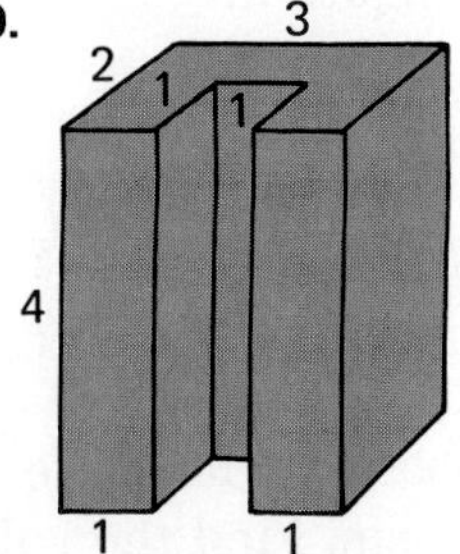

10.

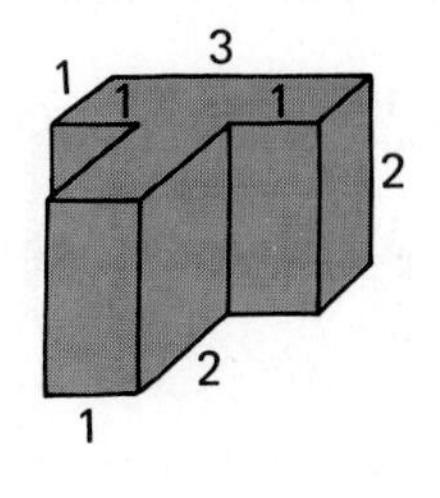

Exercises 11–14 relate to the aquarium shown.

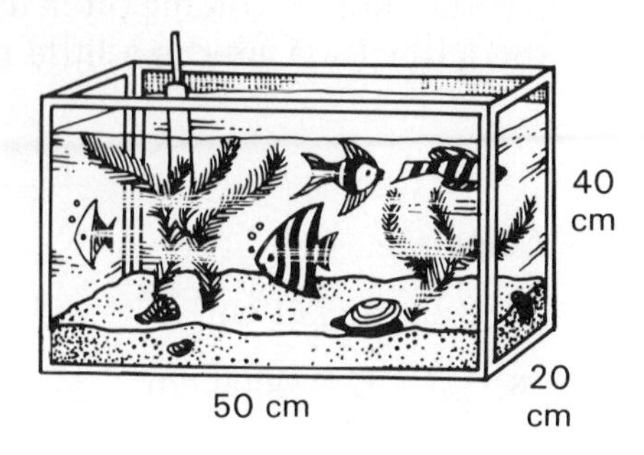

B 11. Find the volume in cubic centimeters.

12. How many liters of water will it hold?
    (1 liter = 1000 cubic centimeters.)

13. How many grams of water is this?
    (1 cubic centimeter holds 1 gram.)

14. How many kilograms will the water be?
    (1 kilogram = 1000 grams.)

15. Will an aquarium measuring 45 cm by 30 cm by 30 cm hold more or less than the aquarium in the drawing?

Find the volume.

16.

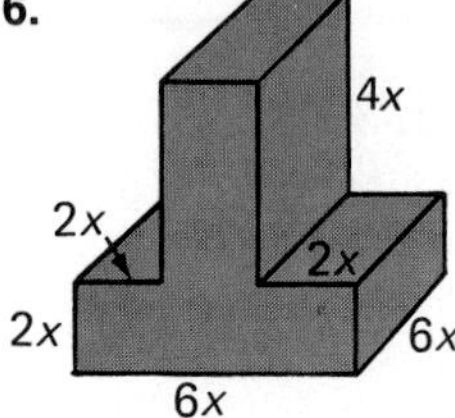

17.

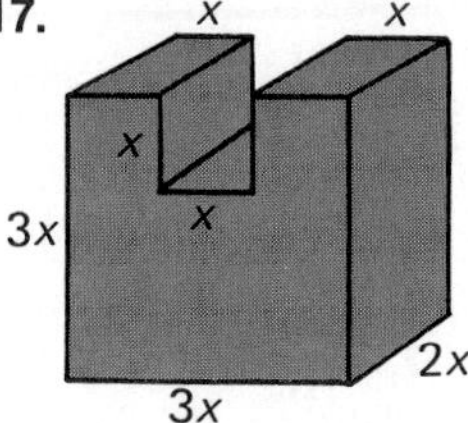

18.

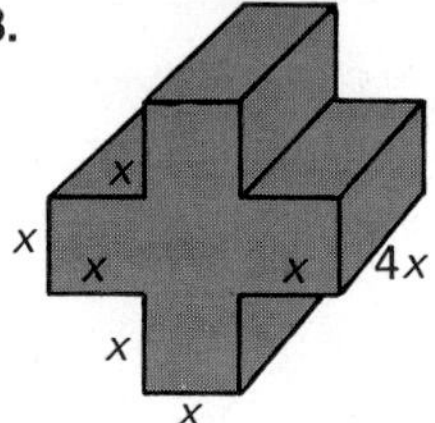

SELF-TEST

Vocabulary

perimeter (p. 112) area (p. 115)
formula (p. 112) volume (p. 119)
circumference (p. 112) prism (p. 119)

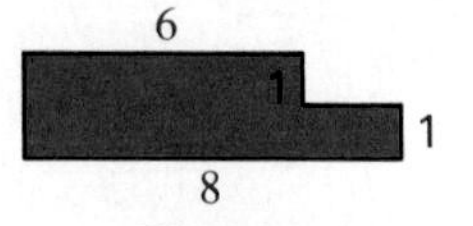

Figure 1

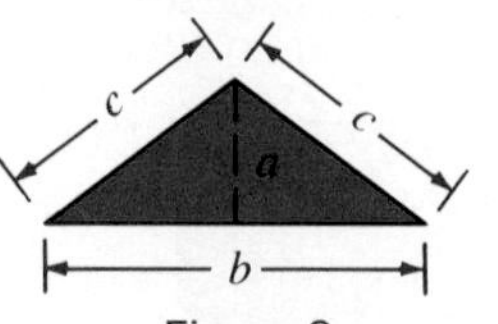

Figure 2

1. What is the perimeter of Figure 1? *(4-1)*

2. What is the perimeter of Figure 2?

3. What is the area of Figure 1? *(4-2)*

4. What is the area of Figure 2?

5. A box has rectangular bases that measure 12 cm by 9 cm. The height of the box is 6 cm. Find the volume. *(4-3)*

CONSUMER APPLICATIONS

WRITING A CHECK

It isn't safe to carry much cash with you. It may be lost or stolen. Many people put their money in a checking account. Then they can pay their bills by check or they can cash a check for themselves when they need money.

Look at the sample below to see how to write a check correctly.

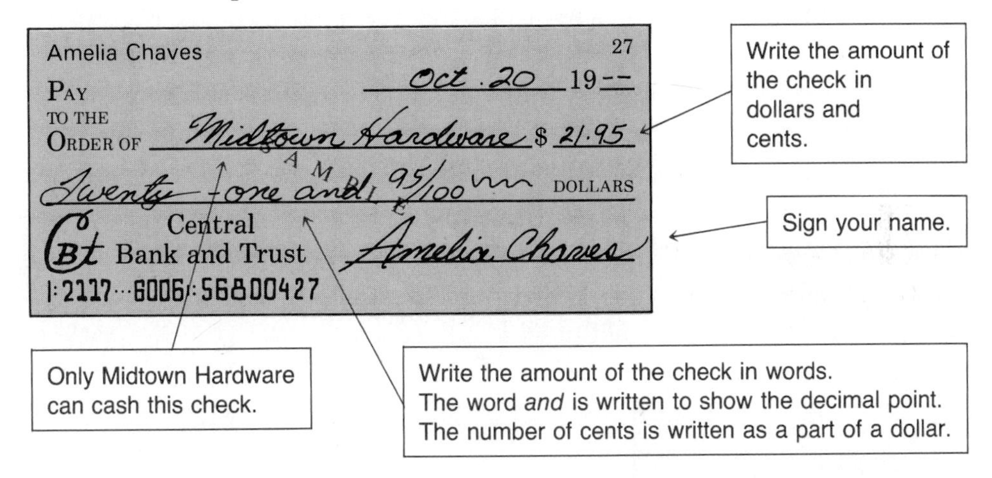

Suppose you wish to get some cash for yourself. You can just go to your bank and make out a check for *Cash*.

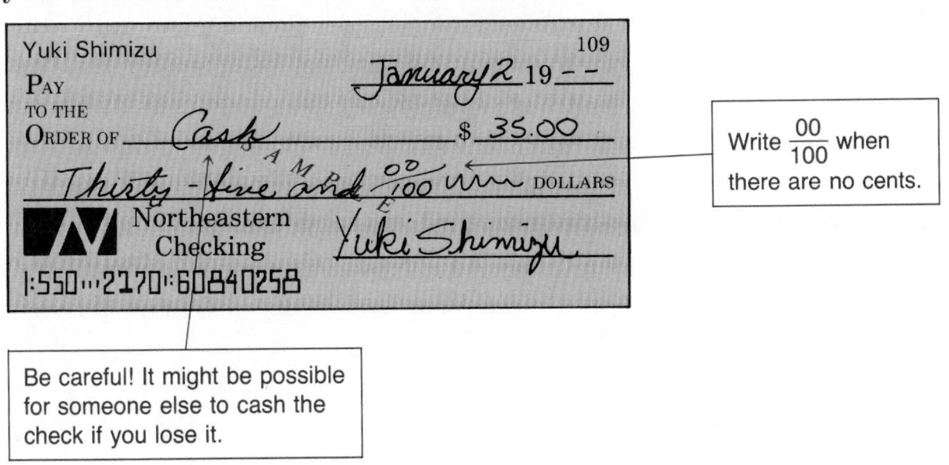

On pages 167 and 286 you'll learn how to keep track of the money in your checking account.

4 Writing Formulas

A formula is a general rule for solving a problem. It tells you *how* to solve the problem, even if you do not know all the numbers.

90 km/h means 90 kilometers per hour.

Suppose you drive 90 km/h for 3 hours. How would you find the distance you traveled? You would *multiply*.

$$90 \times 3 = 270$$

Answer: 270 km

Here's a formula to find distance.

$$\textit{Distance} = \textit{rate} \times \textit{time}$$

$$D = rt$$

Now think of some cost problems. How would you figure the cost of 3 cans of tomatoes at 60¢ each? Multiply!

$$60 \times 3 = 180$$

Answer: $1.80

Here's a formula to find cost.

$$\textit{Cost} = \textit{price per item} \times \textit{number of items}$$

$$C = pn$$

Classroom Practice

Complete.

1. You travel at 75 km/h for h hours. Your distance D = ___?___.

2. You travel at g mph for k hours. Your distance D = ___?___. (mph: miles per hour)

3. You buy t shirts at $6 each. The cost C = ___?___.

4. You buy l liters of gas at m cents a liter. The cost C = ___?___.

Written Exercises

Complete.

A

1. You fly at q km/h for k hours.
   Your distance $D =$ __?__.

2. You paddle for p hours at x mph.
   Your distance $D =$ __?__.

3. You buy 2 tires at t dollars each.
   The cost $C =$ __?__.

4. You buy r magazines at y cents each.
   The cost $C =$ __?__.

5. A touchdown in football scores 6 points.
   Making t touchdowns scores __?__ points.

6. A field goal scores 3 points.
   Making f field goals scores __?__ points.

7. A team makes t touchdowns and f field goals.
   The total $T =$ __?__ points.

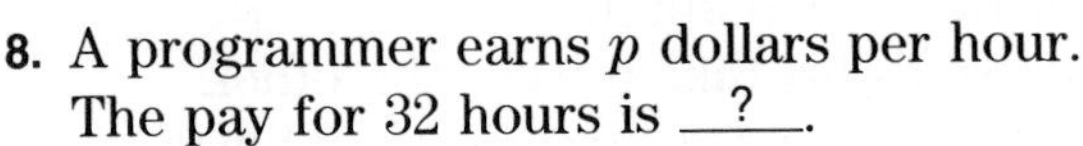

8. A programmer earns p dollars per hour.
   The pay for 32 hours is __?__.

9. You earn \$6 per hour.
   Your pay for z hours is __?__.

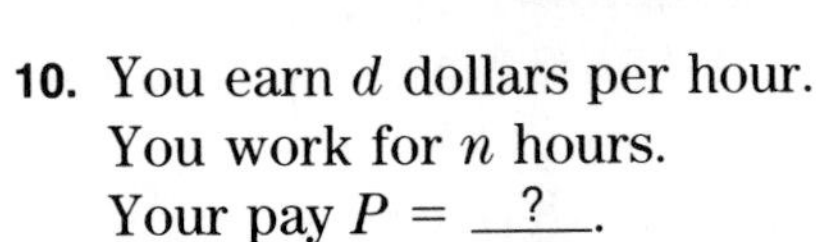

10. You earn d dollars per hour.
    You work for n hours.
    Your pay $P =$ __?__.

11. You buy something for b dollars.
    You sell it for s dollars.
    You make a profit.
    Your profit $P =$ __?__.

12. You buy something for b dollars.
    You sell it for s dollars.
    You take a loss.
    Your loss $L =$ __?__.

B

13. Buy 2 sweaters at u dollars each.
    Buy 3 shirts at w dollars each.
    Total cost $C =$ __?__.

14. Regular wages: \$8 per hour for r hours.
    Overtime: \$12 per hour for s hours.
    Total pay $P =$ __?__.

15. You get the newspaper for a week.
    Each daily paper costs d cents.
    The Sunday paper costs s cents.
    The total cost is __?__.

16. You make a down payment of a dollars, and monthly payments of b dollars each.
    At the end of a year you have paid a total of __?__.

5 Motion Formulas

In the next two sections you will be using formulas to solve motion problems and cost problems. In all of these, you will use basically the same problem-solving methods you used in Chapter 2.

A GUIDE FOR PROBLEM SOLVING
1. *Read the problem. Read it more than once, if you wish.* 2. *Let a variable stand for one of the unknowns.* *Show the other unknowns in terms of that variable.* 3. *Write an equation and solve it.* 4. *Answer the question.* 5. *Check your answer.*

EXAMPLE 1 Tony and Dwight live 81 km apart.
At noon each boy rides his bike toward the other.
Tony travels at 15 km/h. Dwight travels at 12 km/h.
At what time will they meet?

Let t = number of hours until they meet.

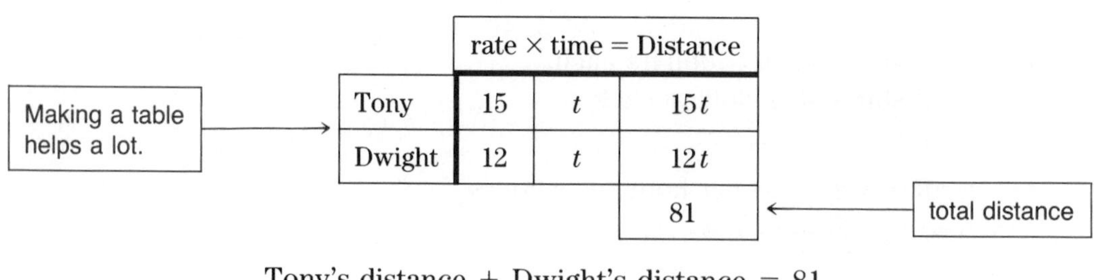

	rate	× time	= Distance
Tony	15	t	$15t$
Dwight	12	t	$12t$
			81

$$\begin{aligned} \text{Tony's distance} + \text{Dwight's distance} &= 81 \\ 15t + 12t &= 81 \\ 27t &= 81 \\ t &= 3 \end{aligned}$$

Answer: The boys will meet 3 hours after they start, at 3:00 P.M.

Check: Tony goes $15 \cdot 3 = 45$ km
Dwight goes $12 \cdot 3 = \underline{36 \text{ km}}$
81 km ✓

EXAMPLE 2

Terry can ride her bike 2 km/h faster than Gina.
They travel from school in opposite directions for 2 hours.
They are then 84 km apart.
How fast does each girl travel?

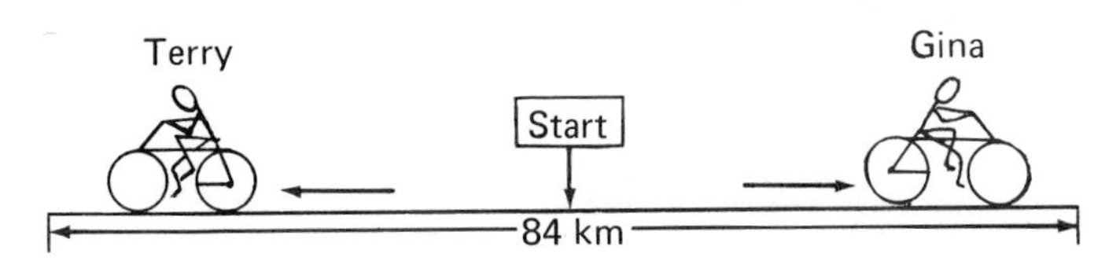

Let r = Gina's speed.
Then $r + 2$ = Terry's speed.

	rate	× time	= Distance
Gina	r	2	$2r$
Terry	$r + 2$	2	$2(r + 2)$
			84 ← total distance

$$\underbrace{\text{Gina's distance}} + \underbrace{\text{Terry's distance}} = 84$$

$$\begin{aligned} 2r \quad + \quad 2(r + 2) &= 84 \\ 2r + 2r + 4 &= 84 \\ 4r + 4 &= 84 \\ 4r &= 80 \\ r &= 20 \end{aligned}$$

Answer: Gina's speed: $r = 20 \longrightarrow 20$ km/h
Terry's speed: $r + 2 = 22 \longrightarrow 22$ km/h

Check: Gina goes $2 \cdot 20 = 40$ km
Terry goes $2 \cdot 22 = 44$ km
84 km ✓

Classroom Practice

Solve.

1. Two cars are 360 km apart.
   They travel toward each other.
   One travels at 80 km/h.
   The other travels at 100 km/h.
   How long will it take them to meet?

	rate	× time	= Distance
First car	80	t	$80t$
Second car	?	?	?
			360

2. Gloria walks 1 km/h faster than Mary Lou.
They walk in opposite directions for 2 hours.
They are then 30 km apart.
How fast do they walk?

	rate × time = Distance		
Gloria	?	?	?
Mary Lou	x	2	?
			?

Written Exercises

Solve.

A

1. Maya and Mike are 56 mi apart.
Each travels toward the other.
Maya travels at 20 mph.
Mike travels at 8 mph.
After how long do they meet?

	rate × time = Distance		
Maya	20	t	$20t$
Mike	?	?	?
			56

2. At noon, a plane leaves Honolulu for San Francisco at 550 km/h.
At the same time a plane leaves San Francisco for Honolulu at 400 km/h.
The distance between the cities is about 3800 km.
At what time do the planes pass each other?

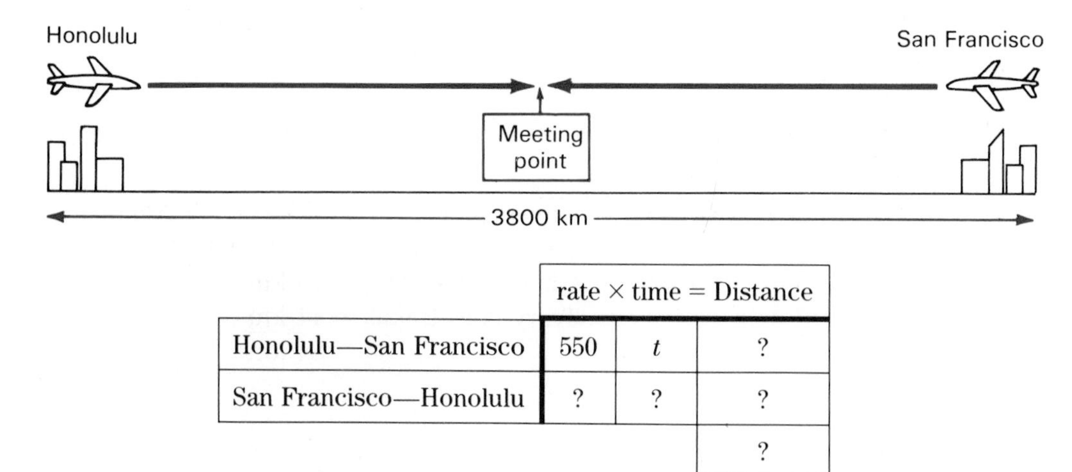

	rate × time = Distance		
Honolulu—San Francisco	550	t	?
San Francisco—Honolulu	?	?	?
			?

3. Lucia bikes 4 km/h faster than Kim.
Lucia travels east for 2 hours.
Kim travels west for 2 hours.
They are then 80 km apart.
How fast is each traveling?

	rate × time = Distance		
Kim	r	?	?
Lucia	$r + 4$	?	?
			?

4. Maria and Tom start at the park.
They drive in opposite directions for 3 hours.
They are then 510 km apart.
Maria's speed is 80 km/h.
What is Tom's speed?

	rate × time = Distance		
Maria	80	?	?
Tom	r	?	?
			?

5. Franklin bikes from A to B at 25 km/h.
He returns by car at 75 km/h.
The bike trip is 2 hours longer than the car trip.
How far is it from A to B?

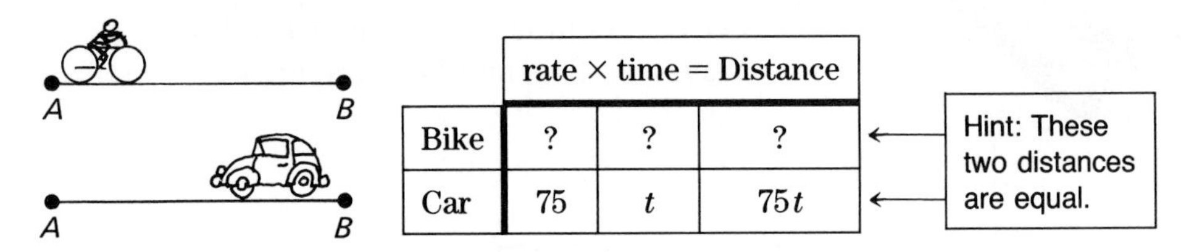

	rate × time = Distance		
Bike	?	?	?
Car	75	t	$75t$

6. Kara floats from A to B at 6 km/h.
She returns by motorboat at 18 km/h.
The float trip is 4 hours longer than the motorboat trip.
How far is it from A to B?

	rate × time = Distance		
Float	6	?	?
Motorboat	?	?	?

B

7. Curtis runs twice as fast as Alvin walks.
In $\frac{1}{2}$ hour Curtis can go 6 km farther than Alvin.
How fast can Alvin walk?

	rate × time = Distance		
Alvin	x	$\frac{1}{2}$	?
Curtis	$2x$	$\frac{1}{2}$	?

6 Cost and Money Problems

In section 4 of this chapter you worked with a formula to find total cost.

$$\text{Cost} = \text{price per item} \times \text{number of items}$$
$$C = pn$$

There are many problems which can make use of this formula.

Irene bought some pencils at 10¢ each.
She kept 2 of them and sold the rest at 15¢ each.
Her profit was \$1.50.
How many pencils did she buy?

Let n = number of pencils Irene bought.
Then $n - 2$ = number of pencils Irene sold.

	p ×	n =	C	
Sell	15	$n - 2$	$15(n - 2)$	← Selling price
Buy	10	n	$10n$	← Buying price
			150	← Profit

$$\underbrace{\text{Selling price}} - \underbrace{\text{Buying price}} = \text{Profit}$$

$$15(n - 2) - 10n = 150$$
$$15n - 30 - 10n = 150$$
$$5n - 30 = 150$$
$$5n = 180$$
$$n = 36$$

Since the prices are given in cents, you write \$1.50 as 150¢.

Answer: Irene bought 36 pencils.

Check: Sold 34 at 15¢ each ⟶ 34 × 15¢ = 510¢
Bought 36 at 10¢ each ⟶ 36 × 10¢ = 360¢ ← Subtract.
Profit = 150¢ ✓

Classroom Practice

Solve.

1. Ramona bought some pears at 40¢ each.
   She sold all but 4 of them at 50¢ each.
   Her profit was $2.00.
   How many pears did she buy?

	price	× number	= Cost	
Sell	50	?	?	← Selling price
Buy	40	n	?	← Buying price
			?	← Profit

2. I bought several notebooks at $.80 each.
   I kept 2 and sold the rest at $1.00 each.
   I didn't make any money, but I didn't lose any either.
   How many notebooks did I buy?

	price	× number	= Cost	
Sell	?	?	?	← Selling price
Buy	80	x	?	← Buying price
			?	← Profit

3. A cycle shop bought 16 helmets at $30 each.
   They sold some of them at $60 each.
   They sold the rest at a sale price of $45 each.
   Their profit was $375.
   How many helmets were sold at the sale price?

Written Exercises

Solve.

A

1. Arnie bought some bagels at 20¢ each.
   He ate 4, and sold the rest at 30¢ each.
   His profit was $2.40.
   How many bagels did he buy?

	price	× number	= Cost
Sell	30	$n - 4$	?
Buy	20	n	?
			240

2. The Video Center bought several video recorders.
Each recorder cost them $220.
They gave away one recorder, and sold the others for $290 each.
Their profit was $1250.
How many recorders did they sell?

	price × number = Cost		
Sell	290	$n - 1$	?
Buy	220	n	?
			?

3. Hank bought several ball-point pens at 40¢ each.
He kept 2 and sold the rest at 60¢ each.
He neither gained nor lost any money.
How many pens did he buy?

	price × number = Cost		
Sell	?	?	?
Buy	40	x	?
			0

4. Mr. and Mrs. Wolfe both make $12 an hour at work.
One week, Mr. Wolfe worked 4 hours more than Mrs. Wolfe.
Their combined pay for the week came to $912.
How many hours did each work?

	pay per hour × number of hours = Weekly pay		
Mrs. Wolfe	12	x	?
Mr. Wolfe	?	?	?
			?

Combined pay

5. Maureen earns $1 more per hour than her husband Karl.
After working a 40-hour work week, their combined pay totaled $840.
How much did each earn per hour?
(Make a chart like the one in Exercise 4.)

6. Admission to the fair was $1 for students and $2 for adults.
There were 80 more student tickets sold than adult tickets.
The ticket sales totaled $980.
How many student tickets were sold?

	price	× number	= Cost
Student tickets	?	?	?
Adult tickets	?	x	?
			980

B

7. Ticket sales for the Spring Fling totaled $1540.
Advance tickets were $3 each. At the door, tickets were $4 each.
There were twice as many advance tickets sold as were sold at the door.
How many of each kind of ticket were sold?

8. During the telethon, people pledged $25 or $40 donations.
Twice as many people pledged $25 as those who pledged $40.
Altogether, $15,300 was pledged.
How many people pledged $25?

SELF-TEST

Complete.

1. You walk for k hours at m km/h.
Your distance D = __?__.

2. You buy x pears at y cents each. *(4-4)*
The cost C = __?__.

3. You earn s dollars per hour.
You work for t hours.
Your pay P = __?__.

4. You buy something for d dollars.
You sell it for c dollars.
Your profit P = __?__.

Solve.

5. Shannon and Len start at the same place. *(4-5)*
They drive in opposite directions for 2 hours.
They are then 310 km apart.
Len's speed is 75 km/h.
What is Shannon's speed?

6. Louisa and Bill Peters each earn $9 an hour at work. *(4-6)*
One week, Bill worked 6 hours more than Louisa.
Their combined pay that week was $684.
How many hours did each person work?

7 New Formulas from Old

Suppose a race car in the Indianapolis 500-mile race finished the course in 3 hours. You could figure the average rate, or speed, in two ways.

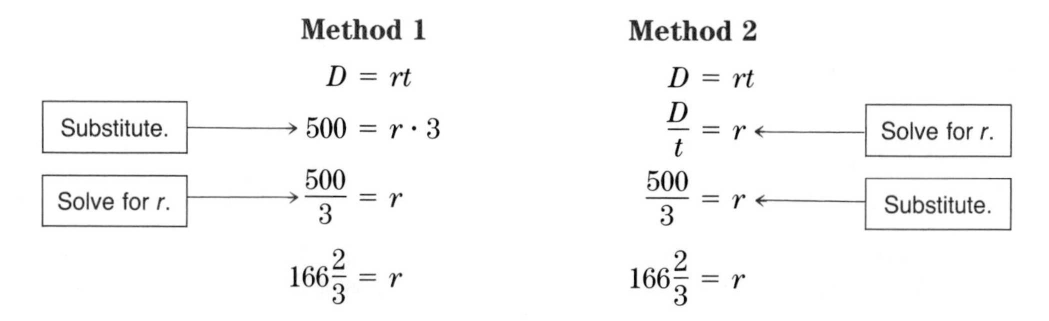

The car went $166\frac{2}{3}$ miles per hour!

In the first step of Method 2, we solved the formula $D = rt$ for r. Here are some other examples that show how to solve a formula.

EXAMPLE 1 Solve $A = lw$ for w.

$$A = lw$$
$$\frac{A}{l} = \frac{lw}{l}$$
$$\frac{A}{l} = w$$

Of course, you can also write $w = \frac{A}{l}$.

EXAMPLE 2 Solve $P = S - B$ for S.

$$P = S - B$$
$$P + B = S - B + B$$
$$P + B = S$$

Or, $S = P + B$.

EXAMPLE 3 Solve $4y + x = 11$ for x.

$$4y + x = 11$$
$$4y + x - 4y = 11 - 4y$$
$$x = 11 - 4y$$

EXAMPLE 4 Solve $2x + y = 9$ for y.

$$2x + y = 9$$
$$2x + y - 2x = 9 - 2x$$
$$y = 9 - 2x$$

EXAMPLE 5 Solve $3x - y = 7$ for y.

$$3x - y = 7$$
$$3x - y - 3x = 7 - 3x$$
$$-y = 7 - 3x$$
$$-1(-y) = -1(7 - 3x)$$
$$y = -7 + 3x$$

Multiply both sides by -1 so that you have solved the equation for y, not $-y$.

Classroom Practice

Solve for x.

1. a. $5x = 15$ b. $5x = y$ c. $bx = y$

Solve for y.

2. a. $y + 2 = 9$ b. $y + 2 = a$ c. $y + x = a$

Solve for s.

3. a. $8 + s = 11$ b. $8 + s = t$ c. $r + s = t$

4. Solve $A = lw$ for l.
5. Solve $bx = y$ for b.
6. Solve $F = ma$ for m.
7. Solve $F = ma$ for a.
8. Solve $C = pn$ for p.
9. Solve $V = Bh$ for B.

Written Exercises

Solve for *y*.

A

1. a. $y - 5 = 7$ b. $y - x = 7$ c. $y - x = t$
2. a. $y + 5 = 7$ b. $y + x = 7$ c. $y + x = t$
3. $2 + y = 5$
4. $2x + y = 4$
5. $3 + y = 11$
6. $4s + y = 9$
7. $y - 2x = 5$
8. $y + 3x = 7$
9. $7 - y = 5$
10. $3x - y = 2$

Solve for *x*.

11. a. $8x = 24$ b. $8x = y$ c. $ax = y$
12. a. $3x = 45$ b. $3x = y$ c. $kx = y$
13. $3 + x = 11$
14. $y + x = 9$
15. $2y + x = 4$
16. $x + 3y = 9$
17. $4y + x = 6$
18. $x - 3y = 5$
19. $y - x = 5$
20. $5y - x = 10$
21. Solve $A = 8w$ for w.
22. Solve $A = Pr$ for r.
23. Solve $D = rt$ for t.
24. Solve $E = 360n$ for n.
25. Solve $N = a + b$ for b.
26. Solve $P = a + b + c$ for c.

Some common formulas are given in Exercises 27–29. You don't have to know how they are used, but their meanings are given, just for interest.

27. Solve for d: $W = fd$ ← work = force × distance

28. Solve for V: $W = AV$ ← watts = amperes × volts

29. Solve for H: $\frac{H}{T} = A$ ← hits ÷ times at bat = batting average

Complete.

30.

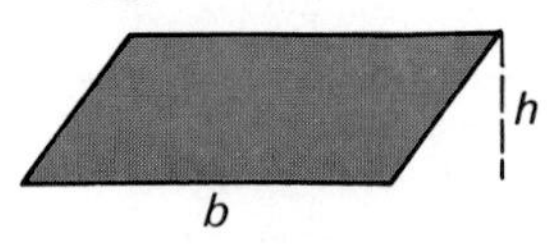

$A = bh$
$A = 84, b = 14$
$h = \underline{\ ?\ }$

31.

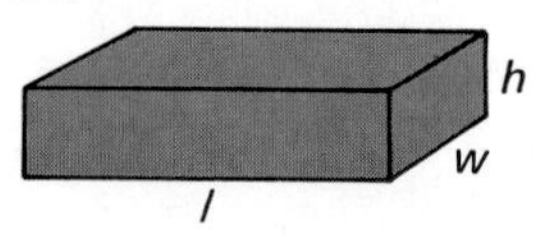

$V = lwh$
$V = 100, l = 10$
$w = 5, h = \underline{\ ?\ }$

32.

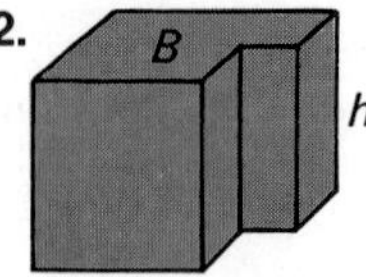

$V = Bh$
$V = 450, B = 50$
$h = \underline{\ ?\ }$

For Exercises 33–34 you'll want to use the formula $D = rt$.

33. The Grand Prix of Monaco is a 256 kilometer race. Suppose the winning car one year took exactly 2 hours. What was the average speed?

34. The famous Le Mans race is won by the automobile which goes the greatest distance in 24 hours. The winning distance for the first Le Mans race in 1923 was 2208 km. What was the average rate?

B **35.** Suppose a golf ball is in orbit around Earth. Suppose the orbital distance of the golf ball is 48,000 km. Suppose it took the golf ball 4 hours to orbit. What was its rate?

$$\text{orbital distance} = \text{rate} \cdot \text{time}$$

SELF-TEST

Solve for x. *(4-7)*

1. a. $9x = 45$ **b.** $9x = y$ **c.** $ax = y$

2. a. $x + 2 = 20$ **b.** $x + 2 = y$ **c.** $x + k = y$

Solve for y.

3. $y - x = 10$

4. $2x + y = 8$

5. $2x - y = 8$

6. $y - 8 = 2x$

7. $y + 3x = 9$

8. $xy = 12$

9. $4 - y = 12$

10. $x + y = z$

PUZZLE ◆ PROBLEMS

How many different combinations of coins have a sum of exactly 27¢? Complete this table to find out.

Coin	25¢	10¢	5¢	1¢
Combinations	1	0	0	2
	0	2	1	2
	and so on			

CAREER NOTES

TRANSPORTATION

Ours is a very mobile society. The field of transportation offers a wide variety of opportunities. Among these are careers in research, design, construction, testing, operations, and maintenance.

Before a new transit system is built, research is done to determine the type of system a city or state needs. Planners research such things as traffic patterns and predicted population figures.

Engineers then design and develop safe and economical transportation equipment. They supervise the construction of the transportation vehicles, such as buses, cars, or subway trains, and the construction of highways, railway systems, and so on.

To make sure that the transportation equipment is safe, technicians thoroughly test new automobiles, train cars, planes, and so on (and test old equipment periodically) to be sure that they meet all standards.

Together the planners, engineers, and testers build safe and efficient means of transportation.

8 Formulas from Mechanics (Optional)

Did you know that there is a formula relating the speeds of bicycle gears? Example 1 explains this formula.

EXAMPLE 1

The pedal gear A on a bicycle has 48 teeth. The wheel gear B has 24 teeth. A chain links both gears so that when you turn gear A, gear B will turn. If A turns at 1 turn per minute, how fast will B turn?

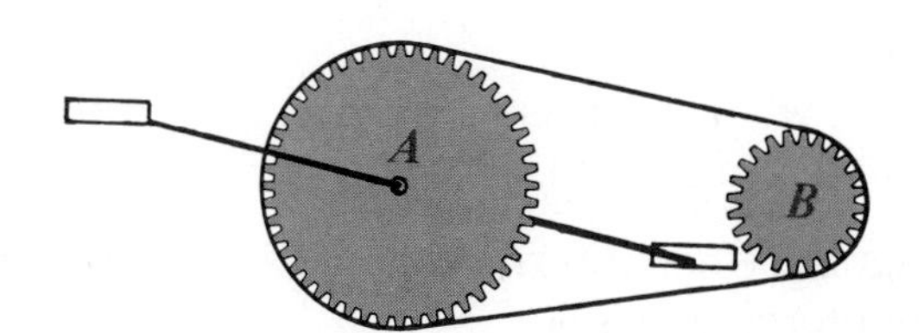

Gear Formula $\longrightarrow$ speed of A × number of teeth in A = speed of B × number of teeth in B

$$1 \times 48 = s \times 24$$
$$48 = 24s$$
$$2 = s$$

Answer: 2 turns per minute

The gear formula is like the lever formula. The following ideas will not seem surprising to you if you have ever been on a seesaw.

EXAMPLE 2

Study the lever (seesaw) below. Then find x.

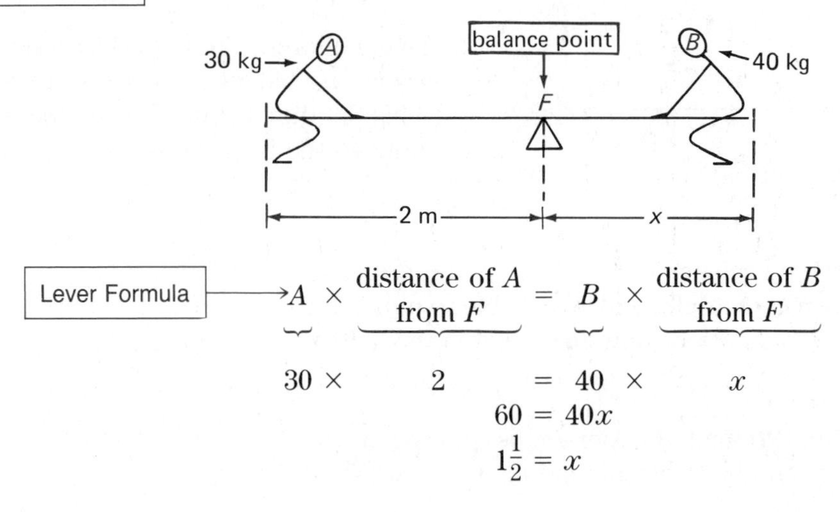

Written Exercises

The diagrams show gear *A* driving gear *B*.

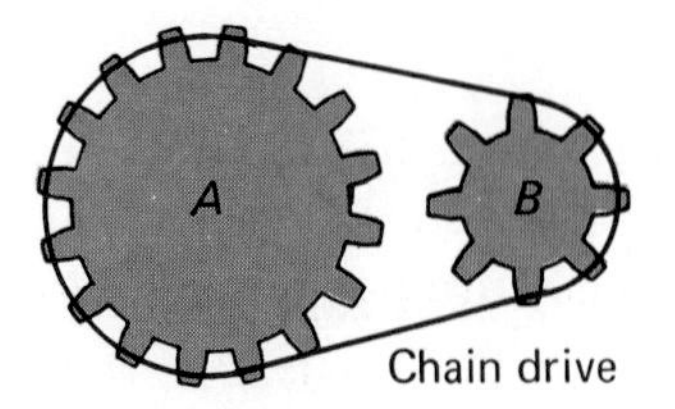

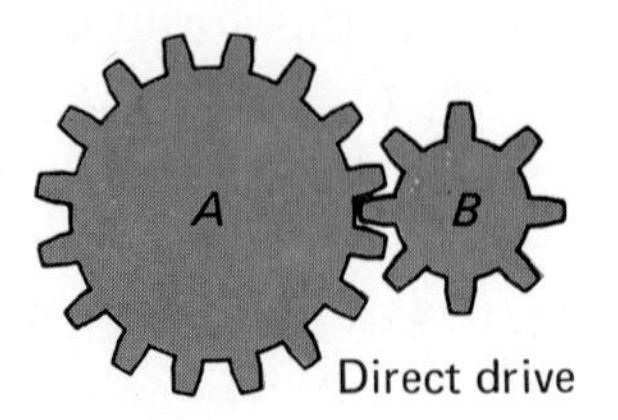

A

1. In the first diagram, when A turns clockwise, gear B turns __?__ (clockwise or counterclockwise).

2. In the second diagram, when A turns clockwise, then B turns __?__ (clockwise or counterclockwise).

3. In the second diagram, count the number of teeth for both gears. When A turns at a speed of 2 turns per second, then how fast does B turn?

4. On a certain five-speed bike, the pedal gear has 52 teeth. The wheel gear used for going up hills has 13 teeth. Suppose you turn the pedal gear at a speed of 1 turn per second. How fast will the first gear turn?

5. Study the lever shown and find the value of d.

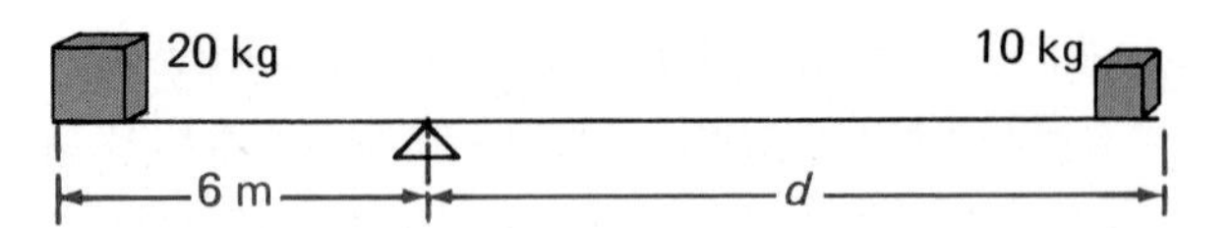

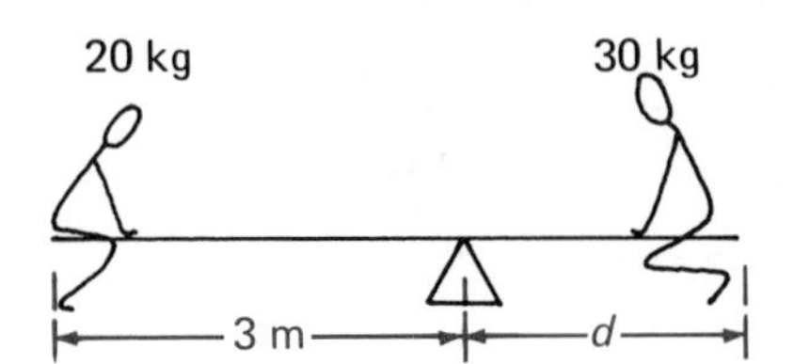

6. Two children, 20 kilograms and 30 kilograms, are balanced on a seesaw as shown. What is the distance from the balancing point to the bigger child?

B

7. A person pushes on the end of the crowbar shown at the right, and is just able to move the large rock. About how many kilograms is the rock?

   (*Hint:* Pretend the person is putting a 75-kilogram object on the end of the crowbar.)

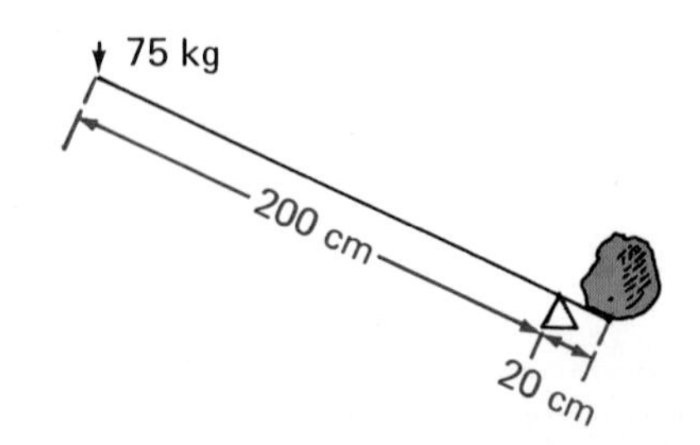

READING ALGEBRA

PROBLEM SOLVING

To be a good problem solver you need to be a careful reader. When you come to a lesson involving problems, do not try to solve the problems until you have read and understood the discussion part of the lesson. Then read each problem slowly and think about the relationships that are given. Use the five steps that are outlined on page 58 and again on page 126. If you come to a word you do not understand, look up its meaning in a dictionary or in the Glossary at the back of the book. Sometimes you may want to make a table or draw a diagram to show what the problem tells you and what you are asked to find. After you have written and solved an equation, be sure to look at the problem again. You may need to do something more in order to answer the problem question. Ask yourself, "Does my answer make sense?"

If your answer is not correct, find out why. You can become a better problem solver by learning to discover and correct your own errors. Read the problem again—did you miss something important? Go over your computation—did you make a careless error? "Talking through" your solution with another person may help you discover where you went wrong.

EXERCISES

Read the following problem and then answer the questions. Ben and Christina started from Ben's house at 1:15 one afternoon and walked in opposite directions. Ben walked at a rate of 4 km/h, and Christina walked at a rate of 5 km/h. At what time were they 27 km apart?

1. What does the problem say about the directions in which Ben and Christina walked? What were their rates of walking?
2. What does the problem tell you about distance?
3. Suppose you use n as the variable. What will you let n stand for? In terms of n, how far did Ben walk? How far did Christina walk?
4. Write an equation for the problem. Solve your equation.
5. Did you get 3 as the value of n? Is 3 the answer to the problem question? If not, what else must you do?
6. Check your solution. *Does* $(3 \times 4) + (3 \times 5)$ equal 27? *Is* 4:15 three hours after 1:15?

PROBLEM SOLVING STRATEGIES

USING DIAGRAMS

You may have noticed that diagrams are often used in this textbook to illustrate explanations and examples. Diagrams can help you better understand mathematical properties and relationships by giving you a visual picture of the known facts.

Would a diagram help you answer the question in this example?

EXAMPLE

Carlos needs to walk six blocks in twenty minutes. In that time, he needs two minutes to stop at a bank and withdraw some money.
How long does Carlos have to walk each block?

You might decide to draw a picture of Carlos, the street, and the bank to illustrate the problem. You may find, however, that a chart of the blocks and minutes alone will be more helpful.

blocks b b b b b b

minutes m m m m m m m m m m m m m m m m m m (m m)

The two minutes needed for the visit to the bank are grouped at the right. Divide the remaining 18 minutes into six groups so that each of the six blocks gets the same number of minutes.

You should find that Carlos will have 3 minutes to walk each block.

This example shows how a simple diagram, or chart, can be used to solve a problem. The following exercises will give you the chance to use diagrams to solve other kinds of word problems.

EXERCISES

1. You need to place five stones into three boxes such that each box gets at least one stone. In how many different ways can this be done?

   a. Draw a picture of the five stones.

   ○ ○ ○ ○ ○

b. Now place two lines between the stones so the stones are divided into three groups. Here is one possibility:

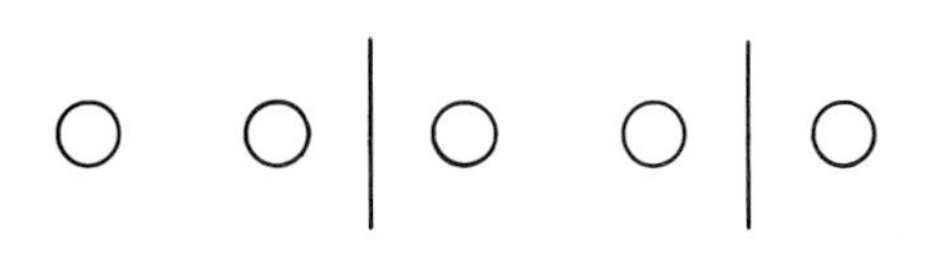

First box	Second box	Third box
2	2	1

c. Finish the problem. Draw a diagram and include a row in the table for each possibility.

2. In a homeroom class of 30 students, 12 students take accounting, 14 take Spanish, and 6 take neither accounting nor Spanish. How many students take *both* accounting and Spanish?

a. Draw two overlapping circles. The circle marked "A" contains all the students who take accounting. The circle marked "S" contains those who take Spanish. Let x be the number of students who take *both* accounting and Spanish. The circle marked "N" contains the students who take *neither* accounting nor Spanish.

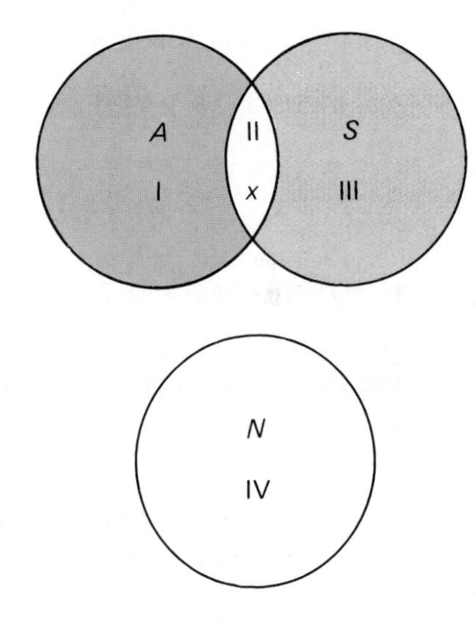

b. Write, in terms of x, expressions for the two shaded regions. (*Hint:* There are 12 students in A, and x of these are not in the shaded region.)

c. Remember that for the four regions shown, the total number of students is 30. Write an equation.

d. Solve the equation in part (c). Answer the question in the problem.

3. In a certain class, 13 students take history, 10 students take science, 3 take both history and science, and 2 take neither history nor science. How many students are in the class? (Use a diagram like the one in Exercise 2.)

SKILLS REVIEW

ORDER OF OPERATIONS

When you have to simplify an expression like $24 \div (-6 + 4) \cdot 3 + 1$, it is important to remember these steps.

Step 1 Do operations inside parentheses.

Step 2 Do all multiplications and divisions in order from left to right.

Step 3 Do all additions and subtractions in order from left to right.

$$24 \div (-6 + 4) \cdot 3 + 1$$
$$24 \div (-2) \cdot 3 + 1$$
$$-12 \cdot 3 + 1$$
$$-36 + 1$$
$$-35$$

Which answer is correct, (a), (b), or (c)?

		a.	b.	c.
1.	$3 \times 5 - 7$	-6	8	6
2.	$4 + 12 \div 2$	8	9	10
3.	$20 - 12 \div (6 - 2)$	16	17	2
4.	$24 \div (-4) + 2$	-12	-4	14
5.	$-3(-5 + 1) + (-2)$	16	10	-14
6.	$5 + 15 \div 3 + 2$	4	12	8
7.	$3 \cdot 4 + 18 \div 2 + 7$	14	40	28
8.	$-7 - (-15) \div 5 + 6$	-2	-4	2

Find the value of the expression.

9. $15 \div (-3) + 2$

10. $-15 \div (-3 + 2)$

11. $-(15 \div (-3)) + 2$

12. $12 \cdot 6 - 4 + 3$

13. $(12 \cdot 6) - (4 + 3)$

14. $12(6 - 4) + 3$

15. $21 \div 7 - (4 - 1)$

16. $4 \cdot 5 + 9 \div 3$

17. $-(15 \cdot 3 - 2)$

18. $7 - 8 + 24 \div 8$

19. $2 + 6 \cdot 10 \div 2$

20. $84 - 16 \div 2 + 2$

21. $17 - 12 \cdot 2 + 3$

22. $6 \cdot 7 + 28 \div 4$

23. $-36 \div (-9) - 5 - (-4)$

24. $216 \div 24 \div (-3) \div (-1)$

CHAPTER REVIEW

CHAPTER SUMMARY

1. Some important formulas are shown below.

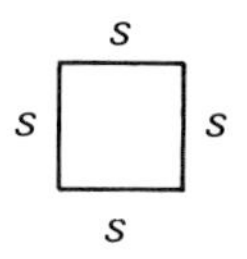

$P = 4s$
$A = s^2$

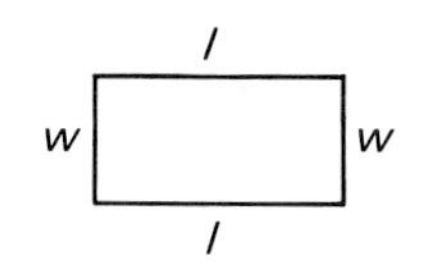

$P = 2l + 2w$
$A = lw$

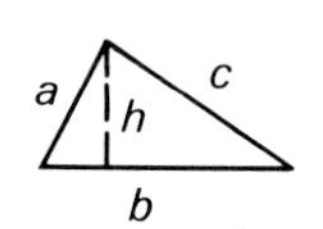

$P = a + b + c$
$A = \frac{1}{2}bh$

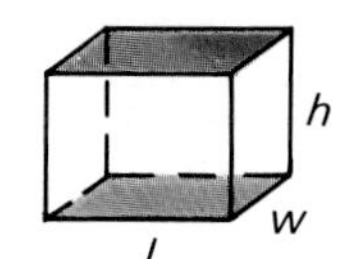

$V = Bh$
$= lwh$

2. Distance formula: Distance = rate × time, or $D = rt$

3. Cost formula: Cost = price per item × number of items, or $C = pn$

REVIEW EXERCISES

Find the perimeter. *(See pp. 112–114.)*

1. 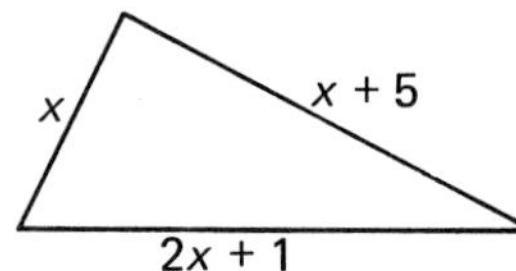

2. 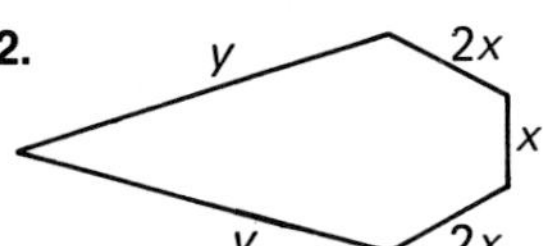

3. 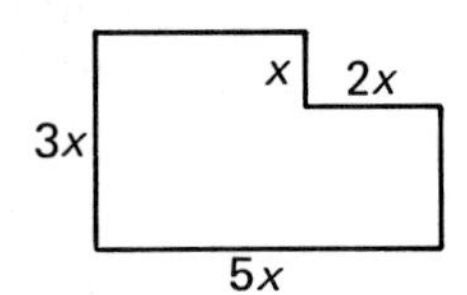

Find the area. *(See pp. 115–118.)*

4. 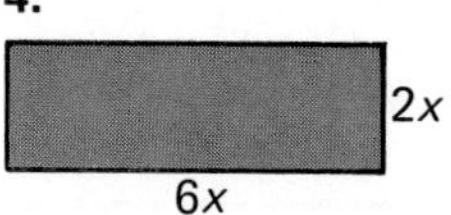

5. 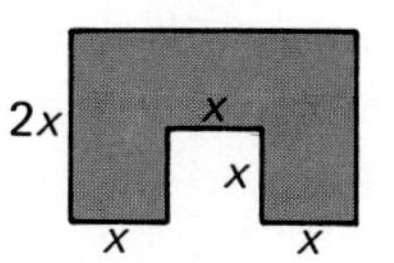

6. 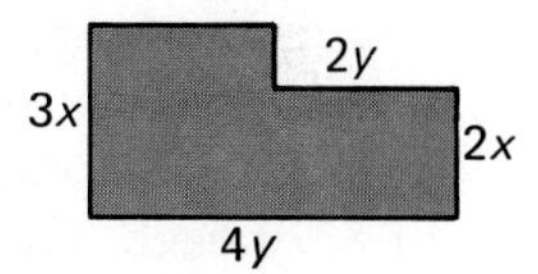

7. A rug is priced at \$11 per square meter. How much would it cost to buy a rug 4 meters long and 3 meters wide?

Find the volume. *(See pp. 119–122.)*

8. 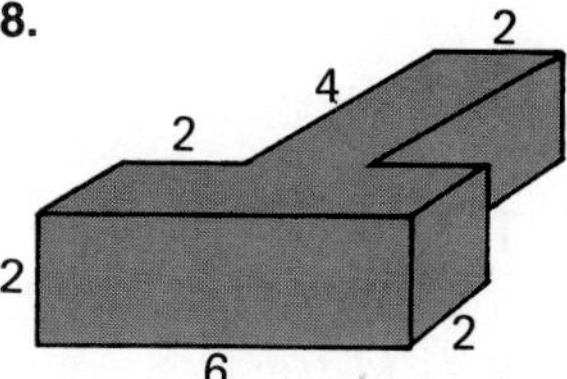

9. 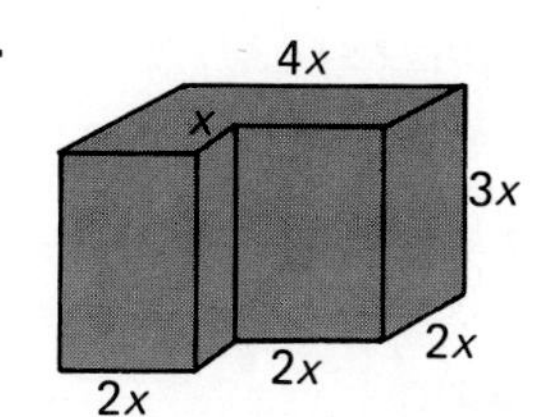

10. A box measures 50 cm by 30 cm by 20 cm.
Its volume is __?__ cm^3, or __?__ liters.

Complete. ***(See pp. 124–125.)***

11. You travel at 90 km/h for h hours.
Your distance D = __?__.

12. You buy s salads at \$$t$ each.
The cost C = __?__.

Solve. ***(See pp. 126–133.)***

13. Shawn bikes for 3 hours at 20 km/h.
Then he rides by car for 2 hours.
His total distance is 246 km.
How fast does the car travel?

	rate	× time	= Distance
Bike	?	?	?
Car	?	?	?
			246

14. Marsha can walk 2 km/h faster than Cammie.
They travel in opposite directions for 4 hours.
They are then 48 km apart.
How fast is each girl traveling?

	rate	× time	= Distance
Marsha	r	?	?
Cammie	?	?	?
			?

15. Compu Hut bought several computer monitors.
Each monitor cost them \$150.
They gave 2 monitors away and sold the rest
for \$210 each. Their profit was \$1440.
How many monitors did they sell?

	rate	× number	= Cost
Sell	?	?	?
Buy	150	n	?
			?

Solve for x. ***(See pp. 134–137.)***

16. a. $7x = 21$ **b.** $kx = 21$ **c.** $kx = a$

17. a. $x + y = 7$ **b.** $x - y = 7$ **c.** $y - x = 7$

18. Solve $\frac{t}{3} = 8$ for t.

19. Solve $W = fd$ for f.

CHAPTER TEST

Find the perimeter.

(4-1)

1. 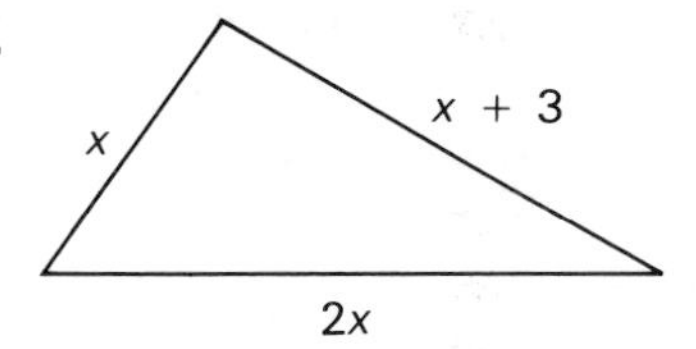

2. 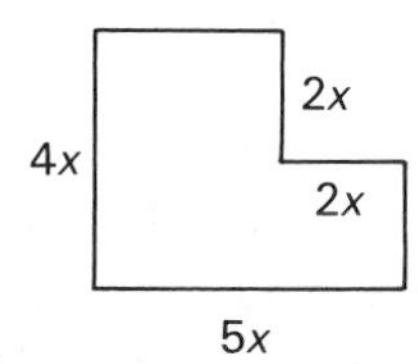

3. The length of a rectangle is 3 times as long as the width. The perimeter is 56 in. Find the length and width.

Find the area of the shaded figure. Give your answer in terms of the variables.

(4-2)

4. 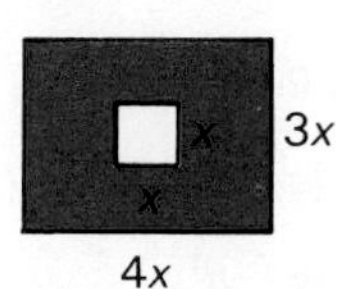

5. 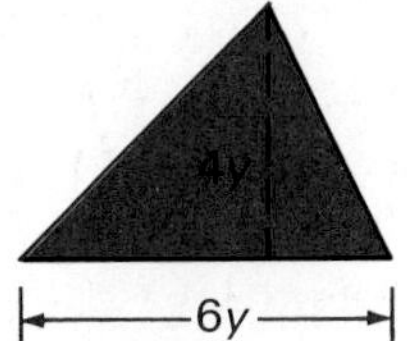

6. One can of paint will cover 300 square feet of area. How many cans of paint are needed to paint both sides of a fence that is 90 ft long and 5 ft high?

7. A box has a rectangular base that measures 10 cm by 6 cm. The height of the box is 8 cm. Find the volume. *(4-3)*

Complete.

(4-4)

8. You jog for x hours at r km/h. Your distance $D =$ __?__.

9. You have x cents. You buy 5 apples at c cents each. Your change $=$ __?__.

Solve.

10. Two cars are 640 km apart. The two cars travel toward each other and meet in 4 hours. One car travels 10 km/h faster than the other. How fast did each travel? *(4-5)*

11. Stan's Store bought some irons at \$26 each. Stan kept two irons and sold the rest at \$35 each. His profit was \$92. How many irons did Stan sell? *(4-6)*

(4-7)

12. Solve $3y - x = 4$ for x.

13. Solve $C = pn$ for n.

UNIT B CUMULATIVE REVIEW

Compare the numbers. Write > or <.

1. $1 \underline{\ ?\ } -1$
2. $-12 \underline{\ ?\ } -10$
3. $0 \underline{\ ?\ } -4$
4. $-8 \underline{\ ?\ } -9$

Graph the solution(s) of the equation or inequality on the number line.

5. $x - 3 = 4$
6. $y + 1 = 1$
7. $x > -5$
8. $x < 5$

Add or subtract.

9. $3 + (-3)$
10. $-3 + (-3)$
11. $-7 + 5$
12. $-11 + (-3)$
13. $8 - 4$
14. $-8 - 4$
15. $8 - (-4)$
16. $-8 - (-4)$

Multiply or divide.

17. $7 \cdot 20$
18. $-4 \cdot 16$
19. $-5(-9)$
20. $-(-12)^2$
21. $3a(-7y)$
22. $-(-1 - x)$
23. $t - (3t - 5)$
24. $(2z - y)(-8)$
25. $24 \div -8$
26. $-16 \div 2$
27. $-54 \div -9$
28. $26 \div -2$

Solve.

29. $x + 5 = -1$
30. $4z = -60$
31. $4t + 2 = t - 4$

Find the perimeter.

32.

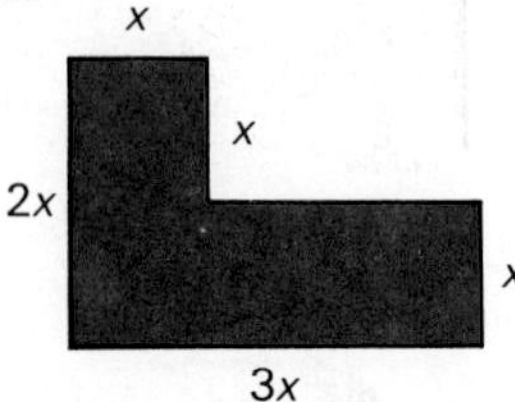

Find the area.

33.

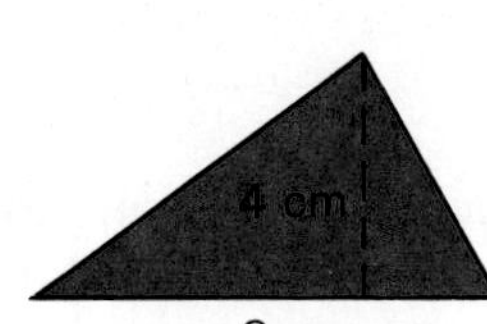

Find the volume.

34.

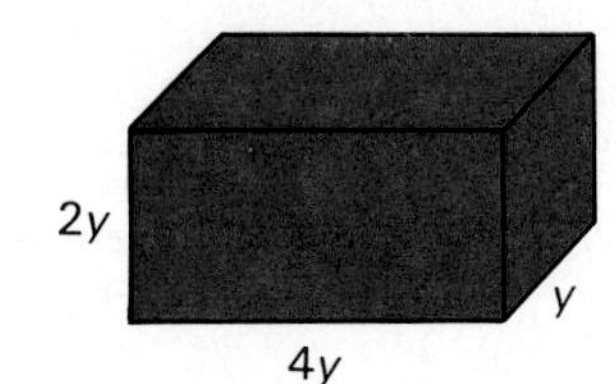

35. A team gets 2 points for a win. It gets 1 point for a tie. The team earned w wins and t ties during the season. The total points earned $P = \underline{\ ?\ }$.

36. A plane flies from A to B at 750 km/h. The return trip at 600 km/h takes 1 hour longer. How far is it from A to B?

37. A shop bought pies at \$2 each. They sold all but 4 at \$3 each. They made a profit of \$56. How many pies did they buy?

UNIT C

Here's what you'll learn in this chapter.

To add and subtract polynomials.

To multiply polynomials.

To multiply binomials at sight.

To square a binomial.

To divide by a monomial.

Notice how rectangular shapes have been combined in this building to give each apartment some window space. Without mathematics, it would not be possible to plan and build such a complicated structure.

Chapter 5

Working with Polynomials

1 Addition

The three expressions below are **polynomials.** A polynomial may have any number of terms. Notice the terms are connected by + or − signs.

$$3xy$$
$$4a + 2b$$
$$5x^2 - 12xy + y^2$$

Special names are given to polynomials with one, two, or three terms.

one term	*two terms*	*three terms*
$3xy$	$4a + 2b$	$5x^2 - 12xy + y^2$
monomial	**binomial**	**trinomial**

It's easy to remember these words if you compare them with words you know.

monocycle

bicycle

tricycle

You can add polynomials in a vertical form or in a horizontal form.

EXAMPLE 1

$$\begin{array}{rrrr} 2a^2 & + 3b^2 & + ab & - 3 \\ 6a^2 & - b^2 & & + 1 \\ \hline 8a^2 & + 2b^2 & + ab & - 2 \end{array}$$

EXAMPLE 2

$$\begin{aligned} &(2xy + 2x + 3y) + (4xy - 2x + 5y) \\ &= 2xy + 4xy + 2x - 2x + 3y + 5y \\ &= 6xy + 8y \end{aligned}$$

Classroom Practice

Is the expression a monomial, a binomial, or a trinomial?

1. $a + 2$
2. x
3. $4ab$
4. $w + 4$
5. $x^2 + 2xy + y^2$
6. -1
7. $t^3 - 3$
8. $a^2 + 4ab + 4$

Add.

9. $\begin{array}{r} 2x + 4 \\ \underline{5x - 1} \end{array}$

10. $\begin{array}{r} 3a + 2b \\ \underline{a + 5b} \end{array}$

11. $\begin{array}{r} a^2 - 3ab - b^2 \\ \underline{6a^2 + 2ab + 5b^2} \end{array}$

12. $(3a + 4ab) + (6a + ab)$

13. $(x^2 + 3xy) + (4x^2 - xy)$

Written Exercises

Add.

A

1. $\begin{array}{r} 2y + 4 \\ \underline{y + 7} \end{array}$

2. $\begin{array}{r} 2x^2 + 8 \\ \underline{7x^2 + 7} \end{array}$

3. $\begin{array}{r} m^2 - n^2 \\ \underline{m^2 + n^2} \end{array}$

4. $\begin{array}{r} -3x + 4y \\ \underline{-8x + 9y} \end{array}$

5. $\begin{array}{r} 12ab + b^2 \\ \underline{-7ab + b^2} \end{array}$

6. $\begin{array}{r} a^2b - 3ab^2 \\ \underline{2a^2b + 5ab^2} \end{array}$

7. $\begin{array}{r} x^2 + xy + y^2 \\ \underline{x^2 \qquad + y^2} \end{array}$

8. $\begin{array}{r} -n^2 + 2n - 1 \\ \underline{n^2 + 8n - 12} \end{array}$

9. $\begin{array}{r} 4y^4 - 4y^2 + 1 \\ \underline{-4y^4 - y^2 \qquad} \end{array}$

10. $(2x - y) + (7x + y)$

11. $(n + 1) + (11n + 1)$

12. $(3y + z + 1) + (18y + 9z + 16)$

13. $(m + 3n + 8) + (9m + 9n + 7)$

14. $(n^2 + 2n + 1) + (3n^2 + 4n - 3)$

15. $(-m^2 + 4m) + (m^2 - 8m + 8)$

16. $(a^2 - 2ab + b^2) + (-2a^2 + b^2)$

17. $(x^2 - 10x + 5) + (9x^2 - 10x - 3)$

18. $(3a^2 + 3a - 6) + (a^2 - 1)$

19. $(x^2 + 2x + 4) + (4x^2 - 5)$

20. $(n^3 - 4) + (n^3 - n^2 + 10)$

21. $(k^3 - 1) + (k^2 - 2k + 7)$

22. $(n^2 - n) + (n^3 - 2n)$

23. $(2x^2 + 3) + (x^3 + 4x^2 - 6)$

24. $(2a^3 - ab + 1) + (a^3 - 2ab)$

25. $(4a^2 - 2ab) + (-3a^2 + ab - b^2)$

26. A city park has the dimensions shown.
    a. Write an expression for its perimeter.
    b. Evaluate if $n = 5$.

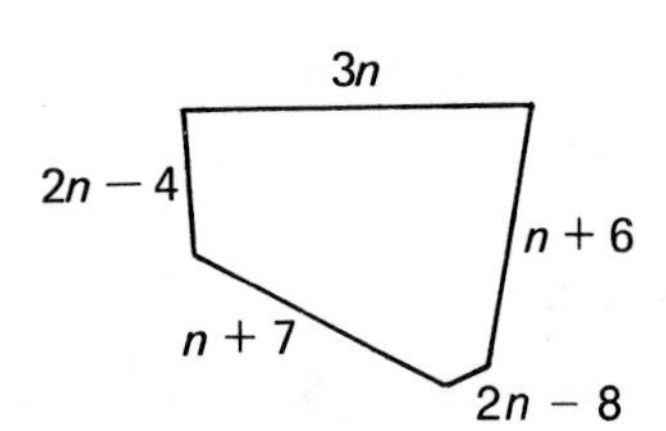

2 Subtraction

You know that when you want to subtract, you add the opposite.

$$4 - 7 = 4 + (-7)$$
$$= -3$$

The same is true when you're subtracting binomials or trinomials. The vertical form is sometimes easier to use.

EXAMPLE 1 $(3x - 1) - (2x + 2)$

$$\begin{array}{r} 3x - 1 \\ -(2x + 2) \\ \hline \end{array} \longrightarrow \boxed{\text{Rewrite and add.}} \longrightarrow \begin{array}{r} 3x - 1 \\ -2x - 2 \\ \hline x - 3 \end{array}$$

EXAMPLE 2 $(5x^2 + x + 1) - (4x^2 - x + 1)$

$$\begin{array}{r} 5x^2 + x + 1 \\ -(4x^2 - x + 1) \\ \hline \end{array} \longrightarrow \boxed{\text{Rewrite and add.}} \longrightarrow \begin{array}{r} 5x^2 + x + 1 \\ -4x^2 + x - 1 \\ \hline x^2 + 2x \end{array}$$

Of course, the idea of changing signs and adding can be used in the horizontal form too.

EXAMPLE 3

$$\begin{aligned}(2a + 4) - (a - 6) &= (2a + 4) + (-a + 6) \\ &= 2a - a + 4 + 6 \\ &= a + 10\end{aligned}$$

EXAMPLE 4

$$\begin{aligned}&(x^2 + 4x + 2) - (2x^2 - x + 1) \\ &= (x^2 + 4x + 2) + (-2x^2 + x - 1) \\ &= x^2 - 2x^2 + 4x + x + 2 - 1 \\ &= -x^2 + 5x + 1\end{aligned}$$

Classroom Practice

Name the opposite.

1. $(4x^2 - 1)$ **2.** $(-x + 4)$ **3.** $(x^2 + 2x - 1)$ **4.** $(x^2 - 2xy + y^2)$

Subtract.

5. $\begin{array}{r} 2a - 4 \\ -(3a + 2) \\ \hline \end{array}$ **6.** $\begin{array}{r} x^2 - 3y \\ -(x^2 - 6y) \\ \hline \end{array}$ **7.** $\begin{array}{r} a^2 - 2b + 5b^2 \\ -(3a^2 - b + 4b^2) \\ \hline \end{array}$ **8.** $\begin{array}{r} 2x^2 + 3x - y^2 \\ -(x^2 - 3x + y^2) \\ \hline \end{array}$

Written Exercises

Subtract.

A

1. $\begin{array}{r} 2x + y \\ \underline{-(x - 3y)} \end{array}$

2. $\begin{array}{r} 3a - 4b \\ \underline{-(a + 5b)} \end{array}$

3. $\begin{array}{r} -x^2 + y^2 \\ \underline{-(x^2 - y^2)} \end{array}$

4. $\begin{array}{r} x^2 - 2x - 1 \\ \underline{-(x^2 + 2x + 1)} \end{array}$

5. $\begin{array}{r} 3x^2 - 4x + 5 \\ \underline{-(x^2 - 4x + 1)} \end{array}$

6. $\begin{array}{r} n^2 - 4n + 5 \\ \underline{-(3n^2 + 7n + 3)} \end{array}$

7. $\begin{array}{r} x^2 - 2xy + y^2 \\ \underline{-(x^2 + 2xy + y^2)} \end{array}$

8. $\begin{array}{r} y^3 + y^2 + 6 \\ \underline{-(y^3 \qquad + 4)} \end{array}$

9. $\begin{array}{r} x^2y + xy^2 + 4 \\ \underline{-(-3xy^2 + 2)} \end{array}$

10. $(a + 1) - (2a - 4)$
11. $(2b - 2) - (b - 4)$
12. $(x^2 + 1) - (x^2 + 1)$
13. $(5m - 16) - (m + 2)$
14. $(3n - 2) - (-n - 2)$
15. $(y + 6) - (-3y - 8)$
16. $(x^2 + 3x + 2) - (x^2 - 4x + 1)$
17. $(x^2 + 2x + 1) - (4x^2 - 3x + 7)$
18. $(a^2 + 5ab - 2c) - (3a^2 + ab - 4c)$
19. $(2ab - 4c) - (a^2 - 3ab + 6c)$
20. $(x^2 - 8x + 7) - (5x^2 + 9)$
21. $(4y^2 - 8) - (2y^2 - 5y + 10)$

Solve for the variable.

22. $3x - (-24 - x) = 60$
23. $(13n + 5) - (3n + 6) = 99$
24. $(2x - 4) - (x + 8) = 24$
25. $(12n - 40) - (10n + 30) = n + 10$

B

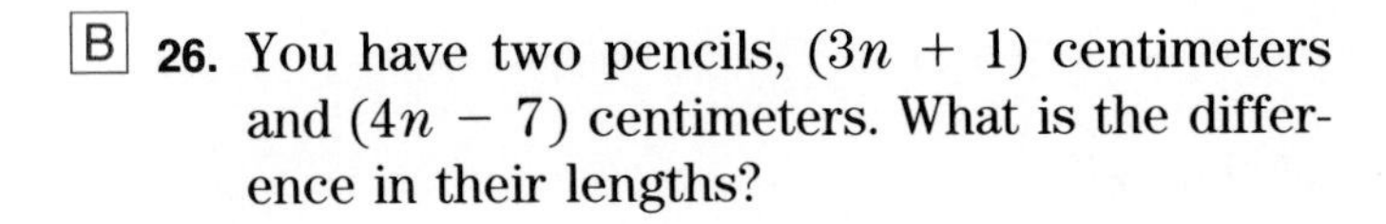

26. You have two pencils, $(3n + 1)$ centimeters and $(4n - 7)$ centimeters. What is the difference in their lengths?

SELF-TEST

Vocabulary

polynomial (p. 152)
monomial (p. 152)
binomial (p. 152)
trinomial (p. 152)

Add or subtract.

1. $(3x - 4) + (4x + 9)$
2. $(5m + n + 1) + (-3m - n + 3)$ *(5-1)*
3. $(b^3 - 2b + 1) + (b^2 + 5b)$
4. $(-a^2 + 3b) + (2a^2 - 4b + 1)$
5. $(x + 4) - (3x - 2)$
6. $(4a - 17) - (-a + 3)$ *(5-2)*
7. $(3b - 4) - (3b^2 + 2b + 5)$
8. $(m^2 + 2m + 1) - (m^2 - 3m + 2)$

3 Multiplying Monomials

You will recall from your work with exponents that $x^2 = x \cdot x$, $x^3 = x \cdot x \cdot x$, and so on. Therefore, you can do the following.

$$\begin{aligned} x^2 \cdot x^3 &= (x \cdot x)(x \cdot x \cdot x) \\ &= x^5 \end{aligned}$$

You can use this idea to multiply any two monomials.

EXAMPLE 1

$$\begin{aligned} x^4 \cdot x^3 &= (x \cdot x \cdot x \cdot x)(x \cdot x \cdot x) \\ &= x^7 \end{aligned}$$

EXAMPLE 2

$$\begin{aligned} (3a^2)(a^3) &= (3 \cdot a \cdot a)(a \cdot a \cdot a) \\ &= 3(a \cdot a \cdot a \cdot a \cdot a) \\ &= 3a^5 \end{aligned}$$

Once you have done a few problems like these, you'll probably see a short cut.

When you multiply two powers of the same number, add the exponents.

$$x^a \cdot x^b = x^{a+b}$$

EXAMPLE 3

$$x^5 \cdot x^7 = x^{5+7} = x^{12}$$

EXAMPLE 4

$$n^6 \cdot n = n^{6+1} = n^7$$

Remember, $n = n^1$.

EXAMPLE 5

$$\begin{aligned} (2x^2)(3x^3) &= (2 \cdot 3)(x^2 \cdot x^3) \\ &= 6x^{2+3} \\ &= 6x^5 \end{aligned}$$

EXAMPLE 6

$$\begin{aligned} (4a^2b)(-3ab^2) &= (4 \cdot -3)(a^2 \cdot a)(b \cdot b^2) \\ &= -12a^{2+1}b^{1+2} \\ &= -12a^3b^3 \end{aligned}$$

Classroom Practice

Multiply.

1. $x^4 \cdot x^2$
2. $n \cdot n^2$
3. $y^2 \cdot y^3$
4. $-2x \cdot x^5$
5. $(x^3)(4x^2)$
6. $(2a)(4a^3)$
7. $(-9y^2)(y^5)$
8. $(6n^4)(2n^4)$
9. $(5ab)(a^2b)$
10. $(2x^2)(4x^2y)$
11. $(-5x^2y)(-2y^2)$
12. $(8a^2b^5)(-4ab^2)$

Written Exercises

Multiply.

A

1. $a \cdot a^2$
2. $b^2 \cdot b^2$
3. $x^2 \cdot 3x$
4. $x^4 \cdot x$
5. $n^3 \cdot n^4$
6. $a^6 \cdot a$
7. $(2x)(2x^2)$
8. $(3x)(-2x^4)$
9. $(c^2)(-5c^3)$
10. $(ab)(a^2b)$
11. $(3x^2)(-2x^5)$
12. $(-y^2)(-y^7)$
13. $(3x^2)(4x^4)$
14. $(-x^2)(-4x)$
15. $(5a)(-ab^2)$
16. $(xy)(-2x)$
17. $(cd)(-3d^3)$
18. $(2mn)(-8m^2)$
19. $(5x^2y)(4xy^2)$
20. $(-5xy)(2xy^2)$
21. $(-r^2s)(-10r^2s^2)$
22. $(-6a^2)(4ab^5)$
23. $(-x^4)(-3xyz^2)$
24. $(-6a^2b^5)(abc^3)$
25. $(a^2b)(-5a^2b^2)$
26. $(-x^3)(-5x^2y)$
27. $(xy^3)(-2x^3y^2)$
28. $(-a^4b^3)(a^2bc^5)$
29. $(m^3n)(-4m^3n^2p^4)$
30. $(r^2s^3t^4)(r^5st^3)$

PUZZLE ◆ PROBLEMS

If you live in Parry Sound, then you live in Ontario.
If you live in Ontario, then you live in Canada.
Therefore, if you live in Parry Sound, you live in Canada.

Crackers are better than nothing.
Nothing is better than peanuts.
Therefore, crackers are better than peanuts.

What happened?

4 Powers of Monomials

Study the examples below and see if you can discover another important rule of exponents.

$$(x^3)^2 = x^3 \cdot x^3 = x^6$$
$$(a^4)^2 = a^4 \cdot a^4 = a^8$$
$$(y^5)^3 = y^5 \cdot y^5 \cdot y^5 = y^{15}$$

To find the power of a power of a number, multiply the exponents.

$$(x^a)^b = x^{ab}$$

EXAMPLE 1 $(x^4)^3 = x^{4 \cdot 3} = x^{12}$

EXAMPLE 2 $(y^8)^2 = y^{8 \cdot 2} = y^{16}$

Suppose you have an expression like $(xy)^2$. You can rewrite it in the following way.

$$(xy)^2 = (xy)(xy) = x^2y^2$$

Make a note of this rule of exponents.

$$(xy)^n = x^ny^n$$

EXAMPLE 3 $(ab)^3 = a^3b^3$

EXAMPLE 4 $(2a)^2 = 2^2 \cdot a^2 = 4a^2$

WARNING! $2a^2$ and $(2a)^2$ are not the same.

$2a^2$	$\stackrel{?}{=}$ $(2a)^2$
$2a^2$	$2^2 \cdot a^2$
	$4a^2$ ← $2a^2 \neq (2a)^2$

EXAMPLE 5 $(-4y)^2 = (-4)^2 \cdot y^2 = 16y^2$

Classroom Practice

1. In 5^3, 5 is used as a factor ___?___ times.
2. In $(x^3)^2$, x^3 is used as a factor ___?___ times.
3. Does $(3x)^2$ equal $3x^2$, or does it equal $9x^2$?
4. Does $(2x)^3$ equal $2x^3$, or does it equal $8x^3$?

Simplify.

5. $(a^4)^2$
6. $(c^2)^5$
7. $(ab)^4$
8. $(5a)^2$
9. $(x^2)^3$
10. $(2a)^2$
11. $(4n^2)^2$
12. $(-3xy)^2$

Written Exercises

Simplify.

A

1. $(x^2)^3$
2. $(a^3)^4$
3. $(b^6)^2$
4. $(x^2)^5$
5. $(c^3)^5$
6. $(n^4)^{10}$
7. $(2x)^2$
8. $(4a)^2$
9. $(ab)^4$
10. $(xy)^6$
11. $(6ax)^2$
12. $(-2xy)^3$

Find the area. Use $A = s^2$.

13.

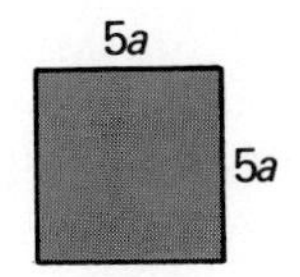

14.

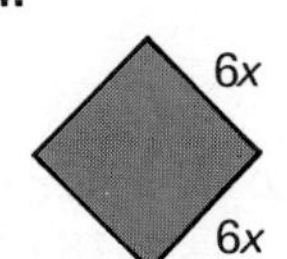

15.

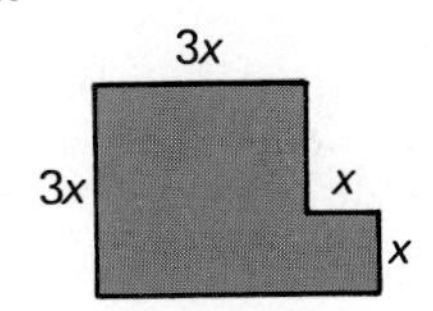

Use both of the exponent rules discussed in the lesson. Simplify.

SAMPLE $(4x^3)^2 = 4^2(x^3)^2 = 16x^6$

16. $(5x^3)^2$
17. $(3a^2)^2$
18. $(2b^3)^2$
19. $(5x^5)^2$
20. $(-2a^5)^2$
21. $(2a^2b)^2$
22. $(3xy^3)^2$
23. $(-x^3y^3)^2$
24. $2(ab^4)^2$
25. $3(x^2y)^2$
26. $-(2n^2)^3$
27. $-(4n^2)^2$

B

28. $(xy)^2(-xy)$
29. $(-xy^3)(-xy)^2$
30. $(3xy^4)^2(-4x^2)^2$
31. $(2mn)^3(3n)^2$
32. $-(x^2y^2)(x^2)$
33. $-(r^4s^3)^2(rs^2)^3$

5 Polynomials Times Monomials

Now that you can multiply monomials, you can put the distributive property to work.

You know that $\longrightarrow 5(3 + 4) = (5 \cdot 3) + (5 \cdot 4)$

In the same way $\longrightarrow$

$$5(3x + 4y) = (5 \cdot 3x) + (5 \cdot 4y)$$
$$= 15x + 20y$$

EXAMPLE 1

$$3(a + 3b) = (3 \cdot a) + (3 \cdot 3b)$$
$$= 3a + 9b$$

EXAMPLE 2

$$-1(2x - 3y) = (-1 \cdot 2x) - (-1 \cdot 3y)$$
$$= -2x + 3y$$

EXAMPLE 3

$$a(a^2 + 2ab + b^2) = (a \cdot a^2) + (a \cdot 2ab) + (a \cdot b^2)$$
$$= a^3 + 2a^2b + ab^2$$

EXAMPLE 4

$$-2x(9x^3 + 3x^2 + x) = (-2x \cdot 9x^3) + (-2x \cdot 3x^2) + (-2x \cdot x)$$
$$= -18x^4 - 6x^3 - 2x^2$$

Classroom Practice

Multiply.

1. $2(a + 4)$
2. $3(x + y)$
3. $-1(2x + 3y)$
4. $x(x - 2)$
5. $a(a + 2b)$
6. $2c(a + b)$
7. $-4(x^2 + 1)$
8. $-5(a^2 - 2c)$
9. $x(x^2 + 2x + 4)$
10. $-x(x^2 - 2x + 4)$
11. $x^2(x^2 - 3x + 1)$
12. $a^2b^2(-a - b)$
13. $xy(x^2 - 2y)$
14. $-4x(x + 2y + 3z)$
15. $m(m^2 - 5m - 6)$

Written Exercises

Multiply.

A

1. $2(x + 4)$
2. $3(a - b)$
3. $4(x + y)$
4. $5(a^2 + b)$
5. $-6(n + 2m)$
6. $-1(5a + b^2)$
7. $a(a - b)$
8. $x(x + 3y)$
9. $-c(a + b)$
10. $-ab(2a - 4b)$
11. $-5x(3x + 2y)$
12. $2x(3x - 1)$
13. $4a(a + 2b + 3)$
14. $-4(1 + 5x + x^2)$
15. $-1(2x + y + z)$
16. $2x(x^2 - 2x - 4)$
17. $-4y(y^3 - 2y + 1)$
18. $ab(a^2 + 2ab - 1)$
19. $-x^2(x + 2x^2)$
20. $-3c(2c^2 + 4c - 5)$
21. $-y^2(y^3 - 2y^2 + 4y)$
22. A rectangle is $10n$ centimeters long by $(n + 6)$ centimeters wide. Write its area as a polynomial.

B

23. You have collected $(3n + 1)$ dimes. What is their value in cents?

Solve.

24. $4(2n + 3) - 3(n - 1) = 0$
25. $-(n + 3) + 2(n + 7) = 0$
26. $5x + 2 - 2(2x + 6) = 0$
27. $(2y - 3) - (y + 6) = 63$
28. $2(5x - 6) - 3(2x - 4) = 0$
29. $3(1 - 2a) - (6 - 2a) = -7$
30. $3(x - 4) + 2(2x + 1) = 4$
31. $2(n - 6) + 5(2n + 4) = 32$
32. $6(1 - 3x) - 2(2x + 5) = 40$
33. $4(2a - 3) - 2(a - 8) = 22$
34. $7(m + 3) - 5(2 - m) = -1$
35. $2(2x - 1) + 3(x + 4) = 52$

SELF-TEST

Simplify.

1. $x \cdot x^3$
2. $(4x)(7x^3)$
3. $(-m^2n^2)(2mn^2)$ *(5-3)*
4. $(m^8)^4$
5. $(ab)^3$
6. $(-7x^2y)^2$ *(5-4)*

Multiply.

7. $x(y + 7)$
8. $m(m - n)$
9. $-a(a^2 - b)$ *(5-5)*
10. $-2(3x - 4y)$
11. $5x(4x + x^2)$
12. $m^2(3m^3 - m^2 + 2m)$

6 Multiplying Polynomials

To multiply by a binomial you also use the distributive property. Let's use a vertical form in multiplying the binomials below. Your work is a lot like multiplying in arithmetic.

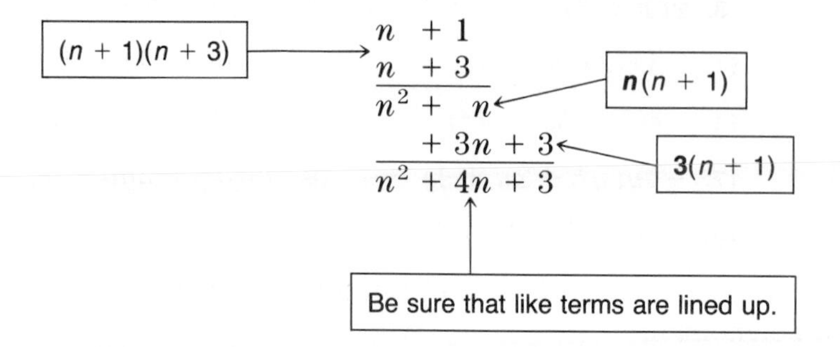

Here are a few more examples to study.

EXAMPLE 1

$$\begin{array}{r} x + 7 \\ x - 2 \\ \hline x^2 + 7x \\ -2x - 14 \\ \hline x^2 + 5x - 14 \end{array}$$

$x^2 + 7x$ ← **x**$(x + 7)$

$-2x - 14$ ← **−2**$(x + 7)$

EXAMPLE 2

$$\begin{array}{r} 2y - 5 \\ y - 4 \\ \hline 2y^2 - 5y \\ -8y + 20 \\ \hline 2y^2 - 13y + 20 \end{array}$$

$2y^2 - 5y$ ← **y**$(2y - 5)$

$-8y + 20$ ← **−4**$(2y - 5)$

EXAMPLE 3

$$\begin{array}{r} a^2 + 2a - 1 \\ a - 3 \\ \hline a^3 + 2a^2 - a \\ -3a^2 - 6a + 3 \\ \hline a^3 - a^2 - 7a + 3 \end{array}$$

$a^3 + 2a^2 - a$ ← **a**$(a^2 + 2a - 1)$

$-3a^2 - 6a + 3$ ← **−3**$(a^2 + 2a - 1)$

Classroom Practice

Multiply.

1. $(x + 2)(x + 4)$
2. $(n + 2)(n + 7)$
3. $(a + 4)(a - 1)$
4. $(x + 3)(x - 5)$
5. $(2x + 1)(x + 5)$
6. $(a + 6)(2a - 4)$
7. $(2a + b)(a + 3b)$
8. $(2x - 4)(5x + 1)$
9. $(3y - 6)(2y - 5)$
10. $(2x - 5)(x + 2)$
11. $(3n + 6)(n - 3)$
12. $(4x - 1)(2x + 3)$

Written Exercises

Multiply.

A

1. $(x + 1)(x + 2)$
2. $(y + 3)(y + 4)$
3. $(n + 4)(n + 6)$
4. $(a + 3)(a + 5)$
5. $(x + 1)(x + 7)$
6. $(y + 9)(y + 3)$
7. $(x + y)(x + y)$
8. $(a + b)(a + c)$
9. $(x + y)(x + 4)$
10. $(x - 1)(x - 2)$
11. $(y - 3)(y - 4)$
12. $(n - 3)(n - 1)$
13. $(x - 1)(x + 2)$
14. $(y + 3)(y - 4)$
15. $(y - 5)(y + 4)$
16. $(2x + y)(x + 2y)$
17. $(4x + 1)(4x + 1)$
18. $(4x + 1)(4x - 1)$
19. $(a + b)(a - b)$
20. $(a + 2b)(a - 2b)$
21. $(2a - 1)(a + b)$
22. $(6 - x)(6 - x)$
23. $(6x - 5)(6x - 5)$
24. $(3y - 2)(2y + 1)$

B

25. $(n + 1)(n^2 + 2n + 1)$
26. $(x + 2)(x^2 + 2x + 1)$
27. $(x - 2)(x^2 - 4x + 4)$
28. $(y + 3)(y^2 + 6y + 9)$
29. $(m - 2)(m^2 - 2m + 4)$
30. $(n - 4)(n^2 - 8n + 16)$
31. $(a + b)(a^2 + 2ab + b^2)$
32. $(a - b)(a^2 - b^2)$

C

33. $(n^2 + 2n + 1)(n^2 + 2n + 1)$
34. $(x^2 - 7x + 12)(x^2 - 3x - 1)$
35. $(2a^2 - ab + 4b^2)(5a^2 + ab + b^2)$
36. $(x - 4)(2x^4 - 5x^3 - 7x^2 + 10)$
37. $(n - 1)(n^4 - 3n^3 + n^2 + 1)$
38. $(a - b)(a^4 + a^3b + a^2b^2 + ab^3 + b^4)$

7 Multiplying at Sight

When you multiplied binomials by using the vertical form, you may have noticed a pattern. The pattern is pointed out at the right.

$$\begin{array}{r} x + 2 \\ x + 3 \\ \hline x^2 + 2x \\ + 3x + 6 \\ \hline x^2 + 5x + 6 \end{array}$$

product of the first terms

sum of $2 \cdot x$ and $3 \cdot x$

product of the last terms

You don't have to write out each step every time you multiply two binomials. You can multiply at sight if you follow three steps.

Step 1: $(x + 3)(x + 2) = x^2 \ldots$ ← Multiply the *first* terms.

Step 2: $(x + 3)(x + 2) = x^2 + 5x \ldots$ ← Multiply the *outer* terms. Multiply the *inner* terms. Add the two products.

Step 3: $(x + 3)(x + 2) = x^2 + 5x + 6$ ← Multiply the *last* terms.

An easy way to remember this is to think of it as the **FOIL method.**

> *FOIL stands for First, Outer, Inner, Last!*

Here's how the FOIL method works.

EXAMPLE 1 Multiply $(x + 4)(x + 5)$.

Step 1: $(x + 4)(x + 5) = x^2 \ldots$ ← F

Step 2: $(x + 4)(x + 5) = x^2 + 9x \ldots$ ← O + I

Step 3: $(x + 4)(x + 5) = x^2 + 9x + 20$ ← L

Be sure to watch the signs!

EXAMPLE 2 Multiply $(x - 4)(x - 2)$.

Step 1: $(x - 4)(x - 2) = x^2 \ldots$ ← F

Step 2: $(x - 4)(x - 2) = x^2 - 6x \ldots$ ← O + I

Step 3: $(x - 4)(x - 2) = x^2 - 6x + 8$ ← L

EXAMPLE 3 Multiply $(x - 4)(x + 2)$.

$$(x - 4)(x + 2) = x^2 - 2x - 8$$

F O + I L

EXAMPLE 4 Multiply $(2x + 3y)(x + y)$.

$$(2x + 3y)(x + y) = 2x^2 + 5xy + 3y^2$$

F O + I L

Classroom Practice

Complete.

1. $(x + 1)(x + 4) = \underline{\ ?\ } + 5x + 4$
2. $(a + 2)(a + 5) = a^2 + 7a + \underline{\ ?\ }$
3. $(y + 3)(y + 2) = y^2 + 5y + \underline{\ ?\ }$
4. $(x + 4)(x + 5) = x^2 + \underline{\ ?\ } + 20$
5. $(n + 1)(n + 7) = n^2 + \underline{\ ?\ } + 7$
6. $(x + 5)(x - 2) = x^2 + 3x - \underline{\ ?\ }$
7. $(a + 8)(a - 2) = a^2 + 6a - \underline{\ ?\ }$
8. $(c - 6)(c + 1) = c^2 - 5c - \underline{\ ?\ }$
9. $(x - 3)(x - 4) = x^2 - 7x + \underline{\ ?\ }$
10. $(a - 5)(a - 9) = a^2 - \underline{\ ?\ } + 45$

Multiply.

11. $(x + 2)(x + 3)$
12. $(y + 5)(y + 1)$
13. $(a + 6)(a + 2)$
14. $(x - 4)(x - 3)$
15. $(y - 6)(y - 1)$
16. $(x + 3)(x - 7)$
17. $(c - 9)(c + 6)$
18. $(2x + 1)(x + 3)$
19. $(4y + 1)(2y - 6)$
20. $(5x + 2)(3x - 1)$
21. $(2a - b)(3a + 4b)$
22. $(7r - s)(r + 4s)$

Written Exercises

Multiply.

A

1. $(x + 4)(x + 3)$
2. $(x + 3)(x + 5)$
3. $(x + 6)(x + 2)$
4. $(x + 2)(x + 1)$
5. $(x - 1)(x - 2)$
6. $(x - 6)(x - 3)$
7. $(x - 4)(x - 3)$
8. $(a - 3)(a - 5)$
9. $(y - 4)(y + 1)$
10. $(a - 1)(a + 3)$
11. $(n + 2)(n - 1)$
12. $(x + 5)(x - 2)$
13. $(y + 6)(y - 2)$
14. $(a + 5)(a - 3)$
15. $(2x + 1)(x + 4)$
16. $(x + 2)(x - 2)$
17. $(a + b)(a - b)$
18. $(2x + 5)(2x - 5)$
19. $(6a + 7)(6a - 7)$
20. $(2y - 3)(2y + 3)$
21. $(3x + 4)(3x - 4)$
22. $(5n - 1)(2n + 3)$
23. $(3x - 1)(2x + 5)$
24. $(3x + 2y)(2x + 3y)$
25. $(3x + y)(7x + 2y)$
26. $(3a - b)(a - 3b)$
27. $(4x - 3y)(x - 5y)$
28. $(3a - b)(7a + 4b)$
29. $(6x - y)(4x + 2y)$
30. $(3n + 2p)(7n - p)$

Find the area. Use $A = lw$.

31.

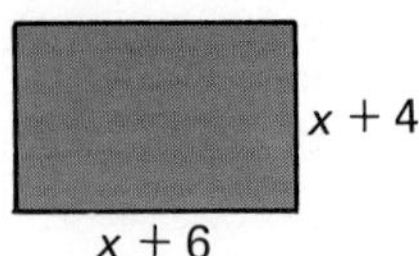

32.

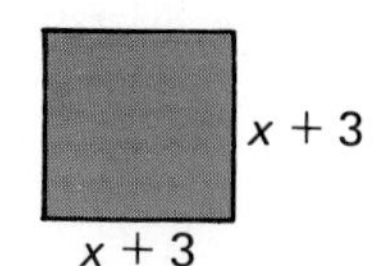

33.

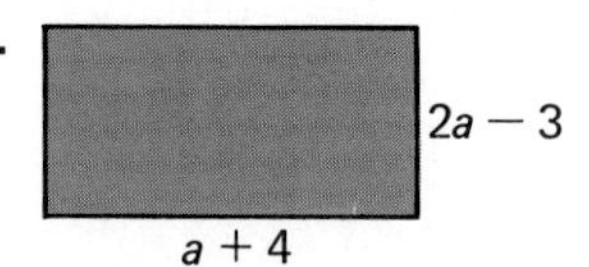

34. A rectangle measures $(17 - x)$ centimeters by $(x - 4)$ centimeters. Find its area.

35. A rectangle measures $(2y + 3)$ centimeters by $(3y + 1)$ centimeters. Find its area.

Find the area. Use $A = \frac{1}{2}bh$.

B

36.

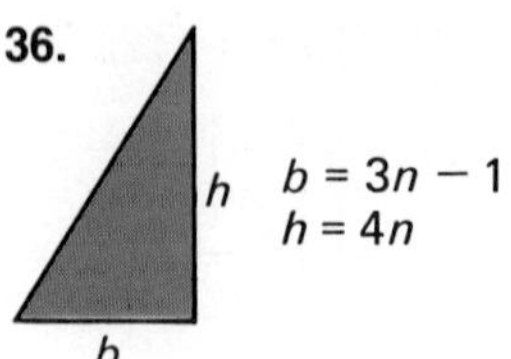

37.

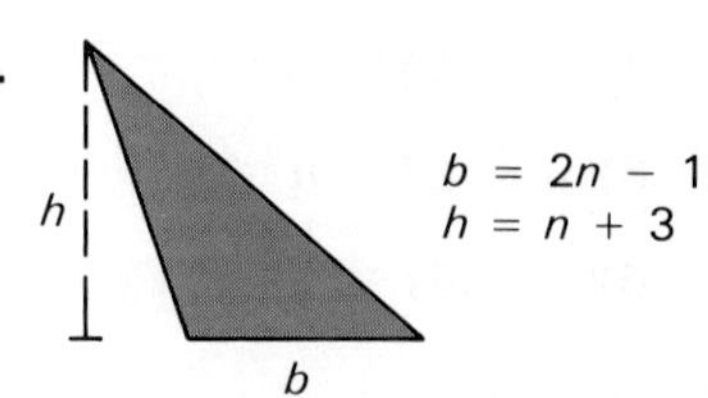

38. Find the area of the side of the house shown at the right.

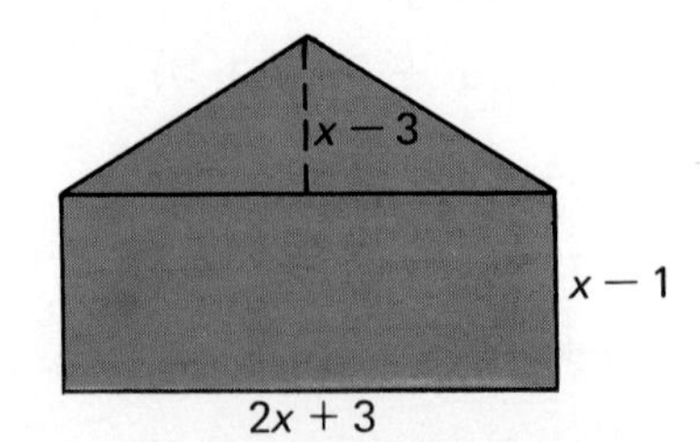

C **39.** A trapezoid is shown at the right. Its area may be found by using the formula $A = \frac{1}{2}h(a + b)$.

Find the area of the trapezoid if $h = 2x$, $a = 3x - 1$, and $b = 3x + 1$.

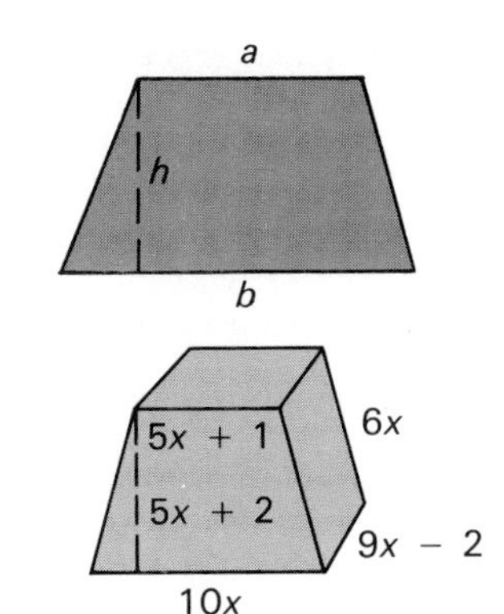

40. The box shown at the right is designed to hold a cup and saucer. The top, the bottom, and two sides are rectangles. Two sides are trapezoids. Find the area of the whole surface of the box.

CONSUMER APPLICATIONS

USING A CHECK REGISTER

When you have a checking account you have to be careful to keep track of how much money you have spent and how much money you have in your account. You can use a check register to keep track of your transactions and your account balances. See the sample below.

CHECK NO.	DATE	DESCRIPTION OF TRANSACTION	AMOUNT OF CHECK	AMOUNT OF DEPOSIT	BALANCE
		BALANCE BROUGHT FORWARD →			406 19
161	9/21	Jones Cleaners	12 15		394 04
162	9/30	Statewide Electric Co.	34 83		359 21
	10/6	Deposit		125 62	484 83
163	10/7	Southern Savings + Loan	152 30		332 53

Be sure to record all checks when you write them. It's just too easy to forget about the check in only a few hours.

When you deposit money into your account, remember to save the deposit slips. Record your deposit in the register as well.

After writing a check or making a deposit, be sure to record the amount of money you have in your account in the Balance column.

8 Square of a Binomial

You know that when you square a number you multiply it by itself.

$$6^2 = 6 \cdot 6 \qquad x^2 = x \cdot x$$

The same is true when you square a binomial. All you do is to write the two factors, then multiply as you did in the last section.

$$\begin{aligned}(a + b)^2 &= (a + b)(a + b)\\ &= a^2 + 2ab + b^2\end{aligned}$$

The square of a binomial is called a **trinomial square.**

Study some more examples.

EXAMPLE 1

a. $(x + 3)^2 = (x + 3)(x + 3) = x^2 + 6x + 9$

b. $(x - 4)^2 = (x - 4)(x - 4) = x^2 - 8x + 16$

Once you have done a few multiplications like those above, you might notice a pattern in squaring a binomial.

SQUARING A BINOMIAL

1. *Square the first term.*
2. *Add or subtract twice the product of the two terms.*
3. *Add the square of the second term.*

$$(a + b)^2 = a^2 + 2ab + b^2$$
$$(a - b)^2 = a^2 - 2ab + b^2$$

Here are a couple of examples. Be sure to watch the signs too.

EXAMPLE 2

a. $(x + 5)^2 = x^2 + 10x + 25$

square of x — twice $5 \cdot x$ — square of 5

b. $(n - 3)^2 = n^2 - 6n + 9$

square of n — twice $3 \cdot n$ — square of 3

Of course, if you find this short cut difficult to remember, you can multiply as in Example 1.

Written Exercises

Express as a trinomial.

A

1. $(x + 4)^2$
2. $(m + 1)^2$
3. $(r + 2)^2$
4. $(y - 3)^2$
5. $(x + 3)^2$
6. $(n + 6)^2$
7. $(x - 2)^2$
8. $(m - 6)^2$
9. $(x - 4)^2$
10. $(m + 5)^2$
11. $(y - 7)^2$
12. $(a + 10)^2$
13. $(x + 9)^2$
14. $(y - 8)^2$
15. $(m + 7)^2$
16. $(n + 8)^2$
17. $(n - 5)^2$
18. $(y - 1)^2$
19. $(3a + 2)^2$
20. $(2x + 3)^2$
21. $(4x - 2)^2$
22. $(2a + b)^2$
23. $(3x + y)^2$
24. $(6n - 1)^2$
25. $(10a - b)^2$
26. $(2m + 3n)^2$
27. $(4x - 3y)^2$
28. $(3x - 4y)^2$
29. $2(a + 1)^2$
30. $2(a + 5)^2$
31. $3(x - 2)^2$
32. $-1(x + 3)^2$
33. $-2(x - 2)^2$
34. $-1(2n + 3)^2$
35. $2(2x + y)^2$
36. $-1(a - 2b)^2$

B

37. Study the figure. Then complete the equation below.

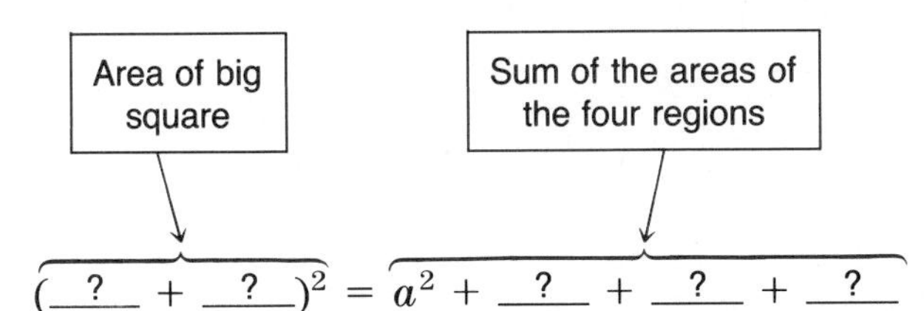

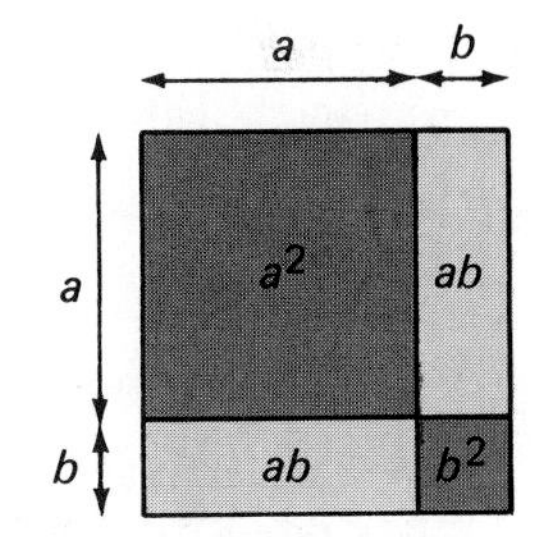

SELF-TEST

Vocabulary

FOIL method (p. 164)   trinomial square (p. 168)

Multiply using the vertical form.

1. $(x + 2)(x + 4)$
2. $(a - 1)(a - 5)$
3. $(a - 1)(a^2 - 2a + 1)$ *(5-6)*

Multiply using the FOIL method.

4. $(m - 5)(m + 1)$
5. $(2a - 7b)(2a + 3b)$
6. $(3m + 1)(2m + 5)$ *(5-7)*

Express as a trinomial.

7. $(a + 4)^2$
8. $(n - 3)^2$
9. $(6x + y)^2$
10. $(4a - 7b)^2$ *(5-8)*

9 Division with Monomials

You know that a division is often shown as a fraction. Here is one way you can simplify.

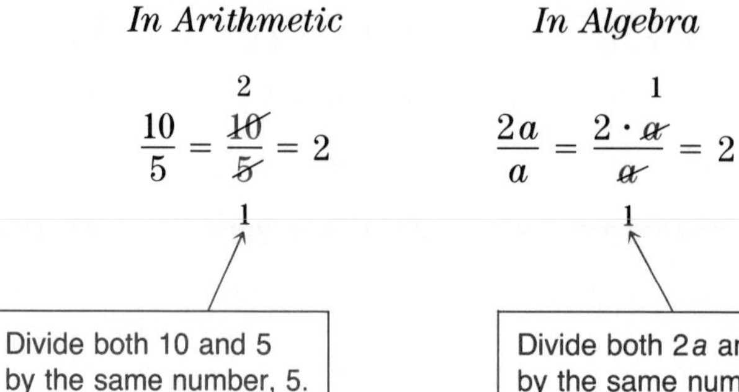
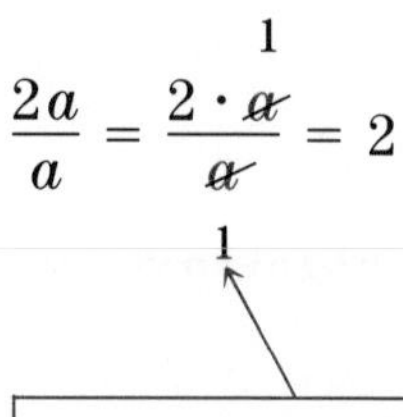

This method works for all divisions.

EXAMPLE 1 $\dfrac{x^4}{x} = \dfrac{\overset{1}{\cancel{x}} \cdot x \cdot x \cdot x}{\underset{1}{\cancel{x}}} = x^3$

EXAMPLE 2 $\dfrac{2x^3}{x^2} = \dfrac{2 \cdot \overset{1}{\cancel{x}} \cdot \overset{1}{\cancel{x}} \cdot x}{\underset{1}{\cancel{x}} \cdot \underset{1}{\cancel{x}}} = 2x$

EXAMPLE 3 $\dfrac{-4ab^4}{2ab} = \dfrac{\overset{-2}{\cancel{-4}} \cdot \overset{1}{\cancel{a}} \cdot \overset{1}{\cancel{b}} \cdot b \cdot b \cdot b}{\underset{1}{\cancel{2}} \cdot \underset{1}{\cancel{a}} \cdot \underset{1}{\cancel{b}}} = -2b^3$

Classroom Practice

Divide.

1. $\dfrac{x^6}{x}$
2. $\dfrac{x^4}{x^3}$
3. $\dfrac{a^7}{a^4}$
4. $\dfrac{2a}{2}$
5. $\dfrac{5x}{x}$
6. $\dfrac{3c^2}{c}$
7. $\dfrac{4ab}{a}$
8. $\dfrac{6a^3}{a^2}$
9. $\dfrac{8a^2b}{ab}$
10. $\dfrac{-7x^2}{x^2}$
11. $\dfrac{-9m^2r^3}{3mr^3}$
12. $\dfrac{24rs^4}{-8rs^3}$

Written Exercises

Divide.

A

1. $\frac{2a}{2a}$
2. $\frac{-6a}{a}$
3. $\frac{a^2}{a}$
4. $\frac{6c}{2}$
5. $\frac{-5ax}{a}$
6. $\frac{x^4}{x^3}$
7. $\frac{c^9}{c^4}$
8. $\frac{2x^2}{x}$
9. $\frac{7n^3}{n^2}$
10. $\frac{10a^5}{2a^3}$
11. $\frac{-6mn}{mn}$
12. $\frac{2xy}{2y}$
13. $\frac{12rs}{-3}$
14. $\frac{4ab}{-2b}$
15. $\frac{-9x^2y}{-3}$
16. $\frac{-8ab^2}{-2a}$
17. $\frac{-56x^4}{8x^2}$
18. $\frac{30m^3}{-6m}$
19. $\frac{-63b^2}{9b}$
20. $\frac{36n^6}{6n^4}$
21. $\frac{-10cd^4}{-5d^2}$
22. $\frac{48bc^2}{-8c^2}$
23. $\frac{42a^2b^2}{-6ab}$
24. $\frac{-72x^4y^3}{9xy^2}$
25. $\frac{50m^3n^2}{2m^2n}$
26. $\frac{48pq^2}{-3pq}$
27. $\frac{5a^5b^3}{ab^2}$
28. $\frac{64r^6s^4}{4r^4s^3}$
29. $\frac{-12x^2y^6}{-xy^4}$
30. $\frac{-32c^9d^6}{8c^3d^2}$

Simplify. Be sure to divide first.

B

31. $\frac{4x^2}{x} - \frac{2x^3}{x^2}$
32. $\frac{2x^4}{x^3} - \frac{3x}{3}$
33. $\frac{5a^2}{5a} + \frac{6a^5}{2a^4}$
34. $\frac{10xy}{5x} + \frac{4x^2y^3}{4x^2y^2}$
35. $\frac{18ab^3}{6b^2} + \frac{5a^2b}{5a}$
36. $\frac{-4x^3y^4}{-2x^2y^2} - \frac{3x^2y^5}{xy^3}$

PUZZLE ◆ PROBLEMS

Here are five squares made from sticks. It is possible to make four squares by moving just two of the sticks. Can you do it?

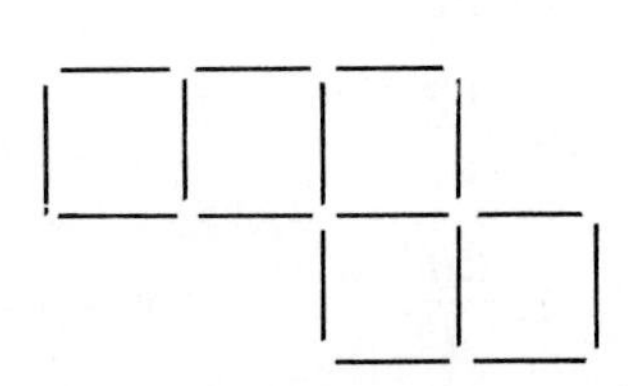

10 Exponent Rules (Optional)

You may have already guessed that you can work with exponents in division in the following way.

$$\frac{x^5}{x^2} = x^{5-2} = x^3$$

Sometimes you may find that the exponent in the denominator is greater than the exponent in the numerator.

$$\frac{x^2}{x^5} = \frac{1}{x^{5-2}} = \frac{1}{x^3}$$

Make a note of the following rules.

$$\frac{x^a}{x^b} = x^{a-b} \quad \textit{when } \text{a} \textit{ is greater than } \text{b.}$$

$$\frac{x^a}{x^b} = \frac{1}{x^{b-a}} \quad \textit{when } \text{b} \textit{ is greater than } \text{a.}$$

$$\frac{x^a}{x^a} = x^{a-a} = x^0 = 1$$

You can use these rules when a and b are positive integers.

EXAMPLE 1 $\frac{3n^6}{n^2} = 3n^{6-2} = 3n^4$

EXAMPLE 2 $\frac{5x^4}{x^6} = \frac{5}{x^{6-4}} = \frac{5}{x^2}$

EXAMPLE 3 $\frac{a^3b^4}{a^3b} = a^{3-3}b^{4-1} = 1 \cdot b^3 = b^3$

Classroom Practice

Divide.

1. $\frac{x^5}{x^3}$ **2.** $\frac{a^7}{a^4}$ **3.** $\frac{c^9}{c^3}$ **4.** $\frac{x}{x^6}$

5. $\frac{2a^4}{a^3}$ **6.** $\frac{5x^4}{5x^2}$ **7.** $\frac{10n^5}{2n^3}$ **8.** $\frac{-8y}{2y^2}$

Written Exercises

Divide.

A

1. $\frac{a^5}{a^2}$
2. $\frac{x^9}{x^4}$
3. $\frac{x^3}{x^7}$
4. $\frac{5a^7}{a^3}$
5. $\frac{9x^6}{x^3}$
6. $\frac{12x^3}{3x^5}$
7. $\frac{-8n^2}{4n^5}$
8. $\frac{14b^{12}}{2b^2}$
9. $\frac{-3a^2b^2}{ab^2}$
10. $\frac{-18a^3b}{-3a^2}$
11. $\frac{16x^4y^3}{-4xy}$
12. $\frac{25x^4y^2}{5xy}$
13. $\frac{-12x^5y^2}{-6x^2y^2}$
14. $\frac{14a^3b^2}{-7a^2b^5}$
15. $\frac{-42m^6n^3}{-7m^4n^4}$
16. $\frac{56x^5y^5}{7xy^4}$
17. $\frac{-20x^6y^7}{5xy^6}$
18. $\frac{-12a^3b^8}{-3a^2b^5}$
19. $\frac{9x^9y^4z}{3xy}$
20. $\frac{15a^2b^4c^3}{-3ac^2}$
21. $\frac{-7a^9bc}{a^5bc^3}$
22. $\frac{49c^2d^{10}x}{-7c^2d^5}$
23. $\frac{-3r^4st^6}{-rst^4}$
24. $\frac{36a^6b^2c^7}{-12abc^6}$
25. $\frac{12x^2y}{-2xy}$
26. $\frac{21a^2b^3c}{-3a^2b}$
27. $\frac{-40a^4b^2c^5}{-10a^3bc^3}$
28. $\frac{30x^2y^3z}{-5xy^2}$
29. $\frac{-42a^5b^2}{6a^3b}$
30. $\frac{36x^4y^2z}{-6x^4}$
31. $\frac{-81a^4b^2c^9}{-3ab^2c^5}$
32. $\frac{54r^2s^4t^{10}}{-9s^3t^4}$
33. $\frac{16ab^4}{2a^2b^3}$
34. $\frac{15m^2n^2s}{-5mn^3s^4}$
35. $\frac{-x^2y^2z^2}{x^4y^2z}$
36. $\frac{-45cd^3}{-9c^3d^6}$

PUZZLE ◆ PROBLEMS

Paula and Peter Piper are picking pickled peppers. They're putting the pickled peppers in a basket. The number of peppers they pick doubles every minute. In one hour the basket is full. When was the basket half full?

11 Polynomials Divided by Monomials

You probably remember that in arithmetic you add fractions in the following way.

$$\frac{20}{4} + \frac{3}{4} = \frac{20 + 3}{4} = \frac{23}{4}$$

If this process is turned around, you have the following.

$$\frac{23}{4} = \frac{20 + 3}{4} = \frac{20}{4} + \frac{3}{4}$$

Now just use this idea in working with division problems.

EXAMPLE 1

$$\frac{3a + 3b}{3} = \frac{3a}{3} + \frac{3b}{3}$$

$$= a + b$$

Divide each term of $3a + 3b$ by 3.

EXAMPLE 2

$$\frac{2rs - 6r^2s^2}{2r} = \frac{2rs}{2r} - \frac{6r^2s^2}{2r}$$

$$= s - 3rs^2$$

EXAMPLE 3

$$\frac{2x^2 + 4x + 8xy}{2x} = \frac{2x^2}{2x} + \frac{4x}{2x} + \frac{8xy}{2x}$$

$$= x + 2 + 4y$$

Classroom Practice

Divide.

1. $\frac{5x + 5y}{5}$
2. $\frac{2x + 3x}{x}$
3. $\frac{4m + 2m}{2m}$
4. $\frac{8ab - 4a}{2a}$
5. $\frac{5a^2b - 3ab}{ab}$
6. $\frac{6mn^2 + 9m^2n}{3mn}$
7. $\frac{2x^2 + 3x - 4xy}{x}$
8. $\frac{3a^2b - 6a + 9a^2}{3a}$
9. $\frac{4xy^2 + 2xy - 6x^2y}{2xy}$
10. $\frac{-2ab + 4b^2 - 6a^3}{-2}$
11. $\frac{4m^2n + 8mn^2 - 4m}{4m}$
12. $\frac{8a^2b - 4ab^2 + 16ab}{4ab}$

Written Exercises

Divide.

A

1. $\frac{5a + 15b}{5}$
2. $\frac{7a - 7b}{7}$
3. $\frac{8x - 4y}{4}$
4. $\frac{3a + 12c}{3}$
5. $\frac{8x - 4y}{2}$
6. $\frac{20x - 16}{4}$
7. $\frac{56m + 63n}{7}$
8. $\frac{42x + 54y}{6}$
9. $\frac{5a^2 + 45a}{5a}$
10. $\frac{6x^2 - 48x}{6x}$
11. $\frac{ab^2 - a^2b}{a}$
12. $\frac{xy - 4x^2y^2}{xy}$
13. $\frac{a^3 + a^2b + ab^2}{a}$
14. $\frac{x^3 + 3x^2 - x}{x}$
15. $\frac{6m^2n^2 - 2m + 4m^2}{2m}$
16. $\frac{3n^4 + 3n^3 - 6n^2}{3n}$
17. $\frac{5x^2 - 15x^3y + 10x}{5x}$
18. $\frac{a^2b^3 - a^3b + 4a^2b^2}{a^2b}$
19. $\frac{6r^2s + 9rs^2 - 18r^3}{3r}$
20. $\frac{n^2 - 3n^4 + 5n^5}{-n^2}$
21. $\frac{24a^2b - 6a^2b^2 + 12a^3}{-6}$

B

22. $\frac{-4x^4 + 12x^3 - 48x^2 + 8x}{4x}$
23. $\frac{32n - 16n + 16n^2 + 32n^2}{-8n}$
24. $\frac{6a^4b^4 + 2a^3b^3 + 4a^2b^2 + 2ab}{-2ab}$
25. $\frac{-18x^3y + 48x^2y^2 - 54xy^3}{-6xy}$
26. $\frac{-24a^6b^3 + 12a^4b^2 - 9a^2b^3}{-3a^2b^2}$
27. $\frac{21x^3b^4 - 28x^5b^3 + 42x^2b^5}{-7x^2b^3}$

SELF-TEST

Divide.

1. $\frac{4x}{2x}$
2. $\frac{2a^6}{a^3}$
3. $\frac{-6mn^2}{-3mn}$
4. $\frac{9x^2y^4}{-3x^2y^3}$

(5-9)

5. $\frac{9m - 12n}{3}$
6. $\frac{4a^2 + 16a}{2a}$

(5-11)

7. $\frac{x^2y^2 - xy^2}{xy}$
8. $\frac{3a^2b - 6ab^2}{3ab}$

12 Dividing Polynomials (Optional)

Dividing by a polynomial is a little like doing a long division problem in arithmetic. Compare the steps.

$$\frac{255}{15} \qquad\qquad \frac{x^2 + 7x + 12}{x + 3}$$

$$15\overline{)255} \qquad\qquad x + 3\overline{)x^2 + 7x + 12}$$

$$\begin{array}{r} 1 \\ 15\overline{)255} \end{array} \qquad\qquad \begin{array}{r} x \\ x + 3\overline{)x^2 + 7x + 12} \end{array}$$

$$\begin{array}{r} 1 \\ 15\overline{)255} \\ \underline{15} \end{array} \qquad\qquad \begin{array}{r} x \\ x + 3\overline{)x^2 + 7x + 12} \\ \underline{x^2 + 3x} \end{array}$$

$$\begin{array}{r} 1 \\ 15\overline{)255} \\ \underline{15} \\ 105 \end{array} \qquad\qquad \begin{array}{r} x \\ x + 3\overline{)x^2 + 7x + 12} \\ \underline{x^2 + 3x} \\ 4x + 12 \end{array}$$

$$\begin{array}{r} 17 \\ 15\overline{)255} \\ \underline{15} \\ 105 \end{array} \qquad\qquad \begin{array}{r} x + 4 \\ x + 3\overline{)x^2 + 7x + 12} \\ \underline{x^2 + 3x} \\ 4x + 12 \end{array}$$

$$\begin{array}{r} 17 \\ 15\overline{)255} \\ \underline{15} \\ 105 \\ \underline{105} \\ 0 \end{array} \qquad\qquad \begin{array}{r} x + 4 \\ x + 3\overline{)x^2 + 7x + 12} \\ \underline{x^2 + 3x} \\ 4x + 12 \\ \underline{4x + 12} \\ 0 \end{array}$$

Check: $15 \times 17 = 255$ ✓ $\qquad (x + 3)(x + 4) = x^2 + 7x + 12$ ✓

Written Exercises

Divide. Check by multiplication.

A

1. $\dfrac{x^2 + 4x + 4}{x + 2}$
2. $\dfrac{a^2 + 6a + 8}{a + 4}$
3. $\dfrac{y^2 + 10y + 24}{y + 6}$
4. $\dfrac{x^2 - 4x + 4}{x - 2}$
5. $\dfrac{x^2 - 4x - 12}{x - 6}$
6. $\dfrac{x^2 - x - 2}{x - 2}$
7. $\dfrac{x^2 - x - 12}{x + 3}$
8. $\dfrac{n^2 - 3n - 4}{n + 1}$
9. $\dfrac{a^2 - 3a - 10}{a + 2}$
10. $\dfrac{s^2 - 7s + 10}{s - 5}$
11. $\dfrac{x^2 - 9x + 18}{x - 3}$
12. $\dfrac{c^2 - 10c + 21}{c - 7}$

B

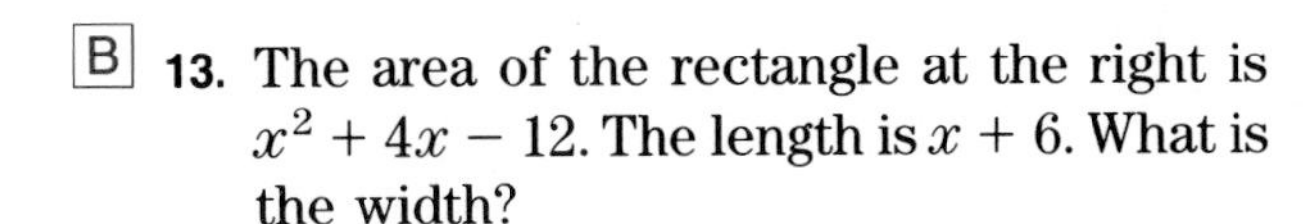

13. The area of the rectangle at the right is $x^2 + 4x - 12$. The length is $x + 6$. What is the width?

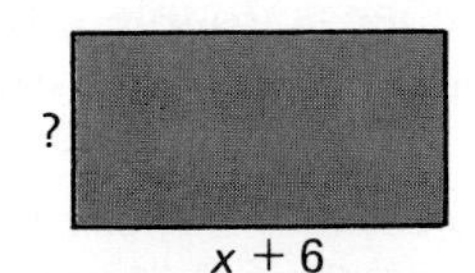

14. The volume of this rectangular prism is $x^3 + 11x^2 + 36x + 36$. Find the height.

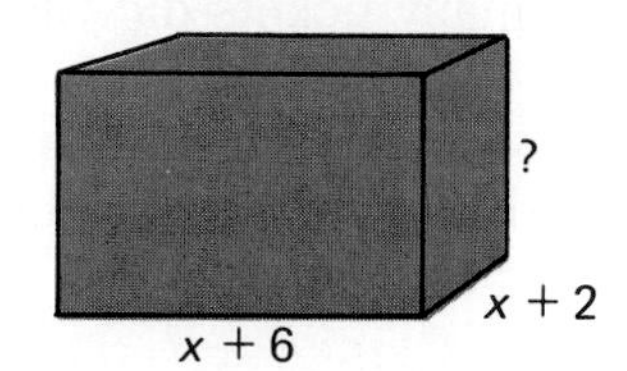

CALCULATOR ACTIVITIES

Does your calculator have a y^x key? This key can be used to find the value of expressions involving powers. Experiment with this key. How would you use it to find the value of 2^3? You will probably need to enter 2 y^x 3 =.

Your calculator may also have an x^2 key. Use this key to find squares. To find 3^2, enter 3 x^2 =, and your calculator will display 9.

Remember that multiplication can also be used to find the value of powers. For example, $2^3 = 2 \cdot 2 \cdot 2$.

If $a = 3$, $b = 4$, and $c = -2$, use a calculator to find the value of each expression. (*Hint:* you may want to simplify the expression first.)

1. $(-b^3)(-b)$
2. $(ab^2)(a^2b)$
3. a^2b^4c
4. $(6a^3)^2$
5. $(8ab^4)(a^2b)$
6. $(-2a^2b)(3ab^4)$

SKILLS REVIEW

PROBLEM SOLVING

One of the most important steps in solving a problem is setting up the proper equation.

Reread the steps on page 126.
Then write an equation to solve each problem.
You do not need to solve the equation.

1. 7 more than twice a number is 11. Find the number.

2. A number is doubled, then 11 is added.
   The result is 51. Find the number.

3. Clifford sold 82 more tickets than Clarisse.
   Together they sold 286 tickets.
   How many tickets did Clarisse sell?

4. A serving of oatmeal has 50 more calories than a half cup of milk.
   Together they have 210 calories.
   How many calories are in a half cup of milk?

5. One number is 8 less than another.
   Twice the larger number is 4 more than 5 times the smaller.
   Find the numbers.

6. The length of a rectangle is 9 cm more than the width.
   The perimeter is 66 cm.
   Find the length and width.

7. Sonya bikes 12 km/h faster than André walks.
   They travel in opposite directions for $\frac{1}{2}$ hour.
   Then they are 13 km apart.
   How fast does André walk?

8. A regular museum ticket costs $5.
   A ticket to a special show costs $7.
   There were 425 more special tickets sold than regular tickets.
   The ticket sales totaled $12,575.
   How many regular tickets were sold?

CHAPTER REVIEW

CHAPTER SUMMARY

1. Some polynomials have special names.

one term	monomial
two terms	binomial
three terms	trinomial

2. To subtract a polynomial, change signs and add.
3. Rules of Exponents:

$$x^a \cdot x^b = x^{a+b} \qquad (x^a)^b = x^{ab} \qquad (xy)^n = x^n y^n$$

4. To multiply by a binomial, use the distributive property.
5. To multiply binomials at sight, use the FOIL method. FOIL stands for First, Outer, Inner, Last.

REVIEW EXERCISES

Add. ***(See pp. 152–153.)***

1. $\begin{array}{r} x + 2 \\ \underline{x - 4} \end{array}$

2. $\begin{array}{r} x^2 - 1 \\ \underline{2x^2 - 5} \end{array}$

3. $\begin{array}{r} 6n - 14 \\ \underline{n + 16} \end{array}$

4. $\begin{array}{r} a^2 + 12 \\ \underline{4a^2 - 13} \end{array}$

5. $(n + 3) + (2n + 5)$

6. $(2x - y) + (9x - 7y)$

7. $(n^2 - 1) + (3n^2 + 1)$

8. $(3x - 7) + (x + 14)$

9. $(a^2 + b^2) + (5a^2 + 7b^2)$

10. $(1 - x^2) + (5 + 9x^2)$

Subtract. ***(See pp. 154–155.)***

11. $\begin{array}{r} x + 2 \\ \underline{-(x - 3)} \end{array}$

12. $\begin{array}{r} 3y - 1 \\ \underline{-(2y + 1)} \end{array}$

13. $\begin{array}{r} 12x - 7y \\ \underline{-(10x + 5y)} \end{array}$

14. $\begin{array}{r} 15m^2 + n \\ \underline{-(11m^2 - 2n)} \end{array}$

15. $(5m - n) - (m + 4n)$

16. $(6a - b) - (7a + b)$

17. $(9x + y) - (2x + 3y)$

18. $(6x - 3y) - (5x - 3y)$

19. $(2x + 1) - (9x + 1)$

20. $(8a + b) - (a - 5b)$

Multiply. ***(See pp. 156–157.)***

21. $a^2 \cdot a^4$
22. $2a^2 \cdot a$
23. $a \cdot a^3$
24. $(5a)(-6a^2)$
25. $(14x)(2x^3)$
26. $(3x^2)(8x)$
27. $(-7y^2)(8y^3)$
28. $(-2n^3)(-9n^2)$

Simplify. ***(See pp. 158–159.)***

29. $(a^2)^3$
30. $-(a^2)^3$
31. $(-a^2)^3$
32. $(-3x^2)^3$
33. $(mn^2)^4$
34. $(3mn^2)^3$
35. $(-x^2y^3)^3$
36. $(3xy^4)^2$

Multiply. ***(See pp. 160–163.)***

37. $5(6 + n)$
38. $7(6x + 3)$
39. $a(a + 4)$
40. $(n - 3)(n - 2)$
41. $(3n - 1)(4n - 1)$
42. $(6x + y)(x - y)$
43. $(r + 4)(2r + 8)$
44. $(2a - 7)(5a + 2)$
45. $(3x + y)(x - y)$

Multiply by the FOIL method. ***(See pp. 164–167.)***

46. $(n - 6)(n - 3)$
47. $(2x - 1)(7x + 2)$
48. $(3n + 1)(10n + 2)$
49. $(1 + 4n)(1 - 2n)$
50. $(y + 4)(3y + 1)$
51. $(x - 5)(x - 7)$

Express as a trinomial square. ***(See pp. 168–169.)***

52. $(n + 4)^2$
53. $(n - 4)^2$
54. $(3m - 2n)^2$
55. $(4a + 3b)^2$

Divide. ***(See pp. 170–171, 174–175.)***

56. $\dfrac{6n^2}{2n}$
57. $\dfrac{50x^3}{5x}$
58. $\dfrac{-10x^3}{x^2}$
59. $\dfrac{24xy^2}{-8xy}$
60. $\dfrac{18a^2b^4}{-3ab}$
61. $\dfrac{40a^4b^3}{8ab^2}$
62. $\dfrac{121mn^2}{11m}$
63. $\dfrac{48x^2y^2}{-4x^2y}$
64. $\dfrac{6n^2 + n}{n}$
65. $\dfrac{4m^2 + 12m}{4m}$
66. $\dfrac{6x^2 - 3x}{3x}$
67. $\dfrac{12a^2b^2 - 9ab}{3ab}$
68. $\dfrac{4a^3 - 2a^2 + 2a}{2a}$
69. $\dfrac{6x^2y + 8xy^2 + 10xy}{2xy}$

(Optional) Divide. ***(See pp. 172–173, 176–177.)***

70. $\dfrac{y^5}{y^9}$
71. $\dfrac{15x^2y^3}{5xy^4}$
72. $\dfrac{18m^8n^6}{-9m^5n^{11}}$
73. $\dfrac{21a^3b^2c}{7ab^2c^3}$
74. $\dfrac{x^2 - 12x + 35}{x - 7}$
75. $\dfrac{y^2 + 11y + 28}{y + 4}$
76. $\dfrac{m^2 - 2m - 15}{m - 5}$

CHAPTER TEST

Add or subtract.

1. $(x^2 - xy + y^2) + (x^2 - y^2)$
2. $(-2m^2 + m - 1) + (4m^2 - m + 5)$ *(5-1)*
3. $(3s - 4t) - (-7t + 8s)$
4. $(6y^2 - y - 4) - (y^2 + 2y - 5)$ *(5-2)*
5. Solve: $(4x + 9) - (5x - 8) = 11$

Multiply.

6. $(-4x^3)(-7x)$
7. $(5b^2)(-2ab^2c^3)$
8. $(-9xz^3)(6x^2y^4)$ *(5-3)*

Simplify.

9. $(-3x)^4$
10. $(2x^2y^3)^3$
11. $-(xy^2)^2(-x^2y)$ *(5-4)*

Multiply.

12. $-4x(x^2 - 3x + 5)$
13. $rs(r^2 - 3s)$
14. $y^3(2y^4 - 7y^2 + 1)$ *(5-5)*

Multiply.

15. $(a + 2b)(a - b)$
16. $(3x - 1)(x - 4)$
17. $(n + 2)(n^2 - 3n + 4)$ *(5-6)*
18. $(r + 7)(r + 2)$
19. $(x + 4y)(x - 4y)$
20. $(3a - 2b)(2a + 5b)$ *(5-7)*

Express as a trinomial.

21. $(n - 9)^2$
22. $(5x + 2y)^2$
23. $-3(2m - 1)^2$ *(5-8)*

Divide.

24. $\dfrac{16n^4}{-2n^2}$
25. $\dfrac{-60r^3s^2}{-6rs^2}$
26. $\dfrac{45c^3d^4}{5d^3}$ *(5-9)*

(Optional) Divide.

27. $\dfrac{15x^4}{-5x^9}$
28. $\dfrac{-8a^3b}{8ab^4}$
29. $\dfrac{-35a^4b^2c}{-7a^2b^2c^2}$ *(5-10)*

Divide.

30. $\dfrac{2a^5 - a^3 + 7a^2}{a^2}$
31. $\dfrac{42x^2y + 14xy - 63xy^2}{7xy}$ *(5-11)*

(Optional) Divide.

32. $\dfrac{a^2 + a - 12}{a - 3}$
33. $\dfrac{b^2 - 12b + 32}{b - 4}$ *(5-12)*

MIXED REVIEW

1. Subtract: $(2x^2 - 5x + 9) - (3x^2 - 7x - 4)$.

2. Jannette and Anne left a parking lot at 2:00.
   They drove in opposite directions for four hours.
   They were then 640 km apart.
   Jannette drove 12 km/h faster than Anne.
   How fast did each travel?

Solve.

3. $27 - x = 3 - 5x$

4. $y - 8 = -2$

5. $3(1 - 2n) = n - 4$

6. Simplify: $5(-9) - 7(-3)$

7. Add: $-9 + 5 + (-8)$

8. Multiply: $(5x + 2y)(3x - 2y)$

9. Find two consecutive whole numbers that total 35.

10. Simplify: $-5a + 3b - 9b - (-2a)$

11. Solve for y: $5x - y = 7$

12. Calvin bought some paint sets for \$3 each.
    He sold all but 3 of them for \$4 each.
    His profit was \$8.
    How many paint sets did he sell?

13. Divide: $\dfrac{16a^4b^4 - 12a^3b^3 + 14a^2b^2}{-2a^2b}$

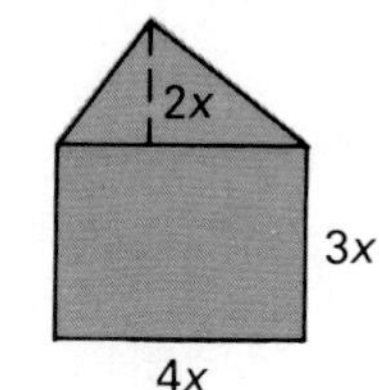

14. Find the area of the shaded figure.

15. Graph the solutions of $x < 5$ on the number line.

16. A number is 17 less than -6. What is the number?

17. Express as a trinomial: $(5a - 1)^2$

18. A taxi ride costs x cents for the first 2 kilometers and y cents for each additional kilometer. A 9-kilometer ride costs __?__ cents.

19. Solve: $4(a - 4) = 9 - 3(2a - 5)$

20. Add: $(5m^2 - 7m - 6) + (4m^3 - 7m^2 + 7m - 2)$

Multiply.

21. $(-3xy^2)(4x^3y^4)$

22. $-x(3 - x)$

23. $(x - 3)(x^2 + 2x - 4)$

24. 7 more than 3 times a number is 3 less than 5 times the number. Find the number.

25. Write an inequality to compare 0, 2, and -7.

26. A square has sides of length $2a + 3b$. Express the area of the square as a trinomial.

27. Find the volume of a box that measures $(x + 2)$ units by x units by x units.

Simplify.

28. $-(4x^2)^3$

29. $(3m^3n)^2 \cdot 5mn^4$

30. $\frac{9x^5}{x^3} - \frac{8x^3}{2x}$

31. Find the perimeter of a rectangle with length 4.5 cm and width 3 cm.

32. If $x = -2$, find the value of $5x + 7$ and $5(x + 7)$.

33. Tell which of the numbers 0, 1, 2, or 3 are solutions of the inequality $8a < 3(a + 6)$.

34. Stu is s years old.
Paul is 4 years older than Stu.
Six years ago Paul was __?__ years old.

35. Compare. Write $<$ or $>$ or $=$: $\frac{-12}{-6}$ __?__ $(-1)^3$

36. You earn x dollars per hour for a 40-hour work week.
You earn $(x + 2)$ dollars per hour for overtime work.
Last week you worked 48 hours and earned $256.
How much did you earn for overtime work?

37. Solve: $7(2a + 5) - (3a - 1) = 58$

38. Find the value if possible. If not, write *impossible.*

a. $\frac{1 \cdot 0}{1 - 0}$ b. $\frac{1 \div 0}{1 + 0}$ c. $\frac{1 + 0}{1 \cdot 0}$ d. $\frac{1 - 0}{1 + 0}$

39. If $A = \frac{1}{2}bh$, $A = 18$, and $b = 4$, then $h =$ __?__.

40. A triangle has sides of lengths $3x - 1$, $2x + 7$, and $x + 8$.
a. Write an expression for the perimeter of the triangle.
b. Evaluate the expression if $x = 6$.
c. If the perimeter is 80, find the value of x.

Here's what you'll learn in this chapter:

- To find a common monomial factor in a polynomial.
- To factor trinomials which are the product of two binomials.
- To multiply binomials which are the sum and the difference of two monomials.
- To factor the difference of two squares.

Most manufactured objects are made up of many small parts such as those shown. In this chapter you will discover that many polynomials are made up of parts called factors.

Chapter 6

Factoring

1 Factoring in Arithmetic

The number 12 can be written as a product of its factors in several ways. The numbers 1, 2, 3, 4, 6, and 12 are all factors of 12.

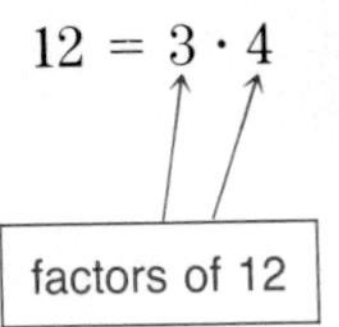

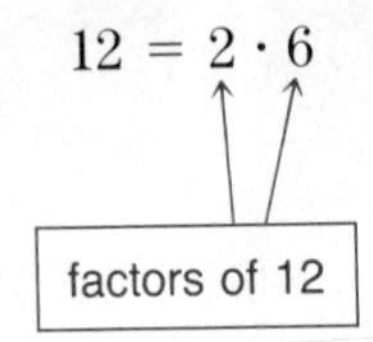

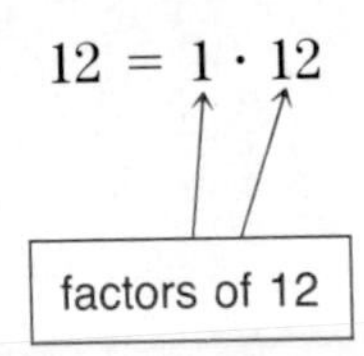

If a number has no factors except itself and 1, it is called a **prime number.** Here are some prime numbers.

prime numbers → 2, 3, 5, 7, 11, 13, 19

Since 2 and 3 are prime numbers and also factors of 12, they are called **prime factors** of 12.

$$12 = 3 \cdot 4 \qquad\qquad 140 = 10 \cdot 14$$

$$= 3 \cdot \overbrace{2 \cdot 2} \qquad\qquad = \overbrace{5 \cdot 2} \cdot \overbrace{2 \cdot 7}$$

$$= 2 \cdot 2 \cdot 3 \qquad\qquad = 2 \cdot 2 \cdot 5 \cdot 7$$

12 and 140 are written as products of prime numbers.

Notice that 2 is a factor of both 12 and 140. We call 2 a **common factor** of 12 and 140.

EXAMPLE 1 Find a common factor of 35 and 28.

$$\left.\begin{aligned} 35 &= 5 \cdot 7 \\ 28 &= 2 \cdot 2 \cdot 7 \end{aligned}\right\}$$ 7 is a common factor.

EXAMPLE 2 Find a common factor of 18 and 45.

$$\left.\begin{aligned} 18 &= 2 \cdot 3 \cdot 3 \\ 45 &= 3 \cdot 3 \cdot 5 \end{aligned}\right\}$$ 3 is a common factor.

In the second example, 3 is a common factor, but $3 \cdot 3$, or 9, is also a common factor. The number 9 is the greatest number that is a common factor of 18 and 45. We call 9 the **greatest common factor** of the numbers 18 and 45.

Study the example below and the clues for finding factors.

EXAMPLE 3 Find the greatest common factor of 28 and 70.

$$\left.\begin{aligned} 28 &= 2 \cdot 2 \cdot 7 \\ 70 &= 2 \cdot 5 \cdot 7 \end{aligned}\right\}$$

$2 \cdot 7$, or 14, is the greatest common factor.

Here are a few clues for finding factors.

1. *A number has 2 as a factor if its last digit is 0, 2, 4, 6, or 8.*
2. *A number has 5 as a factor if its last digit is 0 or 5.*
3. *A number has 3 as a factor if the sum of its digits can be divided evenly by 3.*

SAMPLE 726 is divisible by 3 because $7 + 2 + 6$, or 15, is divisible by 3.

Written Exercises

Write each number as a product of prime numbers.

A

1. 42	**2.** 100	**3.** 36	**4.** 54	**5.** 310
6. 125	**7.** 120	**8.** 625	**9.** 624	**10.** 93

Find the greatest common factor of the pair of numbers.

11. 12, 15	**12.** 16, 22	**13.** 14, 21	**14.** 27, 36
15. 11, 20	**16.** 12, 18	**17.** 22, 33	**18.** 15, 40
19. 16, 64	**20.** 21, 42	**21.** 36, 48	**22.** 14, 49
23. 42, 28	**24.** 18, 24	**25.** 100, 125	**26.** 30, 100
27. 16, 36	**28.** 15, 24	**29.** 24, 56	**30.** 19, 21
31. 12, 20	**32.** 28, 35	**33.** 40, 100	**34.** 15, 35
35. 42, 48	**36.** 50, 60	**37.** 100, 400	**38.** 36, 81

2 Common Monomial Factors

Now you can find common factors of variable expressions.

$$\left.\begin{array}{l} 2a = 2 \cdot a \\ 6a^2 = 2 \cdot 3 \cdot a \cdot a \end{array}\right\} \quad \text{$2 \cdot a$, or $2a$, is the greatest common factor.}$$

Use this idea to find a common monomial factor of the terms of a polynomial. When you do this, you **factor** the polynomial.

$$\begin{aligned} 2a + 6a^2 &= (2a \cdot 1) + (2a \cdot 3a) \\ &= 2a(1 + 3a) \end{aligned}$$

This is a common monomial factor of $2a$ and $6a^2$.

Check: Use the distributive property. → $2a(1 + 3a) = 2a + 6a^2$ ✓

EXAMPLE 1 Factor $3x^2 - ax^2$.

$$\begin{aligned} 3x^2 - ax^2 &= (3 \cdot x^2) - (a \cdot x^2) \\ &= x^2(3 - a) \end{aligned}$$

With practice you will be able to recognize the common monomial factors at sight.

EXAMPLE 2

$$12mx - 6m^2x^2 = 6mx(2 - mx)$$

$\mathbf{6mx} \cdot 2$ $\quad$ $\mathbf{6mx} \cdot mx$

EXAMPLE 3

$$x^3 + x^2 + 2x = x(x^2 + x + 2)$$

$\mathbf{x} \cdot x^2$ $\quad$ $\mathbf{x} \cdot x$ $\quad$ $2 \cdot \mathbf{x}$

EXAMPLE 4

$$y^4 + y^3 + 4y^2 = y^2(y^2 + y + 4)$$

$\mathbf{y^2} \cdot y^2$ $\quad$ $\mathbf{y^2} \cdot y$ $\quad$ $4 \cdot \mathbf{y^2}$

Classroom Practice

Complete.

1. $4xy = 4x(\underline{\ ?\ })$
2. $25x^2 = 5x(\underline{\ ?\ })$
3. $3x^2y^2 = 3xy(\underline{\ ?\ })$
4. $36a^3 = 12a^2(\underline{\ ?\ })$
5. $72b^2 = 8b(\underline{\ ?\ })$
6. $40a^4b^4 = 4a^2b^2(\underline{\ ?\ })$

Find the greatest common factor of the monomial pair.

7. $2a, 6a$
8. $15n, 5n$
9. $n^2, 3n$
10. $3x, 6x^2$

Factor.

11. $3x + 2x^2$
12. $2a^2 + 4a$
13. $7x^3 + 7x^2$
14. $3x^2 + 6xy$

Written Exercises

Factor.

A

1. $9 + 3x$
2. $5 - 15n$
3. $2x - 10$
4. $4y - 16$
5. $3x^2 - x$
6. $5n^2 - 2n$
7. $2x^2 - 6x$
8. $y^2 - 2y$
9. $28n^2 - 7n$
10. $2xy - y^2$
11. $4x^2 - 8x$
12. $21y^2 - 7xy$
13. $25mn - 5m^2n^2$
14. $8a^2b - 24ab^2$
15. $9x^2 - 27x^2y$
16. $12a^2 + 36a^2b$
17. $3x^2 - 6x + 21$
18. $5n^3 + 15n^2 + 25n$
19. $55y^2 + 22y + 44$
20. $2n^2 + 4mn + 80m^2$
21. $4a^2 + 12ab - 16b^2$
22. $3x^2 - 12xy + 9y^2$

B

23. $6x^2 + 6x + 24xy + 42$
24. $-13x + 26x^2 + 39x^3$
25. $-50a^2 + 25b^2 + 75ab$
26. $48mn + 72m^2n^2 + 60m^3n^3$
27. $56x^3y^3 - 72x^2y^2 - 64xy$
28. $32a^2b^4 - 16ab^3 + 48a^3b^5$

C

29. $x(a - b) + 2(a - b)$
30. $y(a + d) - 4(a + d)$
31. $mn - nx + my - xy$
32. $ab - 3ad + b^2 - 3bd$
33. $2a - a^2 + 2b - ab$
34. $3x^2 - 2xy - 3x + 2y$
35. $2x^2 - 8 + x^2y - 4y$
36. $a^3 - ab^2 - a^2b + b^3$

3 Using Factoring

You are probably familiar with the formula $A = \pi r^2$. It is used to calculate the area of a circle. Take a look at this formula and what it means.

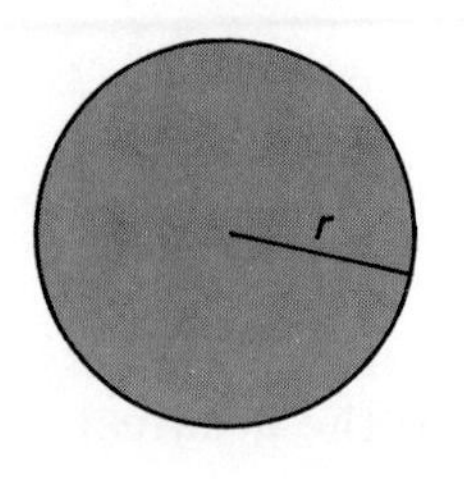

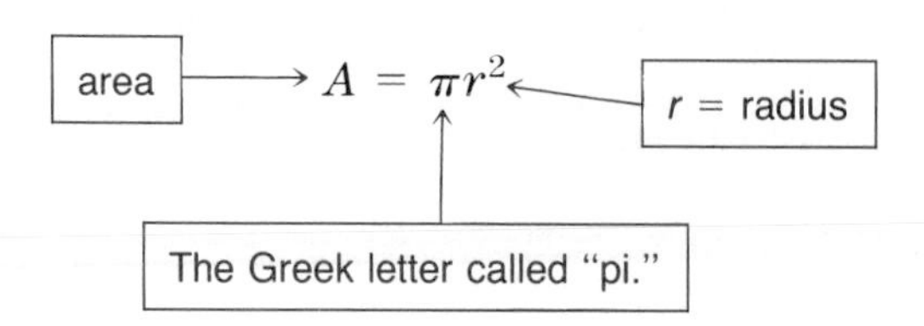

The symbol π in the formula has a value of about 3.14 or $\frac{22}{7}$. Sometimes it's difficult to work with these values, so we often just use the symbol π.

EXAMPLE 1 Find the area of the shaded part in the figure below.

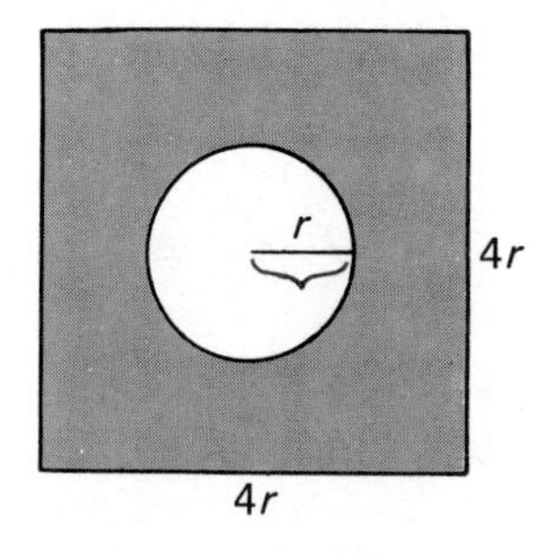

Area of square: $(4r)^2 = 16r^2$
Area of circle: πr^2

Area of shaded part: $16r^2 - \pi r^2$
$= r^2(16 - \pi)$

In doing arithmetic problems with large numbers, be sure to look for common monomial factors as you work. By factoring, you may be able to save yourself some computation.

EXAMPLE 2 Find the area of the shaded part.

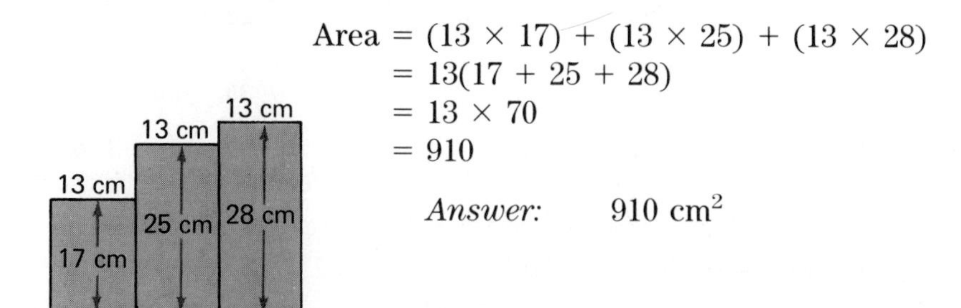

Area $= (13 \times 17) + (13 \times 25) + (13 \times 28)$
$= 13(17 + 25 + 28)$
$= 13 \times 70$
$= 910$

Answer: 910 cm^2

Written Exercises

Find the area of the shaded part. Factor the expression. Leave your answer in terms of π.

A

1.

2.

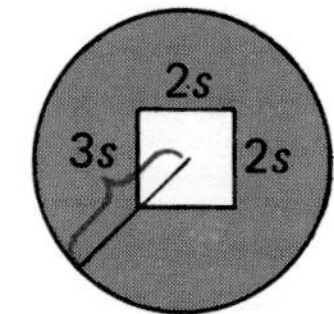

3.

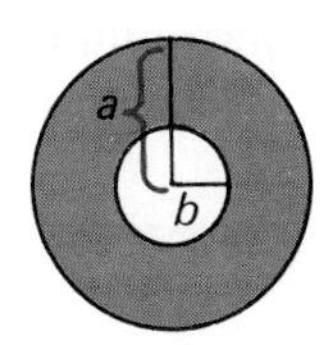

Find the area of the shaded part.

4.

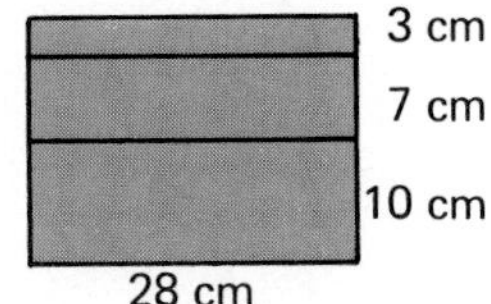

5.

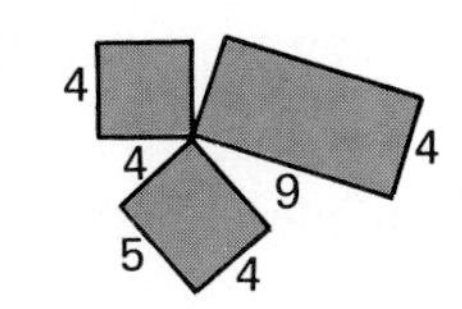

6.

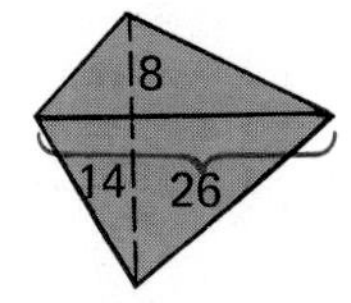

B

7. Find the area of the shaded part. Leave your answer in factored form.

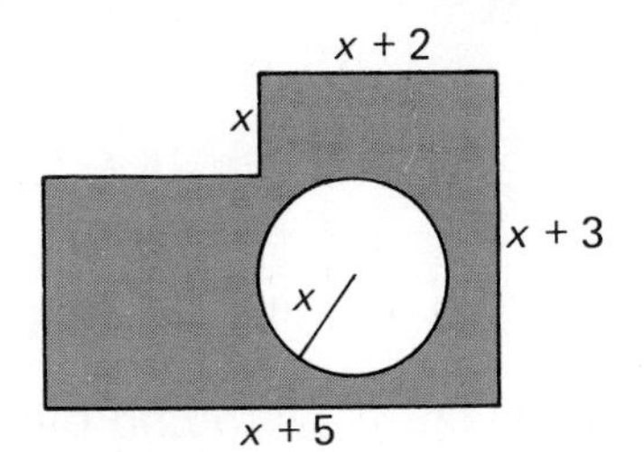

SELF-TEST

Vocabulary

prime number (p. 186)
prime factors (p. 186)
common factor (p. 186)
greatest common factor (p. 187)
factor a polynomial (p. 188)

Find the greatest common factor of the pair.

1. 56, 64 **2.** $24a$, $8b$ **3.** $36x^2$, $3xy$ **4.** $2mn$, $4mn^2$ *(6-1)*

Factor.

5. $15x^2y^2 + 25xy^3$ **6.** $18mn^2 - 9m^2n^2$ **7.** $8a^2b^3 + 56a^2c^3$ *(6-2)*

8. Find the area of the shaded part in the figure at the right. Factor the expression. Leave your answer in terms of π. *(6-3)*

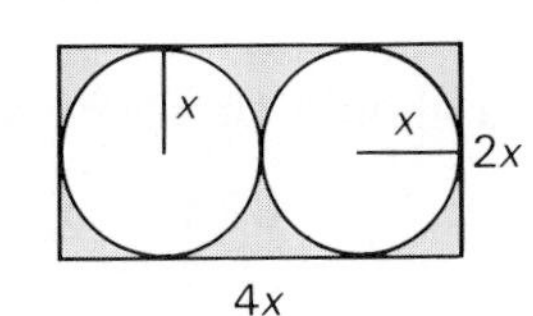

4 Factoring Trinomials—Two Sums

Up to now the only trinomials you have factored are those having a common monomial factor.

$$3x^3 + 2x^2 + x = x(3x^2 + 2x + 1)$$

Common monomial factor

Not all trinomials can be factored, but in the next few lessons you will see that many are the product of two binomials. For example, if $(x + 2)(x + 3) = x^2 + 5x + 6$, then we know that $x^2 + 5x + 6$ has two factors, $x + 2$ and $x + 3$.

Recall how to multiply two binomials at sight.

Step 1: $(x + 2)(x + 3) = x^2 \ldots$

Step 2: $(x + 2)(x + 3) = x^2 + 5x \ldots$

Step 3: $(x + 2)(x + 3) = x^2 + 5x + 6$

Now reverse this process to find the factors.

Step 1: $x^2 + 5x + 6 = (\quad)(\quad)$ ← Draw your parentheses.

Step 2: $x^2 + 5x + 6 = (x \quad)(x \quad)$ ← Your first terms must give you x^2.

Step 3: $x^2 + 5x + 6 = (x \; ? \;)(x \; ? \;)$

What two numbers have a product of 6?

1 and 6 2 and 3

Which of the pairs of factors of 6 have a sum of 5?

2 and 3

Step 4: $x^2 + 5x + 6 = (x + 2)(x + 3)$

Of course, you should check to see if your answer is correct.

$$(x + 2)(x + 3) = x^2 + 5x + 6 \quad ✓$$

Study these two examples.

EXAMPLE 1 Factor $x^2 + 7x + 10$.

Step 1: $x^2 + 7x + 10 = (\quad)(\quad)$

Step 2: $x^2 + 7x + 10 = (x \quad)(x \quad)$

Step 3: $x^2 + 7x + 10 = (x \; ? \;)(x \; ? \;)$

What two numbers have a product of 10?

1 and 10 2 and 5

Which of the pairs of factors of 10 have a sum of 7?

2 and 5

Step 4: $x^2 + 7x + 10 = (x + 2)(x + 5)$

EXAMPLE 2 Factor $a^2 + 10a + 16$.

Step 1: $a^2 + 10a + 16 = (\quad)(\quad)$

Step 2: $a^2 + 10a + 16 = (a \quad)(a \quad)$

Step 3: $a^2 + 10a + 16 = (a \; ? \;)(a \; ? \;)$

What two numbers have a product of 16?

1 and 16 2 and 8

4 and 4

Which of the pairs of factors of 16 have a sum of 10?

2 and 8

Step 4: $a^2 + 10a + 16 = (a + 2)(a + 8)$

Classroom Practice

Name the possible pairs of factors.

1. 10 **2.** 7 **3.** 9 **4.** 6 **5.** 11

6. 12 **7.** 15 **8.** 17 **9.** 16 **10.** 20

11. 21 **12.** 24 **13.** 25 **14.** 30 **15.** 36

Factor.

16. $x^2 + 8x + 7$
17. $x^2 + 4x + 3$
18. $a^2 + 6a + 5$
19. $a^2 + 5a + 4$
20. $x^2 + 4x + 4$
21. $y^2 + 8y + 15$
22. $n^2 + 8n + 12$
23. $x^2 + 6x + 8$
24. $n^2 + 11n + 24$
25. $x^2 + 13x + 42$
26. $x^2 + 13x + 30$
27. $y^2 + 13y + 40$

Written Exercises

Complete.

	If the product of two numbers is: →	and their sum is: →	then the two numbers are:
Sample	18	11	2 and 9
A **1.**	5	6	?
2.	6	5	?
3.	8	6	?
4.	12	8	?
5.	12	7	?
6.	9	6	?
7.	15	8	?
8.	14	9	?
9.	18	9	?
10.	20	9	?
11.	21	10	?
12.	24	14	?
13.	24	11	?
14.	36	13	?
15.	36	12	?

16. $x^2 + 8x + 7 = (x \underline{\ ?\ })(x \underline{\ ?\ })$
17. $a^2 + 7a + 6 = (a \underline{\ ?\ })(a \underline{\ ?\ })$
18. $n^2 + 9n + 20 = (n \underline{\ ?\ })(n \underline{\ ?\ })$
19. $x^2 + 9x + 18 = (x \underline{\ ?\ })(x \underline{\ ?\ })$
20. $y^2 + 9y + 14 = (y \underline{\ ?\ })(y \underline{\ ?\ })$
21. $r^2 + 10r + 21 = (r \underline{\ ?\ })(r \underline{\ ?\ })$

Factor. Check by multiplication.

22. $y^2 + 3y + 2$
23. $n^2 + 17n + 16$
24. $y^2 + 16y + 15$
25. $a^2 + 5a + 4$
26. $m^2 + 7m + 6$
27. $x^2 + 11x + 10$
28. $x^2 + 11x + 18$
29. $n^2 + 6n + 8$
30. $y^2 + 6y + 9$

Factor.

31. $x^2 + 9x + 14$
32. $a^2 + 9a + 18$
33. $y^2 + 12y + 11$
34. $r^2 + 11r + 30$
35. $b^2 + 12b + 27$
36. $m^2 + 11m + 28$
37. $x^2 + 12x + 32$
38. $y^2 + 24y + 23$
39. $x^2 + 13x + 30$
40. $a^2 + 9a + 20$
41. $a^2 + 12a + 20$
42. $n^2 + 12n + 35$

43. $x^2 + 27x + 50$
44. $x^2 + 17x + 72$
45. $m^2 + 28m + 75$
46. $x^2 + 20x + 64$
47. $a^2 + 24a + 63$
48. $y^2 + 24y + 44$

CAREER NOTES

MARKETING

Many people aren't aware of all that is involved in putting a product on the market. First, how do manufacturers know which items to produce? Often, manufacturers ask analysts called market researchers to collect data on consumer needs. Since consumers are the buyers, their needs determine which products will be on the market.

For example, market researchers might decide that there is a market for a new type of athletic shoe or that red cars would sell better than yellow ones.

Market researchers collect data on many separate groups of consumers. For instance, they might study shopping habits of people in California versus people in Missouri or people aged 25–35 years versus people aged 55–65 years. The researchers then analyze the data to forecast sales and give recommendations to the manufacturers on product design, packaging, and advertising.

5 Factoring Trinomials—Two Differences

You know that when you want to factor a positive number like 15, you might also consider the negative factors as well as the positive ones.

$$15 = 3 \times 5 \qquad \text{but also} \qquad 15 = (-3) \times (-5)$$

In this section you will use this idea to factor some trinomials.

EXAMPLE 1 Factor $x^2 - 8x + 15$.

Step 1: $x^2 - 8x + 15 = (\quad)(\quad)$

Step 2: $x^2 - 8x + 15 = (x \quad)(x \quad)$

Step 3: $x^2 - 8x + 15 = (x\ ?\ )(x\ ?\ )$

What two negative numbers have a product of 15?

-1 and -15 $\qquad$ -3 and -5

Which of the pairs of factors of 15 have a sum of -8?

-3 and -5

Step 4: $x^2 - 8x + 15 = (x - 3)(x - 5)$

Check: $(x - 3)(x - 5) = x^2 - 8x + 15$ ✓

EXAMPLE 2 Factor $n^2 - 10n + 16$.

Step 1: $n^2 - 10n + 16 = (\quad)(\quad)$

Step 2: $n^2 - 10n + 16 = (n \quad)(n \quad)$

Step 3: $n^2 - 10n + 16 = (n\ ?\ )(n\ ?\ )$

You have to find two numbers whose product is 16 and whose sum is -10.

-8 and -2

Step 4: $n^2 - 10n + 16 = (n - 8)(n - 2)$

Check: $(n - 8)(n - 2) = n^2 - 10n + 16$ ✓

Classroom Practice

Tell whether the factors will be sums or differences. Factor.

1. $x^2 - 6x + 9$
2. $x^2 + 2x + 1$
3. $y^2 + 3y + 2$
4. $n^2 - 7n + 10$
5. $n^2 - 10n + 24$
6. $n^2 + 5n + 6$
7. $c^2 + 7c + 6$
8. $x^2 - 4x + 4$
9. $a^2 - 9a + 20$

Written Exercises

Complete.

	If the product of two numbers is:	→ and their sum is: →	then the two numbers are:
A 1.	6	−5	?
2.	10	−7	?
3.	18	−9	?
4.	12	−7	?
5.	24	−10	?

Factor.

6. $n^2 - 3n + 2$
7. $x^2 - 7x + 10$
8. $y^2 - 8y + 12$
9. $x^2 - 5x + 6$
10. $n^2 - 10n + 21$
11. $y^2 - 6y + 5$
12. $b^2 - 2b + 1$
13. $x^2 - 11x + 24$
14. $x^2 - 13x + 30$
15. $n^2 + 9n + 18$
16. $n^2 - 11n + 18$
17. $n^2 - 5n + 4$
18. $a^2 - 9a + 14$
19. $x^2 + 6x + 8$
20. $x^2 - 7x + 12$
21. $y^2 - 11y + 28$
22. $x^2 - 12x + 27$
23. $n^2 + 10n + 25$
24. $n^2 - 11n + 24$
25. $y^2 - 11y + 30$
26. $r^2 - 14r + 33$
27. $x^2 - 14x + 49$
28. $a^2 - 12a + 36$
29. $m^2 - 13m + 40$
30. $a^2 - 4a + 4$
31. $z^2 - 18z + 32$
32. $x^2 - 11x + 30$
33. $y^2 - 15y + 36$
34. $b^2 - 12b + 36$
35. $d^2 - 14d + 45$
36. $c^2 - 24c + 80$
37. $x^2 - 52x + 100$
38. $n^2 - 30n + 200$

6 Factoring Trinomial Squares

When the two factors of a trinomial are equal, the trinomial is a *trinomial square.*

This is a trinomial square because it is the product of two equal factors.

$$x^2 + 6x + 9 = (x + 3)(x + 3) = (x + 3)^2$$

Take a look at trinomial squares. Learn how to recognize them.

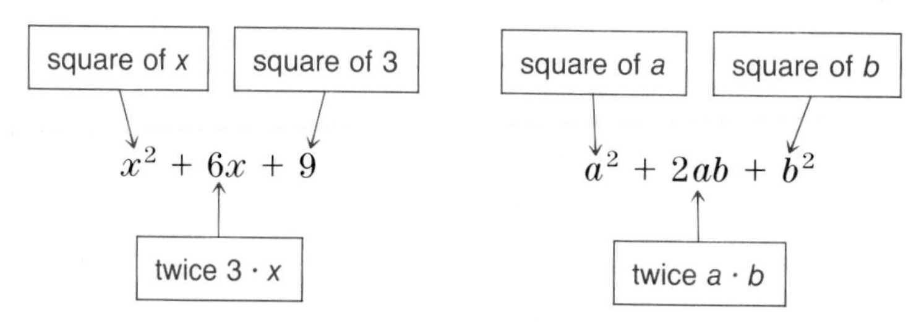

So, $x^2 + 6x + 9 = (x + 3)^2$. So, $a^2 + 2ab + b^2 = (a + b)^2$.

EXAMPLE 1 Is $x^2 + 8x + 16$ a trinomial square? See if the terms meet the test.

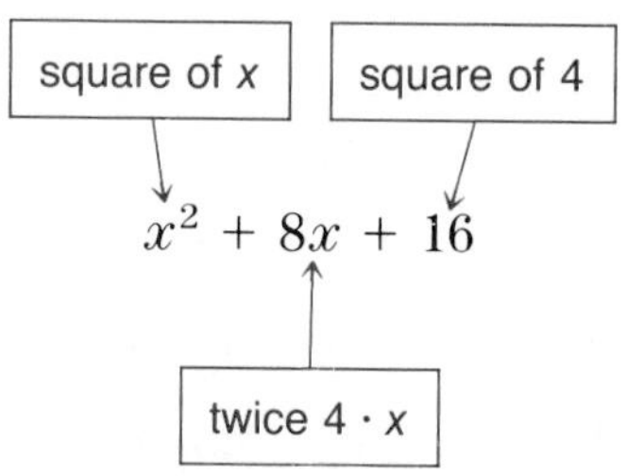

Yes, $x^2 + 8x + 16 = (x + 4)^2$.

EXAMPLE 2 Is $a^2 + 5a + 6$ a trinomial square?

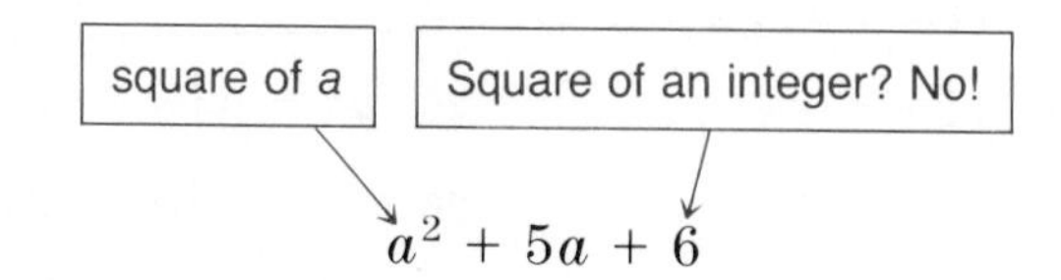

No, $a^2 + 5a + 6$ is not a trinomial square.

Of course, if you forget how to recognize a trinomial square, you can factor using the method you learned in the last two sections.

Sometimes a trinomial is the square of a difference. In that case, the middle term will have a minus sign.

EXAMPLE 3 Is $x^2 - 4x + 4$ a trinomial square?

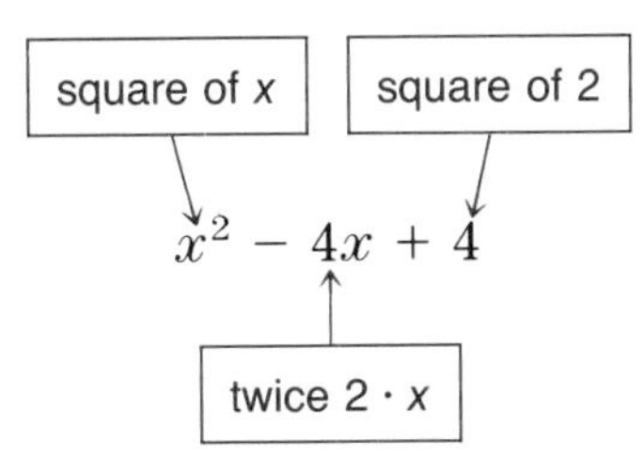

Yes, $x^2 - 4x + 4 = (x - 2)^2$.

Use − because the middle term in the trinomial is negative.

EXAMPLE 4 Is $n^2 - 10n + 25$ a trinomial square?

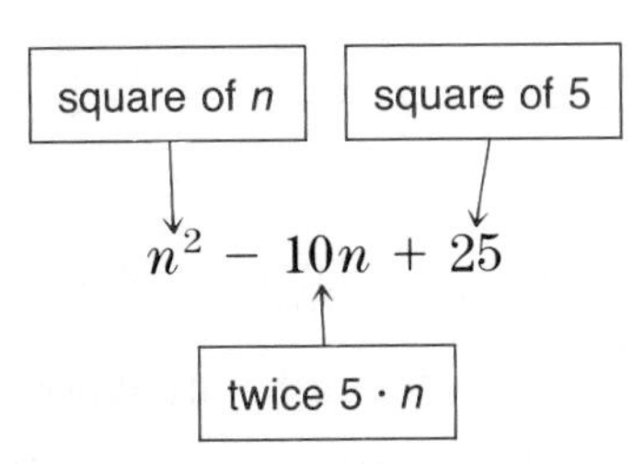

Yes, $n^2 - 10n + 25 = (n - 5)^2$.

Written Exercises

Which of the following are trinomial squares?

A

1. $x^2 + 2x + 1$
2. $y^2 + 4y + 4$
3. $a^2 + 6a + 9$
4. $y^2 + y + 4$
5. $y^2 + 4y + 1$
6. $y^2 + 2y + 4$

Factor. Note that four exercises are not trinomial squares. Factor those as you did in the last sections.

7. $a^2 + 10a + 25$
8. $n^2 + 4n + 4$
9. $x^2 + 8x + 16$
10. $n^2 + 12n + 36$
11. $x^2 - 4x + 4$
12. $a^2 - 9a + 20$
13. $y^2 + 14y + 49$
14. $x^2 - 20x + 100$
15. $n^2 + 18n + 81$
16. $x^2 + 15x + 56$
17. $x^2 - 19x + 90$
18. $b^2 - 15b + 36$
19. $a^2 - 2ab + b^2$
20. $m^2n^2 + 18mn + 81$
21. $1 - 20x + 100x^2$

7 Factoring Trinomials

Here is a trinomial having one factor a sum and one factor a difference. Notice how the signs help you discover the factors.

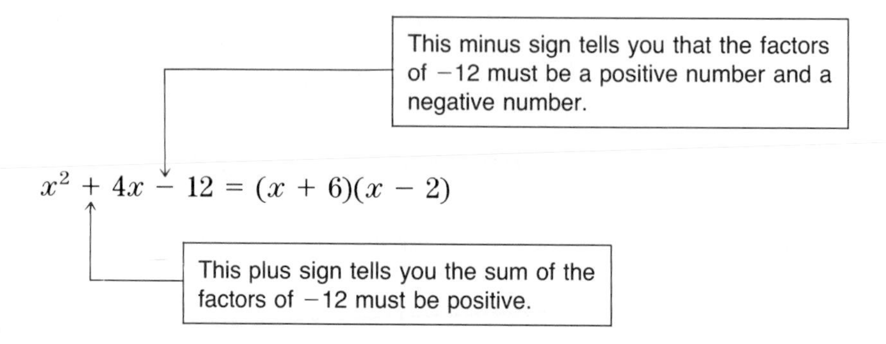

EXAMPLE 1 Factor $n^2 + 4n - 21$.

Step 1: $n^2 + 4n - 21 = (\quad)(\quad)$

Step 2: $n^2 + 4n - 21 = (n \quad)(n \quad)$

Step 3: $n^2 + 4n - 21 = (n\ ?\ )(n\ ?\ )$

Look for factors of -21 whose sum is $+4$.

$3 \times -7 = -21$ and $3 + (-7) = -4$
$-3 \times 7 = -21$ and $-3 + 7 = 4$ ✓

Step 4: $n^2 + 4n - 21 = (n - 3)(n + 7)$

EXAMPLE 2 Factor $x^2 - 2x - 15$.

Step 1: $x^2 - 2x - 15 = (\quad)(\quad)$

Step 2: $x^2 - 2x - 15 = (x \quad)(x \quad)$

Step 3: $x^2 - 2x - 15 = (x\ ?\ )(x\ ?\ )$

Look for factors of -15 whose sum is -2.

$-3 \times 5 = -15$ and $-3 + 5 = 2$
$3 \times -5 = -15$ and $3 + (-5) = -2$ ✓

Step 4: $x^2 - 2x - 15 = (x + 3)(x - 5)$

In the next example, notice that -24 has many different factors. You may have to check quite a few of these before you find the right ones.

EXAMPLE 3 Factor $x^2 - 2x - 24$.

Step 1: $x^2 - 2x - 24 = (\quad)(\quad)$

Step 2: $x^2 - 2x - 24 = (x \quad)(x \quad)$

Step 3: $x^2 - 2x - 24 = (x \ ? \ )(x \ ? \ )$

Look for factors of -24 whose sum is -2.

$2 \times -12 = -24$ and $2 + (-12) = -10$
$3 \times -8 = -24$ and $3 + (-8) = -5$
$4 \times -6 = -24$ and $4 + (-6) = -2$ ✓

Step 4: $x^2 - 2x - 24 = (x + 4)(x - 6)$

It is always wise to check your answer by multiplying.

Classroom Practice

Name the possible pairs of factors.

1. -6
2. -5
3. -10
4. -8
5. -7
6. -12
7. -16
8. -20
9. -36
10. -24

Factor.

11. $a^2 + 5a - 14$
12. $x^2 + x - 6$
13. $y^2 + 2y - 8$
14. $n^2 - 3n - 10$
15. $x^2 - 6x - 16$
16. $n^2 - 4n - 5$
17. $x^2 + 4x - 12$
18. $a^2 - 3a - 40$
19. $x^2 + 5x - 24$
20. $y^2 + 6y - 16$
21. $b^2 - 5b - 24$
22. $m^2 + 8m - 33$
23. $x^2 + x - 30$
24. $y^2 + y - 56$
25. $n^2 - 2n - 63$
26. $x^2 - 2x - 8$
27. $x^2 + 2x - 15$
28. $a^2 + a - 20$
29. $y^2 - 5y - 36$
30. $n^2 + 2n - 35$
31. $x^2 - 4x - 32$
32. $x^2 - 3x - 18$
33. $w^2 + 4w - 12$
34. $z^2 + 10z - 39$
35. $y^2 - 2y - 48$
36. $r^2 + 6r - 55$
37. $m^2 - 4m - 60$

Written Exercises

Complete.

	If the product of two numbers is: →	and their sum is: →	then the two numbers are:
A 1.	−10	−3	?
2.	−18	7	?
3.	−4	−3	?
4.	−21	−4	?
5.	−2	−1	?
6.	−6	1	?
7.	−30	−7	?
8.	−30	−1	?
9.	−28	3	?
10.	−15	2	?
11.	−14	13	?
12.	−24	−2	?

Factor.

13. $x^2 + 4x - 21$

14. $x^2 - 3x - 4$

15. $x^2 + 7x - 18$

16. $x^2 - 6x - 16$

17. $b^2 + b - 12$

18. $b^2 - 4b - 12$

19. $n^2 - 3n - 18$

20. $y^2 - 9y - 10$

21. $x^2 - x - 20$

22. $a^2 + 3a - 10$

23. $y^2 + 14y - 15$

24. $n^2 - 3n - 28$

25. $b^2 + 2b - 24$

26. $x^2 - x - 30$

27. $b^2 - 7b - 30$

28. $a^2 - 9a - 22$

29. $x^2 + 12x - 28$

30. $a^2 + 11a - 26$

31. $y^2 + 4y - 32$

32. $x^2 - 5x - 36$

33. $y^2 + 6y - 27$

34. $x^2 + 9x - 36$

35. $m^2 + 12m - 64$

36. $x^2 - x - 72$

37. $n^2 - 2n - 63$

38. $b^2 + 10b - 24$

39. $y^2 + y - 42$

40. $x^2 - 4x - 45$

41. $y^2 - y - 56$

42. $z^2 + 8z - 65$

43. $c^2 + 16c - 80$

44. $m^2 - 24m - 81$

45. $a^2 + 21a - 100$

Mixed Practice Exercises

Factor.

1. $x^2 + 10x + 9$
2. $x^2 + 8x + 15$
3. $n^2 - 8n + 12$
4. $a^2 - 9a + 20$
5. $y^2 + y - 12$
6. $x^2 + 2x - 15$
7. $y^2 - 4y - 21$
8. $x^2 - 9x - 22$
9. $x^2 + 10x + 21$
10. $a^2 + 4a - 32$
11. $b^2 + 4b - 5$
12. $x^2 - 8x + 16$
13. $b^2 - 8b + 7$
14. $x^2 + 10x - 11$
15. $y^2 + y - 20$
16. $a^2 - 4a - 32$
17. $x^2 + 12x + 35$
18. $y^2 + 5y - 36$
19. $n^2 - 15n + 56$
20. $y^2 - y - 42$
21. $x^2 + 15x + 54$
22. $a^2 + 2a - 48$
23. $n^2 + 14n + 45$
24. $x^2 + x - 72$
25. $n^2 - 5n - 50$
26. $x^2 - 2x - 48$
27. $y^2 + 24y - 52$
28. $x^2 - 9x + 20$
29. $a^2 - 7a - 44$
30. $y^2 + 17y + 52$
31. $c^2 + 4c - 60$
32. $z^2 - 10z - 75$
33. $m^2 - 14m + 40$

34. A rectangular field has an area of $x^2 - 2x - 15$. Find expressions for its length and width.

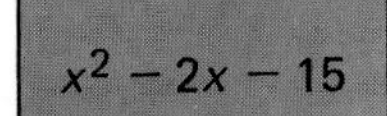

35. What is the perimeter of the field in Exercise 34?

36. A rectangular field has an area of $y^2 - 2y - 48$. Find expressions for its length and width.

37. What is the perimeter of the field in Exercise 36?

SELF-TEST

Factor.

1. $x^2 + 11x + 18$
2. $y^2 + 14y + 24$
3. $m^2 + 15m + 26$ *(6-4)*
4. $a^2 - 6a + 8$
5. $a^2 - 12a + 27$
6. $m^2 - 18m + 56$ *(6-5)*
7. $m^2 + 16m + 64$
8. $x^2 - 12x + 36$
9. $x^2 + 20x + 100$ *(6-6)*
10. $x^2 + 8x - 48$
11. $y^2 + y - 42$
12. $b^2 + 24b - 81$ *(6-7)*

8 A Special Product

When you multiply two binomials, the product isn't always a trinomial. When your factors are the sum and difference of two monomials, look what happens to the product.

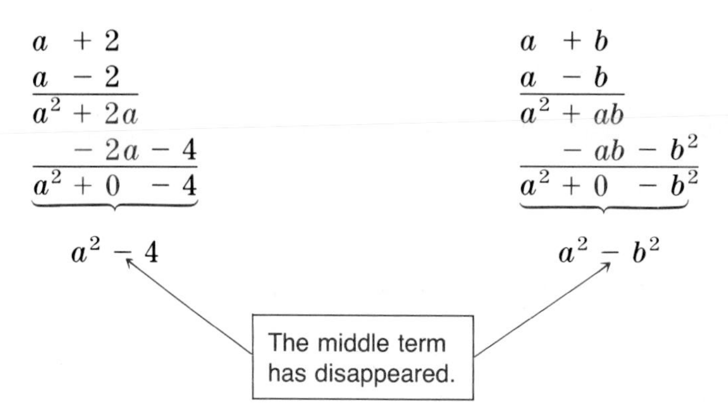

Let's look at what the second example above shows us.

$$(a + b)(a - b) = a^2 - b^2$$

(sum of two numbers) **times** (difference of two numbers) = (square of first number) **minus** (square of second number)

> *To multiply the sum of two monomials by their difference,*
> 1. *square the first term,*
> 2. *then subtract the square of the second term.*

EXAMPLE 1 $(x + 3)(x - 3) = x^2 - 9$

EXAMPLE 2 $(n + 4)(n - 4) = n^2 - 16$

EXAMPLE 3 $(m - n)(m + n) = m^2 - n^2$

EXAMPLE 4
$$\begin{aligned}(2a + b)(2a - b) &= (2a)^2 - b^2 \\ &= 4a^2 - b^2\end{aligned}$$

Classroom Practice

Complete.

1. $(x + 6)(x - 6) = (\underline{\ ?\ })^2 - (\underline{\ ?\ })^2$
2. $(a - 5)(a + 5) = (\underline{\ ?\ })^2 - (\underline{\ ?\ })^2$
3. $(n + 3)(n - 3) = (\underline{\ ?\ })^2 - (\underline{\ ?\ })^2$
4. $(x - 7)(x + 7) = (\underline{\ ?\ })^2 - (\underline{\ ?\ })^2$
5. $(y + 10)(y - 10) = (\underline{\ ?\ })^2 - (\underline{\ ?\ })^2$
6. $(b - 9)(b + 9) = (\underline{\ ?\ })^2 - (\underline{\ ?\ })^2$
7. $(m - 5)(m + 5) = (\underline{\ ?\ })^2 - (\underline{\ ?\ })^2$
8. $(x + 12)(x - 12) = (\underline{\ ?\ })^2 - (\underline{\ ?\ })^2$
9. $(4x + 3)(4x - 3) = (\underline{\ ?\ })^2 - (\underline{\ ?\ })^2$
10. $(2z - y)(2z + y) = (\underline{\ ?\ })^2 - (\underline{\ ?\ })^2$

Written Exercises

Multiply.

A

1. $(n - 7)(n + 7)$
2. $(x - 8)(x + 8)$
3. $(a + 10)(a - 10)$
4. $(y - 9)(y + 9)$
5. $(m + n)(m - n)$
6. $(a + c)(a - c)$
7. $(x + y)(x - y)$
8. $(1 - 3b)(1 + 3b)$
9. $(1 - 2x)(1 + 2x)$
10. $(m + 4n)(m - 4n)$
11. $(3x - 2)(3x + 2)$
12. $(6x + 1)(6x - 1)$
13. $(5y - 2)(5y + 2)$
14. $(7x - 1)(7x + 1)$
15. $(9m - n)(9m + n)$
16. $(4a - b)(4a + b)$
17. $(5x + 6)(5x - 6)$
18. $(10x - 8)(10x + 8)$

B

19. $(5x + 3y)(5x - 3y)$
20. $(2x - 6y)(2x + 6y)$
21. $(10x + 5y)(10x - 5y)$
22. $(x^2 - 1)(x^2 + 1)$
23. $(a^2 - 2)(a^2 + 2)$
24. $(b^3 - 5)(b^3 + 5)$

Find the product by using the rule $(a + b)(a - b) = a^2 - b^2$.

SAMPLE Find (23)(17).

$$\begin{aligned}(23)(17) &= (20 + 3)(20 - 3)\\ &= 400 - 9\\ &= 391\end{aligned}$$

C

25. (21)(19)
26. (51)(49)
27. (52)(48)
28. (18)(22)
29. (32)(28)
30. (38)(42)
31. (93)(87)
32. (83)(77)
33. (24)(36)
34. (94)(86)
35. (55)(65)
36. (95)(85)
37. (84)(76)
38. (59)(61)
39. (48)(52)
40. (73)(67)

9 Factoring the Difference of Squares

In the last section you saw that the following is true.

$$(a + b)(a - b) = a^2 - b^2$$

If this equation is turned around, we have the following.

$$a^2 - b^2 = (a + b)(a - b)$$

The last equation shows the pattern to use when you want to factor the difference of two squares.

EXAMPLE 1 Factor $c^2 - 25$.

Step 1: $c^2 - 25 = (c \quad)(c \quad)$

Step 2: $c^2 - 25 = (c + 5)(c - 5)$

Check: $(c + 5)(c - 5) = c^2 - 25$ ✓

EXAMPLE 2 Factor $a^2 - y^2$.

Step 1: $a^2 - y^2 = (a \quad)(a \quad)$

Step 2: $a^2 - y^2 = (a + y)(a - y)$

Check: $(a + y)(a - y) = a^2 - y^2$ ✓

EXAMPLE 3 Factor $x^2 - 4y^2$.

Step 1: $x^2 - 4y^2 = (x \quad)(x \quad)$

Step 2: $x^2 - 4y^2 = (x + 2y)(x - 2y)$

Check: $(x + 2y)(x - 2y) = x^2 - 4y^2$ ✓

Written Exercises

Which of the expressions are the difference of two squares?

A

1. $n^2 - 9$ **2.** $x^2 - 16$ **3.** $3x^2 - y^2$ **4.** $2n^2 - 9$

5. $25m^2 - n^2$ **6.** $1 - 10x^2$ **7.** $a^2 - 3b^2$ **8.** $36m^2 - 1$

Factor.

9. $x^2 - 16$ **10.** $x^2 - 64$ **11.** $n^2 - 9$ **12.** $y^2 - 81$

13. $a^2 - 9b^2$ **14.** $b^2 - 4a^2$ **15.** $x^2 - 4y^2$ **16.** $r^2 - 16t^2$

17. $b^2 - 64c^2$ 18. $9n^2 - 49$ 19. $4a^2 - 25b^2$ 20. $49m^2 - n^2$

21. $x^2 - 121$ 22. $y^2 - 144$ 23. $64x^2 - 9y^2$ 24. $81a^2 - 49b^2$

Factor. You may need to use the commutative property to rearrange the terms first.

SAMPLES

$$-x^2 + 1 = 1 - x^2 = (1 + x)(1 - x)$$

$$-1 + 4x^4 = 4x^4 - 1 = (2x^2 + 1)(2x^2 - 1)$$

B

25. $-x^2 + 4$ 26. $-1 + 4y^2$ 27. $-x^2 + 64$ 28. $-4x^2 + 9$

29. $-y^2 + x^2$ 30. $4y^4 - 1$ 31. $4x^6 - 9$ 32. $-1 + 49y^4$

33. $144 - 121x^2$ 34. $225a^2 - 1$ 35. $-4a^2 + 36b^2c^2$ 36. $-a^2 + 100b^2$

You may be able to factor more than once in the following exercises.

SAMPLE

$$x^4 - 16 = (x^2 + 4)(x^2 - 4) = (x^2 + 4)(x + 2)(x - 2)$$

C

37. $y^4 - 81$ 38. $16y^4 - 81z^4$ 39. $16a^4 - 1$ 40. $-1 + 81a^4$

41. $256 - x^4y^4$ 42. $a^8 - b^8$ 43. $x^8y^4 - 1$ 44. $81a^8b^4 - c^8$

45. Make a square a units on each side. Cut out a small corner square, b units on a side. Can you cut the shaded region into two pieces and rearrange the pieces to show the area is $(a + b)(a - b)$?

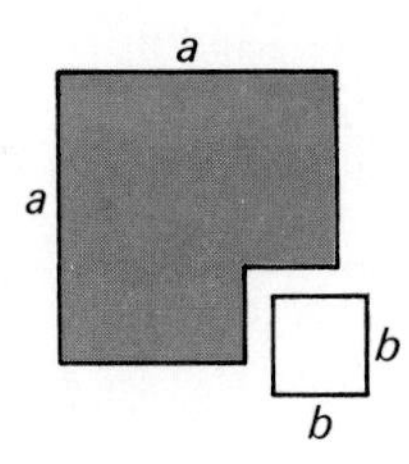

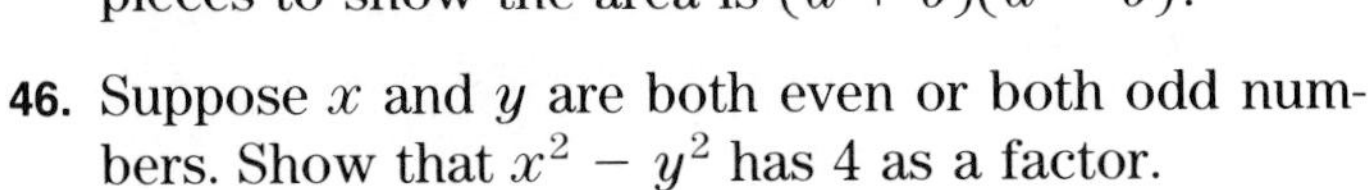

46. Suppose x and y are both even or both odd numbers. Show that $x^2 - y^2$ has 4 as a factor.

PUZZLE ◆ PROBLEMS

The museum director asked a carpenter to put glass panes in the showcase shown here.

Is it possible?

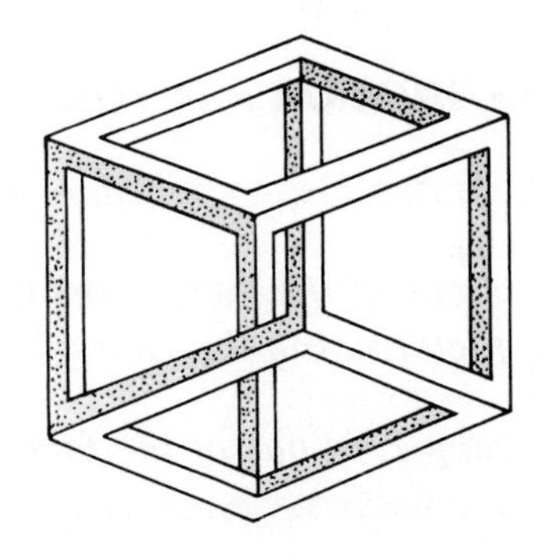

10 Many Types of Factoring

Sometimes you'll see an expression which can be factored several times. In this section you will have to look very closely at the terms of an expression to see what is the easiest way to factor completely.

EXAMPLE 1

Factor $3x^2 + 15x + 18$.

$$\begin{aligned} &3x^2 + 15x + 18 \\ =\ &3(x^2 + 5x + 6) \\ =\ &3(x + 2)(x + 3) \end{aligned}$$

Each term has a common monomial factor, 3.

Next the $x^2 + 5x + 6$ is factored to get $(x + 2)(x + 3)$.

EXAMPLE 2

Factor $x^3 - 9x$.

$$\begin{aligned} &x^3 - 9x \\ =\ &x(x^2 - 9) \\ =\ &x(x + 3)(x - 3) \end{aligned}$$

Each term has a common monomial factor, x.

Next the $x^2 - 9$ is factored to get $(x + 3)(x - 3)$.

EXAMPLE 3

Factor $-x^2 + 6x - 8$.

$$\begin{aligned} &-x^2 + 6x - 8 \\ =\ &-1(x^2 - 6x + 8) \\ =\ &-1(x - 4)(x - 2) \end{aligned}$$

Each term has a common monomial factor, -1.

Next the $x^2 - 6x + 8$ is factored to get $(x - 4)(x - 2)$.

Follow these steps in factoring an expression.

GUIDE FOR FACTORING AN EXPRESSION
1. *Do the terms have a common monomial factor? Factor.*
2. *Now is one of the factors the difference of two squares? If so, factor using the rule* $a^2 - b^2 = (a + b)(a - b)$.
3. *After Step 1, is one of the factors a trinomial? Factor.*

Classroom Practice

Factor.

1. $2a^2 + 14a + 20$
2. $2x^2 - 14x + 24$
3. $3y^2 + 3y - 18$
4. $4x^2 + 4x - 80$
5. $3x^2 - 27$
6. $9a^2 - 4$
7. $x^3 - 16x$
8. $y^3 - x^2y$
9. $2r^2s - 8st^2$
10. $8m^2 - 72n^2$
11. $5x^2 - 25x - 30$
12. $4y^2 - 40y + 100$

Written Exercises

Factor.

A

1. $3y^2 + 18y + 24$
2. $4x^2 + 24x - 64$
3. $2y^2 - 16y + 30$
4. $2b^2 - 4b - 48$
5. $5a^2 - 20a - 60$
6. $3x^2 + 30x + 27$
7. $4x^2 + 8x - 60$
8. $3y^2 - 12y - 63$
9. $2x^2 - 16x + 32$
10. $5a^2 - 5b^2$
11. $8x^2 - 32$
12. $x^2y - y^3$
13. $4x^2 - 36$
14. $3a^2 - 27b^2$
15. $6x^2 - 24y^2$

B

16. $-x^2 - 2xy - y^2$
17. $-x^2 + 4x - 3$
18. $2a^3 - 8a$
19. $-162 + 2x^2$
20. $-48 + 3x^2$
21. $4a - 4a^2 + 8$
22. $-3x^2 + 30x - 75$
23. $-200b^2 + 2a^2$
24. $-2x^2 + 24x - 72$
25. $-12z^2 + 27$
26. $-5x^2 + 30x + 35$
27. $25a^2 + 50a - 200$
28. $x^3 - 4x^2 + 4x$
29. $a^2b^2 + 12a^2b + 36a^2$
30. $x^3y - xy^3$

SELF-TEST

Multiply.

1. $(m - 13)(m + 13)$
2. $(3 + 4x)(3 - 4x)$
3. $(5a - 7)(5a + 7)$ *(6-8)*

Factor.

4. $a^2 - 64$
5. $m^2 - 36n^2$
6. $16x^2 - 9y^2$ *(6-9)*
7. $3x^2 + 15x + 18$
8. $2y^2 - 4y - 30$
9. $2m^2 - 8n^2$ *(6-10)*

11 More Difficult Factoring (Optional)

Trinomials like $2x^2 + 7x + 3$ can often be factored, even though factoring is more difficult. When the coefficient of the first term is not 1, you have to work by trial and error.

EXAMPLE 1 Factor $2x^2 + 7x + 3$.

Step 1: $2x^2 + 7x + 3 = (\quad)(\quad)$

Step 2: $2x^2 + 7x + 3 = (2x \quad)(x \quad)$

Factor the first term.

$2x$ and x

Step 3: $2x^2 + 7x + 3 = (2x\ ?\ )(x\ ?\ )$

What two numbers have a product of 3?

3 and 1 $\quad -3$ and -1

Choose the factors of 3 so that the middle term of the trinomial product will be $+7x$.

$(2x + 3)(x + 1) \qquad (2x + 1)(x + 3)$ ✓
$(2x - 3)(x - 1) \qquad (2x - 1)(x - 3)$

Step 4: $2x^2 + 7x + 3 = (2x + 1)(x + 3)$

Check: $(2x + 1)(x + 3) = 2x^2 + 7x + 3$ ✓

EXAMPLE 2 Factor $3a^2 - 5a - 2$.

Step 1: $3a^2 - 5a - 2 = (\quad)(\quad)$

Step 2: $3a^2 - 5a - 2 = (3a \quad)(a \quad)$

Step 3: $3a^2 - 5a - 2 = (3a\ ?\ )(a\ ?\ )$

What two numbers have a product of -2?

-1 and 2 $\quad$ 1 and -2

Choose the factors of -2 so that the middle term of the trinomial product will be $-5a$.

$(3a - 1)(a + 2) \qquad (3a + 2)(a - 1)$
$(3a + 1)(a - 2)$ ✓ $\qquad (3a - 2)(a + 1)$

Step 4: $3a^2 - 5a - 2 = (3a + 1)(a - 2)$

Check: $(3a + 1)(a - 2) = 3a^2 - 5a - 2$ ✓

Written Exercises

Factor.

A

1. $3x^2 + 4x + 1$
2. $2x^2 + 5x + 2$
3. $2a^2 + 9a + 9$
4. $3y^2 + 7y + 2$
5. $2b^2 + 5b + 3$
6. $2x^2 + 7x + 3$
7. $3a^2 - 8a + 5$
8. $5a^2 + 4a - 1$
9. $3x^2 - 10x + 8$
10. $5x^2 - 7x + 2$
11. $3y^2 - 5y - 2$
12. $3x^2 + x - 2$
13. $2x^2 + 5x - 3$
14. $7x^2 + 13x - 2$
15. $2x^2 - 3x - 9$
16. $4x^2 - 7x - 2$
17. $4a^2 + 5a - 9$
18. $6x^2 - 19x + 3$
19. $6b^2 - 5b - 25$
20. $8x^2 - 14x - 4$
21. $4a^2 - a - 5$
22. $6a^2 - 7a + 2$
23. $4x^2 - 8x - 5$
24. $6y^2 + y - 15$
25. $8r^2 - 6r - 9$
26. $6r^2 - 23r + 7$
27. $8x^2 + 2x - 3$
28. $8c^2 - 10c + 3$
29. $4c^2 + 6c + 2$
30. $6x^2 + 21x - 90$
31. $9y^2 + 15y + 6$
32. $6x^2 + 11x - 10$
33. $6x^2 + 15x + 6$
34. $10a^2 + 14a + 4$
35. $10y^2 - 29y + 10$
36. $12n^2 - 11n + 2$
37. $6x^2 + 11x - 121$
38. $10x^2 - 31x - 14$
39. $12x^2 + 7x - 12$
40. $6x^2 + 19x + 15$
41. $15n^2 + 17n - 4$
42. $20y^2 - 11y - 4$
43. $25x^2 - 5x - 2$
44. $30a^2 + 37a + 9$
45. $10c^2 - 23c + 12$
46. $9y^2 + 14y - 8$
47. $18z^2 + 27z + 10$
48. $36m^2 - 35m - 1$

B

49. $4x^2 + 12x + 9$
50. $9n^2 + 30n + 25$
51. $18b^2 + 13b - 21$
52. $16s^2 - 10s - 21$
53. $16g^2 - 48g + 27$
54. $30x^2 + x - 8$
55. $24x^2 + 2x - 15$
56. $27r^2 - 39r - 10$
57. $48d^2 - 7d - 3$
58. $24y^2 - 73y + 24$
59. $36x^2 + 39x + 10$
60. $36a^2 - 61a + 20$

PUZZLE ◆ PROBLEMS

Some statements that seem logical have built-in contradictions. Such statements are called *paradoxes*. Here's a famous example.

In a certain town the town's only barber states that he shaves all men in the town (and only those men) who don't shave themselves. Does the barber, who is a man, shave himself?

PROBLEM SOLVING STRATEGIES

LOOKING FOR PATTERNS

Some problems that appear to be difficult can be solved by looking for patterns in the solutions of simpler problems. Here is an example.

EXAMPLE Find the sum of the first 50 positive odd integers.

Look at some simpler problems first.

First 2: $1 + 3 = 4 = 2^2$
First 3: $1 + 3 + 5 = 9 = 3^2$
First 4: $1 + 3 + 5 + 7 = 16 = 4^2$
First 5: $1 + 3 + 5 + 7 + 9 = 25 = 5^2$

Use the pattern shown in the statements above to guess the number that is the sum of the first 50 positive odd integers.

First 50: $50^2 = 2500$

Can you give a formula for the sum of the first n positive odd integers? Test your formula by substituting 2, 3, 4, and 5 for n in turn.

When solving a problem directly by computation seems much too complicated and time-consuming, you can try solving the problem by finding patterns in a series of simpler problems. When you reach a conclusion based on specific cases, you are using *inductive reasoning.*

EXERCISES

1. When 4^2 is simplified, the units digit is 6 ($4^2 = 16$). Find the units digit of 7^{22} when it is simplified.

   a. Let's take some simpler problems. Simplify each of the following numbers and find its units digit:

   $$7^1, 7^2, 7^3, 7^4, 7^5, 7^6, 7^7, 7^8, 7^9$$

   b. Use the pattern in the answers in part (a) to find the units digit of 7^{22}.

2. Find the units digit of 3^{15}.

3. How many squares are there in this picture?

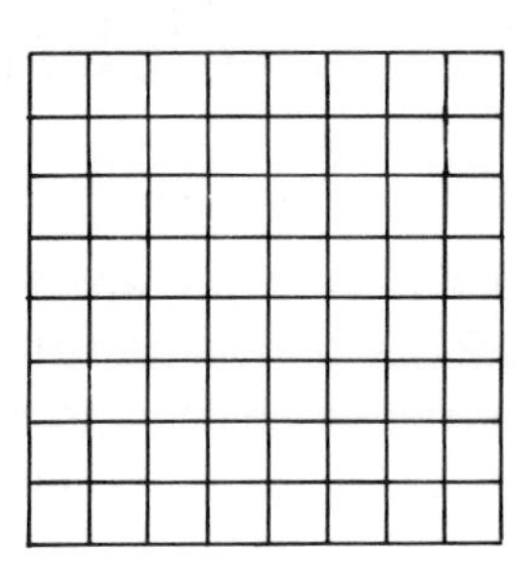

a. Look for a pattern in the problems below to predict the answer to the question.

Number of squares along each edge	Total number of squares
1	1 large 1
2	1 large, 4 small $1 + 4 = 5$
3	1 large, 4 medium, 9 small $1 + 4 + 9 = 14$
4	$1 + 4 + 9 +$ __?__ $=$ __?__
5	$1 + 4 + 9 +$ __?__ $+$ __?__ $=$ __?__

medium square

b. Answer the original question.

COMPUTER ACTIVITIES

THE GROWTH MODEL

Quantities that grow steadily at a certain percent rate are often studied by using a formula called the *growth model.* For example, the growth model is applied to savings accounts with a percent interest rate and to population growth with a percent growth rate.

The growth model uses the formula $A = A_0(1 + r)^t$. In this formula, A_0 is the original amount, r is the percent rate stated as a decimal, and A is the new amount after t years.

The program below will give the amount at the end of a time period. You supply the original amount, the percent rate, and the number of years. INPUT statements appear in lines 40, 70, and 100, to enable you to give these values.

```
10 REM**USING THE GROWTH MODEL
20 PRINT
30 PRINT "ORIGINAL AMOUNT";
40 INPUT A0
50 PRINT "WHAT IS THE GROWTH RATE"
60 PRINT "(EXAMPLE: FOR 5% TYPE 5)";
70 INPUT R
80 LET R1 = R/100
90 PRINT "HOW MANY YEARS";
100 INPUT T
110 REM**FORMULA FOR NEW AMOUNT
120 LET A = A0*(1 + R1) ∧ T
130 PRINT
140 PRINT "NEW AMOUNT: ";A
150 END
```

This symbol indicates that an expression is raised to a power.

EXERCISES

RUN the program to answer the questions. Round money amounts to the nearest dollar, and populations to the nearest million.

1. Find the final value of the investment.
   a. \$500 at 8% for 5 years
   b. \$10,000 at $9\frac{1}{2}$% for 8 years
   c. \$2500 at 12% for 10 years

2. In 1985 the United States population was about 238 million. Find the population in 2010 at the given population growth rate.
   a. 2.5% b. 2% c. 1.5% d. 1%

3. $100 is invested at the given interest rate. Use trial and error to find how long it takes, to the nearest year, for the amount of the investment to double (that is, to reach $200).
   a. 12% b. 10% c. 8% d. 6%

4. Depreciation means the loss in value of an object. For example, cars depreciate year by year. The growth model can be used to study depreciation if you input a negative value for the growth rate. (For depreciation at 5% a year, you would input -5.) The model will compute the final value at the end of a period at a given depreciation rate.
   a. A car that cost $8900 new loses 20% of its value each year. How much will the car be worth when it is 4 years old?
   b. A new combine costs a farmer $100,000. If it loses 15% of its value each year, how long will it take, to the nearest year, until it has lost half its value?

5. Use the program to complete the charts.
   a. Growth of $100 at 5% interest.

years	1	2	3	4	5	6	7	8	9	10
total value	?	?	?	?	?	?	?	?	?	?

   b. Growth of $100 at 10% interest.

years	1	2	3	4	5	6	7	8	9	10
total value	?	?	?	?	?	?	?	?	?	?

Make up your own charts for Exercises 6 and 7. Use the program to complete the charts.

6. Growth of $100 at 8%.

7. Growth of $100 at 13%.

8. Bankers use a rule of thumb called the Rule of 72 to estimate roughly the time required for an investment to double at a given interest rate. Examine the results of Exercise 3. Can you guess what the Rule of 72 claims?

SKILLS REVIEW

ADDING AND SUBTRACTING FRACTIONS

Simplify.

SAMPLE $\frac{4}{6} = \frac{2 \cdot \cancel{2}^{1}}{3 \cdot \cancel{2}_{1}} = \frac{2}{3}$ *or* $\frac{\cancel{4}^{2}}{\cancel{6}_{3}} = \frac{2}{3}$

Divide the numerator and the denominator by 2.

1. $\frac{2}{4}$
2. $\frac{3}{15}$
3. $\frac{4}{16}$
4. $\frac{6}{8}$
5. $\frac{5}{10}$
6. $\frac{3}{9}$
7. $\frac{8}{10}$
8. $\frac{12}{3}$
9. $\frac{16}{12}$
10. $\frac{18}{9}$

Add. Simplify if possible.

11. $\frac{1}{3} + \frac{1}{3}$
12. $\frac{3}{5} + \frac{1}{5}$
13. $\frac{2}{8} + \frac{4}{8}$
14. $\frac{2}{6} + \frac{3}{6}$
15. $\frac{3}{12} + \frac{5}{12}$
16. $\frac{2}{9} + \frac{4}{9}$
17. $\frac{3}{14} + \frac{5}{14}$
18. $\frac{3}{10} + \frac{4}{10}$
19. $\frac{2}{7} + \frac{3}{7}$
20. $\frac{1}{4} + \frac{2}{4}$
21. $\frac{5}{14} + \frac{2}{14}$
22. $\frac{1}{12} + \frac{3}{12}$
23. $\frac{3}{10} + \frac{2}{10}$
24. $\frac{3}{15} + \frac{7}{15}$
25. $\frac{4}{13} + \frac{6}{13}$
26. $\frac{3}{14} + \frac{6}{14}$

Subtract. Simplify if possible.

27. $\frac{4}{5} - \frac{2}{5}$
28. $\frac{7}{8} - \frac{3}{8}$
29. $\frac{7}{9} - \frac{2}{9}$
30. $\frac{9}{11} - \frac{5}{11}$
31. $\frac{3}{4} - \frac{1}{4}$
32. $\frac{9}{10} - \frac{3}{10}$
33. $\frac{11}{12} - \frac{4}{12}$
34. $\frac{9}{14} - \frac{7}{14}$
35. $\frac{8}{13} - \frac{5}{13}$
36. $\frac{7}{12} - \frac{4}{12}$
37. $\frac{5}{6} - \frac{2}{6}$
38. $\frac{6}{10} - \frac{3}{10}$
39. $\frac{5}{8} - \frac{3}{8}$
40. $\frac{10}{14} - \frac{5}{14}$
41. $\frac{5}{12} - \frac{1}{12}$
42. $\frac{11}{13} - \frac{7}{13}$

CHAPTER REVIEW

CHAPTER SUMMARY

1. A number which has no factors except itself and 1 is called a prime number.

2. The greatest common factor of two numbers is the greatest number which is a factor of both.

3. To factor a trinomial like $x^2 + ax + b$ or $x^2 - ax + b$, first write $(x + \quad)(x + \quad)$ or $(x - \quad)(x - \quad)$. Then look for two numbers whose product is b and whose sum is a or $-a$.

4. Recognizing patterns is useful in factoring.

$$a^2 + 2ab + b^2 = (a + b)^2 \qquad a^2 - 2ab + b^2 = (a - b)^2$$
$$a^2 - b^2 = (a + b)(a - b)$$

5. The guide on page 208 will help you factor polynomials completely.

REVIEW EXERCISES

Find the greatest common factor of each pair. ***(See pp. 186–187.)***

1. 6, 9 **2.** 16, 4 **3.** 15, 3 **4.** 24, 32

5. 56, 72 **6.** 30, 27 **7.** $3y$, $9y$ **8.** $16n$, $8n^3$

9. $4x^2$, x^3 **10.** $24x$, $6x^2$ **11.** $20x$, $4y$ **12.** $5x^3$, $30x^2$

Factor. Look for a common monomial factor. ***(See pp. 188–189.)***

13. $2n^2 + 4$ **14.** $3x^2 - 9$ **15.** $ab^2 + a^2b$

16. $7x^2 - 21x$ **17.** $2y^3 + 8y^2$ **18.** $3xy - 27$

19. $2x^2 - 8x^3$ **20.** $x^2 + 4x$ **21.** $36x + 6$

Factor. ***(See pp. 192–199.)***

22. $n^2 + 6n + 9$ **23.** $x^2 + 8x + 7$ **24.** $x^2 - 10x + 25$

25. $x^2 - 9x + 14$ **26.** $n^2 + 9n + 20$ **27.** $n^2 + 12n + 20$

28. $y^2 - 9y + 8$
29. $y^2 - 11y + 30$
30. $n^2 - 13n + 22$
31. $y^2 + 17y + 30$
32. $n^2 + 12n + 35$
33. $y^2 + 12y + 32$
34. $y^2 - 13y + 40$
35. $n^2 - 18n + 32$
36. $y^2 + 9y + 18$
37. $y^2 + 11y + 18$
38. $x^2 + 11x + 30$
39. $h^2 - 24h + 23$
40. $n^2 + 20n + 64$
41. $b^2 - 25b + 100$
42. $x^2 + 12x + 27$

Factor. Watch the signs. ***(See pp. 200–203.)***

43. $x^2 + 3x - 18$
44. $n^2 - 7n - 18$
45. $y^2 - 2y - 24$
46. $m^2 + 6m - 16$
47. $x^2 - 5x - 24$
48. $y^2 - 10y - 24$
49. $x^2 - x - 12$
50. $y^2 - 12y + 36$
51. $n^2 - n - 20$
52. $a^2 + 3a - 10$
53. $a^2 - 9a - 10$
54. $b^2 + 5b - 6$

Multiply. ***(See pp. 204–205.)***

55. $(x - 2)(x + 2)$
56. $(y + 4)(y - 4)$
57. $(n + 5)(n - 5)$
58. $(1 + 2a)(1 - 2a)$
59. $(2y + 3)(2y - 3)$
60. $(4n + 1)(4n - 1)$
61. $(x + y)(x - y)$
62. $(xy + 1)(xy - 1)$
63. $(4x - 3y)(4x + 3y)$

Factor. ***(See pp. 206–207.)***

64. $a^2 - 1$
65. $a^2 - b^2$
66. $x^2 - 4y^2$
67. $25 - n^2$
68. $n^2 - 25$
69. $4x^2 - y^2$
70. $9n^2 - 1$
71. $16a^2 - b^2$
72. $x^2y^2 - 1$

Factor. Look for a common monomial factor first. ***(See pp. 208–209.)***

73. $4x^2 - 16$
74. $3a^2 + 6a + 3$
75. $2n^2 + 6n - 20$
76. $2y^2 - 4y - 48$
77. $3x^2 - 3x - 36$
78. $x^3 - xy^2$
79. $2n^2 + 8n - 42$
80. $4a^2 - 20a - 24$
81. $ab^2 - 4ac^2$

(Optional) Factor. ***(See pp. 210–211.)***

82. $2x^2 + 3x + 1$
83. $3y^2 + 5y - 2$
84. $5z^2 - 3z - 2$
85. $4t^2 + 8t + 3$
86. $10m^2 - 13m - 3$
87. $10k^2 + 17k + 6$

CHAPTER TEST

1. Write 126 as a product of prime numbers. *(6-1)*

Find the greatest common factor of the pair of numbers.

2. 21, 33 3. 50, 63 4. 12, 24 5. 65, 26

Factor.

6. $5x^3 - 10x^2$ 7. $7a^3b - 21a^2b^2 + 7ab^3$ 8. $40y^3 - 5y^4 + 15xy$ *(6-2)*

Find the area of the shaded part. In Question 10, factor the expression and leave your answer in terms of π.

9.

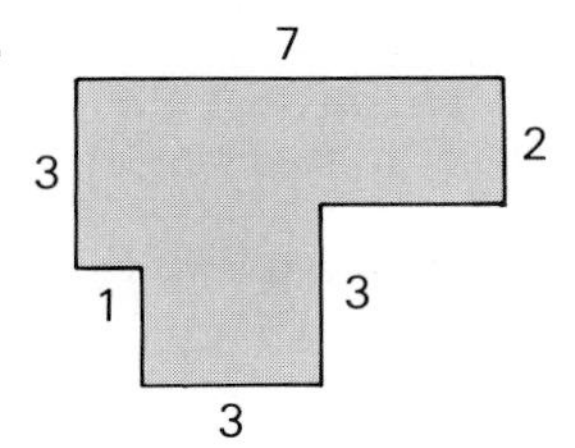

10.

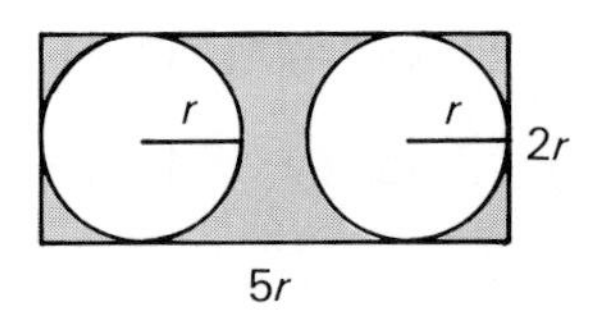

(6-3)

Factor. Check by multiplication.

11. $x^2 + 7x + 10$ 12. $n^2 + 13n + 12$ 13. $a^2 + 15a + 36$ *(6-4)*

14. $y^2 - 12y + 20$ 15. $x^2 - 52x + 100$ 16. $m^2 - 12m + 35$ *(6-5)*

17. $x^2 + 16x + 64$ 18. $r^2 - 8r + 16$ 19. $a^2 - 18ab + 81b^2$ *(6-6)*

20. $n^2 - n - 6$ 21. $y^2 + 5y - 24$ 22. $b^2 - 12b - 45$ *(6-7)*

Multiply.

23. $(2x + 1)(2x - 1)$ 24. $(a + 3b)(a - 3b)$ *(6-8)*

25. $(rs - 7)(rs + 7)$ 26. $(x^2 - 5y)(x^2 + 5y)$

Factor. If necessary, rearrange terms first.

27. $x^2 - 4y^2$ 28. $81 - 16a^2$ 29. $-100 + a^4$ *(6-9)*

30. $b^3 - 5b^2 + 4b$ 31. $98x^2 - 2y^2$ 32. $3m^2 + 12m - 36$ *(6-10)*

(Optional) Factor.

33. $3a^2 + 7a + 4$ 34. $8c^2 + 22c + 9$ 35. $7x^2 + 20x - 3$ *(6-11)*

UNIT C CUMULATIVE REVIEW

Add.

1. $\begin{array}{r} 3x + 2y \\ \underline{4x - 3y} \end{array}$

2. $\begin{array}{r} a^2 - 7a \\ \underline{-3a^2 + 2a} \end{array}$

3. $\begin{array}{r} -7m - 6n \\ \underline{5m - 4n} \end{array}$

4. $\begin{array}{r} 11c^2d - 5c \\ \underline{7c^2d - 5c} \end{array}$

Subtract.

5. $\begin{array}{r} 3x - 4 \\ \underline{-(2x - 1)} \end{array}$

6. $\begin{array}{r} 9z^2 + 2z \\ \underline{-(3z^2 - 2z)} \end{array}$

7. $\begin{array}{r} 2t - 5 \\ \underline{-(-t - 3)} \end{array}$

8. $\begin{array}{r} 4m^2n - 4n^2 \\ \underline{-(6m^2n + 3n^2)} \end{array}$

Multiply.

9. $-3y^3(y^4)$
10. $(x^3y^2)^2$
11. $xy(2x^2 - 3xy + y^2)$
12. $(a + 2)(a + 3)$
13. $(y - 4)(y - 1)$
14. $(2z - 3)(3z + 2)$
15. A rectangle measures $(2x - 3)$ centimeters by $(3x + 4)$ centimeters. Find its area.
16. Express $(3d + f)^2$ as a trinomial.

Divide.

17. $\dfrac{26m^3}{2m}$
18. $\dfrac{-60z^5}{4z^4}$
19. $\dfrac{-56y^8z^4}{-14y^4z^2}$
20. $\dfrac{8y^3 + 12y}{4y}$
21. $\dfrac{10x^5 + 15x^3 - 25x}{5x}$
22. $\dfrac{3a^3b^3 - 9a^2b^4 + 9ab^2}{3ab}$

Find the greatest common factor of each pair.

23. 27, 51
24. 42, 24
25. $3x^3$, $15x$
26. $42ab$, $48a^2$

Factor.

27. $7m^2 - 21$
28. $15y^3 - 9y^2 + 3y$
29. $x^2 + 10x + 16$
30. $y^2 + 7y + 12$
31. $t^2 - 18t + 45$
32. $z^2 - 12z + 35$
33. $1 + 6n + 9n^2$
34. $b^2 - 16b + 64$
35. $x^2 - 8x - 9$
36. $m^2 + 2m - 63$
37. $9 - t^2$
38. $16x^3 - 25xy^2$
39. $2x^2 - 16x - 40$
40. $5r^2 + 50r + 125$
41. $-5a^2 + 20$
42. $-s^2 + 32s - 60$
43. $-2y^2 - 40y - 150$
44. $mn^2 - m^3$

UNIT

D

Here's what you'll learn in this chapter:

To read and draw a bar graph.

To read and draw a broken line graph.

To read and graph ordered pairs of numbers on a coordinate plane.

To draw the graph of an equation when the graph is a line.

To find the slope of a line.

Conveyors such as this one at Lee Ranch Mine in New Mexico are used to move large quantities of ore and other materials. In order to work efficiently, a conveyor must be designed so that it has the correct slope.

Chapter 7

Graphs

1 Graphs You Often See

Many of the graphs in this chapter are like those found in newspapers and magazines. Here is a **pictograph** showing the number of people in the world who speak some of the major languages.

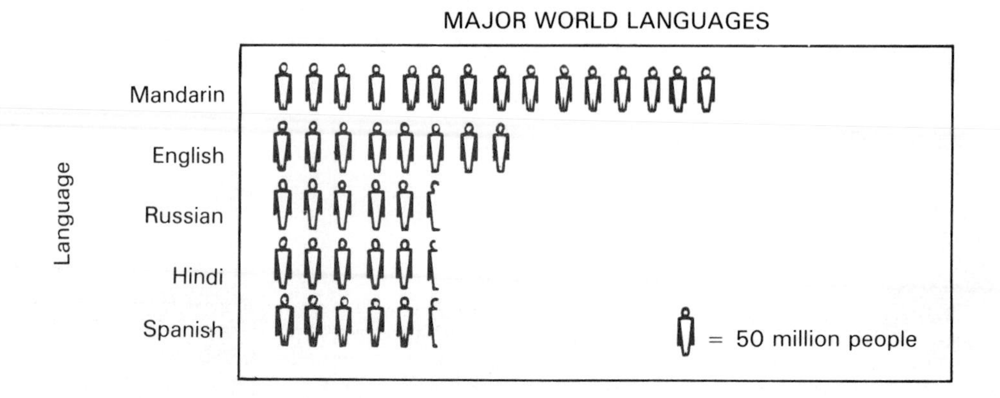

The key to the graph tells you that each symbol represents 50 million people. You can see that about 400 million people speak English.

Here's a type of graph that is closely related to the pictograph. It is called a **bar graph.** Each row of symbols in the graph above is replaced by a bar.

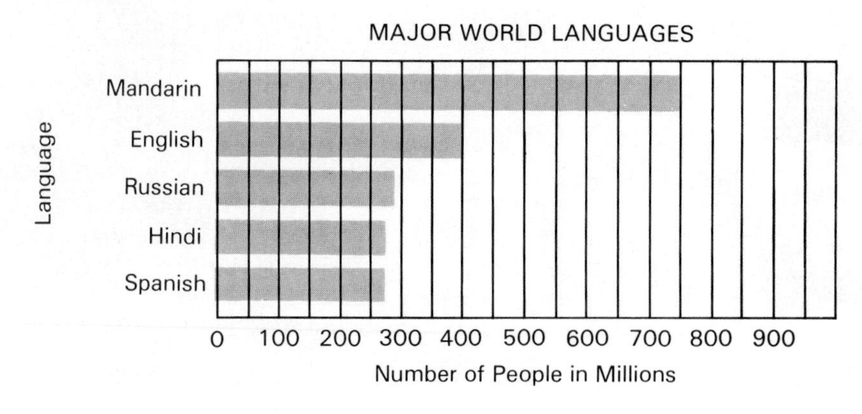

You can use the graph above to find how many more people speak Mandarin than Spanish.

Mandarin = about 750 million
Spanish = about 275 million
about 475 million more speak Mandarin

The bar graph in the advertisement below is misleading. Do you see why?

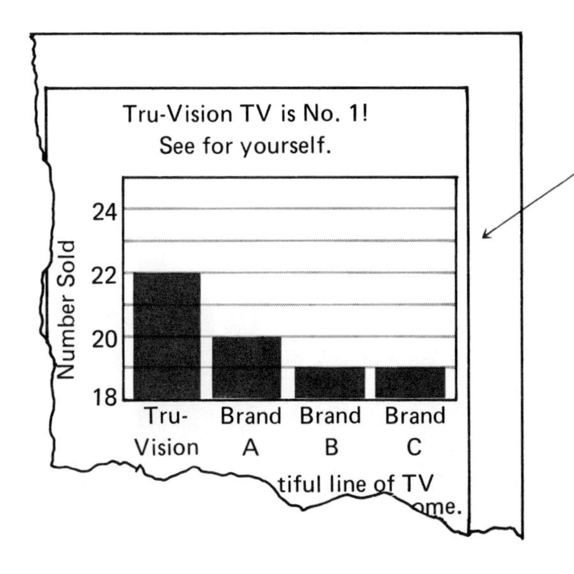

At first glance it looks as if Tru-Vision TV is twice as popular as Brand A. This is because the scale on the side of the graph does not start with 0. Instead, it starts with 18.

Perhaps the graph at the right is more honest.

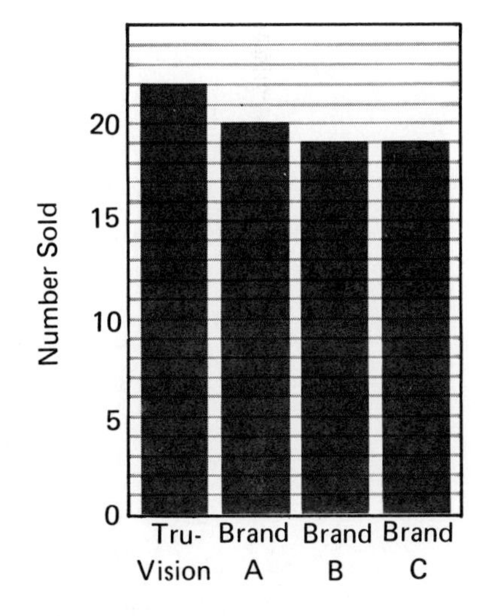

Classroom Practice

1. Study the pictograph on the opposite page. About how many people speak English? About how many speak Russian?
2. Study the bar graph on the opposite page. About how many people speak Hindi?
3. About how many more people speak English than Russian?
4. Study the Tru-Vision TV ad. How many of the TV sets sold were Tru-Vision?
5. How many of the TV sets sold were Brand A?
6. How many more Brand A TV sets were sold than Brand B sets?

Written Exercises

The graph below shows the increase in sales for a newspaper in one year. Use the graph for Exercises 1–3.

A

1. During which months did the sales decrease?
2. The bar for December is twice as long as the bar for January. Does this mean that sales doubled during the year?
3. What was the increase in sales from January to December?

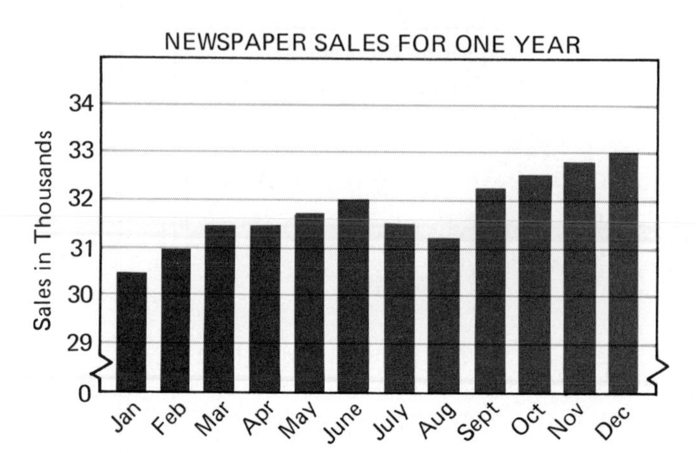

Study the pictograph below. Use it for Exercises 4 and 5.

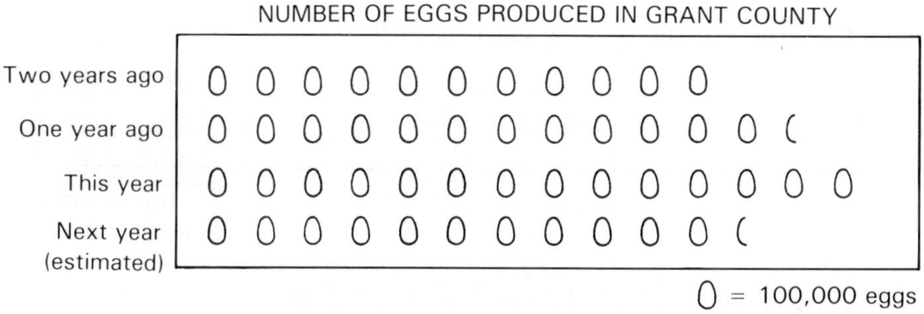

NUMBER OF EGGS PRODUCED IN GRANT COUNTY

Two years ago

One year ago

This year

Next year (estimated)

0 = 100,000 eggs

4. Complete the table.

Time	Number of Eggs
Two years ago	?
One year ago	?
This year	?
Next year (estimated)	?

5. Make a bar graph using the information in your table in Exercise 4.

In a survey, 100 people were asked which sport they participated in most frequently. The results of the survey are shown in the graph below.

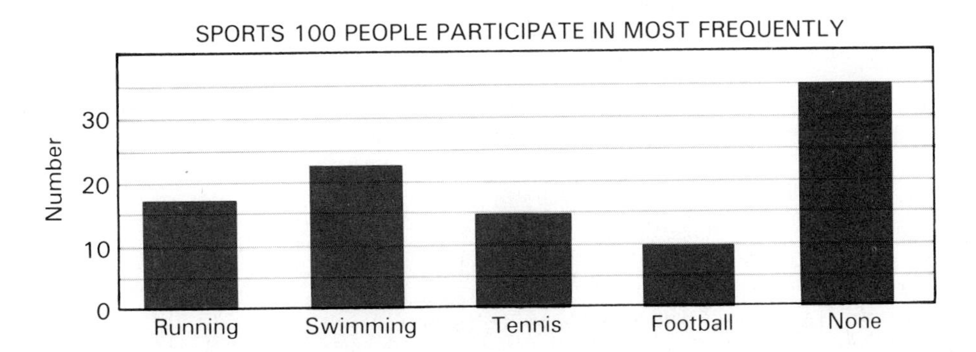

6. How many people do not participate in sports?
7. Which sport was most popular? How many participants?
8. Which sport was least popular? How many participants?
9. Make a bar graph, like the one above, that shows which sports members of your class participate in most frequently.

A consumer magazine used the following graph to show the annual cost of operating several dehumidifiers.

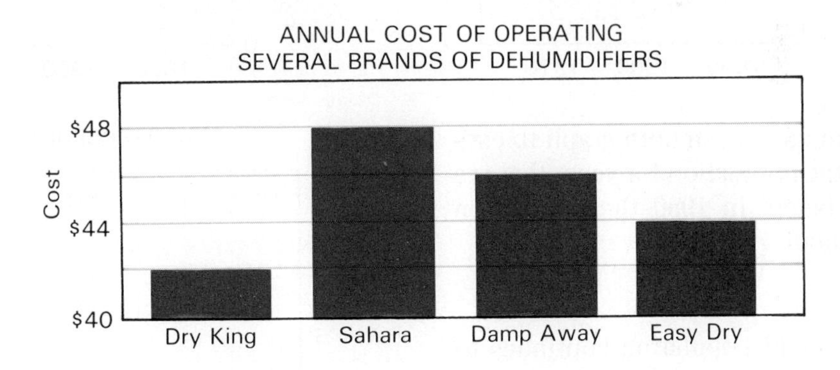

10. Dry King *looks* a lot less expensive. How much less expensive is the Dry King than the Easy Dry?
11. Redraw the graph. This time make the scale at the side of the graph start at $0 instead of $40. Your new graph will better show how the costs of operating the different brands of dehumidifiers compare.

2 Graphs with Lines and Curves

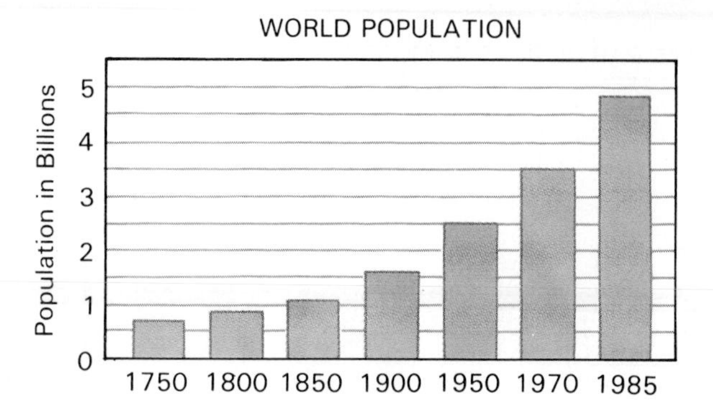

From 1750 to 1985 the world population increased by leaps and bounds! Read the figures on the bar graph at the left.

If you join the tops of the bars, you'll get a **broken line graph** like this.

WORLD POPULATION

If you "smooth out" the broken line graph, you'll get a curve.

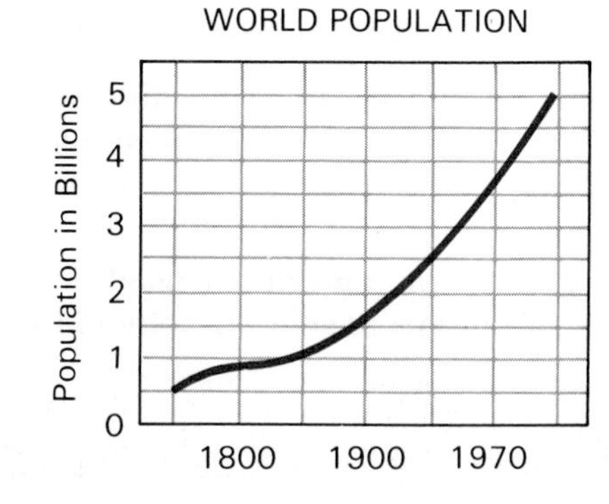

We can use the smooth graph to estimate the population for years that are not labeled. In 1980 the population was about $4\frac{1}{2}$ billion.

If the world population continues to grow at the *same* rate, we can extend the graph and answer questions about the next century.

Question: What will the world population be in the year 2000?

Answer: About 6 billion people.

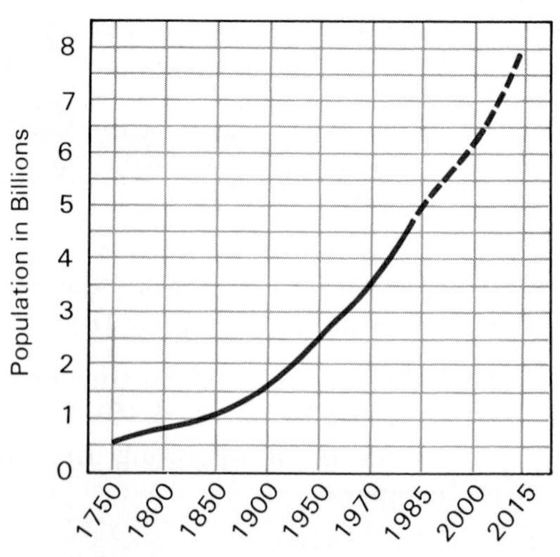

Classroom Practice

Refer to the graphs on the opposite page.

1. The world population was about $3\frac{1}{2}$ billion in the year __?__.
2. The world population in 1985 was about __?__ billion.
3. The world population in 2005 may be about __?__ billion.

Written Exercises

Refer to the graphs on the opposite page.

A

1. The world population in 1800 was about __?__ billion.
2. The estimated world population in 2015 is about __?__ billion.
3. The world population may reach $5\frac{1}{2}$ billion in the year __?__.

The graph at the right shows some population figures for Latin America.

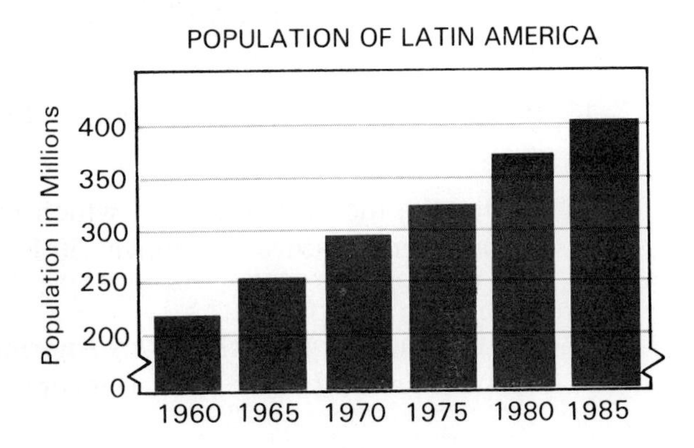

4. Use this bar graph to make a broken line graph of Latin American population.
5. Estimate the population of Latin America in 1990.
6. Estimate Latin America's population in the year 1995.

The amount a manufacturing plant produces changes according to economic conditions. At certain times of the year, a plant may produce 100% of all it can produce. At other times of the year the plant may be partly idle. Examine the graph at the right.

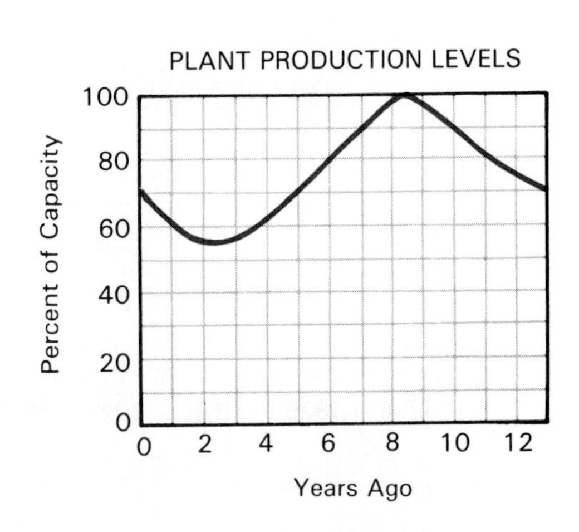

7. Five years ago, the plant was producing __?__ percent of what it was capable of producing.
8. About how many years ago was production the greatest? About how many years ago was production the least?

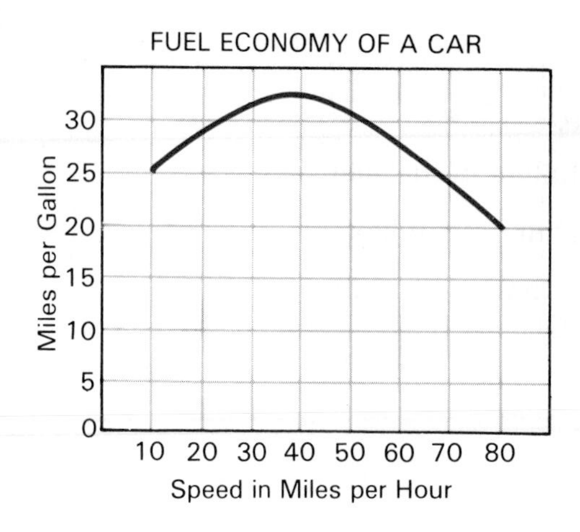

The graph at the left shows the fuel economy of a certain car. Complete the sentences.

9. The car gets the most miles per gallon at about __?__ miles per hour.
10. At 60 miles per hour the car gets about __?__ miles per gallon.
11. The car will get 25 miles per gallon at __?__ miles per hour and also at __?__ miles per hour.

Study the graph at the right.
Complete the following sentences.

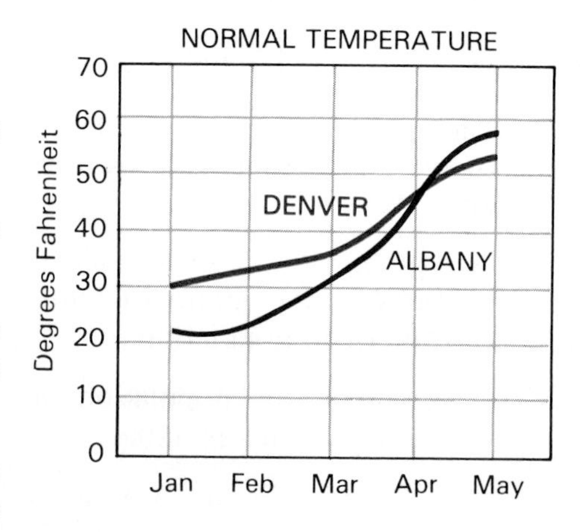

12. In Denver, the normal April temperature is __?__.
13. In Albany, the normal April temperature is __?__.
14. In Denver, the first month in which the normal temperature is above 50°F is __?__.
15. The first month in which Albany's normal temperature is higher than Denver's is __?__.

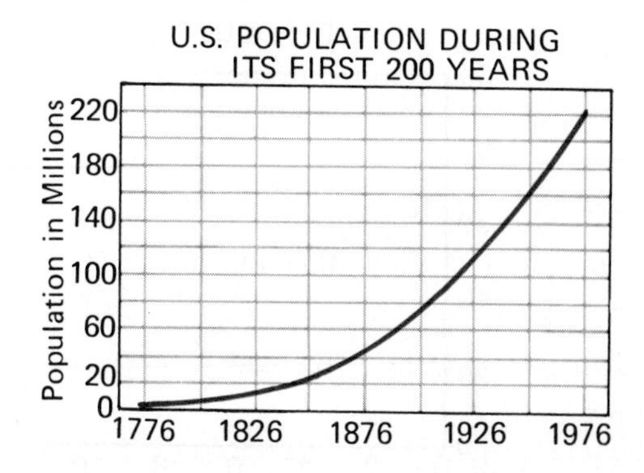

Study the graph at the left.

16. In 1926 the population of the United States was about __?__ million.
17. The population of the United States was 80 million in the year __?__.
18. About what was the increase in population from 1776 to 1976?
19. Estimate the United States' population in the year 2026.

Here are the graphs of three items in a family's yearly budget.

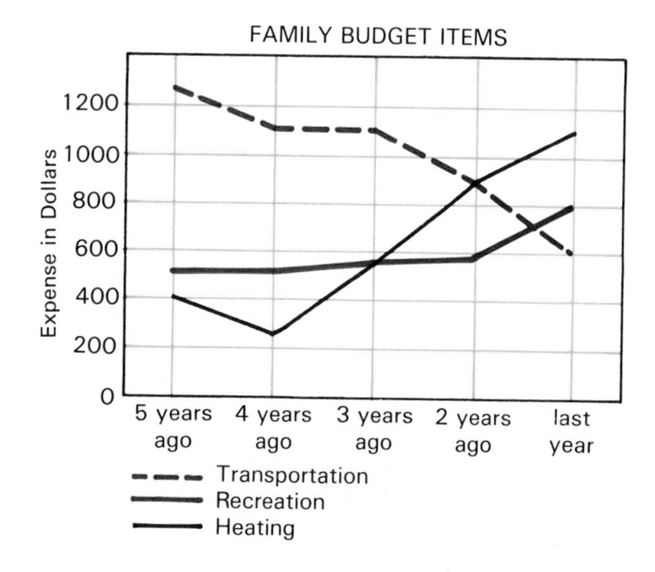

20. Which item cost the most five years ago? How much did it cost?

21. What was the total of these three budget items last year?

22. Which year had the lowest heating cost?

23. When did heating and transportation cost about the same?

24. Which item increased in cost the most?

Suppose gasoline costs $1.25 per gallon and you drive a car that averages 25 miles per gallon.

25. How much does it cost you for gasoline to go 1 mile?

26. Now complete the following table.

Miles traveled per year	10,000	20,000	30,000
Gasoline cost in dollars per year	$500	?	?

27. Use the table in Exercise 26 to draw a graph. Your points should lie along a straight line. Do they?

28. Estimate from your graph the gasoline cost to drive 23,000 miles.

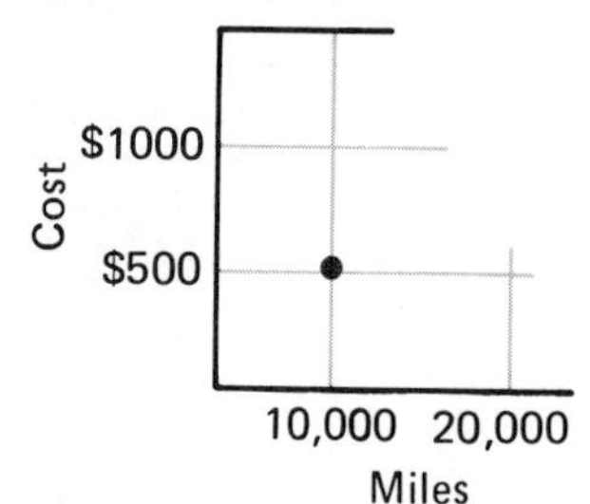

Suppose gasoline costs $1.20 per gallon, but you drive a car that averages 30 miles per gallon.

29. How much does it cost for gasoline to drive 1 mile? How much does it cost for gasoline to drive 2 miles?

30. Make a table and draw a graph like the one in Exercise 27.

SELF-TEST

Vocabulary

pictograph (p. 224) broken line graph (p. 228)
bar graph (p. 224)

The graph below shows the number of sales of four brands of radios in a department store during one month.

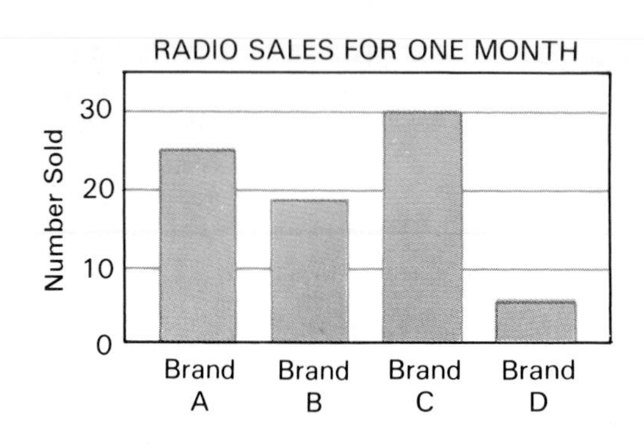

1. Which brand of radio had the most sales during the month? *(7-1)*
2. How many Brand B radios were sold?
3. How many more Brand A radios were sold than Brand D radios?

The graph below shows the temperature in Southville one day, during the hours between midnight and 10 P.M.

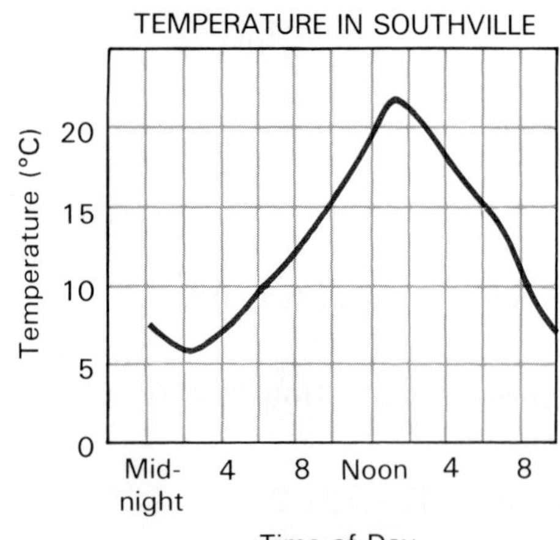

4. The highest temperature of the day occurred at what time? *(7-2)*
5. The lowest temperature of the day occurred at what time?
6. Estimate the temperature at 10 P.M.

3 Points on a Graph

A temperature-time graph is shown below. Notice that the points (2, 5) and (5, 2) are not the same. You can see that the order in which you write the numbers in the pair is very important!

The temperature at 2 o'clock was 5°. We label the point (2, 5) and call it **point (2, 5).**

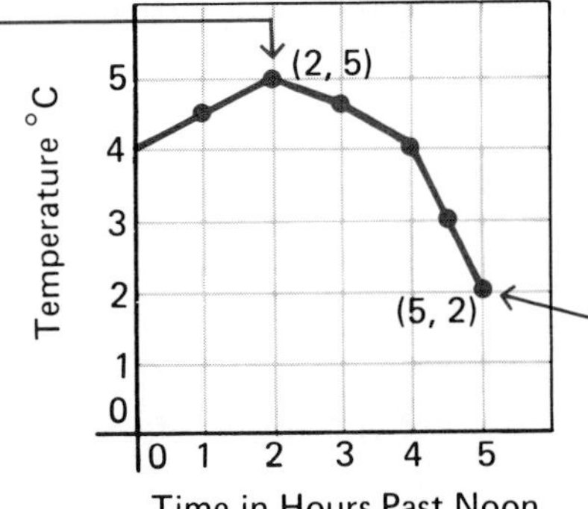

The temperature at 5 o'clock was 2°. We label the point (5, 2) and call it **point (5, 2).**

If the temperature drops below zero, we shall need to extend our graph as shown at the right.

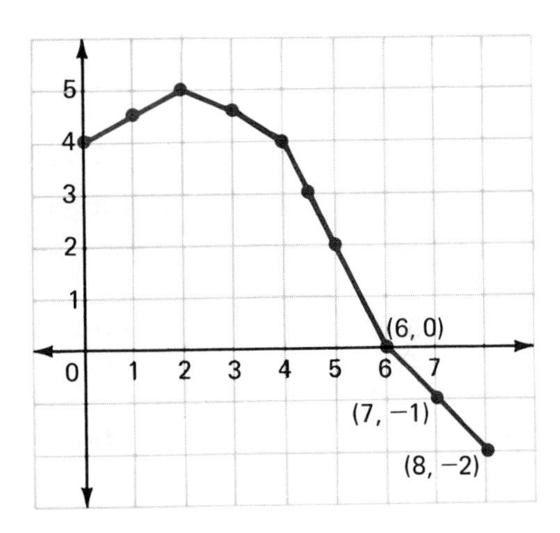

In drawing or reading graphs, you need to know a few terms.

The number lines are called the **horizontal axis** and the **vertical axis.**

The horizontal axis is the ***x*-axis** and the vertical axis is the ***y*-axis.**

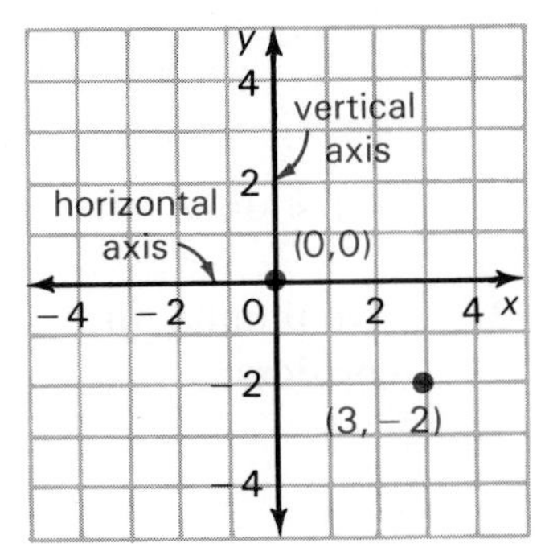

The point (0, 0) is called the **origin.**

The numbers 3 and −2 are the **coordinates** of point (3, −2).

When you want to **plot** (or **locate**) points like $(3, -2)$ and $(-4, 1)$, you follow these steps.

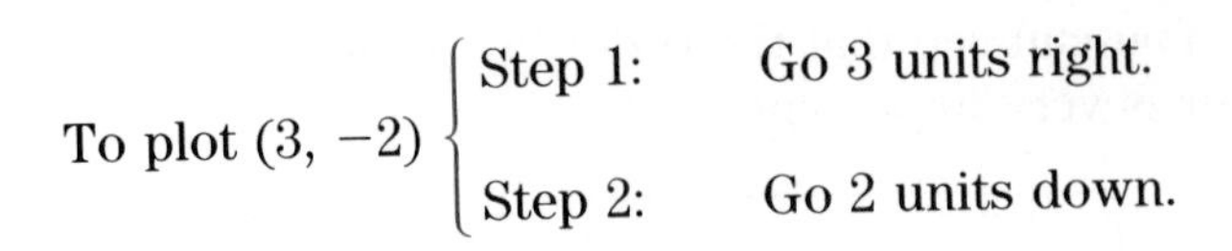

To plot $(3, -2)$
- Step 1: Go 3 units right.
- Step 2: Go 2 units down.

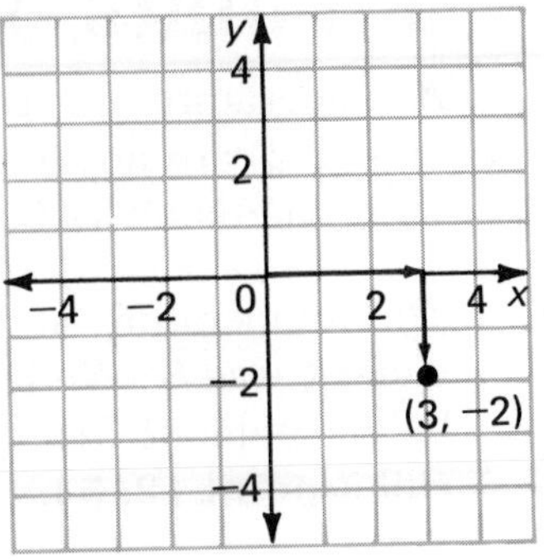

To plot $(-4, 1)$
- Step 1: Go 4 units left.
- Step 2: Go 1 unit up.

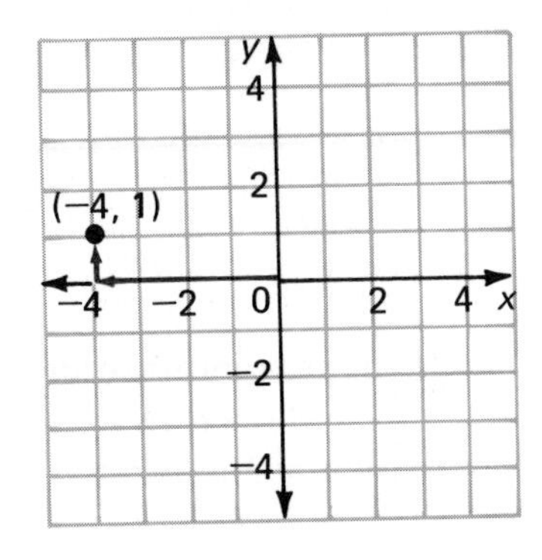

If a point has a 0 coordinate, it lies on one of the axes.

To plot $(0, -3)$
- Step 1: Go 0 units right or left.
- Step 2: Go 3 units down.

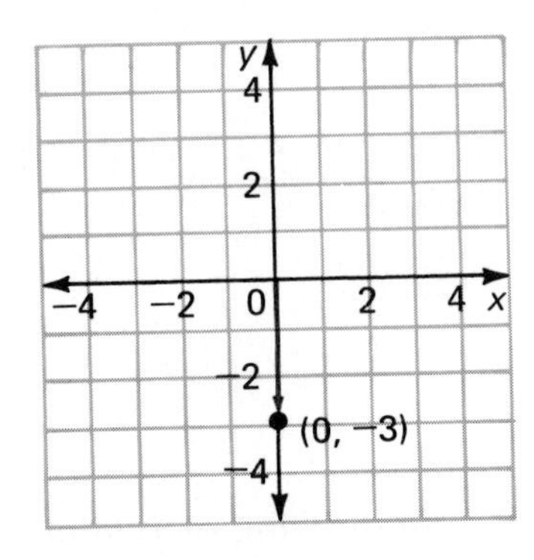

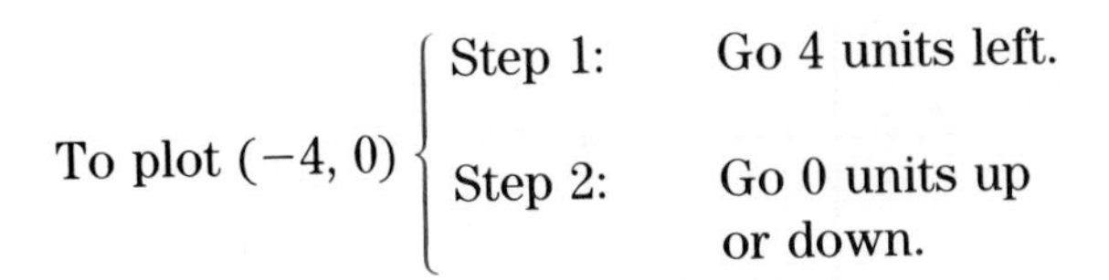

To plot $(-4, 0)$
- Step 1: Go 4 units left.
- Step 2: Go 0 units up or down.

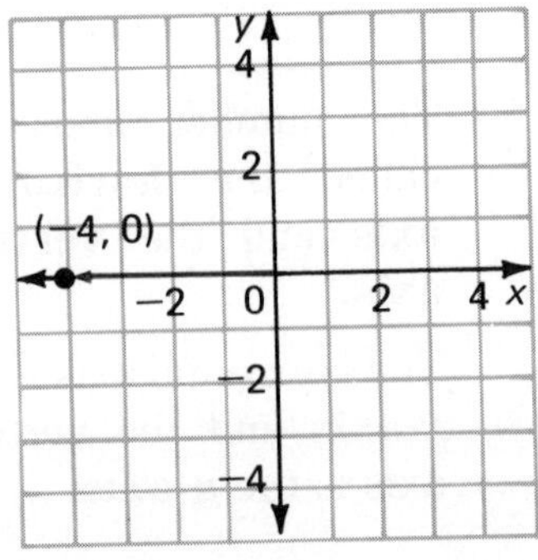

Classroom Practice

Give the coordinates of each point.

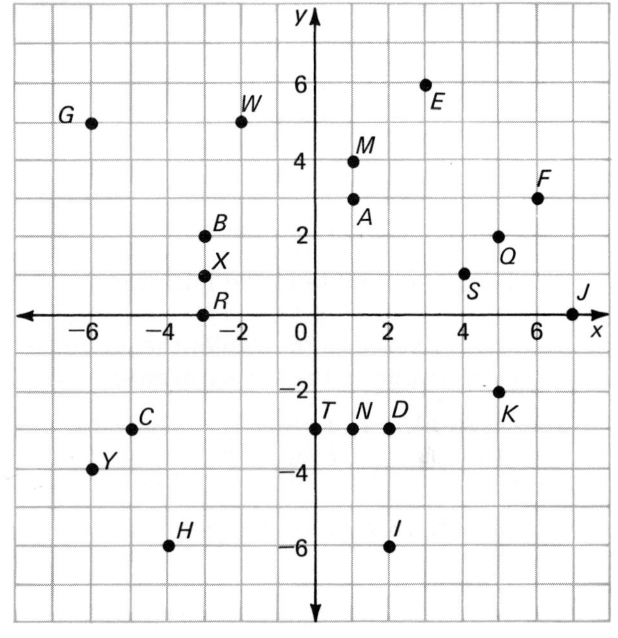

SAMPLE Point A has coordinates (1, 3).

1. B 2. C 3. D
4. E 5. F 6. G
7. H 8. I 9. J

Name the point by its letter.

10. (4, 1) 11. (1, 4)
12. (5, −2) 13. (−2, 5)
14. (−3, 0) 15. (0, −3)
16. (−3, 1) 17. (1, −3)
18. (−6, −4) 19. (5, 2)

Written Exercises

Give the coordinates of each point.

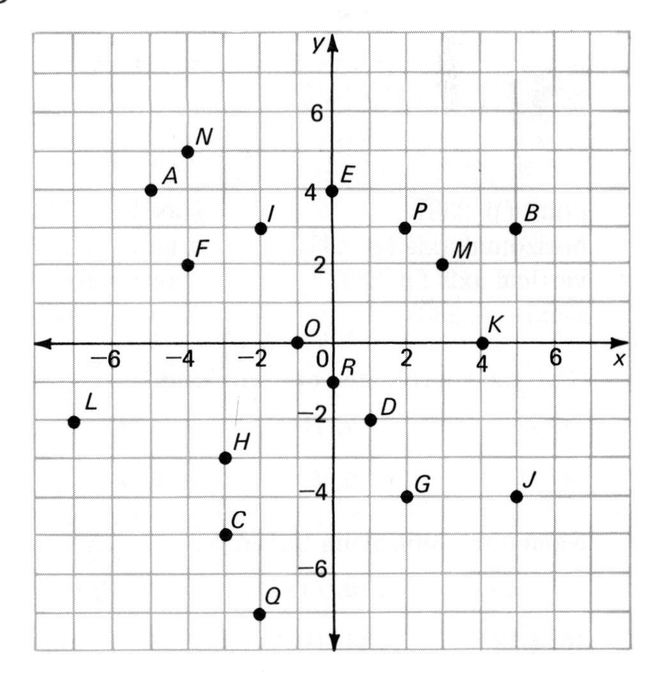

A

1. A 2. B 3. C
4. D 5. E 6. F
7. G 8. H 9. I

Name the point by its letter.

10. (2, 3) 11. (3, 2)
12. (−1, 0) 13. (0, −1)
14. (5, −4) 15. (−4, 5)
16. (−7, −2) 17. (−2, −7)
18. (4, 0)

Refer to the graph at the right.

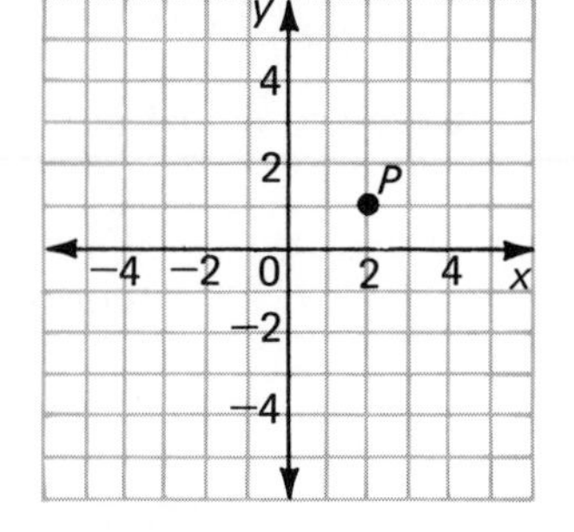

19. How many units is P from the vertical axis?

20. How many units is P from the horizontal axis?

Using the graph at the right, give the coordinates of the following points.

21. The point 3 units above P

22. The point 3 units below P

23. The point 3 units to the right of P

24. The point 3 units to the left of P

In each of Exercises 25–28 plot the points in order. Connect each point to the next one in the list by drawing a line.

25. (4, 2); (4, −4); (−5, −4); (−5, 2); (4, 2) — You should get a rectangle.

26. (5, 0); (0, 5); (−5, 0); (0, −5); (5, 0) — You should get a square.

27. (3, −2); (6, 8); (−4, 6); (−7, −4); (3, −2) — This figure is called a parallelogram.

B 28. (−3, 0); (0, 3); (3, 3); (4, 0); (6, 0); (7, −1); (7, −2); (6, −2); (5, −3); (4, −3); (3, −2); (−1, −2); (−2, −3); (−3, −3); (−4, −2); (−5, −2); (−5, −1); (−4, 0); (−3, 0)

SELF-TEST

Vocabulary

point (p. 233)
horizontal axis (p. 233)
vertical axis (p. 233)
x-axis (p. 233)
y-axis (p. 233)
origin (p. 233)
coordinates (p. 233)
plotting a point (p. 234)

Give the coordinates of each point. *(7-3)*

1. A 2. B 3. C

4. D 5. E 6. F

Name the point by its letter.

7. (2, 1) 8. (1, 2) 9. (3, −2)

10. (−2, 3) 11. (0, 2) 12. (2, 0)

4 Solution Pairs

Some equations have only one variable. Equations of this type have solutions that are single numbers.

$$5x = 10$$
$$x = 2$$

Some equations have two variables. Their solutions are pairs of numbers. In the equation below, $x = 5$ and $y = 1$ is a solution because $5 - 1 = 4$.

$$x - y = 4$$

The solution $x = 5$ and $y = 1$ of the equation above is sometimes called a **solution pair.** It is written as an **ordered pair** of numbers, (5, 1).

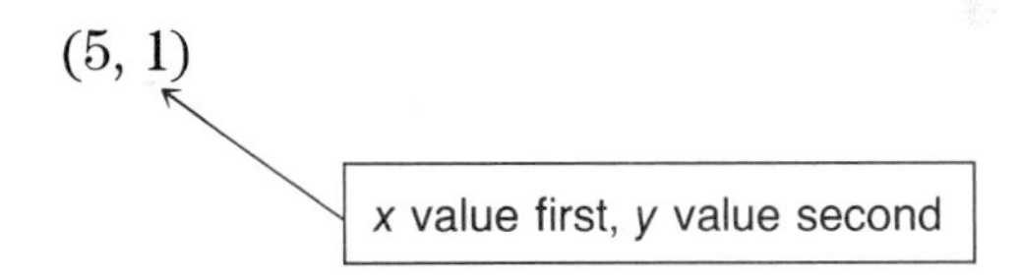

Here are a few more solution pairs of the equation $x - y = 4$.

(6, 2) is a solution because $6 - 2 = 4$
(7, 3) is a solution because $7 - 3 = 4$
(3, −1) is a solution because $3 - (-1) = 4$

(9, 2) is NOT a solution because $9 - 2 \neq 4$

EXAMPLE 1

Find four solution pairs for the equation $x + y = 6$.

$x + y = 6$	
If x is	then y is
0	6
1	5
2	4
4	2

→ (0, 6) is a solution.
→ (1, 5) is a solution.
→ (2, 4) is a solution.
→ (4, 2) is a solution.

(2, 4) and (4, 2) are different solutions.

Some equations are a little more difficult to solve in your head. In that case, just solve for y before you begin to look for solution pairs.

EXAMPLE 2

Find four solution pairs for $y - 2x = 1$.

Step 1: Solve the equation for y.

$$y - 2x = 1$$
$$y - 2x + 2x = 1 + 2x$$
$$y = 1 + 2x$$

Step 2: Replace x with a value. Find the value of y.

$y = 1 + 2x$		
If x is	then y is	
0	1	→ (0, 1)
1	3	→ (1, 3)
−1	−1	→ (−1, −1)
2	5	→ (2, 5)

EXAMPLE 3

Find four solution pairs for $2x + y = 10$.

Step 1: Solve the equation for y.

$$2x + y = 10$$
$$2x + y - 2x = 10 - 2x$$
$$y = 10 - 2x$$

Step 2: Replace x with a value. Find the value of y.

$y = 10 - 2x$		
x	y	
0	10	→ (0, 10)
1	8	→ (1, 8)
−1	12	→ (−1, 12)
2	6	→ (2, 6)

Classroom Practice

Which choices are solution pairs for the equation?

1. $x + y = 4$ a. $(3, 1)$ b. $(6, -2)$ c. $(-2, 2)$
2. $y = 3x + 1$ a. $(2, 7)$ b. $(0, 1)$ c. $(1, 0)$
3. $xy = 6$ a. $(1, 5)$ b. $(-3, -2)$ c. $(6, 0)$
4. $y = x - 2$ a. $(2, 4)$ b. $(-2, 4)$ c. $(1, -1)$
5. $y = 5x$ a. $(0, 5)$ b. $(1, 5)$ c. $(5, 1)$
6. $y - x = 1$ a. $(2, 1)$ b. $(1, 0)$ c. $(1, 2)$

Find four solution pairs for the equation.

7. $y = x + 2$
8. $y = 3x - 1$
9. $2x + y = 8$
10. $y - 2x = 3$
11. $x - y = 2$
12. $y = x + 3$
13. $y = 4 - x$
14. $y = -x + 1$

Written Exercises

Which choices are solution pairs for the equation?

A

1. $x + y = 8$ a. $(5, 3)$ b. $(-5, -3)$ c. $(-4, 12)$
2. $x - y = 8$ a. $(10, 2)$ b. $(-4, 4)$ c. $(-6, -2)$
3. $y = 3x - 2$ a. $(0, 2)$ b. $(-2, 0)$ c. $(2, 4)$
4. $y = 5 - 2x$ a. $(3, 1)$ b. $(1, 3)$ c. $(-1, 7)$
5. $y - 3x = 10$ a. $(0, 4)$ b. $(-1, 7)$ c. $(2, 10)$
6. $2x - y = 8$ a. $(1, -6)$ b. $(0, -8)$ c. $(-8, 8)$

Complete.

7. $y = 2x$

x	y
3	?
5	?
7	?
8	?

8. $y = -3x$

x	y
2	?
−4	?
3	?
−3	?

9. $y = x + 3$

x	y
0	?
2	?
−4	?
5	?

10. $y = 2x - 1$

x	y
0	?
−1	?
1	?
2	?

Complete.

11. $y = -x + 2$

x	y
0	?
1	?
2	?
−2	?

12. $y = 9 - 2x$

x	y
0	?
2	?
3	?
−3	?

13. $s = 3t + 6$

t	s
0	?
−1	?
1	?
−2	?

14. $b = 4a - 1$

a	b
0	?
1	?
−1	?
2	?

Solve the equation for *y*.

15. $x + y = 9$
16. $y + 2x = 4$
17. $y - x = 6$
18. $3x + y = 7$
19. $4x + y = 5$
20. $y - 3x = 7$
21. $y + 4x = 9$
22. $y + 5 = 4x$
23. $7 + y = 9x$
24. $x - y = 9$
25. $2x - y = 5$
26. $3x - y = 6$

Find three solution pairs for the equation.

27. $y = x - 1$
28. $y = 2x + 5$
29. $y + x = 0$
30. $y + x = 7$
31. $y - x = 6$
32. $y - x = 8$
33. $y - 2x = 4$
34. $y - 3x = 2$
35. $y - 4x = 7$
36. $y + 5x = 0$
37. $2x + y = 2$
38. $y - 3x = 9$

Can you guess an equation to go with the table?

B 39.

x	y
1	4
2	8
3	12
4	16

40.

x	y
1	2
2	3
4	5
5	6

41.

x	y
1	5
2	9
3	13
4	17

42.

x	y
0	−1
1	1
2	3
3	5

Which choices are solution pairs for the inequality?

43. $y > 2x$ **a.** (2, 8) **b.** (3, 8) **c.** (4, 8)
44. $y < 3x$ **a.** (2, 5) **b.** (2, 6) **c.** (2, 7)
45. $y < x - 2$ **a.** (5, 1) **b.** (5, 2) **c.** (5, 3)
46. $y > 2x - 3$ **a.** (0, 1) **b.** (0, 0) **c.** (15, 8)

5 Graphs of Equations

The graph at the right shows some of the solutions of the equation $x + y = 3$. Since each solution is an ordered pair of numbers, we can picture each solution as a point.

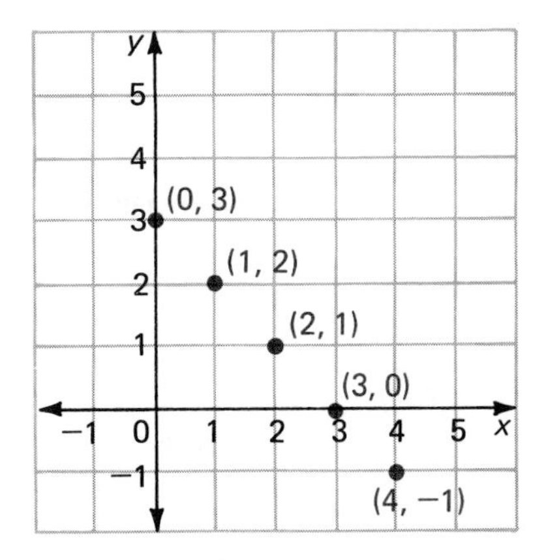

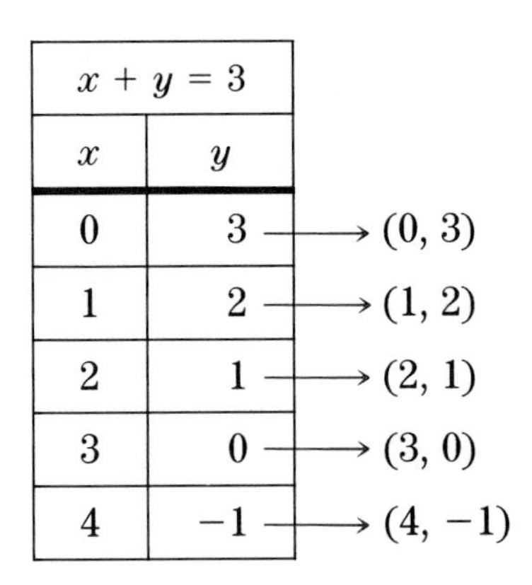

$x + y = 3$		
x	y	
0	3	→ (0, 3)
1	2	→ (1, 2)
2	1	→ (2, 1)
3	0	→ (3, 0)
4	−1	→ (4, −1)

This graph shows some more solutions of the equation $x + y = 3$.

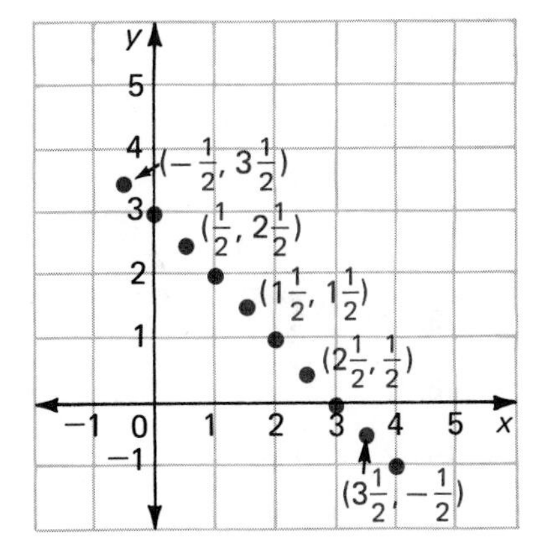

If we could graph *all* the solutions of the equation $x + y = 3$, we would get the line shown. This line is called **the graph of the equation $x + y = 3$.**

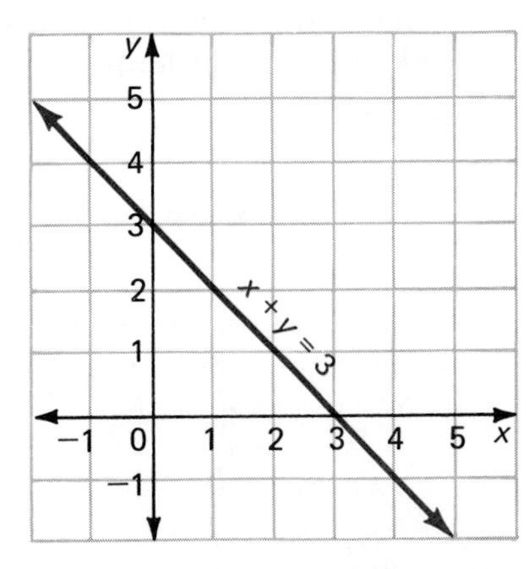

In this section the equations will have graphs that are straight lines. You will find later, however, that not all equations have straight line graphs.

Classroom Practice

Draw the graph of the equation.

1. $y = 2x$
2. $y = x + 2$
3. $y = -x + 3$
4. $y = -2x + 1$
5. $x + y = 3$
6. $y - x = 1$
7. $y - 2x = 2$
8. $2x + y = 7$
9. $x + y = -5$
10. $y - 3x = -2$
11. $3x + y = 4$
12. $y - 4x = -3$

Written Exercises

Draw the graph of the equation.

A

1. $y = 3x$
2. $y = 4x$
3. $y = x + 3$
4. $y = 3x + 3$
5. $y = 4x - 2$
6. $y = -x + 1$
7. $y = 2x + 3$
8. $y = 3x - 4$
9. $y = -2x + 3$
10. $x + y = 4$
11. $x + y = 5$
12. $y - x = 5$
13. $2x + y = 6$
14. $y - 5x = 0$
15. $y - 2x = 7$
16. $x - y = 9$
17. $3x - y = 5$
18. $2x - y = 5$
19. $2x + y = -3$
20. $y - 4x = -1$
21. $3x + y = 0$

Can you guess the equation of the graph? Find some solution pairs and make a chart for help.

B

22.

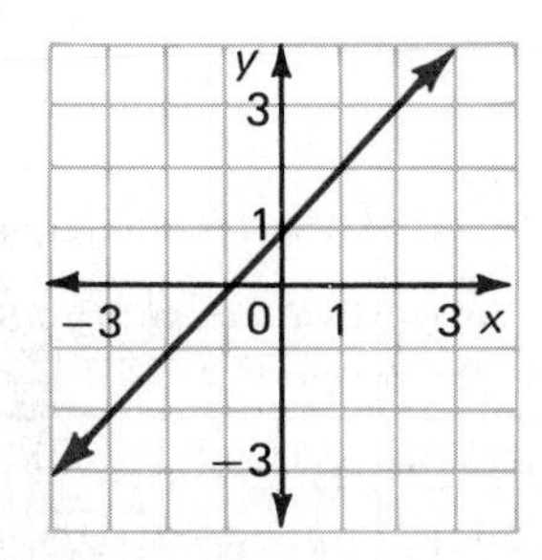

23.

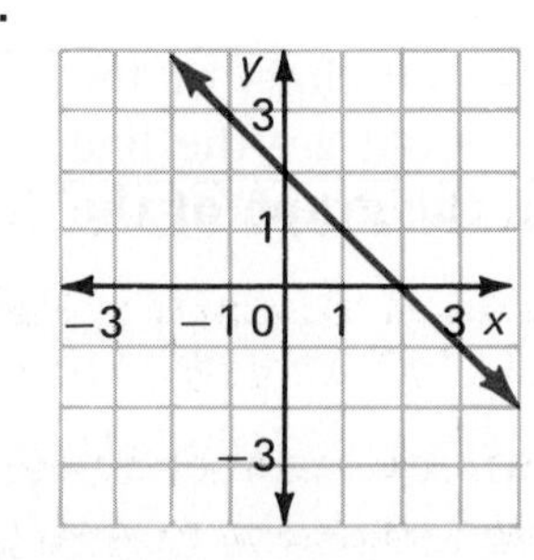

24.

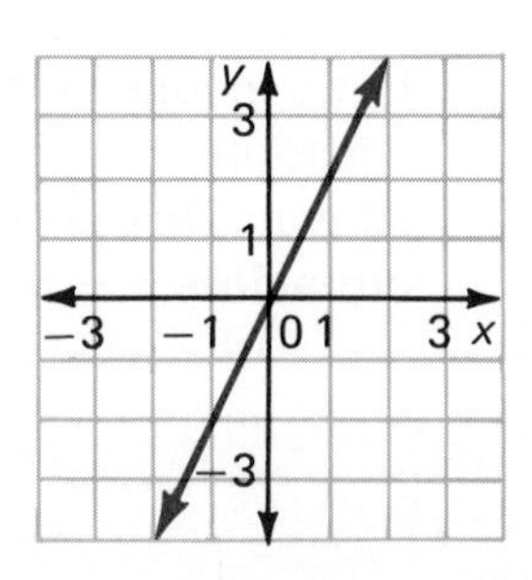

6 Slope of a Line

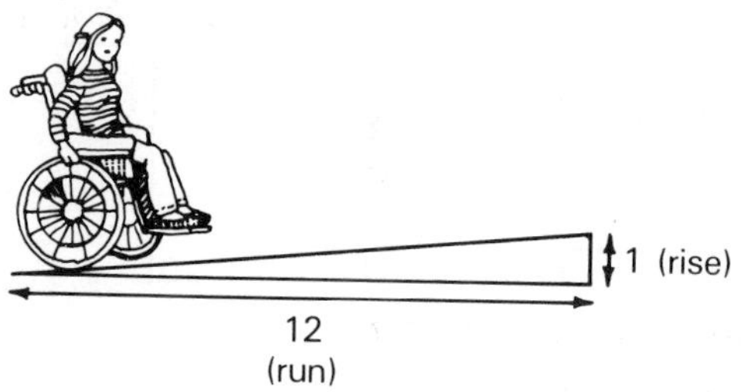

Architects know that wheelchair ramps should not be too steep. They also know that if the ramp is not steep enough, then the ramp might be much too long. The slope of a ramp is very important.

$$\text{slope} = \frac{\text{rise of ramp}}{\text{run of ramp}} = \frac{1}{12}$$

The slope of a line is measured in the same way. It is given as the ratio of the rise to the run. Study the three graphs below. Notice how you can determine the slope of each line. Notice also that you can determine the slope by examining the equation of the graph.

Line 1

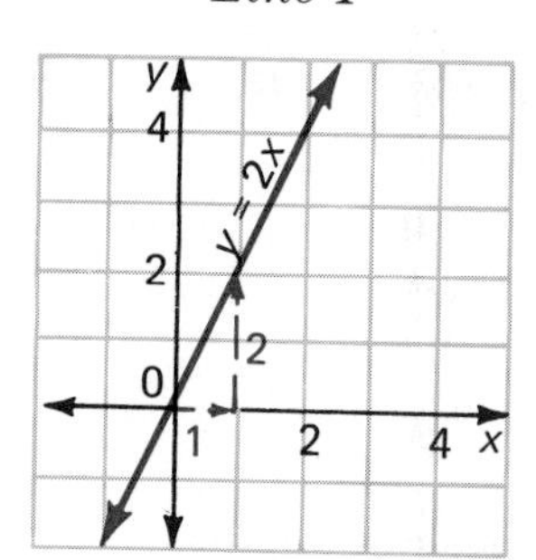

$$\text{slope} = \frac{2}{1} = 2$$

Line 2

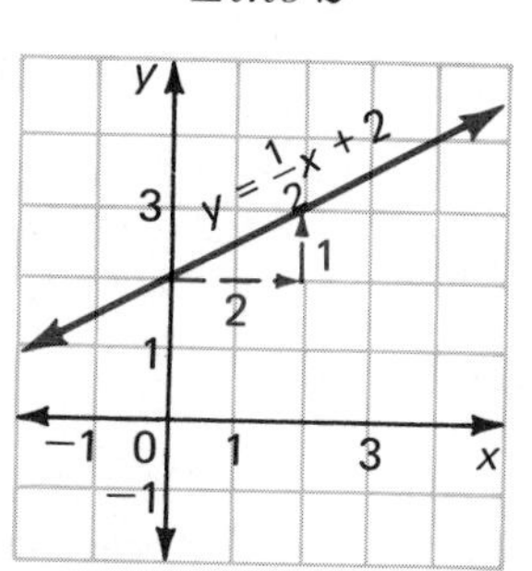

$$\text{slope} = \frac{1}{2}$$

Line 3

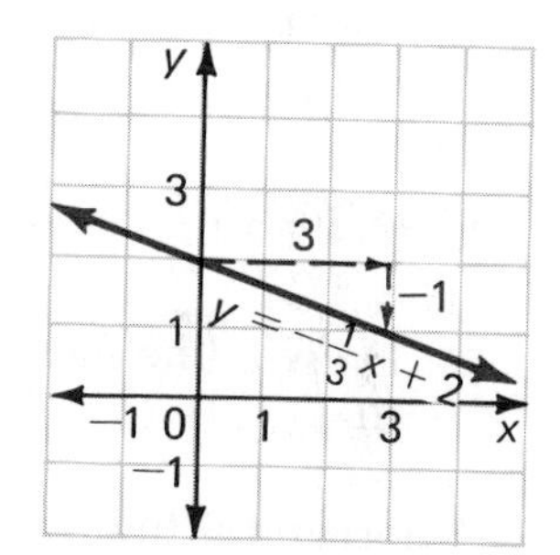

$$\text{slope} = \frac{-1}{3} = -\frac{1}{3}$$

equation: $y = 2x$

slope: 2

equation: $y = \frac{1}{2}x + 2$

slope: $\frac{1}{2}$

equation: $y = -\frac{1}{3}x + 2$

slope: $-\frac{1}{3}$

To find the slope *before* you draw the graph, just look at the coefficient of x after the equation is solved for y. Note that the slope can be a positive or a negative number.

Written Exercises

Find the slope.

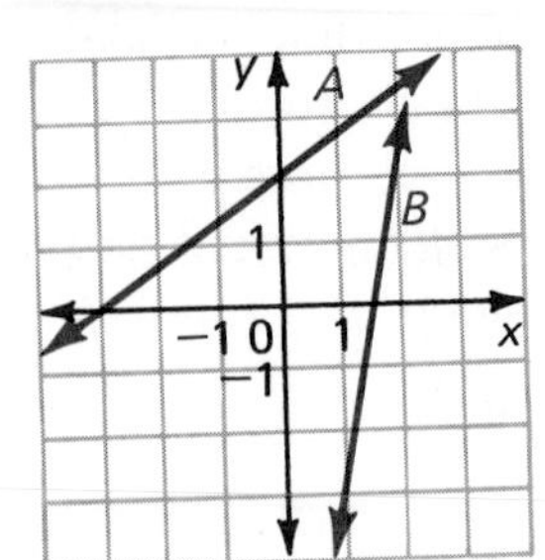

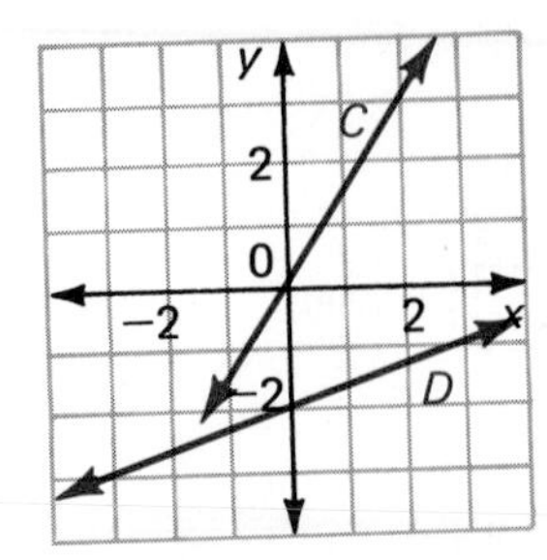

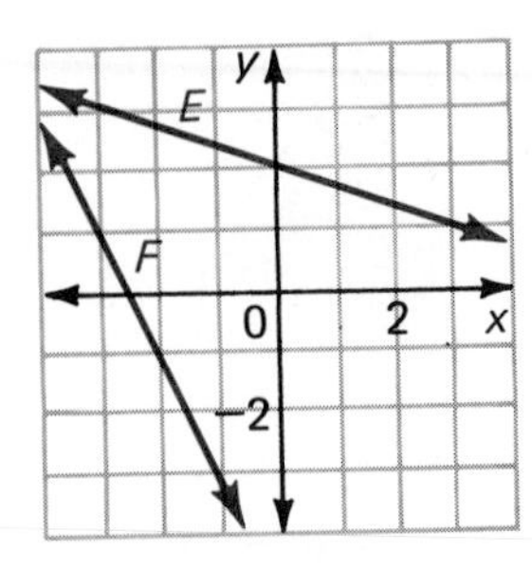

A 1. line A

2. line B

3. line C

4. line D

5. line E

6. line F

What is the slope of the line whose equation is given?

7. $y = 3x$

8. $y = 4x - 2$

9. $y = 2x + 3$

10. $y = 4x + 1$

11. $y = -x + 4$

12. $y = -x - 7$

13. $y = -3x$

14. $y = -5x + 4$

15. $y = \frac{1}{3}x + 7$

16. $y = \frac{3}{4}x$

17. $y = -\frac{1}{5}x + 5$

18. $y = -\frac{2}{3}x + 1$

19. Draw a line having a slope of $\frac{3}{4}$.

20. Draw a line having a slope of $\frac{4}{3}$.

A horizontal line is shown below.

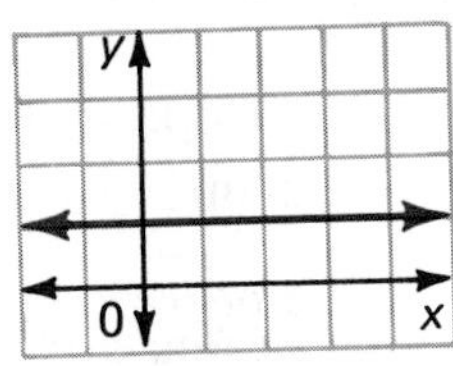

Exercises 21–24

A vertical line is shown below.

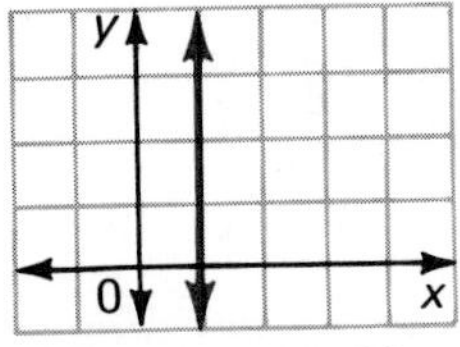

Exercises 25–28

B 21. What is its rise?

22. What is its run?

23. What is its slope?

24. Which axis does this horizontal line cross?

25. What is its rise?

26. What is its run?

27. Does the line have a slope?

28. Which axis does this vertical line cross?

7 Functions

Often in everyday life one quantity depends on another. When this happens we say that the first quantity is a function of the other.

EXAMPLE 1 The distance you can travel in two hours depends on how fast you are going (your rate).

Your distance is a function of your rate.

$$D = 2r$$

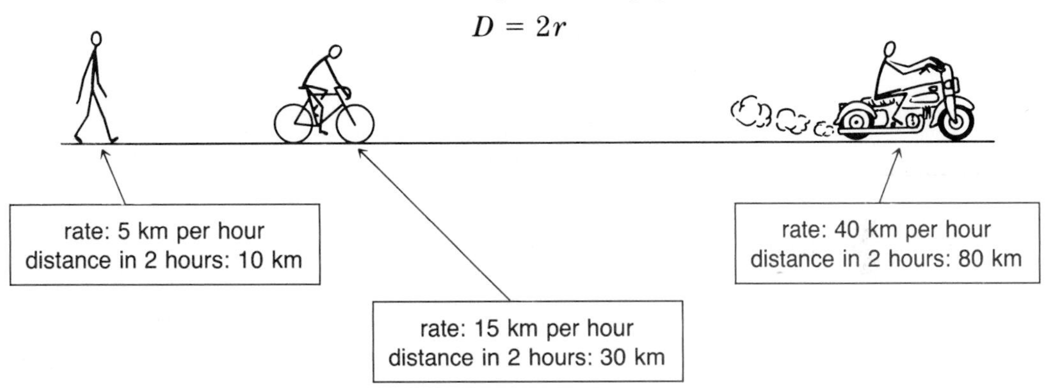

EXAMPLE 2 The value of a car depends upon its age (usually).

The value of a car is a function of its age.

Age of car	new	1 yr	2 yr	3 yr	4 yr
Value of car	\$6000	\$4800	\$3600	\$2700	\$2025

EXAMPLE 3 The temperature outdoors often depends upon the time of day.

The temperature outdoors is a function of the time of day.

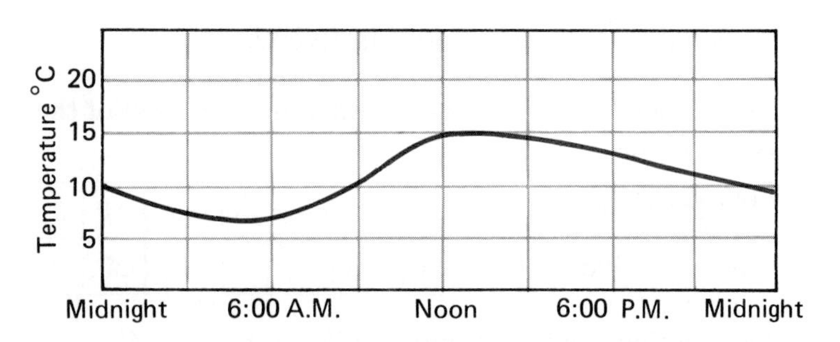

Examples 1–3 show that a function can be described by an equation, a table, or a graph. Let's consider one more function and describe it in all three ways.

EXAMPLE 4 The OUTPUT of a number machine depends upon the INPUT.

The OUTPUT *depends upon the* INPUT.

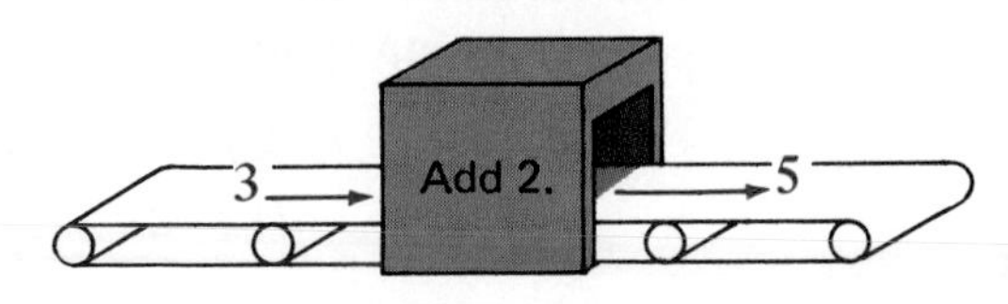

Equation

$y = x + 2$

Table

INPUT x	OUTPUT y
0	2
1	3
2	4
3	5

Graph

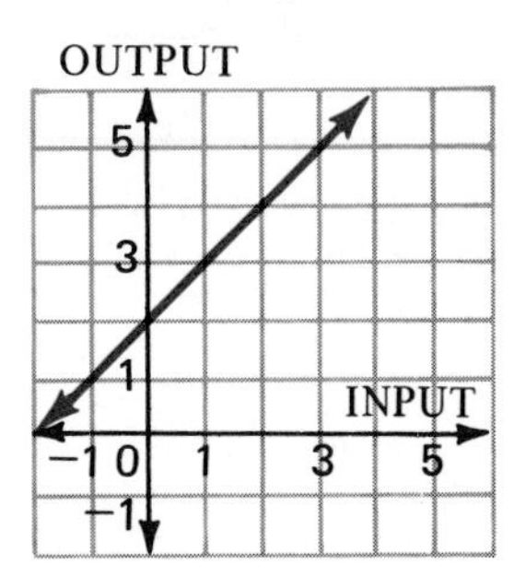

Classroom Practice

The OUTPUT of the machine at the right is a function of the INPUT.

1. Complete the table.

x = INPUT	3	2	0	−2	−4
y = OUTPUT	9	?	?	?	?

2. An equation relating x and y in Exercise 1 is: y = __?__.

The cost of several oranges at 25¢ each is a function of the number of oranges you buy.

3. Complete the table.

n = number you buy	1	2	6	12
C = cost	25	?	?	?

4. An equation relating n and C in Exercise 3 is: C = __?__.

5. If $y = 3x$, then y is a function of __?__.

6. If $D = 2r$, then D is a function of __?__.

7. The time it takes to travel by bus from one part of town to another is a function of which of the following?
   a. How often the buses run
   b. The traffic
   c. Both the traffic and how often the buses run

Written Exercises

In Exercises 1–4 the OUTPUT is a function of the INPUT.

A

1. Complete the table.

x = INPUT	2	3	4	-2	?
y = OUTPUT	4	6	8	?	10

2. An equation relating x and y in Exercise 1 is: y = __?__.

3. Complete the table.

x = INPUT	4	9	-7	?
y = OUTPUT	5	10	-6	-8

4. An equation relating x and y in Exercise 3 is: y = __?__.

The cost of several grapefruit that sell for 48¢ each is a function of the number you buy.

5. Complete the table.

n = number you buy	1	2	3	12	?
C = cost	48	?	?	?	480

6. An equation relating n and C in Exercise 5 is: C = __?__.

The distance you travel in 4 hours is a function of your rate.

7. Complete the table.

Rate (km/h)	Distance (km)
10	40
20	?
30	?
40	?

8. Copy and complete the graph.

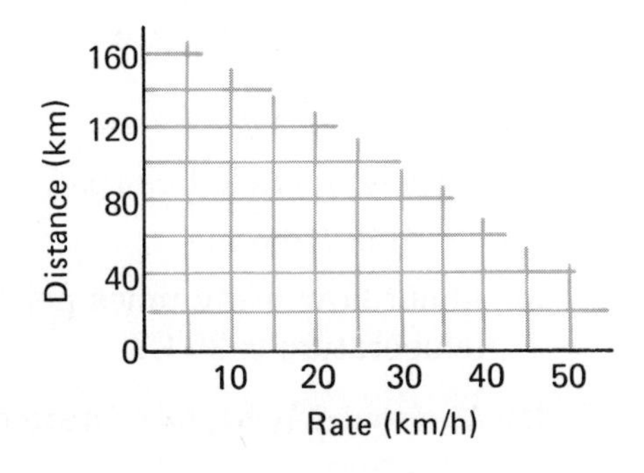

The wind-chill factor is a function of temperature and wind speed.

Wind Speed (km/h)

	10	20	30	40	50	60
0°	−3°	−9°	−13°	−16°	−18°	−19°
−5°	−9°	−17°	−21°	−23°	−26°	−27°
−10°	−14°	−24°	−27°	−31°	−33°	−35°
−15°	−19°	−29°	−35°	−38°	−41°	−42°

If the temperature is −5°C and the wind is blowing 30 km/h, the wind-chill factor is −21°C. This means that it *feels* as though it is −21°C.

9. Find the wind-chill factor when the temperature is −5°C and the wind speed is 20 km/h.

10. Find the wind-chill factor when the temperature is −15°C and the wind speed is 50 km/h.

11. If the temperature is 0°C and the wind-chill factor is −16°, then the wind speed is __?__ km/h.

It is believed that a cricket's rate of chirping is a function of the temperature of the air. Use the equation below to solve Exercises 12–15.

Number of chirps per minute = 7 × Celsius temperature − 30

$$N = 7C - 30$$

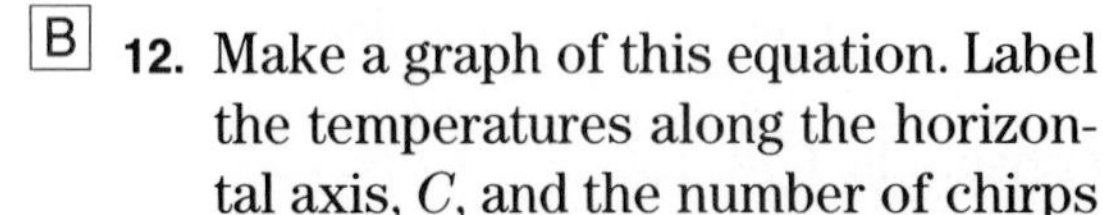

B 12. Make a graph of this equation. Label the temperatures along the horizontal axis, C, and the number of chirps per minute along the vertical axis, N.

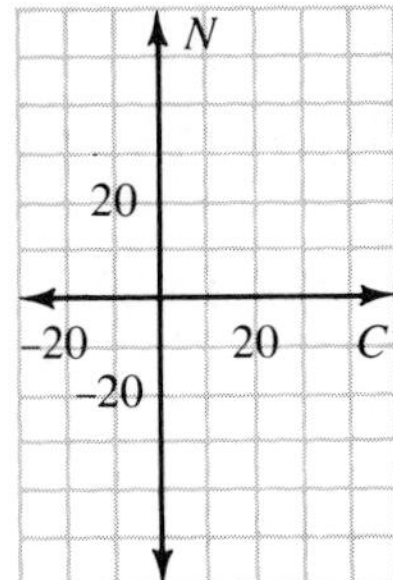
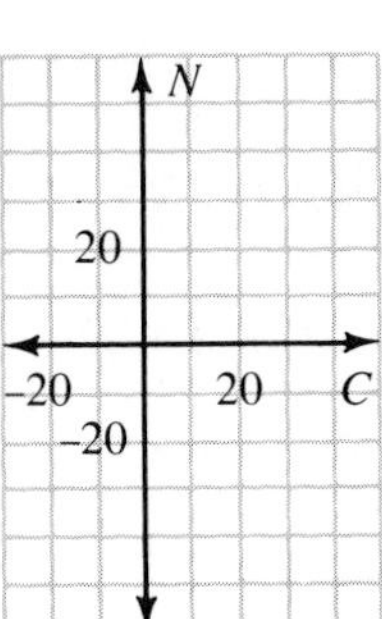

13. Suppose a cricket chirps 75 times a minute. About what temperature is it?

14. About how many times per minute does a cricket chirp if the temperature is 25°C?

15. At about what temperature do crickets stop chirping?

SELF-TEST

Vocabulary

solution pair (p. 237) graph of an equation (p. 241)
ordered pair (p. 237)

Find three solution pairs for the equation.

1. $y = 5x$ **2.** $y = 3x - 6$ **3.** $y = 4x + 1$ *(7-4)*

Draw the graph of the equation.

4. $y = -4x$ **5.** $y - 7x = 0$ **6.** $y = 5x - 6$ *(7-5)*

What is the slope of the line whose equation is given?

7. $y = 9x - 4$ **8.** $y = -3x + 2$ **9.** $y = -x$ *(7-6)*

In Exercises 10 and 11, the OUTPUT is a function of the INPUT.

x = INPUT	-1	2	0	1	3	?
y = OUTPUT	-1	5	1	3	?	9

10. Complete the table. *(7-7)*

11. An equation relating x and y in Exercise 10 is: $y = \underline{\ ?\ }$.

CALCULATOR ACTIVITIES

You can use a calculator to find solution pairs for given equations. Enter values for *x* in each equation to find values for *y*.

Use a calculator to decide which choices are solution pairs for the equation or inequality.

1. $y = 2x + 1$ **a.** (3, 7) **b.** (−2, 1) **c.** (−1, −1) **d.** (0, 3)

2. $y = 2 - 3x$ **a.** (−2, −4) **b.** (4, −10) **c.** (0, −1) **d.** (3, 11)

3. $y = 4x + 6$ **a.** (1, 10) **b.** (−2, −2) **c.** (2, −2) **d.** (−1, 5)

4. $y > x + 5$ **a.** (3, 4) **b.** (−1, 4) **c.** (2, 9) **d.** (−6, 0)

5. $y < 4x - 1$ **a.** (−1, 4) **b.** (3, 0) **c.** (5, 25) **d.** (1, 2)

CONSUMER APPLICATIONS

ADVERTISING

There are lots of gimmicks that advertisers use to promote their products. Don't be mistaken. Gimmicks are not necessarily dishonest. As you probably know, some are really clever! Gimmicks are just methods advertisers use to get attention and to interest the consumer in buying the product. Let's consider a couple of these gimmicks.

If *Sporting Guide* Magazine wants to show its increased sales over a number of years, it could do so with a graph like this.

You can see that the sales have doubled in the last 5 years.

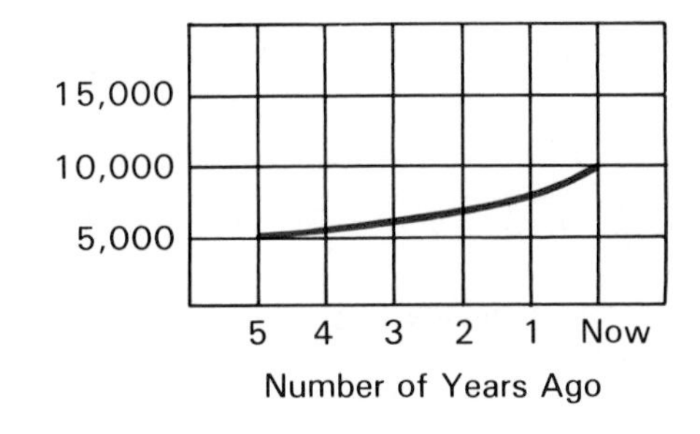

But look how much more impressive this graph is. It looks as though the number of sales has really grown by leaps and bounds!

All you have to do is to change the vertical scale and chop off the bottom of the graph.

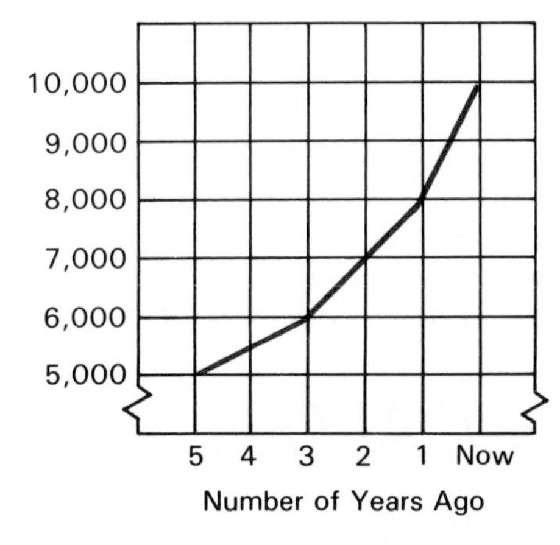

Sales
5 Years Ago

Sales
Today

The advertisers of *Sporting Guide* Magazine could also show the increase in sales in picture form like this.

Since the sales have doubled, the dimensions in the second picture were drawn twice as large as the first.

Since the second area is four times the first, it looks as if sales have more than doubled!

8 Graphs of Inequalities (Optional)

How can you picture, or graph, an inequality like

$$y > -x + 2 \text{ or } y < -x + 2?$$

First, draw the graph of $y = -x + 2$. This line divides the plane, or surface, into two parts.

All the points *above* the line are solutions of $y > -x + 2$. We use shading to show these solutions.

All the points *below* the line are solutions of $y < -x + 2$.

In the graphs of $y > -x + 2$ and $y < -x + 2$, we make the graph of $y = -x + 2$ a dashed line because the points on this line are not solutions of either inequality.

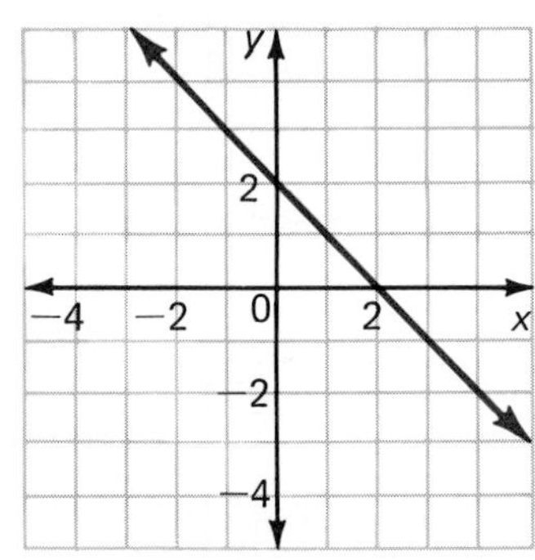

Graph of $y = -x + 2$

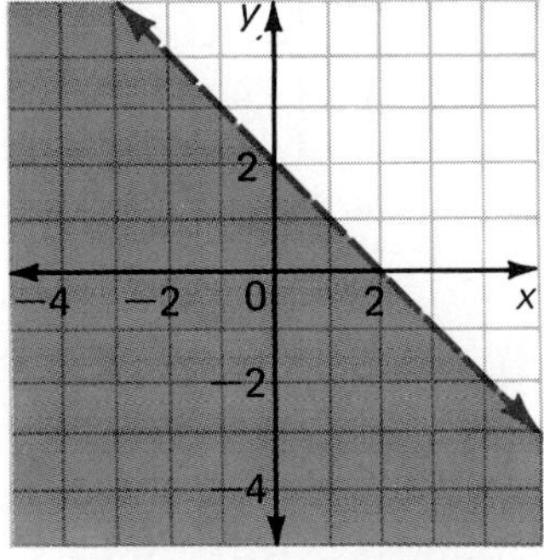

Graph of $y < -x + 2$

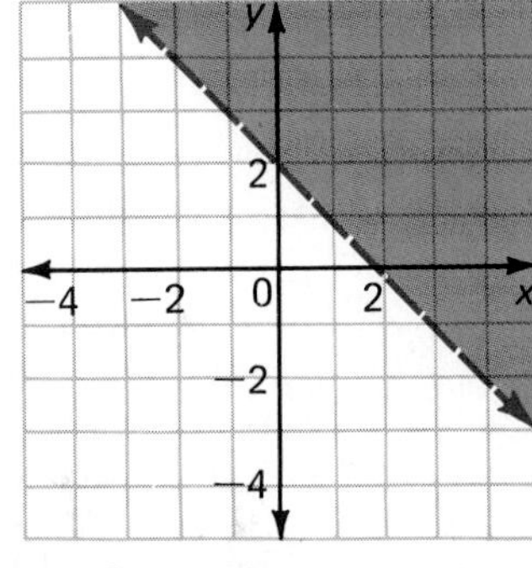

Graph of $y > -x + 2$

If an inequality symbol includes the = sign, as in $y \leq -x + 2$, then we show the graph of the equality $y = -x + 2$ as a solid line.

We call the graph of the equality a *boundary line.*

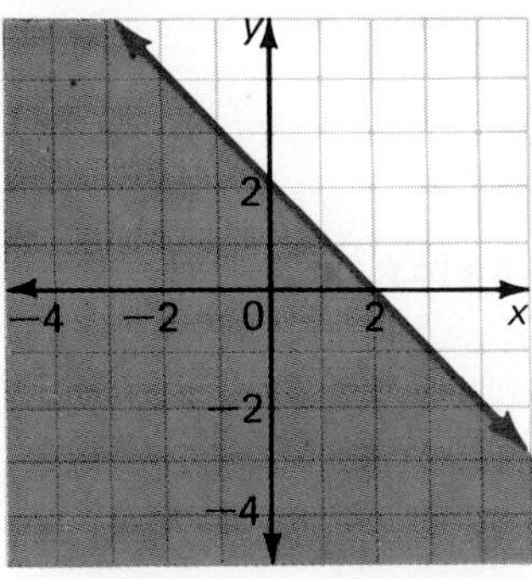

Graph of $y \leq -x + 2$

EXAMPLE 1

Draw the graph of $y > 2x$.

Step 1: Make a table.

Step 2: Graph $y = 2x$. Note that the boundary line should be dashed since the inequality symbol does not include the = sign.

Step 3: Shade the part of the plane (called a half-plane) *above* the line.

$y = 2x$	
x	y
0	0
1	2
−1	−2

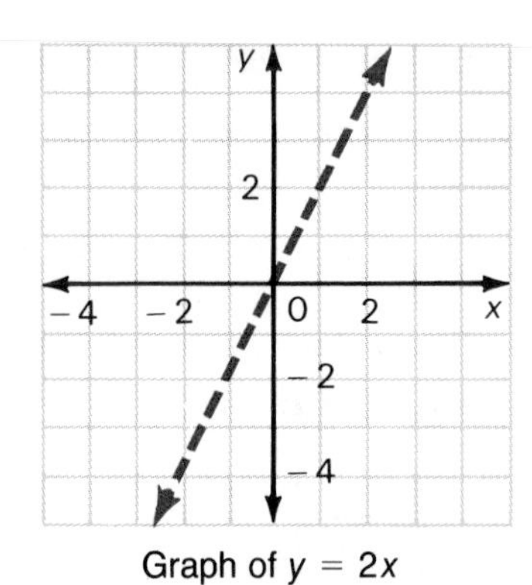

Graph of $y = 2x$

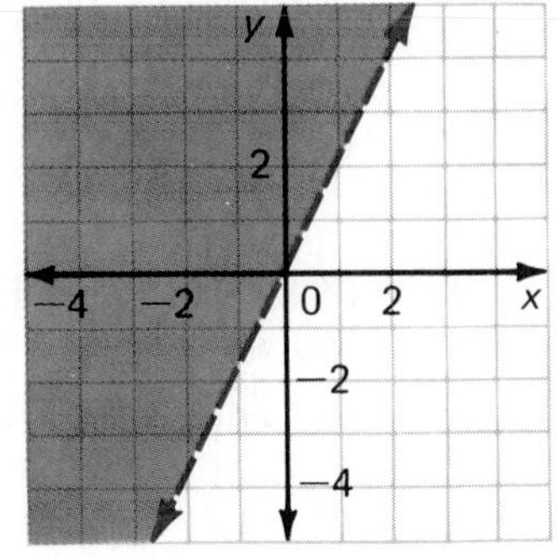

Graph of $y > 2x$

EXAMPLE 2

Draw the graph of

$$y \leq 3x - 2.$$

First, graph $y = 3x - 2$ as a solid line.

Then shade the half-plane *below* the line, since the inequality symbol is $\leq$.

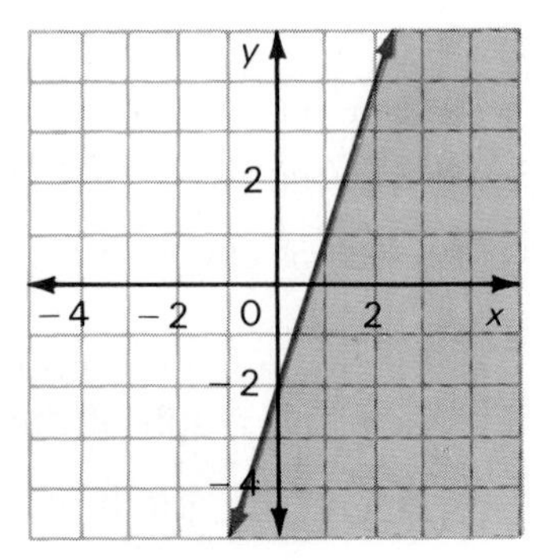

Graph of $y \leq 3x - 2$

Written Exercises

Draw the graph of each inequality.

1. $y > x$

2. $y > -2x$

3. $y < 2x$

4. $y < 2x - 1$

5. $y > 3x + 2$

6. $y > x - 2$

7. $y \leq 4 - x$

8. $y \leq x + 3$

9. $y \geq 4x - 1$

10. $y \geq 5 + 2x$

11. $y \geq 0$

12. $y \leq 4 - 2x$

READING ALGEBRA

THE RIGHT WORD

Suppose you are a nurse working with a surgeon in an operating room. The surgeon would never say, "Hand me that thing over there on the table," because you would have no idea which of the instruments to pick up. Every job has a special vocabulary of its own. In order to make your meaning clear, you have to use these special words along with the common words that hold sentences together—*and, or, with,* and *to,* for example. Your job is easier if you call things by their right names.

Knowing and using the right words is important in algebra, too. Some of the new words in this book may look hard, but they will soon become familiar if you learn their meanings and use them in class discussions. The book has been planned to help you learn new words. Look back at page 10, where the word *coefficient* is used for the first time. The sentence explains the meaning, and colored type is used to pick out the coefficients in the examples. Classroom Practice Exercises 1–3 gave you a chance to test your understanding of the word. On page 14, *coefficient* appears again, with a sentence to help you recall the meaning. Farther along in the book you may have needed to know the meaning of *coefficient* in order to understand some of the explanations. You should try to make each new word a part of your working vocabulary.

EXERCISES

Write a single word or phrase that can be used to replace each of the descriptions below.

1. A little number written above and to the right of a number or letter
2. The distance around a figure
3. An algebraic expression with three terms connected by plus or minus signs
4. The positive and negative whole numbers and zero
5. The numbers in an ordered pair representing a point
6. Numbers that are multiplied together to produce a product

SKILLS REVIEW

RENAMING FRACTIONS

Rename.

SAMPLE $\frac{1}{2} = \frac{?}{10} \longrightarrow \frac{1}{2} = \frac{5}{10}$

1. $\frac{1}{4} = \frac{?}{12}$
2. $\frac{1}{5} = \frac{?}{20}$
3. $\frac{1}{3} = \frac{?}{9}$
4. $\frac{1}{2} = \frac{?}{8}$
5. $\frac{1}{6} = \frac{?}{18}$
6. $\frac{2}{3} = \frac{?}{6}$
7. $\frac{3}{8} = \frac{?}{48}$
8. $\frac{3}{4} = \frac{?}{16}$
9. $\frac{4}{7} = \frac{?}{35}$
10. $\frac{5}{6} = \frac{?}{24}$

In each exercise, rename the fractions so that they have the same denominator.

SAMPLE $\frac{1}{2}, \frac{1}{5} \longrightarrow \frac{5}{10}, \frac{2}{10}$

11. $\frac{1}{3}, \frac{1}{4}$
12. $\frac{1}{2}, \frac{1}{7}$
13. $\frac{1}{4}, \frac{1}{8}$
14. $\frac{1}{3}, \frac{1}{5}$
15. $\frac{1}{4}, \frac{1}{5}$
16. $\frac{2}{5}, \frac{1}{3}$
17. $\frac{1}{6}, \frac{4}{9}$
18. $\frac{2}{3}, \frac{1}{4}$
19. $\frac{3}{7}, \frac{1}{14}$
20. $\frac{1}{8}, \frac{5}{6}$
21. $\frac{3}{4}, \frac{3}{7}$
22. $\frac{4}{7}, \frac{2}{9}$
23. $\frac{7}{10}, \frac{2}{3}$
24. $\frac{5}{6}, \frac{2}{5}$
25. $\frac{4}{9}, \frac{3}{8}$

In each exercise, rename the fractions so that they have the same denominator. Then add or subtract.

26. $\frac{1}{2} + \frac{1}{7}$
27. $\frac{1}{6} + \frac{1}{5}$
28. $\frac{1}{4} + \frac{1}{6}$
29. $\frac{1}{9} + \frac{1}{3}$
30. $\frac{1}{4} + \frac{1}{7}$
31. $\frac{1}{2} - \frac{1}{3}$
32. $\frac{1}{3} - \frac{1}{8}$
33. $\frac{1}{4} - \frac{1}{5}$
34. $\frac{1}{2} - \frac{1}{6}$
35. $\frac{1}{3} - \frac{1}{7}$
36. $\frac{2}{3} + \frac{1}{4}$
37. $\frac{5}{6} + \frac{1}{7}$
38. $\frac{3}{8} + \frac{1}{6}$
39. $\frac{4}{5} + \frac{3}{7}$
40. $\frac{2}{9} + \frac{3}{10}$
41. $\frac{3}{4} - \frac{1}{8}$
42. $\frac{7}{8} - \frac{2}{5}$
43. $\frac{9}{11} - \frac{1}{3}$
44. $\frac{5}{7} - \frac{2}{5}$
45. $\frac{2}{3} - \frac{5}{8}$

CHAPTER REVIEW

CHAPTER SUMMARY

1. Bar graphs, pictographs, and broken line graphs are often used to picture information.
2. A solution for an equation with two variables is an ordered pair.
3. The line that contains all the solutions of an equation is called the graph of the equation.
4. The slope of a line is the ratio of the rise to the run.
5. A quantity that depends on a second quantity is a function of it.

REVIEW EXERCISES

The graph at the right shows the amount of rainfall in East Westburg last year. ***(See pp. 224–227.)***

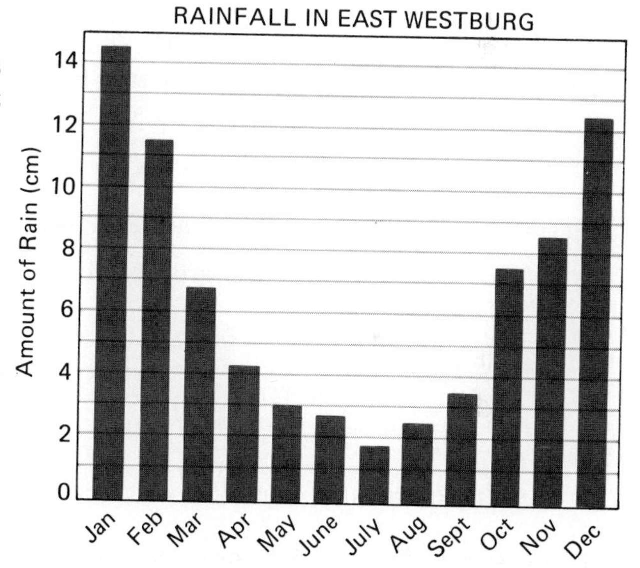

1. About how much rain fell in October?
2. Which month had the greatest amount of rain?
3. Which month had the least amount of rain?
4. About how much more rain fell in February than in September?

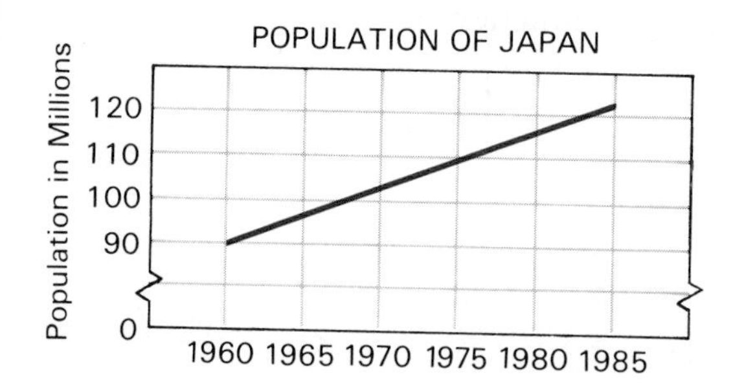

The graph at the left shows the population of Japan from 1960 to 1985. ***(See pp. 228–231.)***

5. Estimate the population in the year 1980.
6. The population was 100 million in the year __?__.

Complete. *(See pp. 237–240.)*

7. $y = -4x$

x	y
0	?
1	?
2	?
−2	?

8. $y = 4x + 3$

x	y
0	?
1	?
2	?
3	?

9. $x + y = 10$

x	y
0	?
1	?
2	?
−1	?

10. $y = 5 - 3x$

x	y
0	?
1	?
2	?
−2	?

Draw the graph of the equation. *(See pp. 241–242.)*

11. $y = x - 6$

12. $y = 3x + 4$

13. $y - 4x = 3$

14. $y + x = 3$

15. $y = -2x + 4$

16. $y + 3x = 2$

Find the slope of the line whose equation is given. *(See pp. 243–244.)*

17. $y = 3x + 4$

18. $y = -x - 4$

19. $y = \frac{3}{4}x - 6$

A certain fabric sells for $4.20 a yard at a local store. *(See pp. 245–248.)*

20. The total cost of several yards of the fabric is a function of ___?___.

21. Complete the table.

n = number of yards	1	2	3	5	?
C = Cost	?	?	?	?	$33.60

22. An equation relating n and C in Exercise 21 is: C = ___?___.

The Booster Club earns 50¢ for each bumper sticker it sells. *(See pp. 245–248.)*

23. The amount of money the club earns is a function of ___?___.

24. Complete the table.

25. Use the table to draw a graph.

n = number sold	10	15	20	50	?
A = Amount earned	?	?	?	?	$55

(Optional) Draw the graph of each inequality. *(See pp. 251–252.)*

26. $y > x + 1$

27. $y < 2x - 1$

28. $y \leq -3x + 2$

CHAPTER TEST

The graph at the right shows the increase in the cost of first-class U.S. postage over a number of years. Use the graph for Questions 1–2.

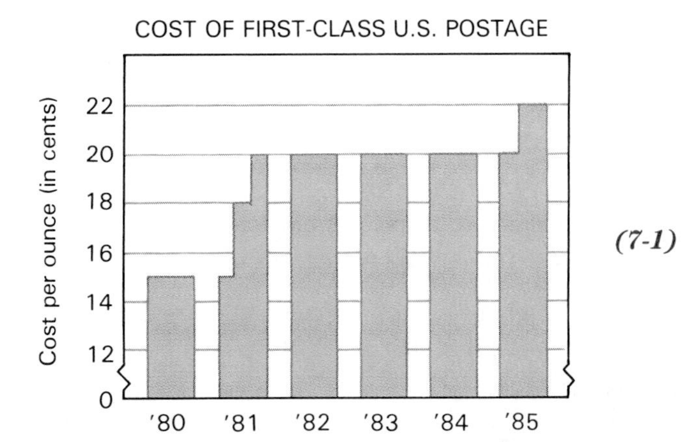

1. In 1983, the cost per ounce was __?__¢.

2. During the year __?__, the cost had the greatest increase.

(7-1)

The graph at the right shows the population of Mayville from 1960 to 1985.

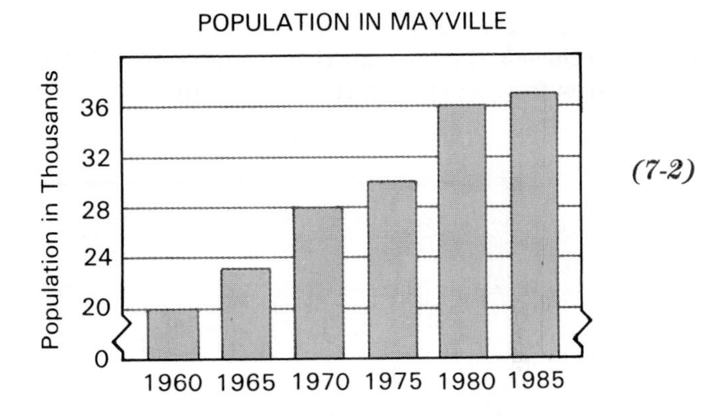

3. Use the bar graph to make a broken line graph of the population of Mayville.

4. Estimate the population of Mayville in 1990.

(7-2)

5. Plot the following points in order: (0, 1), (3, 5), (6, 1), (3, −3), and (0, 1). Connect each point to the next one in the list. *(7-3)*

6. Solve the equation $3x - y = 6$ for y. *(7-4)*

7. Find three solution pairs for the equation $y - 2x = 2$.

Draw the graph of the equation.

8. $x - y = 5$

9. $4x + y = 4$ *(7-5)*

10. Give the slope of the line whose equation is $y = 7x + 2$. *(7-6)*

11. Complete the table. *(7-7)*

x = INPUT	4	10	50	?	100
y = OUTPUT	20	50	250	320	?

MIXED REVIEW

1. Draw a line having a slope of 2.

2. Divide: $\frac{7x^4 - 14x^3y + 35x^2}{-7x^2}$

3. Ramon jogs from his home to the state fair at 10 km/h.
He returns by bus at 50 km/h.
The trip on foot is 4 hours longer than the bus trip.
How far is it from Ramon's home to the state fair?

4. Solve: $2(1 - 4a) - 7(-2 - a) = 1 - 6a$

Multiply.

5. $(2r^3s^2)(-9rs^3)$
6. $(x^2 + 3)(x^2 - 3)$
7. $(n - 2)(n^2 + 2n + 4)$

8. Find the greatest common factor of 75 and 30.

Exercises 9–11 refer to the graph below that shows some net income figures for a certain airline company.

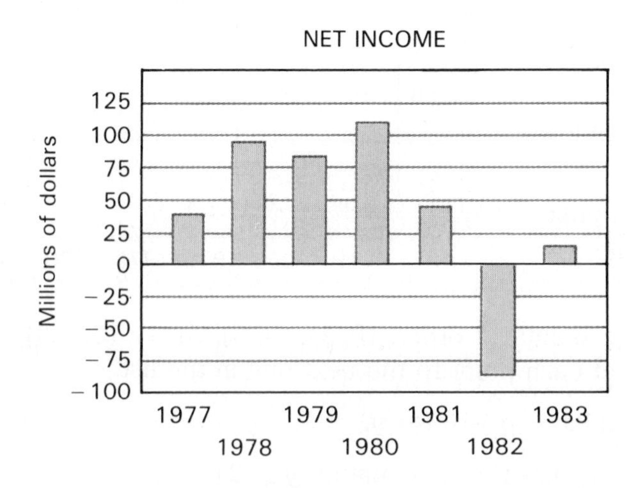

9. In which year was the greatest income earned?

10. What was the decrease in income from 1980 to 1981?

11. Use the bar graph to make a broken line graph showing net income.

12. If P is a point that has coordinates $(2, -3)$, give the coordinates of the point 7 units to the left of P.

13. If a rectangle has length $2n + 3$ and width $n + 4$, give expressions for (**a**) the perimeter and (**b**) the area of the rectangle.

Factor.

14. $b^2 + 2b - 35$

15. $m^2 - 14m + 48$

16. $-9t^2 + r^2s^2$

17. $7x^3 + 21x^2y + 28x$

18. $a^2 + 14a + 24$

19. $x^2 - 8xy + 16y^2$

20. Divide and simplify: $\frac{8mn}{-2n} + \frac{-15m^3}{-5m^2}$

21. Find **(a)** the perimeter and **(b)** the area of the triangle shown.

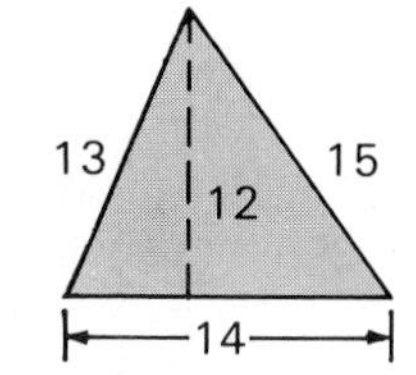

22. Compare. Write < or > or =:
$12 + 6 \div (-2)$ __?__ $-(-3)^2$

23. Graph the solutions of $x < -1$.

24. Find three solution pairs for $y - 3x = 5$.

25. Express $-3(3x - y)^2$ as a trinomial.

26. Factor: $2x^2 - 2x - 40$

27. Draw the graph of $2x + y = 4$.

28. Solve $k = \frac{C}{d}$ for d.

Simplify.

29. $(-x^4y^2)^3(2x)^2$

30. $(y^2 - 3y - 9) - (2y + 6)$

31. $(3a^2 - 5ab) + (ab - b^2)$

A phone call to Aurora costs 36¢ for the first minute and 22¢ for each additional minute. Use this data for Exercises 32–34.

32. An equation relating the cost, C, to the number of minutes, n, is __?__.

33. If $n = 6$, then $C =$ __?__.

34. If $C = \$3.00$, then $n =$ __?__.

35. Solve: $8(3x - 1) - 5(7x + 2) = 4$

36. A shoe store bought some slippers for \$10 per pair.
The store sold all but 8 pairs for \$14 per pair.
The profit earned was \$288.
How many pairs of slippers were sold?

Here's what you'll learn in this chapter:

To find the solution of two equations by graphing.

To determine whether two equations share one solution pair, no solution pairs, or all solution pairs.

To find the solution of two equations by substitution.

To find the solution of two equations by addition or subtraction.

To solve word problems with two equations and two variables.

When there are many choices, as in the fruit market shown, a shopper may need to do some computing in order to decide on the best buy. Algebra can be very useful in comparing costs.

Chapter 8

Equations with Two Variables

1 The Graphing Method

When two or more streets cross, we call the place where they cross their **intersection.**

When two lines cross, the point where they cross is their intersection.

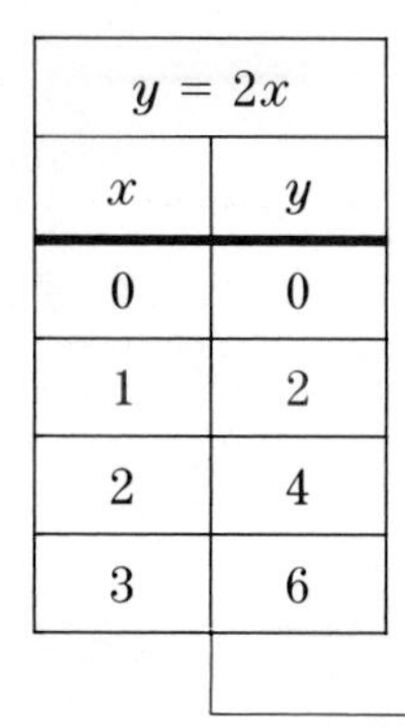

$y = 2x$	
x	y
0	0
1	2
2	4
3	6

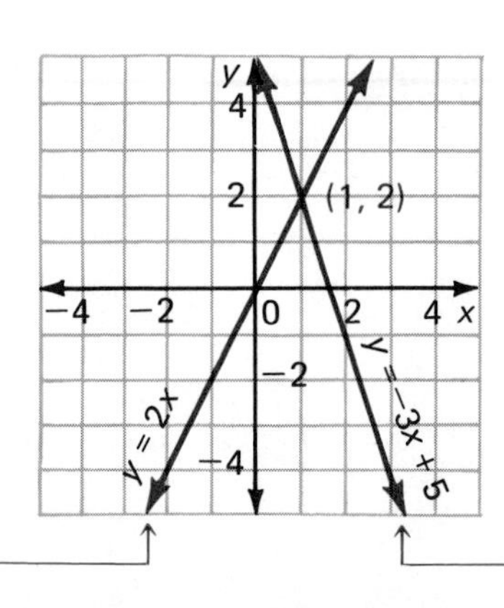

$y = -3x + 5$	
x	y
0	5
1	2
2	−1
3	−4

The graphs of $y = 2x$ and $y = -3x + 5$ intersect at $(1, 2)$. This means that $(1, 2)$ is a solution of *both* equations.

EXAMPLE 1 Find the ordered pair that is a solution of both $y = 3x$ and $y = -x + 4$.

Step 1

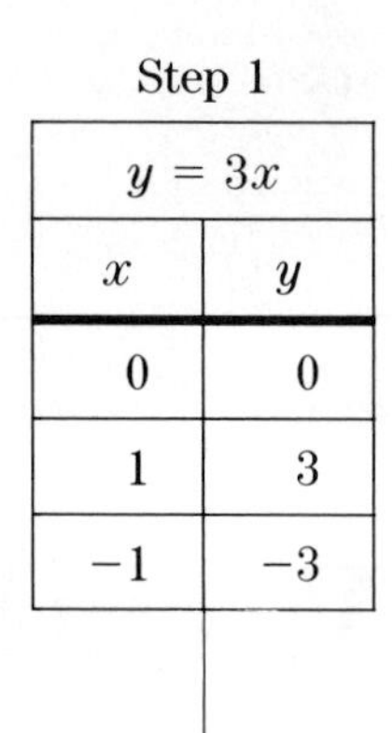

$y = 3x$	
x	y
0	0
1	3
−1	−3

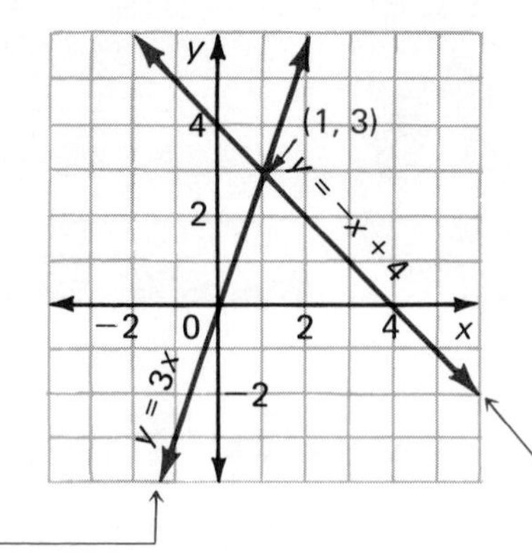

Step 2

$y = -x + 4$	
x	y
0	4
1	3
2	2

Answer: $(1, 3)$

Sometimes you'll have to solve an equation for y before you can draw the graph of the equation.

EXAMPLE 2 Find the ordered pair that is a solution of both $y - 2x = 8$ and $2x + y = 4$.

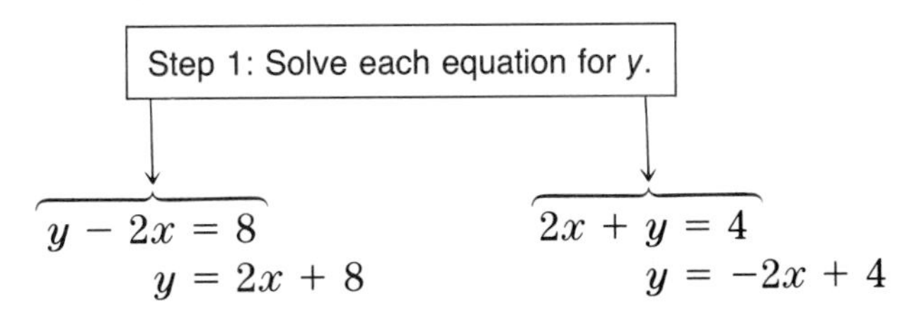

Step 2

$y = 2x + 8$	
x	y
0	8
−1	6
−2	4

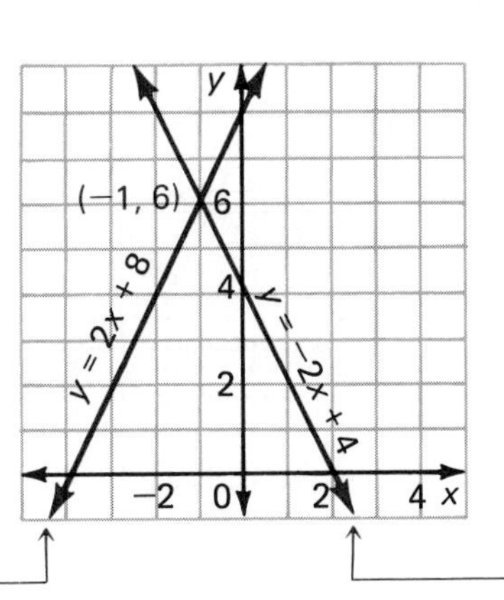

Step 3

$y = -2x + 4$	
x	y
0	4
1	2
2	0

Answer: $(-1, 6)$

Classroom Practice

Name the ordered pair that is a solution of the two equations whose graphs are shown.

1.

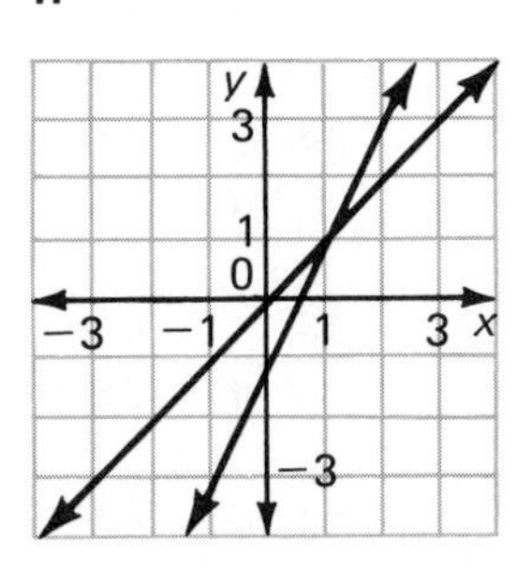

2.

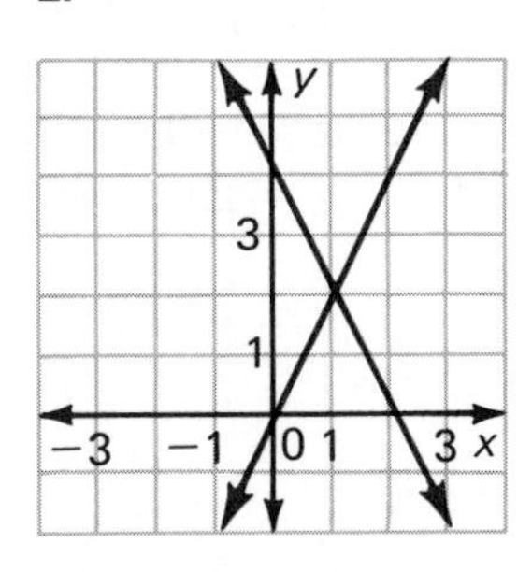

3.

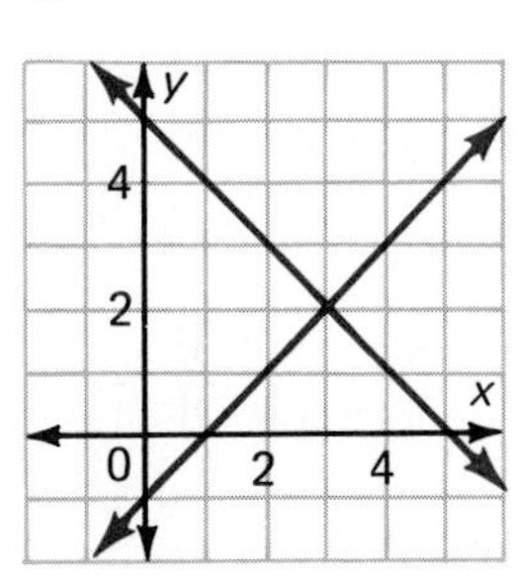

By graphing, find the ordered pair that is a solution of both equations.

4. $y = x$
$y = 2x + 3$

5. $y = x + 4$
$y = -x$

6. $x + y = 2$
$y = 2x + 5$

7. $3x + y = 1$
$y - x = -3$

8. $y = x - 4$
$y + 2x = 2$

9. $y - 2x = 7$
$y = -3x - 3$

10. $y = -x - 3$
$y - 3x = 5$

11. $y + x = 6$
$y - 2x = 3$

Written Exercises

Name the ordered pair that is a solution of the two equations whose graphs are shown.

1.

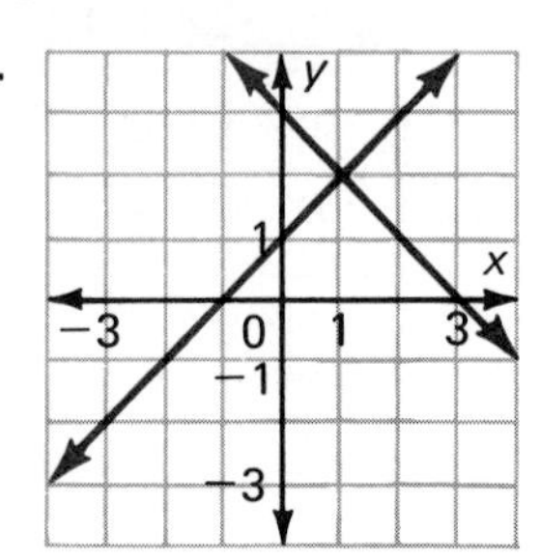

2.

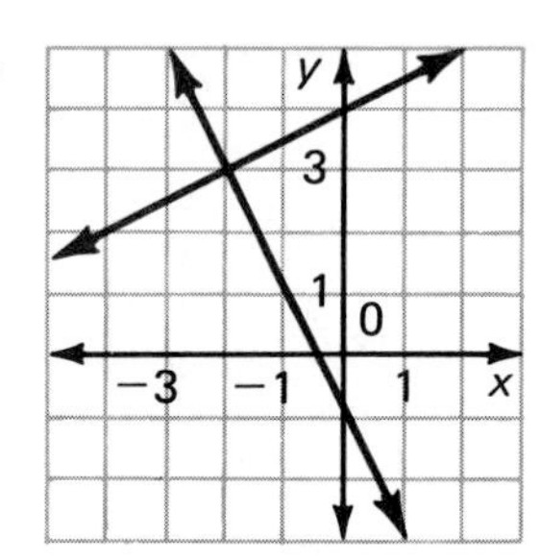

3.

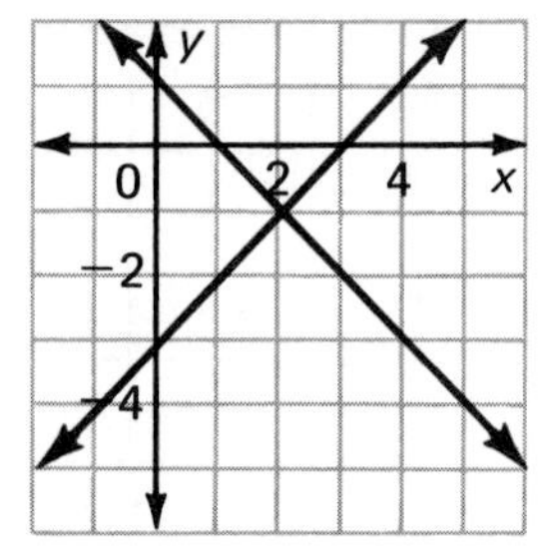

4. Name two points on the graph of $y = 4 - x$.

5. Name two points on the graph of $x + y = 12$.

6. Name two points on the graph of $y - x = 6$.

7. Name two points on the graph of $2x + y = 1$.

By graphing, find the ordered pair that is a solution of both equations.

8. $y = -x$
$y = 3x - 4$

9. $y = x + 1$
$y = -x + 3$

10. $y = -x - 1$
$y = 3x + 7$

11. $y = 2x$
$y = -2x - 4$

12. $y = -3x + 2$
$y = 2x - 3$

13. $y = 2x + 6$
$y = -x - 3$

14. $y - x = 0$
$y + x = 4$

15. $y - x = 3$
$x + y = -1$

16. $x + y = -4$
$y - 2x = 5$

17. $y - x = 0$
$y + 3x = -4$

18. $y - 3x = -1$
$y - 2x = 1$

19. $y - 2x = -5$
$y - x = -3$

Sometimes the solution pair is a pair of fractions instead of a pair of integers. Find the solution pair. Your answer does not have to be exact.

SAMPLE

$y = x + 2$	
x	y
0	2
1	3
−1	1

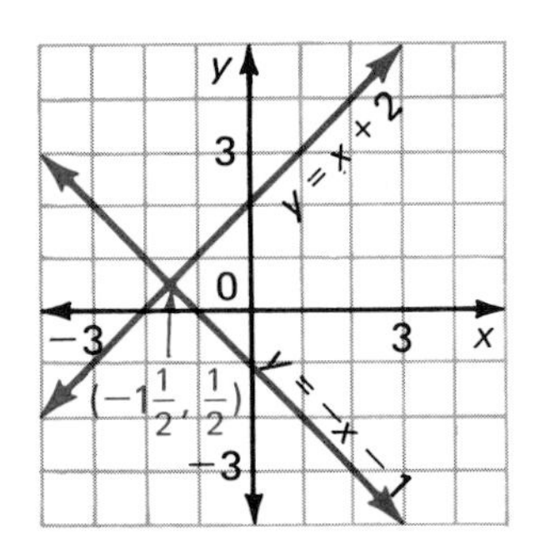

$y = -x - 1$	
x	y
0	−1
1	−2
−2	1

Answer: $\left(-1\frac{1}{2}, \frac{1}{2}\right)$

B

20. $y = -x - 1$
$y = x + 4$

21. $2x + y = 2$
$y - 3x = 9$

22. $y - 2x = 4$
$x + y = 5$

23. $y - 2x = 8$
$x + y = -2$

24. $y = x + 7$
$y = -\frac{1}{5}x + 4$

25. $y = \frac{1}{4}x - 3$
$y = -\frac{1}{4}x - 2$

26. a. Draw the graphs of the equations $y = x + 1$ and $y = x - 1$ on the same set of axes.

b. Do these graphs intersect?

c. Do the two equations share a solution pair?

Graph the pair of inequalities and show their common solutions. (Note: These exercises are for those students who have studied Section 8 in Chapter 7.)

SAMPLE

$y < -x$
$y < 2x + 3$

Graph $y < -x$ and $y < 2x + 3$.

The double-shaded part of the graph is the intersection of the two inequalities. All points in the intersection are solutions of both inequalities.

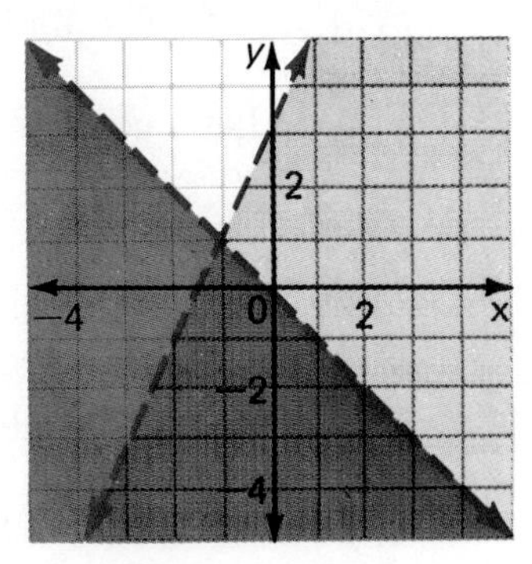

C

27. $y > x + 1$
$y > x - 3$

28. $y > x + 2$
$y \geq -x - 1$

29. $y \leq 2x$
$y \leq -x + 1$

2 No Solution, Many Solutions

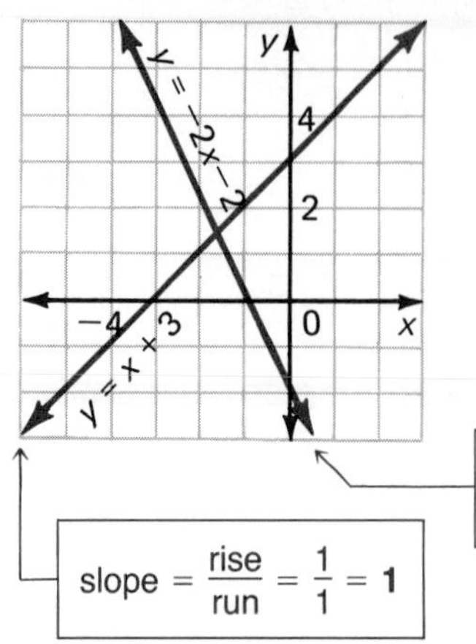

In the last lesson you saw how you could find a solution pair for two equations whose graphs intersect. Look at the graphs at the left. Notice that the two lines slant in different ways. You can see that the **slopes** are not the same, so the lines intersect.

Of course, you don't have to draw the graphs to see whether their slopes are the same or not. You can check by the equations instead.

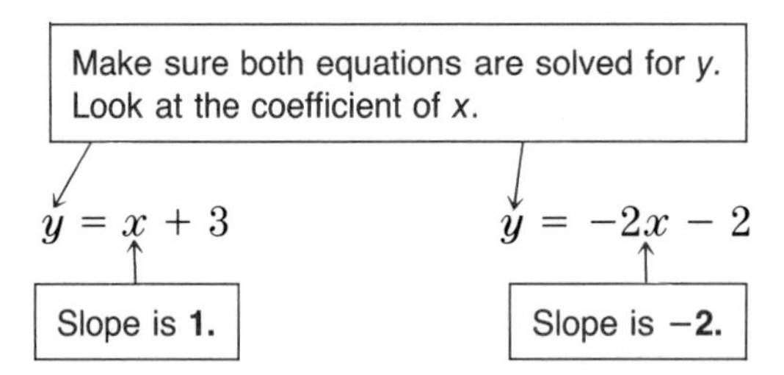

Now let's think of lines whose slopes are the same. There are two possible cases.

CASE 1: The lines will not intersect.

Let's consider $y - 2x = 3$ and $y - 2x = 0$.

$y - 2x = 3$ $\quad$ $y - 2x = 0$

$y = 2x + 3$ $\quad$ $y = 2x$

The slopes are the same.

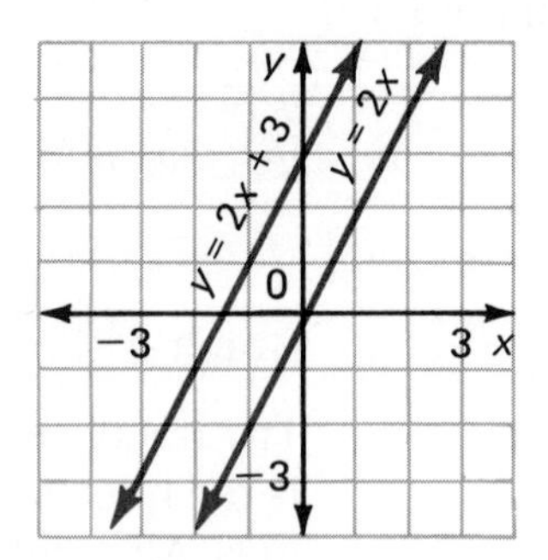

The graphs of these equations will never intersect. They are **parallel lines.**

The equations *do not* share a solution pair.

CASE 2: The lines will overlap.

Consider $y - 1 = x$ and $y - x = 1$.

$y - 1 = x$ $\qquad$ $y - x = 1$

$y = x + 1$ $\qquad$ $y = x + 1$

The slopes are the same. The equations are the same too.

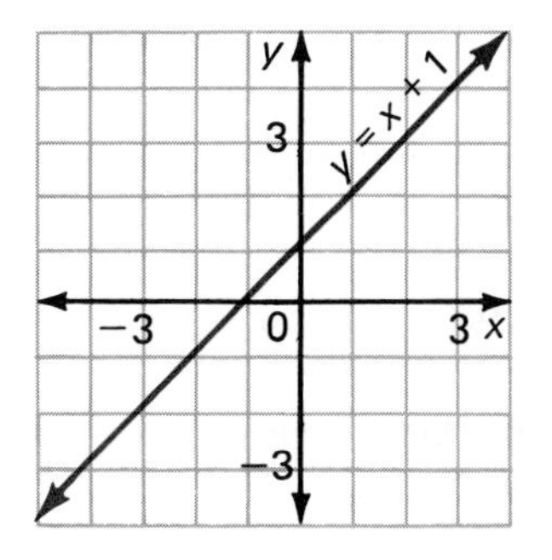

The graph of each equation is the same line.

As you can see, both equations share *all* solutions.

Now let's use these ideas to check equations to see if they share solution pairs.

EXAMPLE 1 Do $x + y = 4$ and $y - 3x = 2$ share any solution pairs?

$x + y = 4$ $\qquad$ $y - 3x = 2$

$y = -x + 4$ $\qquad$ $y = 3x + 2$

Slope is −1. $\qquad$ Slope is 3.

Answer: Yes, the equations share *one* solution pair.

EXAMPLE 2 Do $7 + y = 2x$ and $y - 2x = 5$ share any solution pairs?

$7 + y = 2x$ $\qquad$ $y - 2x = 5$

$y = 2x - 7$ $\qquad$ $y = 2x + 5$

Slope is 2. $\qquad$ Slope is 2.

Answer: No, the equations *do not* share any solution pairs.

EXAMPLE 3 Do $y - 4 = 2x$ and $y - 2x = 4$ share any solution pairs?

$y - 4 = 2x$ $\qquad$ $y - 2x = 4$

$y = 2x + 4$ $\qquad$ $y = 2x + 4$

The slopes are the same, but so are the equations!

Answer: The equations share *all* solution pairs.

Classroom Practice

Find the slope of the line whose equation is given.

1. $y = 3x$
2. $y = 3x - 5$
3. $3x + y = 5$
4. $y - 3x = 7$
5. Which equations in Exercises 1–4 name parallel lines?

Tell whether the equations share *one* solution pair, *no* solution pair, or *all* solution pairs.

6. $y = x + 6$, $y = -2x$
7. $y = 3x - 9$, $y = 3x + 1$
8. $y - 3x = 3$, $y - 2x = 4$
9. $y - 2x = 3$, $-3 + y = 2x$

Written Exercises

Find the slope of the line whose equation is given.

A

1. $y = 4x$
2. $y = 4x + 3$
3. $y = x + 1$
4. $y = \frac{1}{3}x + 2$
5. $y - 3x = 6$
6. $y + 2x = 4$
7. $4x - y = 0$
8. $4x + y = 0$
9. $2x - y = 3$
10. $y + 5x = 2$
11. $y = 8 - 7x$
12. $y = x - 5$

Tell whether the equations share *one* solution pair, *no* solution pair, or *all* solution pairs.

13. $y = 2x$, $y = 2x + 3$
14. $y = \frac{1}{3}x$, $y = \frac{1}{3}x + 2$
15. $y + x = 8$, $y + 2x = 8$
16. $x - y = 4$, $x - y = 6$
17. $x - y = 8$, $x + y = 6$
18. $2x + y = 9$, $y + 4x = 6$
19. $3x + y = 4$, $y - 3x = 2$
20. $4x - y = 6$, $y - 4x = 7$
21. $3x + y = 7$, $y + 3x = 7$
22. $y + 2x = 4$, $2x + y = 8$
23. $y - 4x = 9$, $y - 9 = 4x$
24. $x + y = -3$, $y - 2x = -6$
25. $2x + y = 9$, $y + 2x = 7$
26. $y = -3x + 3$, $y - x = 5$
27. $3x - y = 4$, $y + 4 = -3x$
28. $y + 2x = 4$, $y - 2x = -4$

B

29. a. Will the graphs of these equations intersect?

$$y = 2x + 3$$
$$y = 2x - 1$$

b. Will the graph of $y = -x + 1$ intersect either graph?

c. Can the three equations share a solution pair?

30. a. Are the lines in color at the right parallel? (Check their slopes.)

b. Will they intersect?

c. Do the equations of the lines share a solution pair?

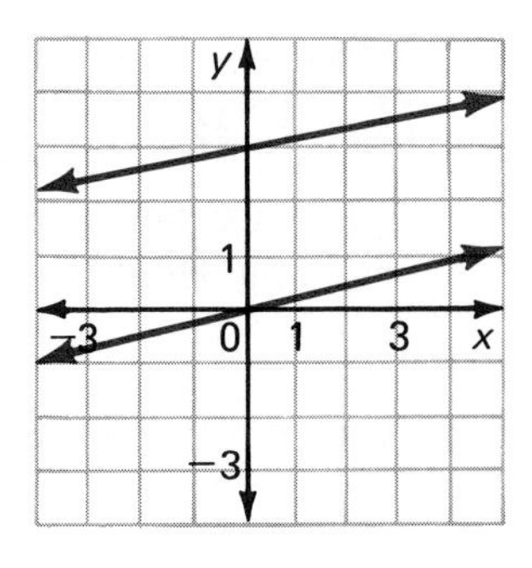

31. a. Are the lines in color at the right parallel? (Check their slopes.)

b. Will they intersect?

c. Do the equations of the lines share a solution pair?

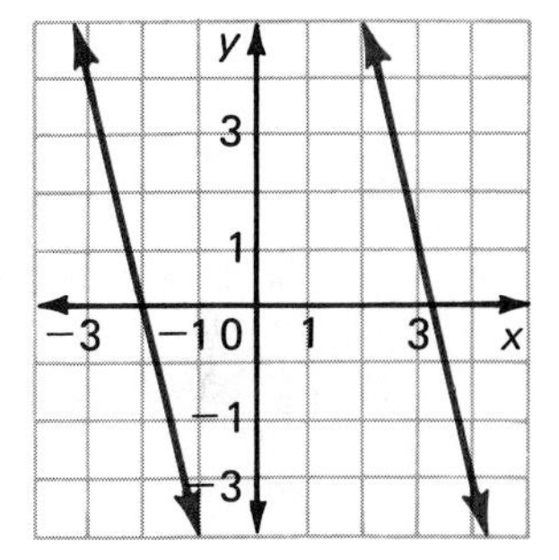

SELF-TEST

Vocabulary

intersection (p. 262) parallel lines (p. 266)
slope (p. 266)

By graphing, find the ordered pair that is a solution of both equations.

1. $y = 2x + 3$
$y = 3x$

2. $y = x + 3$
$y = -2x + 8$

3. $y + x = 0$
$y - x = -2$

(8-1)

Find the slope of the line whose equation is given.

4. $y = 3x + 7$

5. $y - 3x = 0$

6. $4 + y = -x$

(8-2)

Tell whether the equations share *one* solution pair, *no* solution pair, or *all* solution pairs.

7. $y = 3x - 4$
$y - 3x = 0$

8. $y - x = 2$
$y = 2x - 1$

9. $y - 6 = 3x$
$y - 3x = 6$

3 The Substitution Method

There are more exact ways of solving a pair of equations than by graphing. Here is one method.

EXAMPLE 1 Solve $y + 1 = 3x$ and $2x + y = 9$.

Step 1: Solve one equation for y.

$$y + 1 = 3x$$
$$y = 3x - 1$$

Step 2: Substitute the value of y in the other equation. Solve for x.

$$2x + y = 9$$
$$2x + (3x - 1) = 9$$
$$5x - 1 = 9$$
$$5x = 10$$
$$x = 2$$

Step 3: Substitute your x value in the equation in Step 1. Find the value of y.

$$y = 3x - 1$$
$$y = 3 \cdot 2 - 1$$
$$y = 5$$

Answer: The solution pair is (2, 5).

Check: Is (2, 5) a solution of *both* equations? Yes!

$y + 1$	$= 3x$	
$5 + 1$	$3 \cdot 2$	
6	6	✓

$2x + y$	$= 9$	
$2 \cdot 2 + 5$	9	
$4 + 5$		
9		✓

EXAMPLE 2 Solve $x = 2y$ and $x + y = 6$.

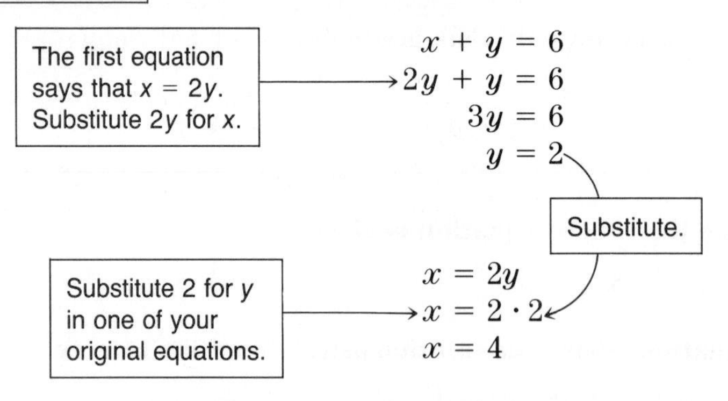

Answer: The solution pair is (4, 2). You may check it.

Classroom Practice

Solve by the substitution method.

1. $y = 4x$
$x + y = 10$

2. $y = 3x$
$x + y = 20$

3. $y = x - 2$
$x + y = 12$

4. $x = 3y$
$5y - x = 8$

5. $x = 2y + 1$
$x + 3y = 16$

6. $x - y = 2$
$3y + x = 14$

7. $x + y = 3$
$5x = 3y - 1$

8. $x - 2y = 5$
$3x + y = -6$

Written Exercises

Solve by the substitution method.

A

1. $y = 2x$
$x + y = 9$

2. $y = 5x$
$y - x = 8$

3. $x = y + 3$
$x + y = 7$

4. $y = x + 1$
$x + y = 5$

5. $y = x + 4$
$x + y = 22$

6. $x = 2 - y$
$2y + x = 9$

7. $y = 3x + 1$
$4x + y = 8$

8. $y = 2x - 3$
$4x + y = 9$

9. $x + y = 2$
$3x + y = 8$

10. $2x + y = 5$
$4x - y = 1$

11. $3x - y = 13$
$2x + 3y = 16$

12. $2x - y = 9$
$5x + 2y = 27$

13. $x = 5y$
$3x = 7y + 16$

14. $x - 5y = 8$
$4x + 2y = 10$

15. $a + 2b = 7$
$2a = 3b$

16. $p - 5q = 6$
$3p - 2q = 5$

Solve.

SAMPLE

$D = rt$
$r = 5t$
Find D in terms of t.

Answer:
$D = rt$
$D = (5t)t$ ← Substitute.
$D = 5t^2$

B

17. $A = bh$
$b = 2h$
Find A in terms of h.

18. $V = Bh$
$B = 2h^2$
Find V in terms of h.

19. $V = lwh$
$l = 3h$
$w = 2h$
Find V in terms of h.

Solve for x, y, and z.

C

20. $x + y + z = 180$
$y = 3x$
$z = 5x$

21. $x + y + z = 62$
$x = 2z - 5$
$y = 3z - 5$

22. $x + 2y + 3z = 0$
$2x + y = 6$
$3x + z = 8$

4 The Add-or-Subtract Method

When solving a pair of equations, you can often add or subtract the equations to get a new equation with just one variable.

EXAMPLE 1 Solve: $\begin{cases} 2x + y = 7 \\ 3x - y = 8 \end{cases}$

Step 1 Add. This makes the y term drop out.

$$\begin{array}{r} 2x + y = 7 \\ 3x - y = 8 \\ \hline 5x + 0 = 15 \end{array}$$

Step 2 Solve the new equation.

$$\begin{aligned} 5x &= 15 \\ x &= 3 \end{aligned}$$

Step 3 Substitute 3 for x in one equation. Find the value of y.

$$\begin{aligned} 2x + y &= 7 \\ 2 \cdot 3 + y &= 7 \\ 6 + y &= 7 \\ y &= 1 \end{aligned}$$

Answer: The solution pair is (3, 1). You may check it.

EXAMPLE 2 Solve: $\begin{cases} 5x + 6y = 3 \\ 5x + 2y = 11 \end{cases}$

Step 1 Subtract to make the x term drop out.

$$\begin{array}{r} 5x + 6y = 3 \\ 5x + 2y = 11 \\ \hline 0 + 4y = -8 \end{array}$$

Step 2 Solve the new equation.

$$\begin{aligned} 4y &= -8 \\ y &= -2 \end{aligned}$$

Step 3 Substitute -2 for y in one equation. Find the value of x.

$$\begin{aligned} 5x + 6y &= 3 \\ 5x + 6(-2) &= 3 \\ 5x - 12 &= 3 \\ 5x &= 15 \\ x &= 3 \end{aligned}$$

Answer: The solution pair is (3, −2). You may check it.

Classroom Practice

Would you add or would you subtract the two equations? Does the x term or does the y term drop out?

1. $x + y = 6$
$x - y = 2$

2. $2x + y = 5$
$2x - y = 3$

3. $5x + 2y = 9$
$3x + 2y = 7$

4. $x + 4y = 5$
$x + y = 2$

5. $2x + 3y = 6$
$2x + y = 2$

6. $x + 2y = 7$
$3x - 2y = 5$

7. $x + 3y = 0$
$2x - 3y = 9$

8. $4x + 2y = 6$
$4x - y = 3$

9–16. Solve the pairs of equations in Exercises 1–8.

Written Exercises

Solve by the addition method.

A

1. $x + y = 2$
$x - y = 10$

2. $5x + 4y = 1$
$3x - 4y = 7$

3. $4x - y = 8$
$2x + y = -2$

4. $5p + 3q = 10$
$2p - 3q = 4$

5. $2a - b = 3$
$4a + b = 9$

6. $2x + y = 10$
$3x - y = 5$

7. $3x - 2y = 8$
$x + 2y = 8$

8. $4r - 7s = 13$
$4r + 7s = -29$

Solve by the subtraction method.

9. $x + 6y = 10$
$x + 2y = 2$

10. $x - y = 20$
$x - 3y = 10$

11. $3x + y = 7$
$-2x + y = -8$

12. $3x - 4y = 21$
$2x - 4y = 18$

13. $3x + 2y = 18$
$x + 2y = 14$

14. $2x - 5y = 14$
$2x - 3y = 10$

Solve by either the addition or the subtraction method.

15. $2a + 3b = 7$
$2a - b = -5$

16. $2s - 5r = 17$
$6s - 5r = 1$

17. $x + y = 9$
$x - y = 5$

18. $x + 2y = 7$
$3x - 2y = 5$

19. $2x + 3y = 8$
$2x - y = -8$

20. $a + 3b = 6$
$2a + 3b = 9$

21. $3x - 2y = 13$
$4x + 2y = 8$

22. $4x - 3y = 9$
$2x - 3y = 3$

23. $3x + y = 0$
$6x - y = 18$

24. $4a - 7b = 13$
$2a - 7b = 3$

25. $4x - 2y = 10$
$-x - 2y = 0$

26. $-4x + y = 7$
$4x + 3y = 5$

Solve by either the addition or the subtraction method.

27. $8x + 3y = 5$
    $x - 3y = 4$

28. $5x - 2y = 7$
    $5x + 3y = 2$

29. $3a - 12b = 9$
    $a - 12b = 11$

30. $2x - 11y = 18$
    $6x - 11y = 10$

31. $6x - 2y = 8$
    $5x - 2y = 5$

32. $4x - y = 15$
    $-4x + 3y = -5$

Can you work with fractions? The following exercises have fractions in the solutions.

SAMPLE

$$\begin{aligned} 2x + 4y &= 2 \\ 6x - 4y &= 2 \\ \hline 8x \quad &= 4 \\ x &= \frac{1}{2} \end{aligned}$$

Substitute.

$$\begin{aligned} 2x + 4y &= 2 \\ 2 \cdot \frac{1}{2} + 4y &= 2 \\ 1 + 4y &= 2 \\ 4y &= 1 \\ y &= \frac{1}{4} \end{aligned}$$

Answer: $\left(\frac{1}{2}, \frac{1}{4}\right)$

B

33. $6x + 9y = 4$
    $6x + 3y = 0$

34. $5a - b = 3$
    $10a + b = 3$

35. $8a + 6b = -1$
    $8a - 4b = 4$

36. $3x + 4y = 5$
    $6x - 4y = 1$

37. $x + 6y = 0$
    $x - 3y = 3$

38. $4m - 3n = 2$
    $8m + 3n = 13$

PUZZLE ◆ PROBLEMS

Here is a stunt you might like to try with one of your friends.

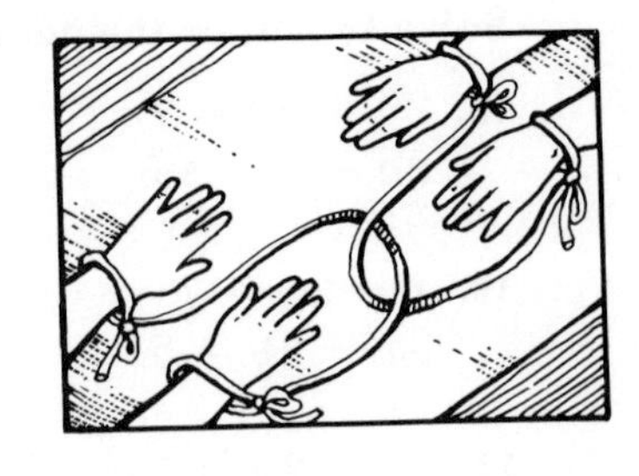

Tie a piece of string to your wrists. Tie another piece of string to your friend's wrists so that the two pieces of string interlock as shown. Now try to separate yourself from your friend without cutting the string, untying the knots, or taking the string off your wrists. It can be done!

Topology is the branch of mathematics that can explain why this stunt can be done. Topology deals with figures and how they can be bent and stretched.

Mixed Practice Exercises

Solve by the graphing method.

1. $y = -x + 3$
$y = 2x - 1$

2. $x + y = 0$
$3x + y = -8$

3. $y + 3x = -7$
$y - 4x = 7$

4. $x + y = 4$
$2x + y = 4$

Solve by the substitution method.

5. $x = 3y$
$x + y = 12$

6. $y = x + 1$
$x + y = 5$

7. $x + 2y = 2$
$x + 3y = 13$

8. $y - 2x = 4$
$y + 4x = 16$

Solve by either the addition or the subtraction method.

9. $x + y = 7$
$2x - y = 5$

10. $2a + 3b = 5$
$2a + b = 3$

11. $3x - 2y = 1$
$-3x + 4y = 7$

12. $4x - 7y = 9$
$6x - 7y = 3$

Solve by the method that seems easiest.

13. $y = x + 2$
$2x + y = 11$

14. $x + y = 8$
$x - 3y = 4$

15. $2x + y = 9$
$3x - y = 6$

16. $x + 2y = 5$
$3x + y = 10$

17. $2x - 3y = 8$
$x + 3y = 7$

18. $x = 2y$
$x + 3y = 5$

19. $y - 2x = 7$
$y + 3x = 2$

20. $y - x = 3$
$y = 2x + 4$

21. $3x - y = 8$
$x + 2y = -2$

22. $2a - 4b = 6$
$-a - 3b = 7$

23. $x + y = -2$
$2x + y = 4$

24. $x - 5y = 2$
$2x + y = 4$

SELF-TEST

Solve by the substitution method. *(8-3)*

1. $y = 2x - 1$
$x + 3y = 11$

2. $x = 3y$
$2x = y + 5$

3. $4p = 3q$
$p - q = 2$

Solve by the addition method. *(8-4)*

4. $3x - y = 2$
$x + y = 6$

5. $2a + 3b = -1$
$a - 3b = 4$

6. $3r + 2s = 4$
$4r - 2s = 10$

Solve by the subtraction method.

7. $x + y = 2$
$x - 3y = -6$

8. $s + 2t = 8$
$3s + 2t = 4$

9. $3x - y = 10$
$x - y = 4$

5 Word Problems, Two Variables

You know how to solve problems by using one variable. Now you can use two variables in problem solving.

EXAMPLE 1 Dot and Ed sold magazines to raise money for a classroom computer.
Dot sold \$30 more in magazines than Ed.
Their combined sales were \$330.
Find each person's sales.

Let a = Ed's sales.
Let b = Dot's sales.

Dot's sales were \$30 more than Ed's.

$$b = 30 + a$$

Their combined sales were \$330.

$$a + b = 330$$

Now solve the two equations. We'll use the substitution method.

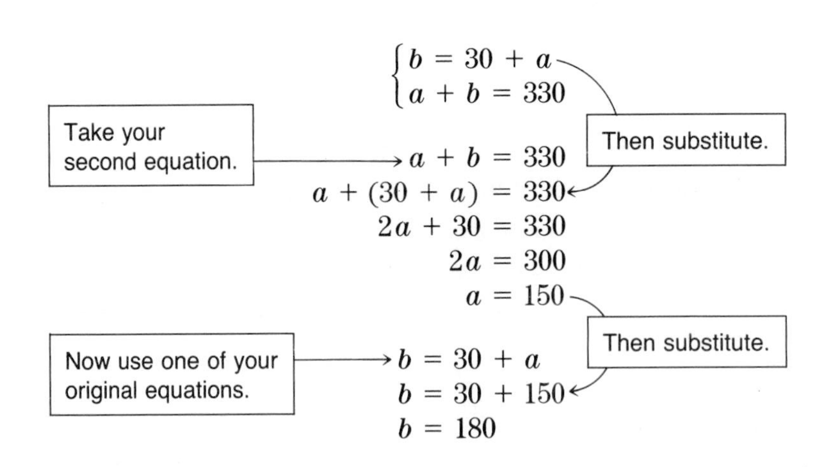

Answer: Ed's sales were \$150. Dot's sales were \$180.

Check: $180 - 150 = 30$ *and* $180 + 150 = 330$ ✓

EXAMPLE 2 The sum of two numbers is 10. ← Fact 1

One number is 4 more than twice the other number. ← Fact 2

Find the numbers.

Let x = one number.
Let y = the other number.

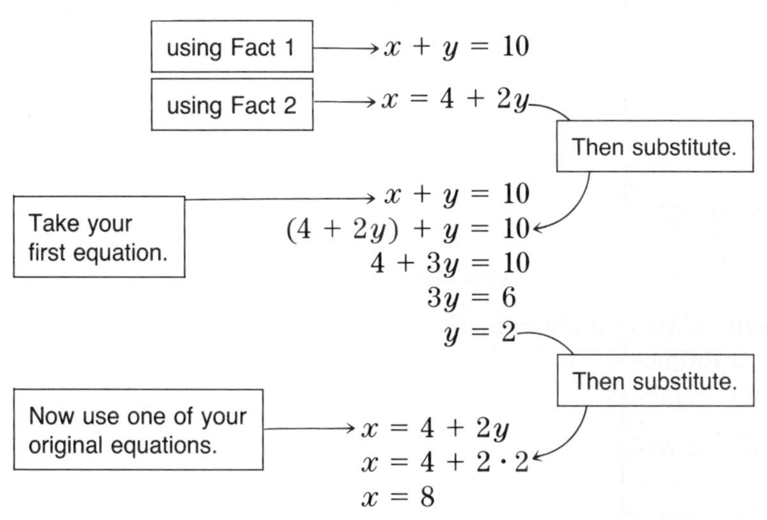

Answer: 8 and 2

Check: $8 + 2 = 10$ *and* $8 = 4 + 2 \cdot 2$ ✓

Classroom Practice

Solve. Use two variables and two equations.

1. One number is 6 more than another.
   The sum of the numbers is 30.
   Find the numbers.

2. The sum of two numbers is 24.
   The difference of the numbers is 16.
   Find the numbers.

3. The length of a rectangle is twice the width.
   The perimeter is 36.
   What are the length and the width?

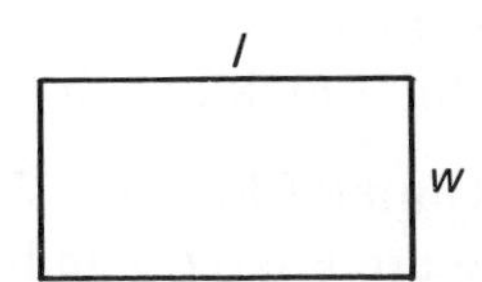

4. Cindy is 2 years older than Marie.
Their ages total 30.
How old is each person?

5. The Jets scored 4 points more than the Vets.
The total of their scores was 38.
How many points did each team score?

6. Mac has a total of 15 dimes and nickels
in his pocket.
He has 7 more dimes than nickels.
How many dimes does he have?

Written Exercises

Solve. Use two variables and two equations.

A

1. The sum of two numbers is 35.
Their difference is 13.
Find the numbers.

2. The sum of two numbers is 48.
One number is 3 times the other.
Find the numbers.

3. One number is 9 more than another number.
The sum of the two numbers is 53.
Find the numbers.

4. One number is 6 times another number.
It is also 12 more than twice the other number.
Find the numbers.

5. A coat costs 3 times as much as a pair of pants.
Together they cost $180.
How much does each cost?

6. Arlene is 7 years older than Todd.
Their ages total 43.
How old is each person?

7. The 1986 rainfall in one city was 18 cm more than in 1985.
The combined rainfall for the two years was 150 cm.
Find the rainfall each year.

8. The Outing Club has 35 members.
There are 3 more climbers than there are skiers.
How many climbers are there?

Solve, using one variable. Then solve again using two variables.

B **9.** The length of a rectangle is three times the width.
The perimeter is 48 cm.
Find the length and the width.

$3x$ / x / one variable

l / w / two variables

10. The Kickers beat the Astros by 3 goals.
There was a total of 13 goals scored in the game.
How many goals did each team score?

11. Norm read 12 more books than Sam did last year.
Altogether, they read a total of 56 books.
How many books did Norm read?

12. I'm thinking of two numbers whose difference is 50.
Twice one number equals three times the other.
Find the numbers.

13. A serving of cereal has 18 g less protein than carbohydrates.
Altogether, there are 32 g of the two nutrients in the serving of cereal.
How much protein is there?

PUZZLE ◆ PROBLEMS

1 ball and 1 cup
balance 12 pennies.

4 balls and 2 pennies
balance the cup.

How many pennies will balance the cup?

6 Using Multiplication

In solving pairs of equations, when the terms in one variable are the same, you know that you can use the addition-or-subtraction method of solving.

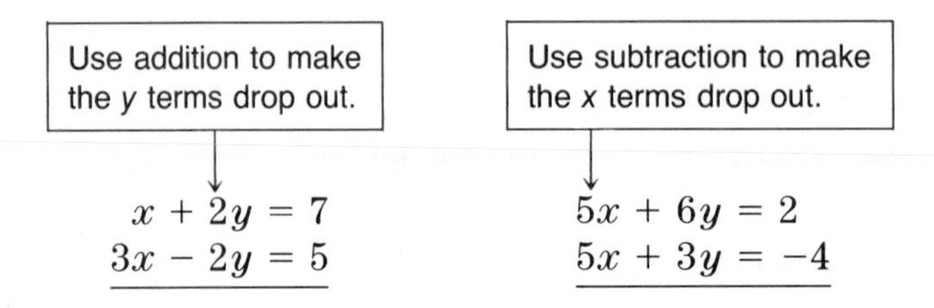

Here's what to do if the terms aren't alike. Just use multiplication to rename one or both equations.

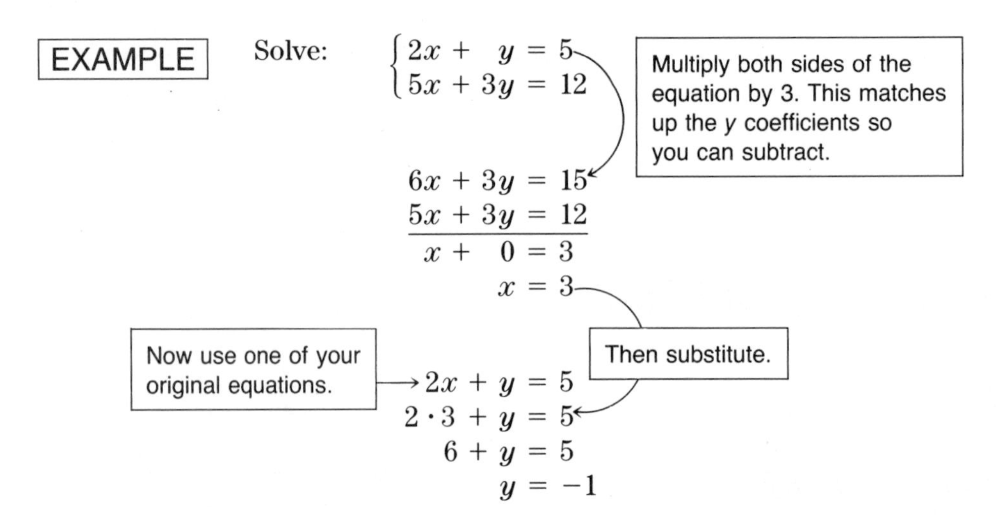

Answer: The solution pair is $(3, -1)$.

Check: Is $(3, -1)$ a solution of *both* equations?

$2x + y = 5$	
$2 \cdot 3 + (-1)$	5
$6 - 1$	
5 ✓	

$5x + 3y = 12$	
$5 \cdot 3 + 3(-1)$	12
$15 - 3$	
12 ✓	

Classroom Practice

Solve.

1. $2x + y = 8$
$3x - 2y = 5$

2. $3x + 5y = 3$
$x + 2y = 13$

3. $2x + y = 5$
$3x - 2y = 4$

4. $a - 3b = 5$
$3a + 2b = 4$

5. $3a + b = 4$
$a - 2b = 6$

6. $x + y = 7$
$3x - 2y = 11$

Written Exercises

Solve.

A

1. $a + b = 9$
$2a - 3b = 3$

2. $x - 4y = 5$
$2x - 7y = 9$

3. $y - 3x = 9$
$2y + x = 4$

4. $a + 3b = 16$
$2a - b = 4$

5. $3m + n = 10$
$m + 2n = 10$

6. $3x - y = 5$
$x + 2y = 11$

7. $3x - 4y = 1$
$-x - 2y = 3$

8. $x - 4y = 3$
$-2x + y = 8$

9. $2x - 3y = 6$
$x - 2y = 3$

10. $6x - 5y = 7$
$2x + 2y = -16$

11. $9x + 6y = 0$
$3x + 5y = -9$

12. $5m - 2n = 14$
$3m + 4n = -2$

13. $3a + 2b = 6$
$-4a + 6b = 44$

14. $5a + 4b = 7$
$3a + 2b = 3$

15. $5x - 9y = -3$
$4x - 3y = 6$

Solve. You'll need to multiply *both* equations, instead of one.

SAMPLE

$2x + 7y = 17$ → Multiply by 3. → $6x + 21y = 51$

$3x + 5y = 9$ → Multiply by 2. → $6x + 10y = 18$

You may finish the solution.

B

16. $2c - 3d = -1$
$3c - 4d = -3$

17. $3r - 2s = 15$
$7r - 3s = 15$

18. $2p + 3q = -1$
$3p + 5q = -2$

19. $5a - 2b = 1$
$-2a + 3b = 4$

20. $2x + 4y = 2$
$3x + 5y = 2$

21. $3x - 2y = -5$
$-4x + 3y = 8$

7 Cost Problems

The cost problems in Chapter 4 involved buying or selling just one item, like tapes. Now that you can work with 2 variables, you can solve cost problems involving 2 items, like tapes and albums.

EXAMPLE 1

Jack buys 2 tapes and an album for \$21.25.
Nancy buys 3 tapes and 2 albums for \$36.00.
How much does each item cost?

Let t = cost of a tape, in cents.
Let a = cost of an album, in cents.

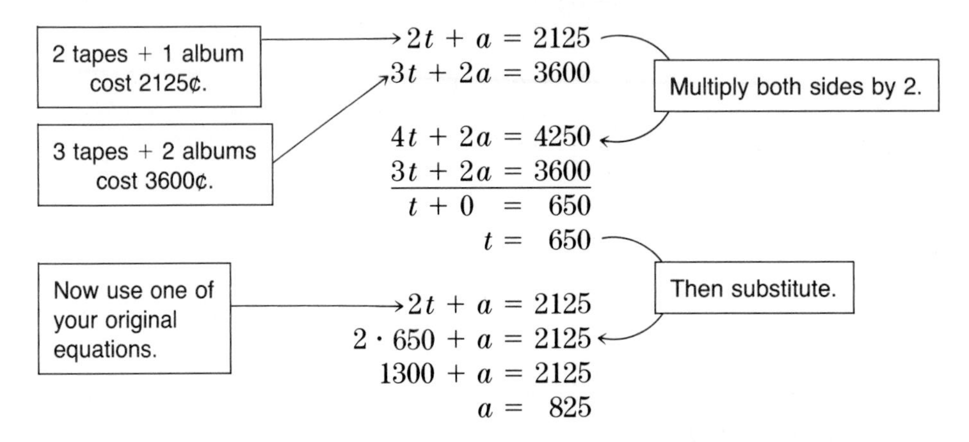

Answer: A tape costs \$6.50. An album costs \$8.25.

Check: 2 tapes + 1 album: \$13.00 + \$8.25 = \$21.25 ✓
3 tapes + 2 albums: \$19.50 + \$16.50 = \$36.00 ✓

Written Exercises

Solve. Use two variables and two equations.

A

1. 2 plates and 1 glass cost \$3.70. ⟶ $2p + g = 370$
   1 plate and 2 glasses cost \$3.35. ⟶ ___?___
   How much does each item cost?

2. 3 chairs and a table cost \$149. ⟶ $3c + t = 149$
   4 chairs and a table cost \$177. ⟶ ___?___
   How much does each item cost?

3. Two adult tickets and 3 student tickets cost \$12. ⟶ ___?___
   One adult ticket and 2 student tickets cost \$7. ⟶ ___?___
   How much does each kind of ticket cost?

4. Three bars of soap and 2 tubes of toothpaste cost \$4.05.
   Two bars of soap and 3 tubes of toothpaste cost \$5.20.
   Find the price of each item.

5. Four cans of tuna and 2 boxes of rice cost \$7.40.
   Six cans of tuna and 2 boxes of rice cost \$9.70.
   Find the cost of each item.

B

6. Bea earned \$285 for 40 hours at regular wages and
   5 hours at the overtime rate.
   Frank earned \$303 for 40 hours at the same regular wage and
   7 hours at the same overtime rate as Bea.
   What is the regular wage rate?
   What is the overtime pay rate?

7. A rugby team bought 20 rugby balls.
   Some cost \$16 each, and some cost \$28 each.
   The total cost was \$380.
   Complete the table.
   Find x and y.

	price	× number	= cost
Expensive balls	28	x	?
Cheaper balls	16	y	?
		20	?

SELF-TEST

Solve. Use two variables and two equations.

1. One number is 17 more than another number. *(8-5)*
   The sum of the two numbers is 25.
   Find the numbers.

2. Yvette is 3 years younger than Lauren.
   Their ages total 33.
   How old is each person?

Solve.

3. $4x + 4y = 8$
   $2x + 3y = 4$

4. $-12m + 3n = 18$
   $2m + 2n = 0$

5. $x + 2y = 0$
   $-3x + 4y = 10$

(8-6)

Solve. Use two variables and two equations.

6. Four brushes and 2 combs cost \$8.20. *(8-7)*
   Two brushes and 3 combs cost \$5.10.
   How much does each item cost?

8 Boat and Aircraft Problems (Optional)

You probably know that a boat may move downstream on a river more quickly than it moves upstream. This is because of the moving water.

EXAMPLE A motorboat went 18 km downstream in 1 hour. The return trip took 3 hours. Find the rate of the flowing water.

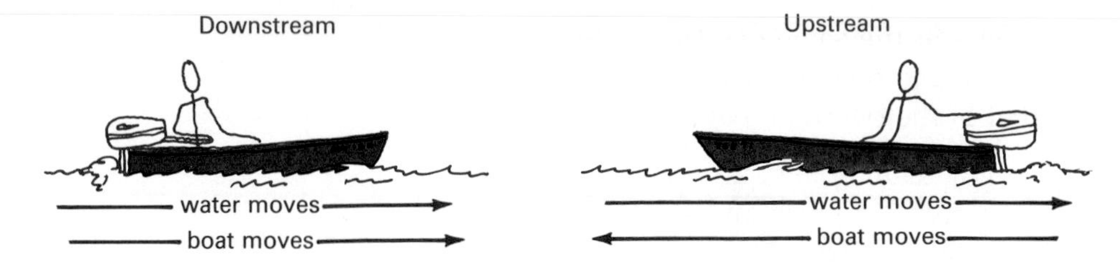

Let x = rate of the boat in still water.
Let y = rate of the flowing water.

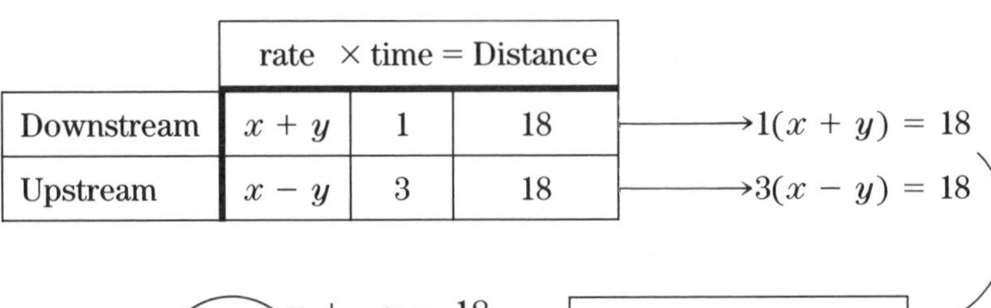

	rate	× time	= Distance
Downstream	$x + y$	1	18
Upstream	$x - y$	3	18

$\rightarrow 1(x + y) = 18$

$\rightarrow 3(x - y) = 18$

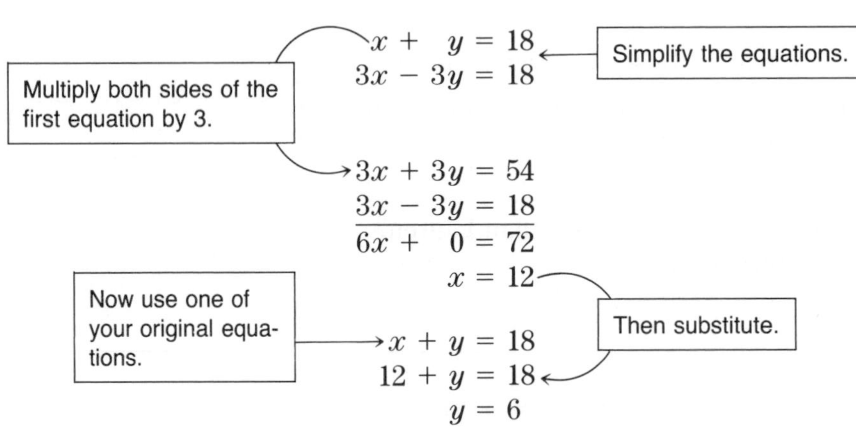

$$x + y = 18$$
$$3x - 3y = 18$$

$$3x + 3y = 54$$
$$3x - 3y = 18$$
$$6x + 0 = 72$$
$$x = 12$$

$$x + y = 18$$
$$12 + y = 18$$
$$y = 6$$

Answer: The water was flowing at 6 km/h.

Check:

downstream rate:
$x + y = 12 + 6 = 18$
or 18 km in 1 hour ✓

upstream rate:
$x - y = 12 - 6 = 6$
or 6 km in 1 hour
or 18 km in 3 hours ✓

Written Exercises

Solve.

A

1. A motorboat can go 12 kilometers downstream in 2 hours.
   The return trip takes 3 hours.
   Find the speed of the boat in still water.

2. A motorboat can go 30 kilometers downstream in 3 hours.
   The return trip takes 5 hours.
   Find the speed of the boat in still water.

Aircraft problems are like motorboat problems. Aircraft can fly *with* the wind more quickly than they can fly *into* the wind.

3. An airplane takes 3 hr flying 1200 km into the wind.
   With the same wind, the return trip takes 2 hr.
   What is the speed of the wind?

   Let x = airplane's rate with no wind.
   Let y = the wind speed.

	rate	× time	= Distance
Flying *with* the wind	$x + y$	2	1200
Flying *into* the wind	?	?	?

4. An airplane can go 1560 km in 5 hours flying into the wind.
   With the same wind, the return trip takes 4 hours.
   What is the speed of the wind?

   Let x = airplane's rate with no wind.
   Let y = the wind speed.

	rate	× time	= Distance
Flying *with* the wind	$x + y$	?	?
Flying *into* the wind	?	?	?

5. An airplane flies 1656 km into the wind in 4 hours.
   With the same wind, the return trip takes 3 hours.
   Find the airplane's rate with no wind.

   Let x = airplane's rate with no wind.
   Let y = the wind speed.

	rate	× time	= Distance
Flying *with* the wind	$x + y$	?	?
Flying *into* the wind	?	?	?

CONSUMER APPLICATIONS

BANK STATEMENTS

If you have a checking account, you will receive a statement from the bank each month. With the bank statement, you will usually get back all those checks that have been processed and cancelled by the bank. The statement will show a record of these processed checks and the deposits you made during the statement period.

Below is a sample of a bank statement.

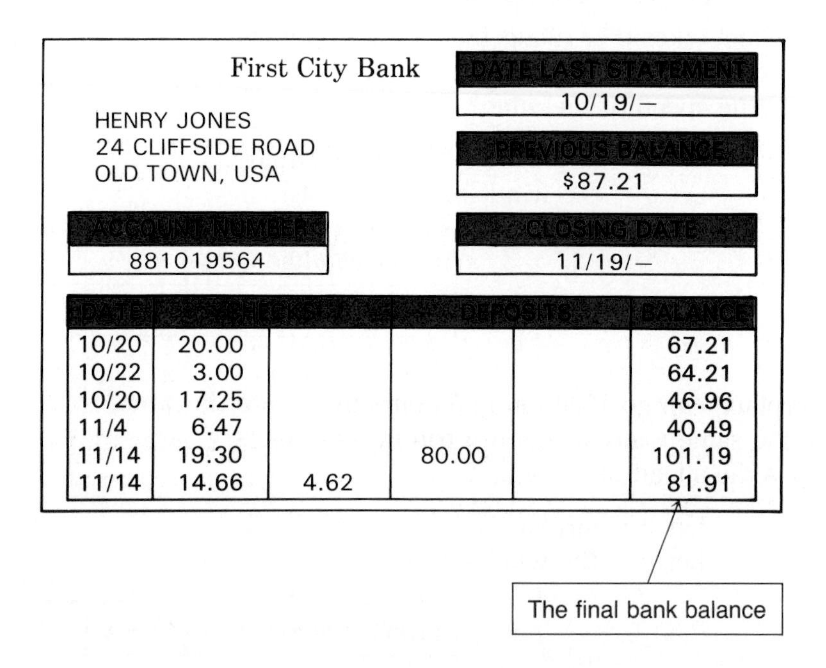

First City Bank

HENRY JONES
24 CLIFFSIDE ROAD
OLD TOWN, USA

DATE LAST STATEMENT: 10/19/–

PREVIOUS BALANCE: $87.21

ACCOUNT NUMBER: 881019564

CLOSING DATE: 11/19/–

DATE	CHECKS		DEPOSITS		BALANCE
10/20	20.00				67.21
10/22	3.00				64.21
10/20	17.25				46.96
11/4	6.47				40.49
11/14	19.30		80.00		101.19
11/14	14.66	4.62			81.91

Here's what you can do to make sure your checkbook balances.

1. Did you make any deposits that are not on the statement? Add that amount to the balance on the statement.
2. Arrange the returned cancelled checks by number. Mark in your register those checks that have been processed.
3. Have you written some checks that have not been processed? Subtract the total of these checks from the balance you found in step 1.
4. Does the final figure check with the balance you have on the last line in your register? It should!

READING ALGEBRA

WORDS, TABLES, AND DIAGRAMS

Often there are several different ways of stating the same facts. This book frequently uses words, tables, and diagrams together to help you gain a better understanding of algebra. If the words of a problem are hard for you, a table or a diagram may make the relationships clear.

Motion problems such as the ones on pages 126–129 become easier to understand when tables and diagrams are used. Notice that direction of motion and distance traveled can be shown very clearly by a diagram. Tables help you match up known and unknown values—time with time, distance with distance, and so on. Then it is easy to set up an equation.

Sometimes a diagram can be used to picture an important concept. For example, see Exercise 37 on page 169. Squaring a binomial is not just something that takes place in your head—it can be illustrated by a diagram that shows the meaning of $a^2 + 2ab + b^2$ in terms of areas.

The idea that an ordered pair of numbers can be pictured by a point in the plane is very valuable in algebra. A table of solution pairs can tell you a lot about an equation, and the graph of the equation can tell you even more. Graphs are especially useful when you are working with two equations in two variables.

EXERCISES

1. Anita made the following table for a problem about ticket sales at a pet show. See if you can work backward from the table and write a problem about the price of the tickets.

	price ×	number =	cost
child's ticket	x	85	?
adult's ticket	$2x$	34	?
			153

2. Look at Example 2 on page 263. What does the graph tell you that you cannot discover by looking at the given tables of ordered pairs?

PROBLEM SOLVING STRATEGIES

GUESSING WITHIN LIMITS

One way to understand a problem is to guess a possible answer and then test your guess. Before you guess, however, you should try to find the least and the greatest possible answers. These values set limits on the choice of values. Consider the following example.

EXAMPLE

Ryan has pigs and chickens on his farm.
He has 22 animals, and the number of feet they have totals 56.
How many pigs and chickens does Ryan have?

Find the greatest possible answers to the problem. What are the greatest number of pigs and the greatest number of chickens that Ryan could have?

If Ryan had *all pigs* and *no chickens* he could have *at most* $56 \div 4$, or 14, pigs (since pigs have 4 feet).

If Ryan had *all chickens* and *no pigs* he could have *at most* $56 \div 2$, or 28, chickens. Since the farm has only 22 chickens and pigs, we know that Ryan could have no more than 22 chickens.

Looking at the greatest possible answers to the problem, you can see that Ryan can have no more than 14 pigs and no more than 22 chickens.

A. Now you can guess an answer to the problem between these two greatest possible answers. You know, for instance, that a guess of 16 pigs and 6 chickens will not answer the problem, since $16 > 14$.

A guess of 12 pigs and 10 chickens, however, might be the answer. Check this guess by testing to see if the answer satisfies all the facts given in the problem. (It does not; there would be too many feet.) While you are testing your guess you may get clues about how to set up the equation(s) needed to solve the problem.

B. If you solve the problem by writing and solving equations, you can use the greatest possible answers to check your final solution. Your solution will make sense if it falls between the limits of the greatest possible answers.

EXERCISES

In Exercises 1–4, show that the guess in parentheses cannot be the answer to the problem. (*Hint:* Show that the guess does not satisfy all the conditions of the problem.)

1. A store bought several wool coats at $70 each.
   The store sold all but 10 coats at $95 each.
   The store's profit was $500.
   How many coats did the store buy?
   (The store bought 60 coats.)

2. The volleyball team at South High won three fourths of the games they played last season.
   They lost nine of the games they played.
   How many games did the team play last season?
   (The team played 24 games.)

3. A cup and a saucer together cost $2.50.
   The cup costs $1.20 more than the saucer.
   Find the cost of the cup. (The cup costs $1.75.)

4. Andrew is 30 years older than his son.
   Four years ago, he was three times as old as his son was.
   How old is Andrew now? (He is 44 years old.)

In Exercises 5 and 6, (a) find the least and greatest possible values for the answer, (b) guess and test a possible answer, and (c) solve the problem, comparing the answer to the limits found in part (a).

5. Altogether Byung Kyu has 51 albums and tapes.
   He has 23 more tapes than albums.
   How many tapes does he have?

6. Marcella has 100 m of fencing for a rectangular garden.
   The length of the garden is 10 m less than twice the width.
   What is the width of the garden?

Solve.

7. Acme Auto Body has 13 cars and motorcycles to repair.
   The vehicles have a total of 42 wheels.
   How many cars need to be repaired?
   a. Find the greatest possible number of cars.
   b. Find the greatest possible number of motorcycles.
   c. Guess and test one possible answer.
   d. Solve. Compare your answer to the limits found in parts (a) and (b).

SKILLS REVIEW

MULTIPLYING AND DIVIDING FRACTIONS

Multiply. Simplify if possible.

SAMPLE $\frac{3}{4} \cdot \frac{5}{8} = \frac{3 \cdot 5}{4 \cdot 8} = \frac{15}{32}$

Multiply the numerators.

Multiply the denominators.

1. $\frac{1}{5} \cdot \frac{2}{3}$
2. $\frac{1}{3} \cdot \frac{1}{2}$
3. $\frac{1}{2} \cdot \frac{5}{6}$
4. $\frac{3}{7} \cdot \frac{1}{2}$
5. $\frac{2}{5} \cdot \frac{3}{4}$
6. $\frac{9}{10} \cdot \frac{1}{4}$
7. $\frac{4}{5} \cdot \frac{2}{3}$
8. $\frac{2}{7} \cdot \frac{2}{3}$
9. $\frac{4}{5} \cdot \frac{1}{2}$
10. $\frac{8}{9} \cdot \frac{1}{4}$
11. $\frac{5}{7} \cdot \frac{4}{15}$
12. $\frac{7}{10} \cdot \frac{5}{6}$
13. $\frac{4}{9} \cdot \frac{3}{4}$
14. $\frac{5}{6} \cdot \frac{2}{3}$
15. $\frac{3}{8} \cdot \frac{4}{9}$
16. $\frac{6}{7} \cdot \frac{14}{15}$

Divide. Simplify if possible.

SAMPLE $\frac{2}{3} \div \frac{3}{4} = \frac{2}{3} \cdot \frac{4}{3} = \frac{2 \cdot 4}{3 \cdot 3} = \frac{8}{9}$

To divide by $\frac{3}{4}$, multiply by $\frac{4}{3}$.

17. $\frac{1}{3} \div \frac{1}{2}$
18. $\frac{1}{2} \div \frac{1}{3}$
19. $\frac{2}{5} \div \frac{3}{4}$
20. $\frac{5}{8} \div \frac{1}{5}$
21. $\frac{4}{5} \div \frac{1}{2}$
22. $\frac{4}{5} \div 2$
23. $\frac{3}{8} \div \frac{1}{5}$
24. $\frac{3}{8} \div 5$
25. $\frac{3}{5} \div \frac{3}{4}$
26. $\frac{7}{10} \div \frac{14}{15}$
27. $\frac{5}{12} \div \frac{2}{3}$
28. $\frac{4}{7} \div \frac{8}{14}$
29. $\frac{7}{8} \div \frac{3}{8}$
30. $\frac{9}{10} \div \frac{3}{5}$
31. $\frac{7}{9} \div \frac{2}{3}$
32. $\frac{15}{16} \div \frac{5}{8}$

CHAPTER REVIEW

CHAPTER SUMMARY

1. If the graphs of two equations intersect at a point (a, b), then (a, b) is a solution of both equations.
2. If the graphs of two equations have different slopes, the equations share one solution pair. If the slopes are the same, the equations share either no solution pair or all solution pairs.
3. The following are methods of solving pairs of equations.

   The Graphing Method

   The Substitution Method

   The Add-or-Subtract Method

   Using Multiplication before Adding or Subtracting
4. You need to use two variables and two equations in solving many word problems.

REVIEW EXERCISES

Solve by the graphing method. ***(See pp. 262–265.)***

1. $y = x + 2$
   $y = 2x$
2. $y + x = 3$
   $y - x = 1$
3. $y = 2x + 5$
   $y = -x - 1$
4. $3x + y = 0$
   $y = 2x + 5$
5. $x + y = -2$
   $y - x = -6$
6. $y = 4x + 1$
   $y - 2x = 1$

Find the slope of the line whose equation is given. ***(See pp. 266–269.)***

7. $y = 9x$
8. $y = -x + 3$
9. $y - x = 9$
10. $y + 2x = 0$
11. $y = \frac{1}{2}x + 4$
12. $\frac{2}{3}x - y = 2$

Tell whether the equations share *one* solution pair, *no* solution pair, or *all* solution pairs. ***(See pp. 266–269.)***

13. $y = 4x$
    $y = 4x + 3$
14. $y = x + 4$
    $y + x = -4$
15. $y + 5x = 3$
    $y - 3 = -5x$
16. $y - 7x = 2$
    $y + 2 = -7x$
17. $y - 6x = 2$
    $y = 6x + 5$
18. $y + 2x = 6$
    $y = 6 + 2x$

Solve by the substitution method. *(See pp. 270–271.)*

19. $y = 3x$
$x + y = 12$

20. $y = -4x$
$x - y = -10$

21. $y = 2x + 1$
$3x + y = 6$

22. $5x - y = 0$
$x + 2y = 22$

23. $3x - y = 1$
$2x + 3y = 8$

24. $y + 4x = 5$
$5x + y = 0$

Solve by either the addition or the subtraction method. *(See pp. 272–274.)*

25. $2x + y = 0$
$x - y = 6$

26. $-3x + 2y = 5$
$3x + y = 7$

27. $x + 4y = 16$
$-x + 3y = -2$

28. $x + 3y = -1$
$-2x - 3y = 5$

29. $5x + 2y = -18$
$3x + 2y = -6$

30. $3x + 2y = 5$
$-3x + 2y = -1$

31. $x + 2y = 7$
$x - y = 1$

32. $2x + 3y = 7$
$x + 3y = 2$

33. $2x - y = 8$
$2x + 2y = 14$

Solve. *(See pp. 280–281.)*

34. $2x + 3y = 12$
$4x + y = 14$

35. $5x - 2y = 11$
$3x - 6y = -3$

36. $3x + 4y = 1$
$6x + 5y = -1$

37. $3x - 2y = 0$
$7x - 8y = 10$

38. $3x - 4y = -1$
$5x + 8y = 13$

39. $9x - 2y = -3$
$-3x + 5y = 27$

Solve. Use two variables and two equations. *(See pp. 276–279, 282–283.)*

40. The sum of two numbers is 27.
Their difference is 13.
Find the numbers.

41. One number is 14 more than another.
The sum of the two numbers is 20.
Find the numbers.

42. One saw and 4 hammers cost $72.
Two saws and 6 hammers cost $114.
How much does each item cost?

43. One adult ticket and 3 student tickets cost $11.50.
Three adult tickets and 2 student tickets cost $17.00.
How much does each kind of ticket cost?

(Optional) Solve. *(See pp. 284–285.)*

44. A motorboat can go 36 km downstream in 2 hours.
The return trip takes 3 hours.
Find the speed of the boat in still water.

CHAPTER TEST

By graphing, find the ordered pair that is a solution of both equations. In Exercise 3, your answer does not have to be exact.

1. $y = -x + 2$
$y = x + 4$

2. $y - 4x = 0$
$y + 2x = -6$

3. $2x - y = -2$
$2x + y = 8$

(8-1)

Tell whether the equations share *one* solution pair, *no* solution pair, or *all* solution pairs.

4. $3x - y = 0$
$y - 3x = 9$

5. $x + y = 8$
$x - y = 3$

6. $y + 1 = 4x$
$y - 4x = -1$

(8-2)

Solve by the substitution method.

7. $y = x - 4$
$3x + y = 12$

8. $3x - y = 1$
$2x - 3y = 10$

9. $a = 2b + 1$
$6a - 7b = 16$

(8-3)

Solve by either the addition or the subtraction method.

10. $3x + 2y = 7$
$3x + y = 11$

11. $2x - 3y = 5$
$x + 3y = 16$

12. $3x + 4y = -2$
$5x + 4y = 2$

(8-4)

Solve. Use two variables and two equations.

13. The sum of two numbers is 16. Their difference is 8. Find the numbers. *(8-5)*

14. Walt has 16 coins in pennies and nickels. The coins are worth 40¢. How many nickels does he have?

Solve.

15. $5x - 3y = 7$
$7x - 6y = 8$

16. $3x + 2y = 2$
$-x + 3y = 25$

17. $2x - 5y = 9$
$5x - 7y = -5$

(8-6)

18. Five albums and 2 tapes cost $57. Four albums and 1 tape cost $42. How much does each item cost? *(8-7)*

19. Two adult tickets and 3 student tickets cost $14. Four adult tickets and 5 student tickets cost $26. Find the price of each kind of ticket.

20. (Optional) A motorboat travels 48 miles downstream in 3 hours. The return trip takes 4 hours. Find the rate of the flowing water. *(8-8)*

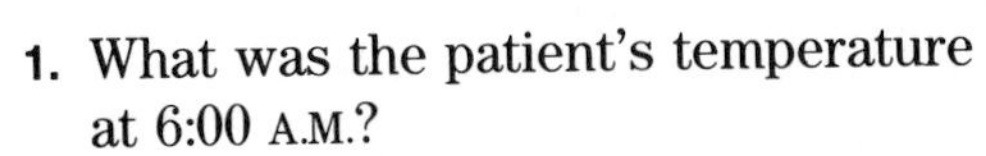

UNIT D CUMULATIVE REVIEW

The graph at the right shows a patient's hourly temperature from midnight to 6:00 A.M.

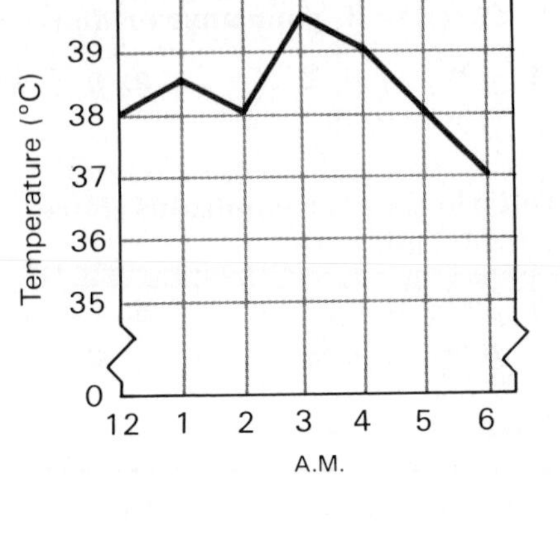

1. What was the patient's temperature at 6:00 A.M.?
2. At what time was it highest?
3. Between which two readings did it rise most?
4. Was it ever below normal (37°C)?
5. Solve $3x - y = 2$ for y. Then complete the table.

x	-2	-1	0	1	2
y	?	?	?	?	?

The cost of several notebooks at 40¢ each is a function of the number of notebooks you buy.

6. Complete the table.

n = number you buy	1	2	5	8
C = cost (in cents)	?	?	?	?

7. An equation relating n and C in Exercise 6 is: $C = \underline{\ ?\ }$.
8. Solve by graphing: $y - 2x = 5$
   $y = x + 3$
9. Find the slope of the line with equation $2x + y = 5$.
10. Tell whether the equations $x - y = 1$ and $x + y = 3$ share *one* solution pair, *no* solution pair, or *all* solution pairs.

Solve.

11. $y = 2x + 1$
    $x + 3y = 10$
12. $x + y = 9$
    $x - y = 1$
13. $5x - 2y = 7$
    $5x + 3y = 2$
14. $5x - 2y = 2$
    $2x - 3y = 3$
15. The sum of two numbers is 56. One number is 8 more than the other. Find the numbers.
16. 3 apples and 2 pears cost \$1.50. 2 apples and 1 pear cost \$.85. How much does each item cost?

UNIT

E

Here's what you'll learn in this chapter:

To use factoring to simplify fractions.

To solve word problems involving ratios and proportions.

To add, subtract, multiply, and divide fractions.

To solve equations with fractions.

To solve work problems.

Ratio and proportion, which you will study in this chapter, were used in the construction of the scale model shown. Each measurement in the model dome is in proportion to the corresponding measurement in the full-sized dome.

Chapter 9

Working with Fractions

1 Fractions in Algebra

Working with fractions in algebra is like working with fractions in arithmetic. You already know that fractions can be used to express division.

In Arithmetic	In Algebra
$\frac{5}{6}$ 5 ← Numerator, 6 ← Denominator	$\frac{2}{y}$ 2 ← Numerator, y ← Denominator The denominator cannot be 0. $y \neq 0$

You can simplify a fraction by dividing both its numerator and denominator by the same number. Keep in mind that when you simplify a fraction, you do not find a new number. You just find a simpler way of writing it.

$$\frac{8}{12} = \frac{2 \cdot \cancel{4}^{1}}{3 \cdot \cancel{4}_{1}} = \frac{2}{3}$$

Both numerator and denominator are divided by **4.**

$$\frac{2a}{6a} = \frac{\cancel{2}^{1} \cdot \cancel{a}^{1}}{3 \cdot \cancel{2}_{1} \cdot \cancel{a}_{1}} = \frac{1}{3}$$

Both numerator and denominator are divided by **2a.**

Study a few more examples. Watch the signs.

EXAMPLE 1

$$\frac{-6x}{12x} = \frac{\cancel{-6}^{-1} \cdot \cancel{x}^{1}}{\cancel{12}_{2} \cdot \cancel{x}_{1}} = \frac{-1}{2} = -\frac{1}{2}$$

negative ÷ positive = negative

$\frac{-1}{2} = -\frac{1}{2}$

EXAMPLE 2

$$\frac{4a}{-20a} = \frac{\cancel{4}^{1} \cdot \cancel{a}^{1}}{\cancel{-20}_{-5} \cdot \cancel{a}_{1}} = \frac{1}{-5} = -\frac{1}{5}$$

positive ÷ negative = negative

$\frac{1}{-5} = -\frac{1}{5}$

EXAMPLE 3

$$\frac{-3xy}{-6xy^2} = \frac{\cancel{-3}^{-1} \cdot \cancel{x}^{1} \cdot \cancel{y}^{1}}{\cancel{-6}_{-2} \cdot \cancel{x}_{1} \cdot \cancel{y}_{1} \cdot y} = \frac{1}{2y}$$

negative ÷ negative = positive

$\frac{-1}{-2y} = \frac{1}{2y}$

Classroom Practice

Simplify.

1. $\frac{6}{8}$
2. $\frac{2}{14}$
3. $\frac{-3}{6x}$
4. $\frac{4xy}{-8x}$
5. $\frac{a}{ab}$
6. $\frac{x^3}{x}$
7. $\frac{abc}{bcd}$
8. $\frac{-9y^2}{-18xy}$
9. $\frac{8xy^2}{24xy}$
10. $\frac{24r^2s}{24r^2s^3}$

Written Exercises

Simplify.

A

1. $\frac{10}{15}$
2. $\frac{9}{12}$
3. $\frac{3}{18a}$
4. $\frac{7}{21g}$
5. $\frac{-5x}{10x}$
6. $\frac{3xy}{9y}$
7. $\frac{7ab}{14ab}$
8. $\frac{4x}{12xy}$
9. $\frac{-8x^2}{6x}$
10. $\frac{12abc}{9bcd}$
11. $\frac{28r}{21r^2}$
12. $\frac{6a}{18a^2}$
13. $\frac{4ab^2}{12b}$
14. $\frac{-9st}{-6s^2}$
15. $-\frac{3a^2b}{15b}$
16. $\frac{5x^2y}{10xy^2}$
17. $\frac{8ab}{4a^2b}$
18. $\frac{9x^2y}{15x}$
19. $\frac{8a^2x}{4ax^2}$
20. $\frac{9mn^2}{7n^2}$
21. $\frac{10x^2y}{25y^2}$
22. $\frac{-6rs}{15r^2s}$
23. $\frac{10m^2k}{8mk}$
24. $\frac{9n^2r}{24nr^2}$
25. $\frac{4pq^2}{32qr}$
26. $\frac{7a}{28ab^2c}$
27. $\frac{6y^2z}{-3y}$
28. $\frac{12a^2b^4}{4a^3b}$
29. $\frac{10x^2w^3}{35xw^4}$
30. $\frac{12y^3t^3}{19y^2t^4}$

B

31. $\frac{125a^2b^3c}{-25a^3bc^2}$
32. $\frac{-76xy^2z^3}{19x^2z}$
33. $\frac{12rst^3}{144r^2s^3t}$
34. $\frac{-21x^2yz^3}{-84xy^2z^2}$
35. $\frac{121ab^2c^3}{-11a^2b^3}$
36. $\frac{-13r^3s^2t}{104rs^3t^2}$
37. $\frac{102x^3y^2}{-18x^2y^3z}$
38. $\frac{-24a^3b^2c^2}{-168a^2b^3c}$

A fraction has no meaning if the denominator is zero. State the value of *x* for which the fraction has no meaning.

39. $\frac{7}{x}$
40. $\frac{4}{x-1}$
41. $\frac{3}{x-2}$
42. $\frac{2x}{x+1}$
43. $\frac{8}{3x-6}$

2 Simplifying Fractions

Fractions having several terms in the numerator or denominator can often be simplified too.

Factor the numerator. →

$$\frac{4x-8}{8} = \frac{\overset{1}{\cancel{4}}(x-2)}{\underset{2}{\cancel{8}}} = \frac{x-2}{2}$$

WARNING! In examples like these, write out the factors first.

Correct →

$$\frac{4x-8}{8} = \frac{\overset{1}{\cancel{4}}(x-2)}{\underset{2}{\cancel{8}}} = \frac{x-2}{2}$$

Not Correct →

$$\frac{4x - \overset{1}{\cancel{8}}}{\underset{1}{\cancel{8}}} = \frac{4x-1}{1} = 4x - 1$$

Here are a few more examples to study.

EXAMPLE 1

$$\frac{x^2-3x}{2x-6} = \frac{x(x-3)}{2(x-3)}$$

$$= \frac{x\overset{1}{\cancel{(x-3)}}}{2\underset{1}{\cancel{(x-3)}}}$$

Divide the numerator and denominator by **(x − 3).**

$$= \frac{x}{2}$$

EXAMPLE 2

$$\frac{3t-15}{t^2-25} = \frac{3(t-5)}{(t-5)(t+5)} = \frac{3}{t+5}$$

EXAMPLE 3

$$\frac{a^2-2a+1}{a^2-a} = \frac{(a-1)(a-1)}{a(a-1)}$$

$$= \frac{a-1}{a}$$

Classroom Practice

Simplify.

1. $\frac{3a + 3b}{3a}$
2. $\frac{x + 1}{2x + 2}$
3. $\frac{y - 4}{2y - 8}$
4. $\frac{x^2 - 1}{x - 1}$
5. $\frac{x^2 + x}{x + 1}$
6. $\frac{a - b}{2a - 2b}$
7. $\frac{a^2 - 4}{a + 2}$
8. $\frac{x^2 + 2x + 1}{x^2 - 1}$
9. For what values of x does the fraction $\frac{5}{x^2 - 4}$ have no meaning?

Written Exercises

Simplify.

A

1. $\frac{x + 1}{3x + 3}$
2. $\frac{x - 1}{2x - 2}$
3. $\frac{3p + 3q}{7p + 7q}$
4. $\frac{2a - 2b}{2a + 2b}$
5. $\frac{x^2 + 2x}{x + 2}$
6. $\frac{a + 1}{a^2 + a}$
7. $\frac{3y}{9y^2 - 6y}$
8. $\frac{6a + 6b}{9a + 9b}$
9. $\frac{2c + 6}{c^2 - 9}$
10. $\frac{a^2 - 4}{a - 2}$
11. $\frac{2x + 10}{x^2 - 25}$
12. $\frac{x^2 - y^2}{x + y}$
13. $\frac{n^2 - 9}{n^2 + 3n}$
14. $\frac{y^2 - 4y}{2y - 8}$
15. $\frac{2xy - 6x}{y^2 - 3y}$
16. $\frac{2x - 4xy}{1 - 2y}$
17. $\frac{5k + 10}{(k + 2)^2}$
18. $\frac{(m - 4)^2}{3m - 12}$
19. $\frac{6u + 6v}{u^2 - v^2}$
20. $\frac{g^2 - h^2}{5g - 5h}$
21. $\frac{12x^2y}{3xy^2 + 6x^2y}$
22. $\frac{3x - 9xy}{1 - 3y}$
23. $\frac{x^2 + 6x + 9}{x^2 - 9}$
24. $\frac{y^2 - 4y + 4}{y^2 - 4}$

B

25. $\frac{2x^2 - 50}{2x + 10}$
26. $\frac{2ab^2 + 2a^2b}{4a^2 - 4b^2}$
27. $\frac{3x^2y}{6x^2y + 3xy^2}$
28. $\frac{x^2 - 8x + 15}{x^2 - 2x - 15}$
29. $\frac{x^2 - x - 6}{x^2 + 5x + 6}$
30. $\frac{x^2 - 6x + 8}{x^2 - x - 2}$
31. $\frac{y^2 - 4y - 5}{y^2 - 2y - 3}$
32. $\frac{t^2 + 7t + 10}{t^2 - t - 6}$
33. $\frac{z^2 - 12z + 20}{z^2 + 4z - 12}$

3 The −1 Factor

You know that 1 is a factor of every number, even though the factor 1 is not written very often.

$$2 = 1 \cdot 2 \qquad x = 1 \cdot x \qquad a - 1 = 1(a - 1)$$

You can also write −1 as a factor of every number.

$$2 = -1(-2) \qquad x = -1(-x) \qquad a - 1 = -1(-a + 1)$$
$$\text{or} \quad a - 1 = -1(1 - a)$$

Now see how useful this −1 factor can be in simplifying fractions.

EXAMPLE 1 $\dfrac{a-1}{1-a} = \dfrac{a-1}{-1(a-1)} = \dfrac{\overset{1}{\cancel{a-1}}}{-1\underset{1}{\cancel{(a-1)}}} = \dfrac{1}{-1} = -1$

EXAMPLE 2 $\dfrac{b-a}{2a-2b} = \dfrac{b-a}{2(a-b)} = \dfrac{-1(a-b)}{2(a-b)} = \dfrac{-1}{2} = -\dfrac{1}{2}$

EXAMPLE 3 $\dfrac{x^2-9}{3-x} = \dfrac{(x-3)(x+3)}{3-x} = \dfrac{(x-3)(x+3)}{-1(x-3)} = \dfrac{x+3}{-1} = -x-3$

Classroom Practice

Complete.

1. $3 = -1(\underline{\ ?\ })$

2. $7 = -1(\underline{\ ?\ })$

3. $17 = -1(\underline{\ ?\ })$

4. $2x = -1(\underline{\ ?\ })$

5. $-x - y = -1(\underline{\ ?\ })$

6. $-x + y = -1(\underline{\ ?\ })$

7. $a + b = (\underline{\ ?\ })(-a - b)$

8. $1 - x = (\underline{\ ?\ })(x - 1)$

9. $2 - y = -1(\underline{\ ?\ })$

10. $a - b = -1(\underline{\ ?\ })$

11. $3 + x = -1(\underline{\ ?\ })$

12. $x - 2 = -1(\underline{\ ?\ })$

13. $2a - b = -1(\underline{\ ?\ })$

14. $-x - 4y = -1(\underline{\ ?\ })$

15. $x^2 - 4 = -1(\underline{\ ?\ })$

16. $9 - a^2 = -1(\underline{\ ?\ })$

17. $-b^2 + 1 = -1(\underline{\ ?\ })$

18. $m^2 + 2 = -1(\underline{\ ?\ })$

Written Exercises

Write the expression with a factor of −1.

A

1. $-2x$
2. $-3y$
3. $-a - b$
4. $-7 - 4x$
5. $2 - 3y$
6. $6 - 5b$
7. $3 - 8m$
8. $16 - r$
9. $25 - v$
10. $3 - x^2$
11. $-4 - y^2$
12. $a^2 - b^2$

Simplify.

13. $\frac{a - b}{b - a}$
14. $\frac{c - 2}{2 - c}$
15. $\frac{x - y}{y - x}$
16. $\frac{a - c}{c - a}$
17. $\frac{1 - a}{a - 1}$
18. $\frac{b - 1}{1 - b}$
19. $\frac{m - n}{n - m}$
20. $\frac{a - 6}{6 - a}$
21. $\frac{2x + 2y}{-y - x}$
22. $\frac{3b + 6a}{-b - 2a}$
23. $\frac{5x - 5y}{5y - 5x}$
24. $\frac{1 - b}{2b - 2}$
25. $\frac{a^2 - 1}{1 - a}$
26. $\frac{m^2 - 1}{1 - m}$
27. $\frac{1 - r}{r^2 - 1}$
28. $\frac{a^2 - 9}{3 - a}$
29. $\frac{2 - c}{c^2 - 4}$
30. $\frac{3y^2 - 3}{3 - 3y}$
31. $\frac{2x^2 - 50}{10 - 2x}$
32. $\frac{-a - b}{b^2 - a^2}$

B

33. $\frac{8 - 4t}{t^2 + t - 6}$
34. $\frac{5 - 5x}{x^2 + 3x - 4}$
35. $\frac{p^2 - 4p - 21}{14 - 2p}$
36. $\frac{z^2 - 6z + 9}{6 - 2z}$
37. $\frac{r^2 - 7r + 12}{9 - 3r}$
38. $\frac{8 - 4a}{a^2 - 5a + 6}$

SELF-TEST

Simplify.

1. $\frac{24}{21x}$
2. $\frac{-6a}{18a}$
3. $\frac{-12r^2s}{-9r^2s^2}$ *(9-1)*
4. $\frac{5a - 5b}{a - b}$
5. $\frac{9 - m^2}{3 + m}$
6. $\frac{6x + 6y}{9x + 9y}$ *(9-2)*
7. $\frac{x - 3y}{3y - x}$
8. $\frac{4m - 4n}{2n - 2m}$
9. $\frac{2r - 8s}{-r + 4s}$ *(9-3)*

4 Ratio

One way to compare numbers is by a **ratio.** The ratio of two numbers is their quotient. We often write a ratio as a fraction.

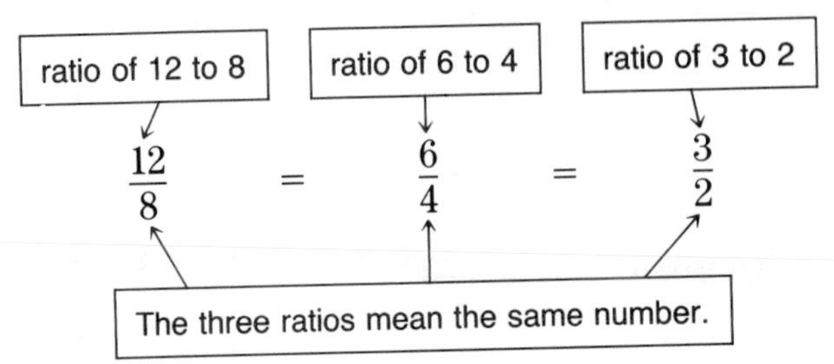

Suppose in your school there are 500 students and 25 teachers. You want to compare the number of students and teachers by a ratio in simplest form.

$$\text{ratio of students to teachers} = \frac{\text{number of students}}{\text{number of teachers}} = \frac{500}{25}$$

$$\frac{500}{25} = \frac{100}{5} = \frac{20}{1}$$

There are 20 students for every 1 teacher. The ratio is 20 to 1.

Classroom Practice

Give the ratios in simplest form.

1. 12 centimeters to 16 centimeters
2. 40 grams to 25 grams
3. 6 liters to 18 liters
4. 15 minutes to 60 minutes
5. A school has 506 students and 23 teachers. Find the ratio of students to teachers.
6. In a small town last year, there were 20 births and 12 deaths. Find the ratio of births to deaths.
7. A recipe for trail mix says to mix dried fruit, oatmeal, and nuts in a 3 to 2 to 1 ratio. What does this mean?

Written Exercises

Give the ratios in simplest form.

A

1. 25 centimeters to 40 centimeters
2. 30 millimeters to 20 millimeters
3. 50¢ to 75¢
4. 4 months to 12 months
5. 12 buckets to 20 buckets
6. 6 parts to 8 parts
7. One part of cement is mixed with two parts of sand in order to make concrete. What is the ratio of sand to cement?

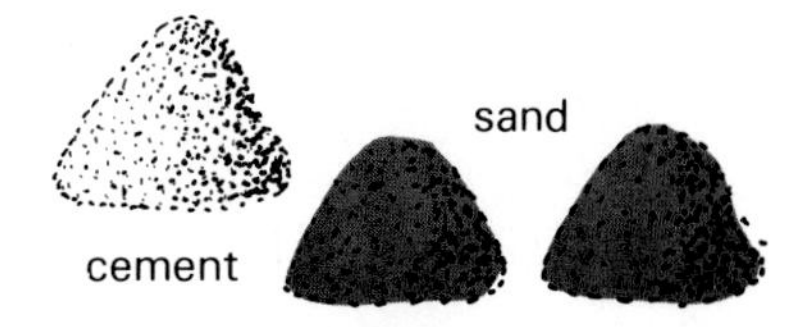

8. A school has 1000 students and 40 teachers. What is the ratio of teachers to students?
9. The Tigers won 9 games and lost 6 games last season. What is the ratio of wins to losses?
10. A school has 2100 students. 1200 students are girls. What is the ratio of girls to boys?
11. A commuter drove 40 km, driving 24 km on the freeway and the rest on city streets. What is the ratio of the distance on the freeway to the distance on city streets?
12. **a.** The slope of a line is the ratio of its __?__ to its __?__.

    b. The slope of the line shown is __?__.

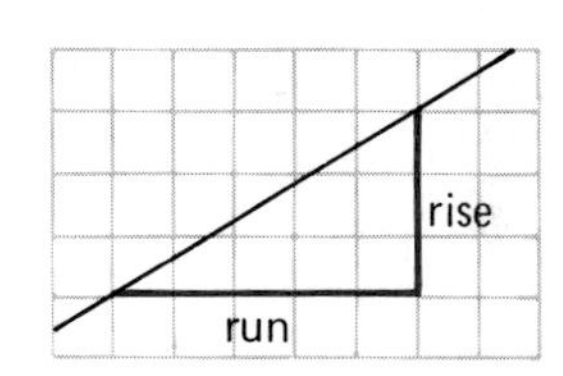

Write the ratio using the same units. Give the ratio in simplest form.

SAMPLE

30 minutes to 2 hours $= \dfrac{30 \text{ min}}{120 \text{ min}} = \dfrac{1 \text{ min}}{4 \text{ min}}$ *or* $\dfrac{1}{4}$

2 hours = 120 minutes

13. 3 days to 3 weeks
14. 30 minutes to 3 hours
15. 400 grams to 2 kilograms
16. 9 months to 2 years
17. 2 minutes to 40 seconds
18. 1 century to 4 decades

5 Problems Involving Ratio

If you know that the ratio of two numbers is 2 to 3, you can represent these numbers by $2x$ and $3x$.

$$\frac{2x}{3x} = \frac{2}{3} \longleftarrow \boxed{\text{Ratio of } 2x \text{ to } 3x = \text{Ratio of 2 to 3}}$$

Now let's use this idea in solving some problems.

EXAMPLE 1

The ratio of girls to boys in a club is 5 to 4.
There are 45 members in all.
How many are girls? How many are boys?

Let $5x$ = number of girls, $4x$ = number of boys.

$$\underbrace{\text{number of girls}}_{5x} + \underbrace{\text{number of boys}}_{4x} = 45$$

$$9x = 45$$

$$x = 5$$

Answer: Number of girls $= 5x = 5 \cdot 5 = 25$
Number of boys $= 4x = 4 \cdot 5 = 20$

Check: $\frac{25}{20} = \frac{5}{4}$ ✓ $25 + 20 = 45$ ✓

EXAMPLE 2

The ratio of tin to copper in a metal mixture is 3 to 5.
We need to make 400 kg of the mixture.
How much of each metal should be used?

Let $3x$ = kg of tin, $5x$ = kg of copper.

$$\underbrace{\text{kg of tin}}_{3x} + \underbrace{\text{kg of copper}}_{5x} = 400$$

$$8x = 400$$

$$x = 50$$

Answer: Amount of tin $= 3x = 3 \cdot 50 = 150 \longrightarrow$ 150 kg
Amount of copper $= 5x = 5 \cdot 50 = 250 \longrightarrow$ 250 kg

Check: $\frac{150}{250} = \frac{3}{5}$ ✓ $150 + 250 = 400$ ✓

Written Exercises

A

1. There are 35 students at a game.
The ratio of boys to girls is 4 to 3.
How many boys are there?

2. A marching band has 80 members.
The ratio of members who play wind instruments to all others is 2 to 3.
How many members play wind instruments?

3. A coin was tossed 100 times.
The ratio of heads thrown to tails thrown was 13 to 12.
How many tails were thrown?

4. A large bookstore took in $4,000,000 in sales last year.
The ratio of book sales to other sales was 5 to 3.
How much money in book sales did the store take in?

5. An 80-meter cable is cut into two pieces.
The lengths of the pieces have a ratio of 11 to 5.
How long is the shorter piece?

6. A store spends $7 for TV ads for every $2 spent on newspaper ads.
The advertising budget is $360,000.
How much is spent on newspaper ads?

7. Pat pays 26 cents out of every dollar of her income in taxes.
Her yearly income is $30,000.
How much does she have left after taxes?

8. At Tidewater Tech, full-time teachers outnumber part-time teachers by a ratio of 5 to 3.
There are 72 teachers on the faculty.
How many part-time teachers are there?

9. In a certain salad dressing, the ratio of vinegar to oil is 4 to 3.
How much of each is in 735 milliliters of dressing?

B

10. One town had $64,000 to spend on streets, parks, and the library.
The money was divided in the ratio of 4 to 3 to 1.
How much money did each get?

11. Three people in an election split the vote in the ratio 6 to 5 to 5.
The number of votes cast was 56,000.
How many votes did each person receive?

12. A small garden has daffodils, hyacinths, and tulips.
The plants are in a ratio of 7 to 5 to 3.
If there are 60 plants in all, how many tulips are there?

6 Proportion

A **proportion** is a statement that two ratios are equal.

$$\frac{4}{6} = \frac{2}{3} \leftarrow \boxed{\text{4 to 6 equals 2 to 3.}}$$

Note what happens in the proportion if you cross-multiply.

$$\boxed{4 \cdot 3 = 12} \rightarrow \frac{4}{6} = \frac{2}{3} \leftarrow \boxed{6 \cdot 2 = 12}$$

The products are the same!

If you know three of the four numbers in a proportion, you can use cross-multiplication to find the fourth number.

EXAMPLE 1

$$\frac{5}{4} = \frac{30}{x}$$

$$5x = 4 \cdot 30$$
$$5x = 120$$
$$x = 24$$

Check: $\frac{5}{4} = \frac{30}{x}$

$\frac{5}{4}$	$\frac{30}{24}$
	$\frac{5}{4}$ ✓

EXAMPLE 2

$$\frac{2x}{5} = \frac{x+1}{3}$$

$$2x \cdot 3 = 5(x + 1)$$
$$6x = 5x + 5$$
$$6x - 5x = 5x + 5 - 5x$$
$$x = 5$$

Check: $\frac{2x}{5} = \frac{x+1}{3}$

$\frac{10}{5}$	$\frac{6}{3}$
2	2 ✓

Classroom Practice

Solve.

1. $\frac{4}{12} = \frac{x}{9}$
2. $\frac{a}{3} = \frac{16}{12}$
3. $\frac{x}{4} = \frac{9}{12}$
4. $\frac{5}{3} = \frac{a}{6}$
5. $\frac{1}{6x} = \frac{2}{36}$
6. $\frac{3x}{5} = \frac{12}{5}$
7. $\frac{4n}{3} = \frac{8}{6}$
8. $\frac{5n}{3} = \frac{25}{5}$

Written Exercises

Solve.

A

1. $\frac{3}{5} = \frac{x}{15}$
2. $\frac{2}{9} = \frac{6}{x}$
3. $\frac{8}{6} = \frac{x}{9}$
4. $\frac{10}{a} = \frac{5}{10}$
5. $\frac{3x}{4} = \frac{9}{6}$
6. $\frac{2x}{7} = \frac{4}{1}$
7. $\frac{6}{1} = \frac{48}{x}$
8. $\frac{3a}{2} = \frac{15}{5}$
9. $\frac{3a}{5} = \frac{12}{10}$
10. $\frac{15}{5} = \frac{3a}{4}$
11. $\frac{4}{5} = \frac{2x}{25}$
12. $\frac{4r}{8} = \frac{3}{6}$
13. $\frac{x}{4} = \frac{x + 2}{5}$
14. $\frac{x}{5 - x} = \frac{2}{3}$
15. $\frac{a - 2}{a + 1} = \frac{1}{2}$
16. $\frac{x + 4}{x} = \frac{5}{3}$
17. $\frac{x - 1}{3} = \frac{x + 1}{9}$
18. $\frac{x + 1}{3} = \frac{x}{2}$
19. $\frac{2a + 3}{4} = \frac{a - 6}{8}$
20. $\frac{y + 2}{y - 2} = \frac{5}{3}$
21. $\frac{3n - 5}{8n} = \frac{1}{4}$
22. $\frac{x + 3}{5} = \frac{x + 1}{4}$
23. $\frac{y + 2}{9} = \frac{y - 1}{18}$
24. $\frac{a + 4}{6} = \frac{a + 6}{4}$
25. $\frac{2x + 3}{4x} = \frac{3}{4}$
26. $\frac{3a - 4}{1} = \frac{2a}{2}$
27. $\frac{5m + 2}{3} = \frac{m - 1}{2}$

B

28. $\frac{x}{x - 2} = \frac{x + 2}{x + 1}$
29. $\frac{y}{y - 2} = \frac{y + 2}{y + 4}$
30. $\frac{a - 1}{a} = \frac{a + 1}{a + 3}$
31. $\frac{x - 3}{x} = \frac{x + 3}{x + 8}$
32. $\frac{n - 3}{n + 5} = \frac{n - 4}{n + 3}$
33. $\frac{c + 5}{c - 1} = \frac{c - 3}{c - 5}$

34. You can find the height of the cliff in the drawing by using this proportion.

$$\frac{\text{height of cliff}}{\text{height of pole}} = \frac{\text{length of cliff's shadow}}{\text{length of pole's shadow}}$$

Find the height of the cliff.

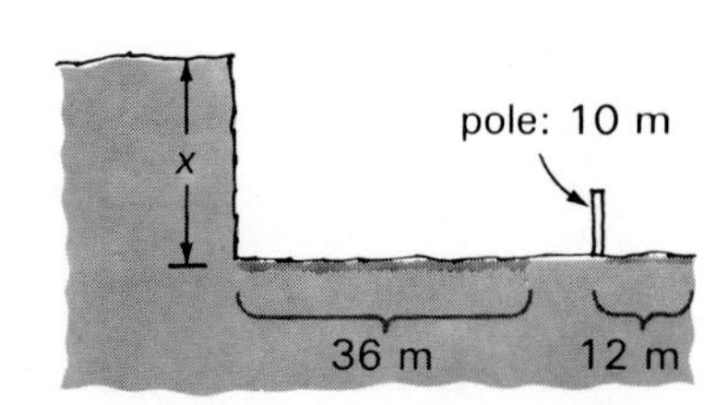

7 Applying Proportions

Proportions give an easy way to solve many real-life problems.

EXAMPLE 1

One large apartment house has 170 tenants.
Sixty tenants were surveyed.
Forty-two said they had at least one pet.
About how many of the tenants have pets (if the ratio of the survey holds true)?

	Survey	Real
Tenants with pets	42	x
Total	60	170

$$\frac{42}{60} = \frac{x}{170}$$

$$60x = 42 \cdot 170$$

$$60x = 7140$$

$$x = \frac{7140}{60}$$

$$x = 119$$

Answer: About 119 tenants

EXAMPLE 2

One family drinks 14 liters of milk in one week.
They usually drink milk at the same rate.
How many liters of milk does this family drink in 10 days?

Number of liters	14	x
Time	7 days	10 days

$$\frac{14}{7} = \frac{x}{10}$$

$$7x = 14 \cdot 10$$

$$7x = 140$$

$$x = 20$$

Answer: 20 liters

EXAMPLE 3

The weekly requirements for a flock of 15 chickens are shown below.
How much feed would be needed for a flock of 24 chickens?
How much water would be needed for a flock of 24 chickens?

FEED AND WATER
10 kg feed
20 L water

Number of chickens	Feed	Water
15	10	20
24	x	y

Feed: $\frac{15}{24} = \frac{10}{x}$

$15x = 240$

$x = 16$

Water: $\frac{15}{24} = \frac{20}{y}$

$15y = 480$

$y = 32$

Answer: 16 kg of feed and 32 L of water

Written Exercises

A

1. Three cans of cat food cost 93¢.
Five cans will cost how much?

Number	3	5
Cost	93¢	?

2. Six oranges cost 90¢.
How much will 9 oranges cost?

Number	6	?
Cost	90¢	?

3. Four bottles of vinegar cost $2.76.
How much will 10 bottles cost?

Number	?	?
Cost	?	?

4. Three loaves of bread cost $2.67.
How much will 5 loaves cost?

Number	?	?
Cost	?	?

5. An airplane can fly 1800 km in 4 hours.
How long would it take to fly 3150 km?

Distance	1800 km	?
Time	4	?

6. A metal pipe 160 cm long is 4 kg.
How long is a pipe that is 7 kg?

Centimeters	?	?
Kilograms	?	?

7. An herbal tea contains 12 g of chamomile in each 30 g package of tea. How much chamomile is there in a pot of tea made with 20 g of the tea?

Tea	30 g	?
Chamomile	12 g	?

8. A pump can fill a 2000-liter tank in 1 hour and 20 minutes. How long will it take to fill a 1500-liter tank?

Quantity	2000 L	?
Time	?	?

9. A photograph measures 20 cm by 15 cm. It is enlarged so that the larger measure is 28 cm. What does the shorter measure become?

20 cm
15 cm

10. A 600 g box of cereal contains 21 servings. How many servings does an 800 g box contain?

11. A map is drawn so that 5 cm represents 2 km. On the map, two towns measure 20 cm apart. Find the actual distance between the towns.

12. A car worth $8400 is taxed $476. At that rate, what is the tax on a car worth $9600?

13. A telephone pole casts a shadow of 660 cm. A girl who is 160 cm tall is standing nearby. Her shadow is 120 cm long. How tall is the telephone pole?

	Girl	Telephone pole
Height of object	?	?
Length of shadow	?	?

14. A man 180 cm tall casts a shadow 150 cm long. The shadow of a flagpole is 540 cm long. How tall is the flagpole?

B 15. The Binets pay \$660 a year for homeowner's insurance.
Last year they sold their home.
They had owned it only 3 months of the year.
How much should the refund on the unused part of the policy be?

16. The annual tax on the Carey home is \$1452.
On October 1 the Careys sold the house to the Robinsons.
How much tax must the Careys pay for that year?

17. The Bergmans want to put an addition onto their house.
Their neighbors' addition cost \$12,000 for an area of 18 m^2.
The Bergmans want to add about 27 m^2.
How much money should the Bergmans expect to pay?

SELF-TEST

Vocabulary

ratio (p. 304) proportion (p. 308)

1. A bookshelf contains 24 hardcover books and 32 paperback books. What is the ratio of hardcover books to paperback books? *(9-4)*

Solve.

2. A 65-centimeter string is cut into two pieces. The lengths of the pieces have a ratio of 3 to 2. How long is each piece? *(9-5)*

3. $\frac{7}{9} = \frac{x}{36}$ 4. $\frac{3}{4} = \frac{6}{x}$ 5. $\frac{a + 2}{a} = \frac{5}{3}$ 6. $\frac{x}{x + 4} = \frac{2}{1}$ *(9-6)*

7. Daryl drove 75 km on 9 L of gasoline. How far can he drive on 12 L of gasoline? *(9-7)*

8. A pump can fill a 900-liter tank in 30 minutes. How long will it take to fill a 1200-liter tank?

PUZZLE ◆ PROBLEMS

In how many different ways can 4 postage stamps be attached to each other? Here is one way.

8 Multiplying Fractions

You can multiply fractions in algebra in the same way as in arithmetic.

In Arithmetic	In Algebra
$\frac{2}{3} \cdot \frac{4}{5} = \frac{2 \cdot 4}{3 \cdot 5} = \frac{8}{15}$	$\frac{3}{4} \cdot \frac{x}{y} = \frac{3 \cdot x}{4 \cdot y} = \frac{3x}{4y}$

Sometimes you can simplify the product of two fractions.

EXAMPLE 1

$$\frac{2}{x^2} \cdot \frac{x}{2} = \frac{2 \cdot x}{x \cdot x \cdot 2} = \frac{1}{x}$$

EXAMPLE 2

$$\frac{r+s}{r-s} \cdot \frac{r-s}{3} = \frac{(r+s)(r-s)}{3(r-s)} = \frac{r+s}{3}$$

EXAMPLE 3

$$\frac{a^2-1}{12} \cdot \frac{6}{a-1} = \frac{6(a^2-1)}{12(a-1)} = \frac{6(a+1)(a-1)}{2 \cdot 6(a-1)} = \frac{a+1}{2}$$

Classroom Practice

Simplify.

1. $\frac{3}{5} \cdot \frac{3}{4}$
2. $\frac{a}{2} \cdot \frac{b}{3}$
3. $\frac{x}{3} \cdot \frac{2x}{5}$
4. $\frac{2a}{3} \cdot \frac{1}{4a}$
5. $\frac{7}{19} \cdot \frac{19}{20}$
6. $\frac{n}{3} \cdot \frac{6}{2n^2}$
7. $\frac{1}{a-b} \cdot \frac{a-b}{2}$
8. $\frac{a^2-b^2}{a} \cdot \frac{a^2}{a-b}$
9. $\frac{2m^2}{5} \cdot \frac{5n}{4n^3}$
10. $\frac{6}{5} \cdot \frac{10}{2x-4}$
11. $\frac{9}{3x+3} \cdot \frac{7x+7}{14}$
12. $\frac{x^2-4}{3} \cdot \frac{x}{x^2-2x}$

Written Exercises

Simplify.

A

1. $\frac{1}{3} \cdot \frac{1}{4}$
2. $\frac{1}{5} \cdot \frac{1}{8}$
3. $\frac{2}{3} \cdot \frac{4}{7}$
4. $\frac{x}{3} \cdot \frac{x}{4}$
5. $y \cdot \frac{y}{2}$
6. $m \cdot \frac{m}{8}$
7. $\frac{a}{b} \cdot \frac{2b}{3a}$
8. $\frac{xy}{5} \cdot \frac{10}{x}$
9. $\frac{6x}{5y} \cdot \frac{y^2}{3x}$
10. $\frac{5a}{3b^2} \cdot \frac{6b^3}{a^2}$
11. $\frac{4x}{3y} \cdot \frac{3}{8xy}$
12. $\frac{16x^2}{5} \cdot \frac{15}{4y^2}$
13. $\frac{9r}{7s} \cdot \frac{14rs}{3r}$
14. $\frac{x^2}{3y} \cdot \frac{5xy}{x}$
15. $\frac{3a^2}{4} \cdot \frac{20b}{12b^2}$
16. $\frac{12a}{5a^2} \cdot \frac{a^2}{6a^2}$
17. $\frac{3}{7} \cdot \frac{7x - 7}{6}$
18. $\frac{5}{4} \cdot \frac{4x - 4}{15}$
19. $\frac{2}{3} \cdot \frac{3x - 12}{4}$
20. $\frac{5}{4x + 16} \cdot \frac{4}{25}$
21. $\frac{a^2 - a}{5} \cdot \frac{25}{a}$
22. $\frac{x}{b} \cdot \frac{b^2 - b}{x^2y}$
23. $\frac{a^2b}{2a - 2} \cdot \frac{2}{ab^2}$
24. $\frac{9}{x + 2} \cdot \frac{x^2 + 2x}{36}$
25. $\frac{a^2 - b^2}{10} \cdot \frac{5}{a - b}$
26. $\frac{x^2 - 1}{4} \cdot \frac{12}{x - 1}$
27. $\frac{r^2 - 2r}{4} \cdot \frac{2}{r - 2}$
28. $\frac{3x - 3}{6} \cdot \frac{4}{1 - x}$
29. $\frac{6}{x^2 - 4} \cdot \frac{2 - x}{18}$

B

30. $\frac{x^2 - y^2}{x^2 - 49} \cdot \frac{x + 7}{x - y}$
31. $\frac{x^2 - x - 6}{14} \cdot \frac{7}{x - 3}$
32. $\frac{2x}{x + 1} \cdot \frac{2x + 2}{6x^2}$
33. $\frac{x^2 - 3x - 10}{2} \cdot \frac{4}{x^2 + x - 2}$
34. $\frac{2a - 4}{3a + 6} \cdot \frac{2a + 4}{a - 2}$
35. $\frac{3x - 9}{x - 3} \cdot \frac{x + 2}{3x + 12}$
36. $\frac{4a - 1}{a^2} \cdot \frac{a^3 - a^2}{a - 1}$
37. $\frac{n^2 - 1}{3} \cdot \frac{21}{1 - n}$

C

38. $\frac{3 - 3x}{x^2 - 2x - 3} \cdot \frac{x + 1}{x - 1}$
39. $\frac{m^2 - 1}{m^2 + 4m + 3} \cdot \frac{m^2 + m - 6}{m^2 + m - 2}$

9 Dividing Fractions

The numbers 2 and $\frac{1}{2}$ are called **reciprocals** because when they are multiplied, the result is 1. The numbers $\frac{3}{4}$ and $\frac{4}{3}$ are reciprocals also. Dividing by a number is the same as multiplying by its reciprocal.

In Arithmetic	In Algebra
$6 \div 2 = 6 \cdot \frac{1}{2} = 3$	$4 \div \frac{x}{3} = 4 \cdot \frac{3}{x} = \frac{12}{x}$

EXAMPLE 1

$$\frac{2}{3x} \div \frac{x}{4} = \frac{2}{3x} \cdot \frac{4}{x} = \frac{8}{3x^2}$$

EXAMPLE 2

$$\frac{1}{x + 2} \div \frac{1}{x + 1} = \frac{1}{x + 2} \cdot \frac{x + 1}{1} = \frac{x + 1}{x + 2}$$

Sometimes you'll notice the same factors in the numerator and in the denominator.

EXAMPLE 3

$$\frac{x^3}{3} \div \frac{x^2}{4} = \frac{x^3}{3} \cdot \frac{4}{x^2} = \frac{x \cdot x \cdot x \cdot 4}{3 \cdot x \cdot x} = \frac{4x}{3}$$

EXAMPLE 4

$$\frac{t + 3}{t^2 - 25} \div \frac{1}{t - 5} = \frac{t + 3}{t^2 - 25} \cdot \frac{t - 5}{1} = \frac{t + 3}{(t + 5)(t - 5)} \cdot \frac{t - 5}{1} = \frac{(t + 3)(t - 5)}{(t + 5)(t - 5)} = \frac{t + 3}{t + 5}$$

Classroom Practice

Simplify.

1. $\frac{3}{4} \div \frac{9}{8}$
2. $\frac{x}{3} \div \frac{x}{2}$
3. $\frac{a}{b^2} \div \frac{a^2}{b}$
4. $\frac{v^2}{w} \div \frac{v}{w^2}$
5. $\frac{3r^2}{8} \div \frac{9}{4s^2}$
6. $\frac{x^2}{7x} \div \frac{x}{21}$
7. $\frac{c^2}{d^3} \div \frac{c}{d^2}$
8. $\frac{2x}{3y^2} \div \frac{8x^3}{9y^2}$
9. $\frac{21x^2y^3}{5} \div 7x^2y$
10. $\frac{6}{y^2 - 9} \div \frac{3}{y - 3}$
11. $\frac{b + 4}{7} \div \frac{b^2 - 16}{21}$
12. $\frac{x^2 - y^2}{8} \div \frac{3x - 3y}{2}$

Written Exercises

Simplify.

A

1. $\frac{5}{6} \div \frac{1}{3}$
2. $\frac{x}{2} \div \frac{1}{4}$
3. $\frac{a}{5} \div \frac{a}{6}$
4. $\frac{a}{4} \div \frac{a^2}{2}$
5. $\frac{x}{10} \div \frac{6}{5x}$
6. $\frac{2x^2}{9} \div \frac{4x}{3}$
7. $\frac{2a^2}{5b^2} \div \frac{4a^3}{10b}$
8. $\frac{5r}{12t^2} \div \frac{15r^2}{6s^3}$
9. $3m^2 \div \frac{8m}{7}$
10. $\frac{3km}{8} \div 18k^2$
11. $\frac{6}{a^2b} \div \frac{a}{a^3b^2}$
12. $\frac{(xy)^2}{x} \div \frac{xy}{y}$
13. $\frac{(ab)^2}{4} \div 2ab$
14. $\frac{a - b}{7} \div \frac{a - b}{6}$
15. $\frac{h - g}{4} \div \frac{h - g}{8}$
16. $\frac{2x - 2}{4} \div \frac{2}{9}$
17. $\frac{a^3}{3} \div \frac{a^4}{3a^2 + 3}$
18. $\frac{2x + 2}{5} \div \frac{2}{3}$
19. $\frac{6}{y^2 - 9} \div \frac{3}{y - 3}$
20. $\frac{a^2 - b^2}{4} \div \frac{a - b}{8}$
21. $\frac{2a - 2}{2} \div (a - 1)$
22. $\frac{x^2 - 1}{2 - x} \div (x - 1)$
23. $\frac{x^2 - 16}{2x} \div (x - 4)$
24. $\frac{r - 7}{r^2 - 4} \div \frac{1}{r - 2}$

Simplify.

B 25. $\frac{a^2 - a - 6}{a^2 - 9} \div \frac{a + 2}{a - 3}$

26. $\frac{x^2 - 16}{2x} \div \frac{x + 4}{2x^2 - 2x}$

27. $\frac{6y + 12}{3y - 9} \div \frac{4y - 8}{3}$

28. $\frac{a - b}{4ab} \div \frac{b - a}{8a^2}$

29. $\frac{y + 2}{y + 1} \div \frac{2y + 4}{y^2 - 2y - 3}$

30. $\frac{u^2 + u^3}{36 - u^2} \div \frac{u - u^3}{6u + u^2}$

C 31. $\frac{(d - 5)^2}{d + 3} \div \frac{d^2 - d - 20}{d^2 + 7d + 12}$

32. $\frac{h^2 + h - 6}{h^2 - 2h - 15} \div \frac{h^2 + 4h - 5}{h^2 - 25}$

33. $\frac{y^2 - 7y + 12}{y^2 + y - 20} \div \frac{y^2 - 9y + 18}{y^2 + 7y + 10}$

34. $\frac{x^2 + 5x + 6}{x^2 - 4} \div \frac{x^2 + 2x - 3}{x^2 - 3x + 2}$

CAREER NOTES

MANUFACTURING

Products of the manufacturing industry range from a child's toy to a large jetliner. Manufacturing involves many different processes and therefore offers a variety of career opportunities.

About two thirds of the workers in a manufacturing plant are actually involved in production. These workers include tool designers, machine operators, sorters, and packagers. Efficiency experts aid the production team by studying procedures to help make better use of people, time, and equipment.

Other workers in the factory are involved in receiving supplies and distributing products to the market. Inventory controllers record data for ordering, receiving, storing, and shipping goods. Traffic managers must determine the quickest and most economical means of shipping goods to and from the company.

10 Add, Subtract—Same Denominators

In arithmetic, when the denominators of two fractions are the same, the fractions can be added or subtracted easily. This is true in algebra also.

In Arithmetic	In Algebra
$\frac{1}{7} + \frac{4}{7} = \frac{1+4}{7} = \frac{5}{7}$	$\frac{x}{7} + \frac{4x}{7} = \frac{x+4x}{7} = \frac{5x}{7}$
$\frac{4}{5} - \frac{3}{5} = \frac{4-3}{5} = \frac{1}{5}$	$\frac{4}{y} - \frac{3}{y} = \frac{4-3}{y} = \frac{1}{y}$

EXAMPLE 1 $\frac{5}{2a} + \frac{6}{2a} = \frac{5+6}{2a} = \frac{11}{2a}$

EXAMPLE 2
$$\frac{2r}{r-u} - \frac{2u}{r-u} = \frac{2r-2u}{r-u} = \frac{2(r-u)}{r-u} = 2$$

EXAMPLE 3
$$\frac{3t}{7} - \frac{7-2t}{7} = \frac{3t-(7-2t)}{7} = \frac{3t-7+2t}{7} = \frac{5t-7}{7}$$

Be careful. Do not write this: ~~$\frac{3t-7-2t}{7}$~~

Classroom Practice

Simplify.

1. $\frac{5}{8} + \frac{4}{8}$
2. $\frac{3x}{5} + \frac{x}{5}$
3. $\frac{5}{a} - \frac{3}{a}$
4. $\frac{2x}{7} + \frac{3x}{7}$
5. $\frac{y}{11} - \frac{4y}{11}$
6. $\frac{2}{a} - \frac{5}{a}$
7. $\frac{m}{5k} + \frac{9m}{5k}$
8. $\frac{2r}{3s} + \frac{r}{3s}$
9. $\frac{7y}{y+1} - \frac{y}{y+1}$
10. $\frac{3x}{x+2} + \frac{5x}{x+2}$
11. $\frac{2x}{13} + \frac{x+7}{13}$
12. $\frac{5a}{8} - \frac{a-4}{8}$

Written Exercises

Simplify.

A

1. $\frac{4}{5} + \frac{3}{5}$
2. $\frac{4x}{5} + \frac{3x}{5}$
3. $\frac{6}{2a} + \frac{3}{2a}$
4. $\frac{8}{9k} + \frac{10}{9k}$
5. $\frac{8n}{9} - \frac{2n}{9}$
6. $\frac{a}{8} - \frac{4a}{8}$
7. $\frac{2x}{14} + \frac{5x}{14}$
8. $\frac{2s}{5} + \frac{8s}{5}$
9. $\frac{4y}{3} - \frac{7y}{3}$
10. $\frac{8p}{9} - \frac{5p}{9}$
11. $\frac{-3x}{4} - \frac{x}{4}$
12. $\frac{-a}{5} - \frac{3a}{5}$
13. $\frac{x}{y} + \frac{1}{y}$
14. $\frac{3}{a} + \frac{b}{a}$
15. $\frac{x}{3x} + \frac{1}{3x}$
16. $\frac{a}{2a} - \frac{1}{2a}$
17. $\frac{3}{a + b} + \frac{2}{a + b}$
18. $\frac{7}{x - 1} + \frac{2}{x - 1}$
19. $\frac{3}{y - 2} - \frac{1}{y - 2}$
20. $\frac{4}{a + 3} - \frac{5}{a + 3}$
21. $\frac{2x}{x + 1} + \frac{2}{x + 1}$
22. $\frac{3a}{a - 1} - \frac{3}{a - 1}$
23. $\frac{v}{v - 4} - \frac{4}{v - 4}$
24. $\frac{x + y}{3xy} + \frac{2x - y}{3xy}$
25. $\frac{3a + 2b}{4ab} + \frac{a + 2b}{4ab}$
26. $\frac{3x}{x + 2} - \frac{x + 2}{x + 2}$
27. $\frac{2c + 1}{7c} - \frac{4c - 3}{7c}$
28. $\frac{3w + 2}{w + 3} - \frac{w - 4}{w + 3}$

SELF-TEST

Vocabulary

reciprocals (p. 316)

Simplify.

1. $\frac{3x}{4} \cdot \frac{2}{x}$
2. $\frac{3x}{2y} \cdot \frac{y}{x}$
3. $\frac{3}{4} \cdot \frac{4a + 4}{9}$

(9-8)

4. $4m^2 \div \frac{12m}{5}$
5. $\frac{3}{a^2b} \div \frac{9a}{ab^2}$
6. $\frac{3x - 3}{2x^2 + 2} \div \frac{3}{2}$

(9-9)

7. $\frac{5x}{7} + \frac{4x}{7}$
8. $\frac{5}{9a} - \frac{3}{9a}$
9. $\frac{7x}{5y} + \frac{3x}{5y}$

(9-10)

11 Renaming Fractions

There are two basic ideas to use when renaming fractions. The first of these we have been using already in this chapter.

1. You can divide both the numerator and denominator by the same number (except 0).

$$\frac{\overset{3}{\cancel{15}}}{\underset{4}{\cancel{20}}} = \frac{3}{4}$$

2. You can multiply the numerator and denominator by the same number (except 0).

$$\frac{3}{8} = \frac{6}{16} \quad (\times 2 \text{ numerator and denominator})$$

The second idea is used in the examples below.

EXAMPLE 1

$\frac{3}{5} = \frac{?}{10}$ — **5** was multiplied by 2 to get **10.**

$\frac{3}{5} = \frac{6}{10}$ — So, **3** is multiplied by 2 to get **6.**

EXAMPLE 2

$\frac{3}{a} = \frac{?}{4a}$ — ***a*** was multiplied by 4 to get **4*a*.**

$\frac{3}{a} = \frac{12}{4a}$ — So, **3** is multiplied by 4 to get **12.**

EXAMPLE 3

$\frac{2}{3x} = \frac{?}{6x^2}$ — **3*x*** was multiplied by $2x$ to get $6x^2$.

$\frac{2}{3x} = \frac{4x}{6x^2}$ — So, **2** is multiplied by $2x$ to get **4*x*.**

Classroom Practice

Find the missing numerator.

1. $\frac{2}{3} = \frac{?}{12}$
2. $\frac{4}{9} = \frac{?}{18}$
3. $\frac{3a}{7} = \frac{?}{21}$
4. $\frac{x}{2y} = \frac{?}{4y}$
5. $\frac{7}{3a} = \frac{?}{12ab}$
6. $\frac{a}{4a} = \frac{?}{8ab^2}$
7. $\frac{5}{a} = \frac{?}{a^2}$
8. $\frac{2}{3x} = \frac{?}{3x^2}$

Written Exercises

Find the missing numerator.

A

1. $\frac{4}{5} = \frac{?}{10}$
2. $\frac{2}{3} = \frac{?}{18}$
3. $\frac{4}{5} = \frac{?}{20}$
4. $\frac{y}{6} = \frac{?}{18}$
5. $\frac{x}{2} = \frac{?}{8}$
6. $\frac{x}{3} = \frac{?}{9}$
7. $\frac{y}{4} = \frac{?}{12}$
8. $\frac{4a}{3} = \frac{?}{9}$
9. $\frac{3x}{2} = \frac{?}{4}$
10. $\frac{4}{a} = \frac{?}{2a}$
11. $\frac{3}{x} = \frac{?}{3x}$
12. $\frac{2}{b} = \frac{?}{3b}$
13. $\frac{3}{4} = \frac{?}{8a}$
14. $\frac{5}{3} = \frac{?}{9x}$
15. $\frac{2}{3x} = \frac{?}{6xy}$
16. $\frac{4}{5a} = \frac{?}{10ab}$
17. $\frac{5}{x} = \frac{?}{3xy}$
18. $\frac{3}{a} = \frac{?}{4ab}$
19. $\frac{2}{n} = \frac{?}{4nm}$
20. $\frac{1}{6r} = \frac{?}{24rs}$
21. $\frac{5}{x} = \frac{?}{x^2}$
22. $\frac{3a}{b} = \frac{?}{b^2}$
23. $\frac{2}{3b} = \frac{?}{3b^2}$
24. $\frac{3}{4x} = \frac{?}{8x^2}$

PUZZLE ◆ PROBLEMS

A man decided to give his horses to his children. His seventeen horses are to be divided among his children in the following way:

Floyd, $\frac{1}{2}$; Denise, $\frac{1}{3}$; Harriet, $\frac{1}{9}$

How many horses will each person get?

12 Add, Subtract–Different Denominators

When you want to add or subtract fractions with different denominators, you will have to rename your fractions so that the denominators are the same. Let's simplify $\frac{n}{4} + \frac{n}{6}$.

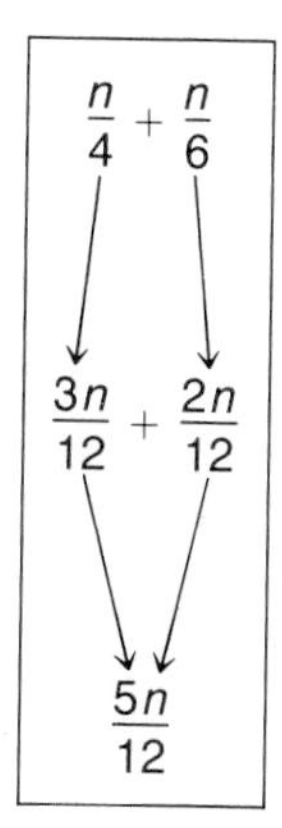

Step 1: Find a new denominator which can be divided by 4 and 6. You can use 12.

Step 2: Now rename your two fractions with the new denominator.

$$\frac{n}{4} = \frac{3n}{12} \qquad \frac{n}{6} = \frac{2n}{12}$$

Step 3: Find the sum.

$$\frac{3n}{12} + \frac{2n}{12} = \frac{5n}{12}$$

In the addition above, you could use 4×6, or 24, for your new denominator, but 12 is a little easier.

EXAMPLE 1

$$\frac{3}{2x} + \frac{1}{2}$$

The new denominator can be $2x$.

$$\frac{3}{2x} + \frac{x}{2x} = \frac{3 + x}{2x}$$

EXAMPLE 2

$$\frac{1}{5} - \frac{2}{y}$$

The new denominator can be $5y$.

$$\frac{y}{5y} - \frac{10}{5y} = \frac{y - 10}{5y}$$

EXAMPLE 3

$$\frac{2}{ab} + \frac{5}{2a}$$

The new denominator can be $2ab$.

$$\frac{4}{2ab} + \frac{5b}{2ab} = \frac{4 + 5b}{2ab}$$

Written Exercises

Simplify.

A

1. $\frac{1}{2} + \frac{1}{3}$
2. $\frac{x}{5} + \frac{x}{3}$
3. $\frac{y}{4} + \frac{y}{8}$
4. $\frac{4x}{9} - \frac{x}{6}$
5. $\frac{2x}{3} + \frac{x}{2}$
6. $\frac{3x}{2} + \frac{x}{4}$
7. $\frac{4x}{3} - \frac{x}{6}$
8. $\frac{a}{2} + \frac{2a}{3}$
9. $\frac{3}{2a} + \frac{1}{a}$
10. $\frac{5}{2x} - \frac{1}{4x}$
11. $\frac{4}{3y} - \frac{7}{2y}$
12. $\frac{4}{3a} + \frac{1}{4a}$
13. $\frac{3x}{4} + \frac{x}{3}$
14. $\frac{a}{3} - \frac{2a}{7}$
15. $\frac{5}{x} + \frac{1}{2x}$
16. $\frac{4}{a} + \frac{3}{2a}$
17. $\frac{5}{n} + \frac{3}{4n}$
18. $\frac{7}{3n} - \frac{4}{9n}$
19. $\frac{3x}{8} - \frac{x}{12}$
20. $\frac{1}{9y} - \frac{1}{15y}$
21. $\frac{1}{5} - \frac{3}{x}$
22. $\frac{3}{4} + \frac{1}{a}$
23. $\frac{4}{x} + \frac{1}{2}$
24. $\frac{1}{a} + \frac{3}{2ab}$
25. $\frac{2}{3x} + \frac{1}{6xy}$
26. $\frac{1}{2x} - \frac{1}{3xy}$
27. $\frac{2}{x} + \frac{3}{y}$
28. $\frac{3}{b} - \frac{1}{2b^2}$

B

29. $\frac{3x}{4} - \frac{2x}{3} + \frac{x}{6}$
30. $\frac{a}{3} + \frac{2a}{6} - \frac{5a}{2}$
31. $\frac{2x}{5} - \frac{1}{4} + \frac{3x}{5}$
32. $\frac{3a}{a} + \frac{a}{a^2} - \frac{2a}{3a}$
33. $\frac{x}{2x} + \frac{3x}{x} - \frac{5}{3x}$
34. $\frac{6}{y^2} - \frac{5x}{3y} + \frac{4x}{6xy^2}$

SELF-TEST

Find the missing numerator. *(9-11)*

1. $\frac{7}{9} = \frac{?}{36}$
2. $\frac{x}{6} = \frac{?}{24}$
3. $\frac{3}{m} = \frac{?}{mn}$
4. $\frac{5}{x} = \frac{?}{x^2}$

Simplify. *(9-12)*

5. $\frac{1}{3} + \frac{1}{5}$
6. $\frac{5x}{7} - \frac{x}{3}$
7. $\frac{1}{9a} + \frac{3}{4a}$
8. $\frac{3}{m} - \frac{4}{3m^2}$

13 More Difficult Fractions (Optional)

When the fractions you add or subtract have more than one term in their numerators, you have a few more steps to do.

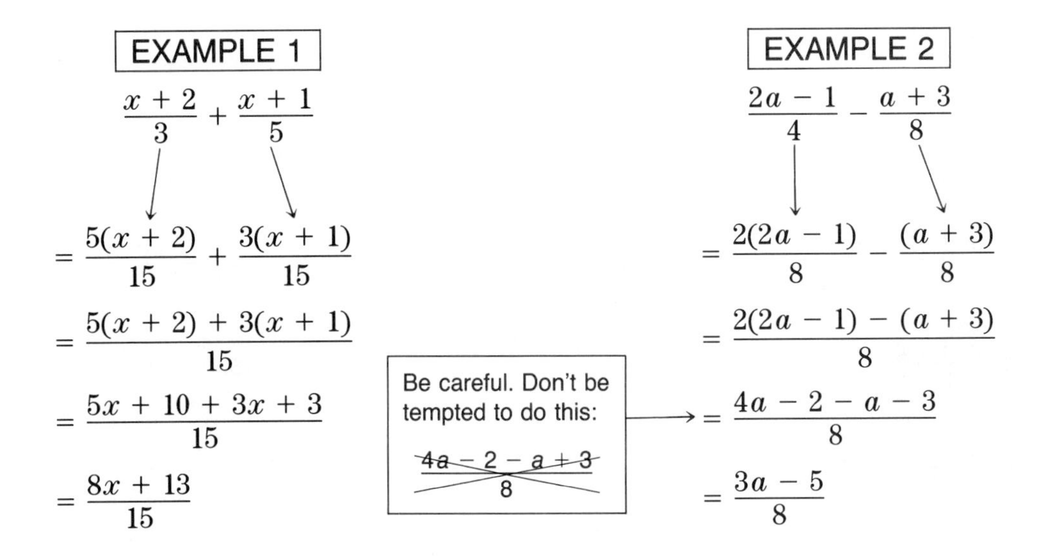

EXAMPLE 1

$$\frac{x+2}{3}+\frac{x+1}{5}$$

$$=\frac{5(x+2)}{15}+\frac{3(x+1)}{15}$$

$$=\frac{5(x+2)+3(x+1)}{15}$$

$$=\frac{5x+10+3x+3}{15}$$

$$=\frac{8x+13}{15}$$

EXAMPLE 2

$$\frac{2a-1}{4}-\frac{a+3}{8}$$

$$=\frac{2(2a-1)}{8}-\frac{(a+3)}{8}$$

$$=\frac{2(2a-1)-(a+3)}{8}$$

$$=\frac{4a-2-a-3}{8}$$

$$=\frac{3a-5}{8}$$

Be careful. Don't be tempted to do this: ~~$\frac{4a-2-a+3}{8}$~~

Classroom Practice

Complete.

1. $\frac{x+4}{2}+\frac{x-1}{3}=\frac{?}{6}+\frac{?}{6}$

2. $\frac{2x+3}{4}-\frac{x+1}{2}=\frac{?}{4}-\frac{?}{4}$

3. $\frac{2a+3}{5}-\frac{a-1}{2}=\frac{?}{10}-\frac{?}{10}$

4. $\frac{x+1}{3}+\frac{x+2}{5}=\frac{?}{15}+\frac{?}{15}$

5. Find the errors in the work below.

$$\frac{2x+5}{3}-\frac{x-3}{4}=\frac{4(2x+5)}{12}-\frac{3(x-3)}{12}$$

$$=\frac{4(2x+5)-3(x-3)}{12}$$

$$=\frac{8x+20-3x-9}{12}$$

$$=\frac{5x-11}{12}$$

Written Exercises

Simplify.

A

1. $\frac{x+1}{8}+\frac{3}{4}$
2. $\frac{2k-1}{6}-\frac{k+1}{9}$
3. $\frac{3t-2}{6}+\frac{2t-1}{4}$
4. $\frac{x+2}{3}+\frac{x}{5}$
5. $\frac{x-5}{4}+\frac{2x}{3}$
6. $\frac{y-9}{4}+\frac{3y}{8}$
7. $\frac{a-2}{4}-\frac{a}{12}$
8. $\frac{a+3b}{8}-\frac{a}{2}$
9. $\frac{x-3}{4}-\frac{5x}{6}$
10. $\frac{x+2}{3}+\frac{x+1}{2}$
11. $\frac{x-1}{2}+\frac{x+3}{4}$
12. $\frac{a-5}{3}+\frac{a-1}{6}$
13. $\frac{x+5}{3}+1$
14. $\frac{2a+1}{5}-2$
15. $\frac{4y-4}{3}+8$
16. $\frac{4y+4}{4}-\frac{y-1}{2}$
17. $\frac{4x+2}{2}-\frac{x-2}{3}$
18. $\frac{2b-2}{4}-\frac{b-4}{3}$

B

19. $\frac{r}{4}+\frac{r-2}{8}-\frac{r-4}{16}$
20. $\frac{4+3a}{6}-\frac{5a}{18}+\frac{3-a}{12}$
21. $\frac{10x+4}{5}-\frac{6x+2}{3}-\frac{2}{15}$
22. $\frac{x+1}{8x}+\frac{3x-1}{12x}-\frac{1}{6x}$

PUZZLE ◆ PROBLEMS

Diophantus was an ancient Greek mathematician. According to a legend, this problem was written on his tombstone:

> HERE LIE THE REMAINS OF DIOPHANTUS. HE WAS A CHILD FOR ONE SIXTH OF HIS LIFE. AFTER ONE TWELFTH MORE, HE BECAME A MAN. AFTER ONE SEVENTH MORE, HE MARRIED. FIVE YEARS LATER HIS SON WAS BORN. THE SON LIVED HALF AS LONG AS HIS FATHER AND DIED FOUR YEARS BEFORE HIS FATHER.

How old did Diophantus live to be?

14 Equations with Fractions

Sometimes you'll want to solve equations with fractions in them. Let's see how these equations can be solved.

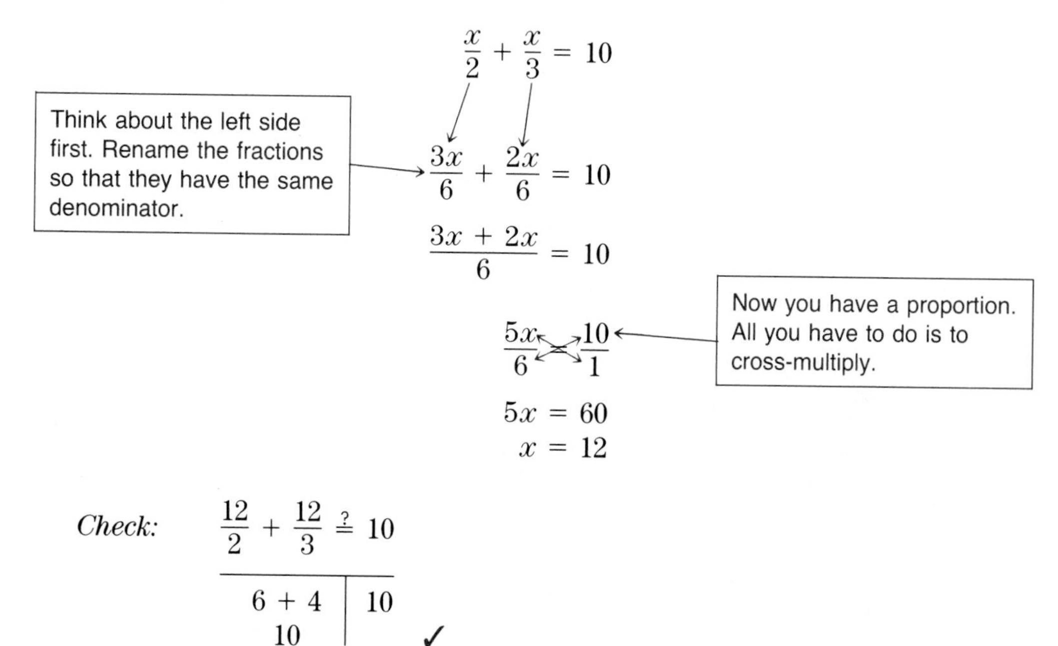

It is easy to do just one step at a time, and simplify one side of the equation at a time. That way you can get a proportion which is not difficult to solve.

EXAMPLES

$$\frac{2}{5} + \frac{3}{10} = \frac{7}{n}$$

$$\frac{4}{10} + \frac{3}{10} = \frac{7}{n}$$

$$\frac{4 + 3}{10} = \frac{7}{n}$$

$$\frac{7}{10} = \frac{7}{n}$$

$$7n = 70$$
$$n = 10$$

$$\frac{a}{4} + 5 = \frac{3a}{2}$$

$$\frac{a}{4} + \frac{20}{4} = \frac{3a}{2}$$

$$\frac{a + 20}{4} = \frac{3a}{2}$$

$$2(a + 20) = 4(3a)$$
$$2a + 40 = 12a$$
$$40 = 10a$$
$$4 = a$$

It is always wise to check your answers.

Classroom Practice

Rename the fractions on the left side of the equation so that they have the same denominator. State the new equation but do not solve it.

1. $\frac{a}{2} - \frac{a}{4} = 5$
2. $\frac{m}{6} + \frac{m}{3} = 1$
3. $\frac{s}{4} + \frac{s}{10} = 2$
4. $\frac{5}{12} + \frac{1}{6} = \frac{7}{a}$
5. $\frac{5}{6} - \frac{2}{3} = \frac{4}{y}$
6. $\frac{1}{4} + \frac{1}{5} = \frac{18}{x}$
7. $\frac{x}{4} + 2 = \frac{x}{2}$
8. $\frac{x}{5} - 4 = \frac{x}{7}$
9. $\frac{m}{3} + 6 = \frac{4m}{3}$

Written Exercises

Solve.

A

1. $\frac{x}{3} - \frac{x}{5} = 4$
2. $\frac{x}{6} + \frac{x}{7} = 13$
3. $\frac{x}{6} + \frac{x}{3} = 6$
4. $\frac{1}{4} + \frac{1}{12} = \frac{x}{3}$
5. $\frac{2}{3} - \frac{1}{6} = \frac{x}{4}$
6. $\frac{4}{5} - \frac{2}{3} = \frac{n}{15}$
7. $\frac{2}{5} + \frac{3}{10} = \frac{7}{n}$
8. $\frac{4}{5} - \frac{2}{15} = \frac{6}{x}$
9. $\frac{3}{4} - \frac{5}{6} = \frac{5}{y}$
10. $\frac{n}{2} + \frac{2n}{3} = 7$
11. $\frac{a}{6} - \frac{2a}{9} = \frac{1}{2}$
12. $\frac{1}{c} + \frac{3}{2c} = \frac{1}{6}$
13. $\frac{3}{4n} + \frac{1}{n} = \frac{7}{8}$
14. $\frac{3n}{7} - \frac{n}{3} = 4$
15. $\frac{2}{x} - \frac{4}{3x} = \frac{2}{9}$

B

16. $\frac{a+2}{3} + \frac{a-1}{6} = 5$
17. $\frac{x-1}{4} - \frac{2x-3}{4} = 5$
18. $\frac{x}{4} - \frac{x+5}{3} = 6$
19. $\frac{x+3}{8} - \frac{x-2}{6} = 1$
20. $\frac{c-1}{2c} + \frac{c+3}{4c} = \frac{5}{8}$
21. $\frac{2-x}{x} - \frac{x+3}{3x} = \frac{-1}{3}$

C

22. $\frac{6x-4}{3} - 2 = \frac{18-4x}{3} + x$
23. $2 - \frac{7x-1}{6} = 3x - \frac{19x+3}{4}$

15 Work Problems

Suppose an office manager wants to figure how long it will take to get a report typed by two typists who work at different speeds.

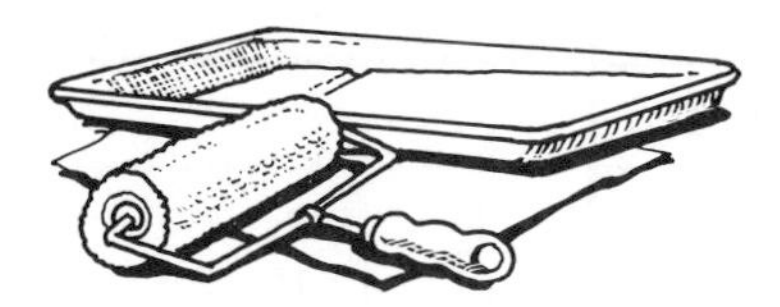

Suppose your mother helps you to paint a room and you wonder how long the job will take.

Algebra can help solve both problems.

Let's consider the painting problem.

EXAMPLE

Ella and her mother plan to paint the living room.
Ella thinks it would take her 10 hours alone.
Her mother says she could do it herself in 5 hours.
How long would it take them to do the job together?

	Ella	Mother	Together
Hours needed	10	5	n
Part done in one hour	$\frac{1}{10}$	$\frac{1}{5}$	$\frac{1}{n}$

$$\underbrace{\text{Part Ella does in one hour}} + \underbrace{\text{Part Mother does in one hour}} = \underbrace{\text{Part of job done in one hour}}$$

$$\frac{1}{10} + \frac{1}{5} = \frac{1}{n}$$

$$\frac{1}{10} + \frac{2}{10} = \frac{1}{n}$$

$$\frac{1+2}{10} = \frac{1}{n}$$

$$\frac{3}{10} \times \frac{1}{n}$$

$$3n = 10$$

$$n = 3\tfrac{1}{3}$$

Look above in the chart. n represents the number of hours needed if they work together. We've solved the problem.

Answer: It will take them $3\frac{1}{3}$ hours working together.

Written Exercises

A

1. Mr. Macy's two grandchildren offer to paint his kitchen. Nathan thinks it would take him 10 hours alone. Meg says it would take her 15 hours alone. How long would the job take if they worked together?

	Nathan	Meg	Together
Hours needed	10	15	n
Part done in one hour	$\frac{1}{10}$	$\frac{1}{15}$	$\frac{1}{n}$

$$\frac{1}{10} + \frac{1}{15} = \frac{1}{n}$$

2. It would take Milo 3 hours to mow the yard. When Jean does it alone it takes 2 hours. How long would it take if they worked together?

	Milo	Jean	Together
Hours needed	3	2	h
Part done in one hour	$\frac{1}{3}$	?	?

3. A carpenter needs 2 hours to replace 3 windows. The apprentice would need 4 hours. If they do the job together, how long will it take?

	Carpenter	Apprentice	Together
Hours needed	?	?	?
Part done in one hour	?	?	?

4. A washing machine can be filled by the cold water faucet in 6 minutes. The hot water faucet can fill it in 9 minutes. How long will it take both faucets to fill the washer together?

	Cold	Hot	Together
Minutes needed	?	?	?
Part done in one minute	?	?	?

5. Corrie and Gary are stacking wood for their stove.
When Corrie stacked the same amount alone last year it took 8 hours.
Gary estimates it would take him 10 hours alone.
How long will it take them together?

	Corrie	Gary	Together
Hours needed	?	?	?
Part done in one hour	?	?	?

6. Eugene is a welder on an assembly line that also uses robots.
Eugene can complete 50 welds in 3 hours.
The robot can do 50 welds in 4 hours.
How long would it take to do 50 welds if they worked together?

	Eugene	Robot	Together
Hours needed	?	?	?
Part done in one hour	?	?	?

7. Ralph, Hiroshi, and Vivian have to type the company annual report.
Ralph thinks it would take him 10 hours, working alone.
Hiroshi says it would take him 10 hours alone also.
Vivian says she could do it in 8 hours.
How long would the job take if they worked together?

	Ralph	Hiroshi	Vivian	Together
Hours needed	10	10	8	x
Part done in one hour	$\frac{1}{10}$	$\frac{1}{10}$	$\frac{1}{8}$	$\frac{1}{x}$

8. Celia, Al, and Julio are going to decorate the gym.
It would take Julio 3 hours, working alone.
It would take Celia and Al each 4 hours, working alone.
How long will the job take if they work together?

	Celia	Al	Julio	Together
Hours needed	?	?	?	?
Part done in one hour	?	?	?	?

B **9.** Pat Marston needs 6 hours to deliver the mail. When a temporary carrier helps, the job is finished in 4 hours. How long would it take the temporary carrier to deliver the mail alone?

	Pat	Temporary	Together
Hours needed	?	x	4
Part done in one hour	?	$\frac{1}{x}$	?

10. Frank can do a typing job in 10 hours. When Paula helps him, it takes only 6 hours. If Paula did the job alone, how long would it take her?

	Frank	Paula	Together
Hours needed	?	?	?
Part done in one hour	?	?	?

11. Working together, Dana and Jack can refinish the living room floor in 4 hours. It would take Dana alone twice as long as it would take Jack alone to complete the job. How long would it take each person alone?

	Dana	Jack	Together
Hours needed	?	?	?
Part done in one hour	?	?	?

C **12.** The boss in the repair shop needs 6 hours to do a special repair job alone. Yesterday her assistant started helping her 2 hours after the job started. They finished the job 3 hours later.
How long would it have taken the assistant to do the job alone?

	Boss	Assistant	Together
Hours needed	?	?	?
Part done in one hour	?	?	?

SELF-TEST

Solve.

1. $\frac{x}{4} - \frac{x}{6} = 11$ 2. $\frac{a}{3} + \frac{a}{2} = 5$ 3. $\frac{x}{3} - \frac{2x}{9} = 4$ *(9-14)*

4. $\frac{1}{n} + \frac{3}{2n} = \frac{1}{2}$ 5. $\frac{2}{x} + \frac{3}{4x} = \frac{1}{8}$ 6. $\frac{1}{2x} + \frac{2}{3x} = \frac{1}{12}$

7. It would take Bruce five hours to clean out the garage. Manuel can clean the garage alone in 3 hours. How long would it take if they worked together? *(9-15)*

	Bruce	Manuel	Together
Hours needed	?	?	?
Part done in one hour	?	?	?

CALCULATOR ACTIVITIES

You can use your calculator to solve proportions. Solve the proportion for the variable and use the calculator to perform the operations shown.

$$\frac{x}{8} = \frac{9}{12}$$
$$x \cdot 12 = 8 \cdot 9$$
$$x = \frac{8 \cdot 9}{12}$$

Use your calculator to find that $x = 6$.

Use a calculator to solve each proportion.

1. $\frac{x}{10} = \frac{13}{20}$ 2. $\frac{4}{x} = \frac{10}{45}$ 3. $\frac{2x}{5} = \frac{18}{15}$

4. $\frac{15}{10} = \frac{3x}{8}$ 5. $\frac{x+1}{3} = \frac{16}{6}$ 6. $\frac{4}{4-x} = \frac{12}{3}$

16 Binomial Denominators (Optional)

When you want to add or subtract fractions with more than one term in the denominator, there will probably be quite a few steps.

EXAMPLE 1

$$\frac{4}{x} + \frac{3}{x+2}$$

Find a new denominator that is divisible by both x and $x + 2$. $x(x + 2)$ will work.

$$= \frac{4(x+2)}{x(x+2)} + \frac{3x}{x(x+2)}$$

$$= \frac{4(x+2) + 3x}{x(x+2)}$$

$$= \frac{4x + 8 + 3x}{x(x+2)}$$

$$= \frac{7x + 8}{x(x+2)}$$

EXAMPLE 2

$$\frac{2}{a-4} + \frac{3}{a+2}$$

Find a new denominator that is divisible by both $a - 4$ and $a + 2$. $(a - 4)(a + 2)$ will work.

$$= \frac{2(a+2)}{(a-4)(a+2)} + \frac{3(a-4)}{(a-4)(a+2)}$$

$$= \frac{2(a+2) + 3(a-4)}{(a-4)(a+2)}$$

$$= \frac{2a + 4 + 3a - 12}{(a-4)(a+2)}$$

$$= \frac{5a - 8}{(a-4)(a+2)}$$

EXAMPLE 3

$$\frac{5}{x-3} - \frac{4}{x+1}$$

Find a new denominator that is divisible by both $x - 3$ and $x + 1$. $(x - 3)(x + 1)$ will work.

$$= \frac{5(x+1)}{(x-3)(x+1)} - \frac{4(x-3)}{(x-3)(x+1)}$$

$$= \frac{5(x+1) - 4(x-3)}{(x-3)(x+1)}$$

$$= \frac{5x + 5 - 4x + 12}{(x-3)(x+1)}$$

$$= \frac{x + 17}{(x-3)(x+1)}$$

Classroom Practice

Simplify.

1. $\frac{1}{z} + \frac{2}{z + 1}$
2. $\frac{5}{x} - \frac{3}{x - 2}$
3. $\frac{4}{y} - \frac{10}{y - 5}$
4. $\frac{5}{m - 1} + \frac{5}{m + 3}$
5. $\frac{m}{m + 4} - \frac{m}{m + 2}$
6. $\frac{4}{r + 1} - \frac{3}{r - 1}$
7. $\frac{2}{x - 4} + \frac{3}{x + 4}$
8. $\frac{6}{y + 1} + \frac{8}{y - 2}$
9. $\frac{1}{a - 4} + \frac{a}{a + 4}$

Written Exercises

Simplify.

A

1. $\frac{5}{k} + \frac{3}{k + 1}$
2. $\frac{2}{t} + \frac{3}{t + 5}$
3. $\frac{2}{x} - \frac{5}{x - 3}$
4. $\frac{r}{r + 5} - \frac{r}{r + 2}$
5. $\frac{a}{a + 4} - \frac{a}{a - 4}$
6. $\frac{9}{b + 3} - \frac{9}{b - 3}$
7. $\frac{6}{y + 3} - \frac{y}{y - 2}$
8. $\frac{c}{c - 4} - \frac{5}{c + 5}$
9. $\frac{7}{x + 3} + \frac{x}{x - 3}$
10. $\frac{6}{a + 4} + \frac{3}{a - 1}$
11. $\frac{b}{b + 6} + \frac{2b}{b - 1}$
12. $\frac{a}{4 - a} - \frac{a}{a - 4}$
13. $\frac{7}{x + y} - \frac{x}{x - y}$
14. $\frac{s}{r - s} + \frac{r}{r + 2}$
15. $\frac{4}{a + b} + \frac{2a}{a - b}$
16. $\frac{3}{x + 6} - \frac{11}{x^2 - 36}$
17. $\frac{8}{(c + d)^2} + \frac{5}{c + d}$
18. $\frac{12}{t + s} - \frac{4}{t^2 + ts}$
19. $\frac{1}{x^2 - y^2} + \frac{3}{x - y}$
20. $\frac{c}{c^2 - 9} + \frac{1}{c + 3}$
21. $\frac{2a + 5}{4 - a^2} + \frac{5}{a + 2}$

Solve for the variable.

B

22. $\frac{2}{a + 1} - \frac{1}{a + 3} = \frac{1}{a}$
23. $\frac{3}{x + 1} - \frac{2}{x - 1} = \frac{1}{x}$
24. $\frac{2}{s + 6} + \frac{1}{s - 3} = \frac{3}{s}$
25. $\frac{1}{m - 1} + \frac{1}{m + 2} = \frac{2}{m}$

COMPUTER ACTIVITIES

FOREIGN EXCHANGE

Suppose you planned to travel in foreign countries. You would want to know about the rate of exchange between U.S. and foreign money. Newspapers often provide information on exchange rates.

The chart below shows how many units of a foreign currency could be exchanged for one U.S. dollar on a certain day.

Britain	0.675
Canada	1.39
France	6.88
Japan	181
Mexico	468
West Germany	2.24

For example, for one U.S. dollar you could get 181 Japanese yen or 0.675 British pounds. (Check a newspaper to find more recent exchange rates. Have they gone up or down from the rates shown here?)

The program on the facing page can be used to print a table of amounts of money in U.S. dollars and the equivalent amount in a foreign currency. The symbol A$ in line 30 is known as a *string variable.* This type of variable allows you to give input in letters and words, rather than in only numbers. In this program, you will be asked to input the name of a country.

Lines 80–100 and 110–130 make up two *loops.* A loop is a sequence of instructions that you wish the computer to repeat a number of times. The first time the computer reads line 80, it assigns C the value 1. At line 100, the computer is told to return to line 80 and assign C the next value. Since 10 is the final value given in line 80, the computer will complete this loop ten times before going on to line 110.

In the first loop (lines 80–100), currency conversions are figured one dollar at a time. In the second loop (lines 110–130), the dollar amounts skip $5 each time. This jump is done by including a STEP 5 in line 110. To skip $10 each time, you would include a STEP 10 in line 110, and so on.

```
10 REM**TO CONVERT MONEY
20 PRINT "WHAT COUNTRY";
30 INPUT A$
40 PRINT "EXCHANGE RATE"
50 PRINT "PER DOLLAR";
60 INPUT R
70 PRINT "U.S.", A$
80 FOR C = 1 TO 10
90 PRINT C, R*C
100 NEXT C
110 FOR C = 15 TO 25 STEP 5
120 PRINT C, R*C
130 NEXT C
140 END
```

EXERCISES

Use the exchange rates shown on the facing page.

1. RUN the program to produce a table for converting U.S. dollars to French francs.

2. RUN the program to produce a table for converting U.S. dollars to Mexican pesos. Change the second loop so the table will go from $25 to $100, skipping $25 each time.

3. Print a table for converting U.S. dollars to Canadian dollars. Use only one loop by removing lines 80–100. (To "erase" a line, type the line number only and hit Enter on the keyboard.) For this table, have the dollar amounts go from $10 to $100, skipping $10 each time.

CONSUMER APPLICATIONS

WHICH IS A BETTER BUY?

At some time you have probably looked at two different brands of an item, and have thought about which one was the better buy. Of course, there are many things like quality and packaging to consider. As a consumer, however, you should be able to recognize which item is less expensive, and therefore be able to make good decisions.

Which brand of shampoo is less expensive?

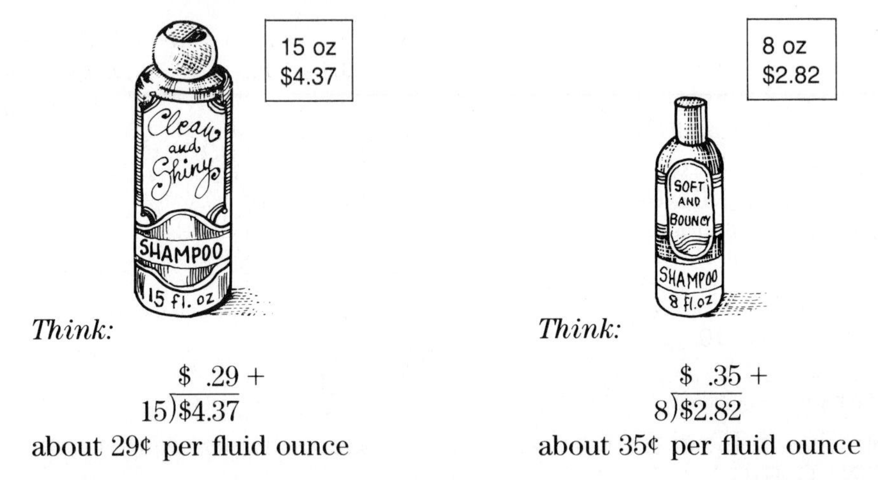

Think:

$$15\overline{)\$4.37}\quad \$\ .29+$$

about 29¢ per fluid ounce

Think:

$$8\overline{)\$2.82}\quad \$\ .35+$$

about 35¢ per fluid ounce

You don't always have to do this kind of arithmetic while you're shopping. Unit pricing labels on store shelves show you the price of the item per pound, per gram, per kilogram, per fluid ounce, or whatever unit seems easiest.

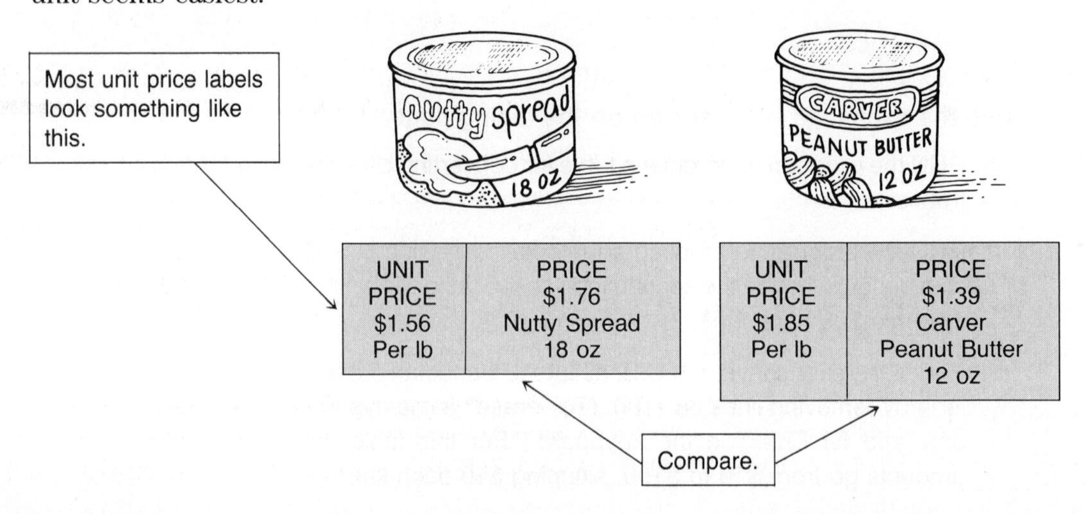

SKILLS REVIEW

DECIMALS

Recall that decimals are another way of writing fractions.

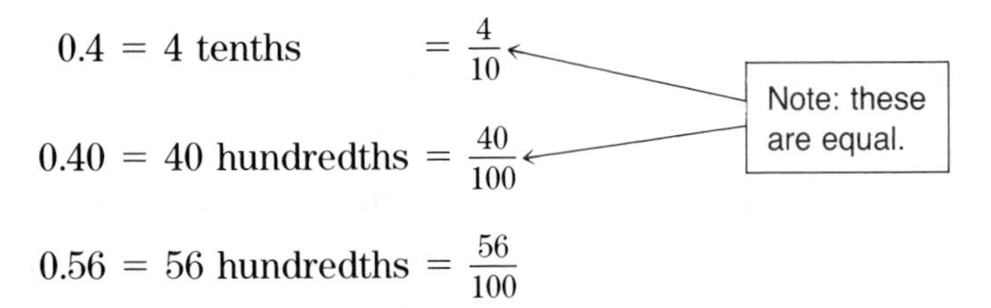

Let's compare 0.6 and 0.45.

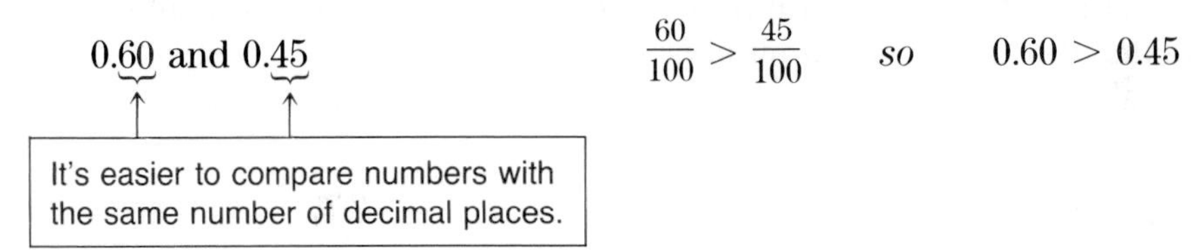

Compare. Write > or <.

1. 0.2 ___?___ 0.9 **2.** 0.72 ___?___ 0.59 **3.** 2.4 ___?___ 3.7 **4.** 5.8 ___?___ 5.08

5. 7.41 ___?___ 7.48 **6.** 6.05 ___?___ 5.21 **7.** 0.03 ___?___ 0.3 **8.** 0.042 ___?___ 0.42

To round decimals, use the same rules as you use to round integers.

SAMPLES

Round 2.346 to one decimal place. *Answer:* 2.3

This digit is less than 5. Round *down*.

Round 6.786 to two decimal places. *Answer:* 6.79

This digit is 5 or greater. Round *up*.

Round to one decimal place.

9. 6.73 **10.** 5.84 **11.** 9.29 **12.** 7.68 **13.** 4.273

Round to two decimal places.

14. 0.651 **15.** 9.207 **16.** 3.415 **17.** 0.078 **18.** 5.924

CHAPTER REVIEW

CHAPTER SUMMARY

1. You can simplify a fraction by dividing both its numerator and denominator by the same number.
2. The ratio of two numbers is their quotient.
3. A proportion is a statement that two ratios are equal.
4. You add, subtract, multiply, and divide fractions in algebra in the same way as in arithmetic.
5. Two numbers are reciprocals if their product is 1.
6. To rename fractions, multiply or divide both the numerator and denominator by the same number (except 0).

REVIEW EXERCISES

Simplify. ***(See pp. 298–303.)***

1. $\frac{8}{24}$
2. $\frac{3}{15x}$
3. $\frac{11x}{22}$
4. $\frac{-4a^2b}{-ab^2}$
5. $\frac{16x^2y^3}{8x^2y}$
6. $\frac{3a - 15}{3}$
7. $\frac{x - 4}{3x - 12}$
8. $\frac{n^2 + 4n + 4}{n + 2}$
9. $\frac{(3 + x)^2}{9 - x^2}$
10. $\frac{3 - a}{a - 3}$
11. $\frac{x - y}{y - x}$
12. $\frac{1 - x}{2x - 2}$
13. $\frac{7 - a}{a^2 - 49}$

Solve. ***(See pp. 304–307.)***

14. The ratio of blue to yellow in a paint mixture is 3 to 2. How much blue paint is needed to make 10 cans of the mixture?
15. Central High's soccer team has 22 members, including coaches. The ratio of coaches to players is 2 to 9. How many players are there?

Solve. ***(See pp. 308–313.)***

16. $\frac{x}{14} = \frac{2}{7}$
17. $\frac{2n}{6} = \frac{12}{9}$
18. $\frac{x - 2}{x + 2} = \frac{1}{3}$
19. $\frac{2b}{b - 7} = \frac{1}{4}$

20. 18 oranges cost \$2.70. How much will 36 oranges cost?

21. 3 avocados cost \$1.95. How much will 2 avocados cost?

Simplify. *(See pp. 314–318.)*

22. $\frac{9}{21} \cdot \frac{7}{63}$ **23.** $\frac{x}{4} \cdot \frac{12}{x^2}$ **24.** $\frac{10m^2}{3m} \cdot \frac{m}{5}$ **25.** $\frac{5}{3x - 3} \cdot \frac{3}{75}$

26. $\frac{3}{4} \div \frac{1}{12}$ **27.** $\frac{n}{3} \div \frac{n}{9}$ **28.** $\frac{5}{a^2b^2} \div \frac{25}{ab}$ **29.** $\frac{6}{mn^2} \div \frac{3}{m^2n}$

30. $\frac{x^2 - 4}{25} \cdot \frac{5}{x + 2}$ **31.** $\frac{a + b}{28} \div \frac{a - b}{7}$ **32.** $\frac{3a - 3}{15} \div \frac{a - 1}{3}$

Simplify. *(See pp. 319–324.)*

33. $\frac{5x}{3} + \frac{x}{3}$ **34.** $\frac{n}{2} - \frac{4}{2}$ **35.** $\frac{7}{a + 1} + \frac{8}{a + 1}$ **36.** $\frac{6a}{a + 2} - \frac{3a}{a + 2}$

37. $\frac{1}{3} + \frac{1}{5}$ **38.** $\frac{n}{3} - \frac{n}{5}$ **39.** $\frac{2x}{7} + \frac{x}{2}$ **40.** $\frac{3a}{4} - \frac{a}{3}$

41. $\frac{1}{x} + \frac{1}{4}$ **42.** $\frac{1}{3} - \frac{3}{x}$ **43.** $\frac{1}{2b} - \frac{4}{3b^2}$ **44.** $\frac{1}{x} - \frac{1}{y}$

(Optional) Simplify. *(See pp. 325–326.)*

45. $\frac{x + 3}{6} + \frac{1}{2}$ **46.** $\frac{2y - 3}{5} - \frac{y + 1}{10}$ **47.** $\frac{3z - 2}{4} + \frac{z + 5}{8}$

Solve. *(See pp. 327–332.)*

48. $\frac{1}{6} + \frac{1}{2} = \frac{x}{3}$ **49.** $\frac{n}{3} - \frac{n}{4} = 1$ **50.** $\frac{3}{4x} - \frac{5}{2x} = 7$ **51.** $\frac{1}{n} + \frac{3}{4n} = \frac{1}{4}$

52. Elton and Paulette are papering their living room. Elton says he could do it in 6 hours alone. Paulette says she could do it in 5 hours alone. If they work together, how long will it take?

(Optional) Simplify. *(See pp. 334–335.)*

53. $\frac{3}{t} + \frac{2}{t - 2}$ **54.** $\frac{2}{x} + \frac{3}{x - 1}$ **55.** $\frac{4z}{z + 1} - \frac{z}{z - 1}$

CHAPTER TEST

Simplify.

1. $\dfrac{18x^3y}{-6xy^2}$
2. $\dfrac{12ab^2}{16bc^2}$
3. $\dfrac{-85a^3b^2c^2}{-34ab^3c^2}$ *(9-1)*

4. $\dfrac{x^2 - 1}{6x + 6}$
5. $\dfrac{k^2 + 4k + 4}{14k + 28}$
6. $\dfrac{x^2 - 7x - 8}{x^2 - 10x + 16}$ *(9-2)*

7. $\dfrac{1 - a}{7a - 7}$
8. $\dfrac{4y - 4x}{x^2 - y^2}$
9. $\dfrac{t^2 + 7t - 18}{-18 - 2t}$ *(9-3)*

Give the ratio in simplest form.

10. 24 meters to 40 meters
11. 54 ounces to 42 ounces *(9-4)*

Write the ratio using the same units. Give the ratio in simplest form.

12. 40 seconds to 2 minutes
13. 1 quarter to 55 cents

14. There are 306 students in the senior class at City High. The ratio of boys to girls in the class is 5 to 4. How many girls are there? *(9-5)*

Solve.

15. $\dfrac{3}{14} = \dfrac{9}{2x}$
16. $\dfrac{y + 1}{5y + 2} = \dfrac{2}{9}$
17. $\dfrac{a - 7}{a - 5} = \dfrac{a + 1}{a + 5}$ *(9-6)*

18. Eight apples cost \$1.50. How much will one dozen apples cost? *(9-7)*

19. You drove 170 km and used 16 L of gas. How far can you drive with 24 L of gas?

Simplify.

20. $\dfrac{6x^2}{25} \cdot \dfrac{10}{9xy^2}$
21. $\dfrac{12}{a^2 + a} \cdot \dfrac{a}{18}$
22. $\dfrac{r - 5}{r + 5} \cdot \dfrac{r^2 - 25}{5 - r}$ *(9-8)*

23. $\dfrac{(2x)^3}{xy} \div \dfrac{4x}{3y}$
24. $\dfrac{3b + 9}{9} \div (b + 3)$
25. $\dfrac{x^2 - x - 2}{2x - 4} \div \dfrac{x + 1}{8}$ *(9-9)*

26. $\dfrac{5y}{8} + \dfrac{7y}{8}$
27. $\dfrac{2x + 1}{x + 1} + \dfrac{1}{x + 1}$
28. $\dfrac{x - y}{2xy} - \dfrac{6x - y}{2xy}$ *(9-10)*

Find the missing numerator.

29. $\frac{5x}{6} = \frac{?}{24}$ 30. $\frac{7}{3r} = \frac{?}{27rs}$ 31. $\frac{3}{4a} = \frac{?}{4a^2}$ *(9-11)*

Simplify.

32. $\frac{2}{3} - \frac{7}{a}$ 33. $\frac{1}{2x} + \frac{1}{6x}$ 34. $\frac{4}{9y} - \frac{3}{y^2} + \frac{1}{2y}$ *(9-12)*

(Optional) Simplify.

35. $\frac{2x + 1}{5} - \frac{x - 4}{7}$ 36. $\frac{a - 3}{6a} - \frac{a - 2}{4a}$ 37. $\frac{x}{6} + \frac{3x - 1}{8} + \frac{x}{12}$ *(9-13)*

Solve.

38. $\frac{2b}{3} - \frac{b}{2} = 3$ 39. $\frac{4}{x} + \frac{8}{3x} = \frac{5}{9}$ 40. $\frac{3}{c} - \frac{5c + 2}{3c} = \frac{8c - 2}{9c}$ *(9-14)*

41. Rae and Yvette are addressing envelopes. *(9-15)*
Rae needs 60 minutes to address one box of envelopes.
Yvette needs 40 minutes to do the job.
How long would the job take if they worked together?

	Rae	Yvette	Together
Minutes needed	60	40	m
Part done in one minute	$\frac{1}{60}$	?	?

42. A pool can be filled by one pipe in 18 hours.
A larger pipe can do it in 12 hours.
How long will it take if both pipes are used?

	Small pipe	Large pipe	Together
Hours needed	?	?	?
Part done in one hour	?	?	?

(Optional) Simplify.

43. $\frac{r}{r + 2} + \frac{3}{r - 3}$ 44. $\frac{1}{t - 4} - \frac{t}{(t - 4)^2}$ 45. $\frac{2}{1 - x} + \frac{4}{x^2 - 1}$ *(9-16)*

MIXED REVIEW

Solve.

1. $7x + 3y = -4$
   $9x + 3y = 0$

2. $4x + 3y = 15$
   $3x - y = 8$

3. $2x + 5y = 20$
   $3x + 2y = 8$

Simplify.

4. $\frac{x}{10} + \frac{x}{8}$

5. $\frac{z^2 + 3z - 28}{z^2 - 5z + 4}$

6. $\frac{a^2 - b^2}{3ab - 3a^2}$

7. Find three solution pairs for the equation $y + 3x = 4$.

8. Four rolls of tape and 3 notebooks cost \$5.30.
   Six rolls of tape and 2 notebooks cost \$5.70.
   Find the price of each of these items.

9. If the First City Bank paid \$12.50 in monthly interest on a \$2000 money market account, find the monthly interest paid on a \$4200 account.

10. Find the volume of the prism shown.

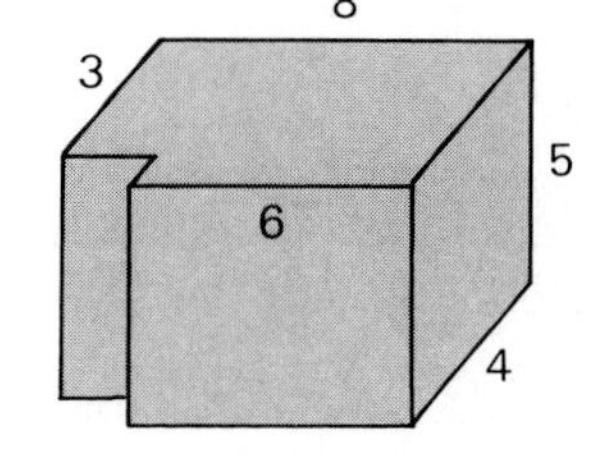

Factor.

11. $x^2 - 16x + 63$

12. $6a^3 + 42a^2b + 12ab^2$

13. Compare. Write < or >:
    $-(-4)^3 \underline{\ \ ?\ \ } (2 - 9)8 - 7$

14. Which choices are solution pairs for the inequality $y > 7 - 2x$?
    a. $(1, 1)$ b. $(4, 0)$ c. $(0, 4)$ d. $(5, -1)$

Solve.

15. $\frac{x - 7}{x - 3} = \frac{x - 1}{x + 6}$

16. $\frac{7}{3n} - \frac{1}{n} = \frac{2}{3}$

17. $5(2x + 1) = -3(1 - 4x)$

18. Solve by the graphing method: $x - y = -2$
    $y = 2x + 5$

19. At the Sunrise Camp the ratio of counselors to campers is 3 to 14. Each day a total of 136 people are at the camp. How many counselors are there?

20. A square has sides of length $x + 9$ and perimeter 68. Find x.

21. Subtract: $(3y^3 - 7y + 4) - (-y^4 + 8y^3 - y^2 + 3y - 5)$

22. Express $(5a - 2b)^2$ as a trinomial.

23. Rose's weekly wage is a function of the number of hours she works. Copy and complete the table.

n = number of hours	40	37	?	22.5
W = weekly wage	400	?	355	225

Simplify.

24. $\dfrac{10r^2}{r^2 + 3r} \cdot \dfrac{2r + 6}{5}$

25. $\dfrac{(2ab^2)^2}{3} \div 6a^3b^3$

26. $\dfrac{-6x^2y}{-8y^3z}$

27. $\dfrac{4a^2 + b^2}{2ab} - \dfrac{b^2 - 5b^3}{2ab}$

28. $(-5xy^3)(4x^2y)^2$

29. $\dfrac{4a^2}{a - 1} \div \dfrac{2a^3}{1 - a}$

30. Use two variables to solve: The sum of two numbers is 32. Their difference is 8. Find the numbers.

Multiply.

31. $(2a - b)(a^2 + ab)$

32. $-3x^2(x^2 + 5x - 1)$

33. $(5y- 7)(3y + 8)$

34. You travel for 2 hours at x km/h.
Then you travel for 3 hours at $(x + 5)$ km/h.
Total distance D = __?__.

35. Tony needs 12 hours to retile the kitchen floor.
His friend can do the job in 15 hours.
How long will the job take if they work together?

36. Graph the solutions of $x > 4$.

37. At 8:00, a car leaves Toronto for Detroit at 92 km/h.
At the same time another car leaves Detroit for Toronto at 97 km/h.
The cities are 378 km apart.
At what time do the cars pass each other?

38. Tell whether the equations share *one* solution pair, *no* solution pair, or *all* solution pairs: $x - y = 2$
$y = x + 2$

Here's what you'll learn in this chapter:

- To solve equations with decimals.
- To use equations with decimals to solve word problems.
- To solve equations with percents.
- To solve word problems involving percents.
- To solve interest, investment, and mixture problems.

The land on a large farm may be used for several different purposes. Percents are useful in calculating the number of acres to be planted with each crop.

Chapter 10

Decimals and Percents

1 Decimals

Many equations contain decimals. First, take a new look at decimals and how to work with them.

Fraction	Decimal	Read
$\frac{7}{10}$	0.7	seven tenths
$\frac{13}{100}$	0.13	thirteen hundredths
$\frac{27}{1000}$	0.027	twenty-seven thousandths

Working with decimals is often much easier than working with ordinary fractions. For instance, in addition and subtraction all you do is line up the decimal points and compute as you would with whole numbers.

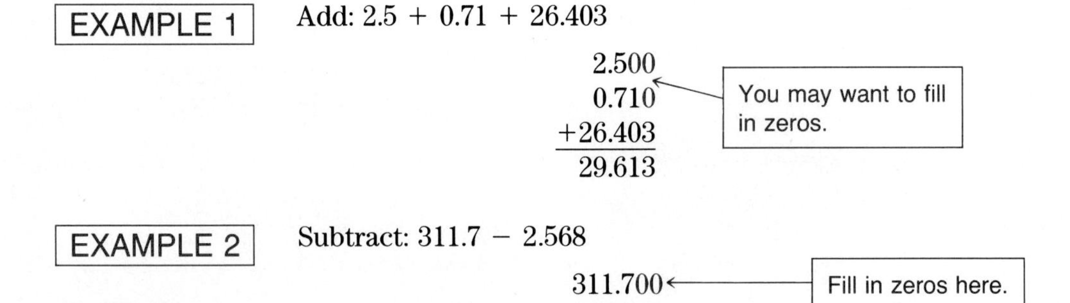

To multiply, don't pay attention to the decimal points at first. Multiply as you would with whole numbers, then put the decimal point in the answer.

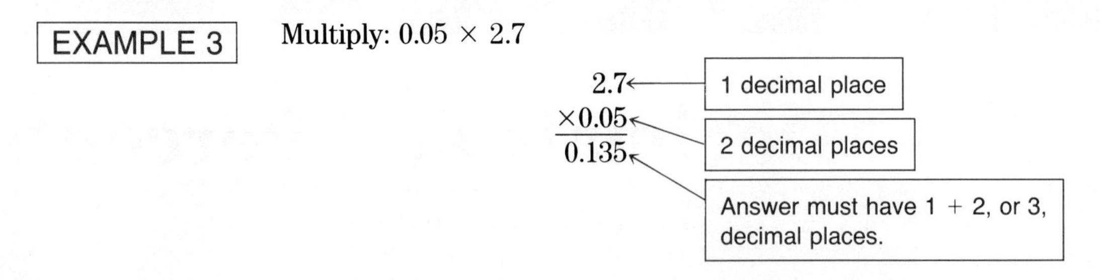

EXAMPLE 4 Multiply: 0.42×6.65

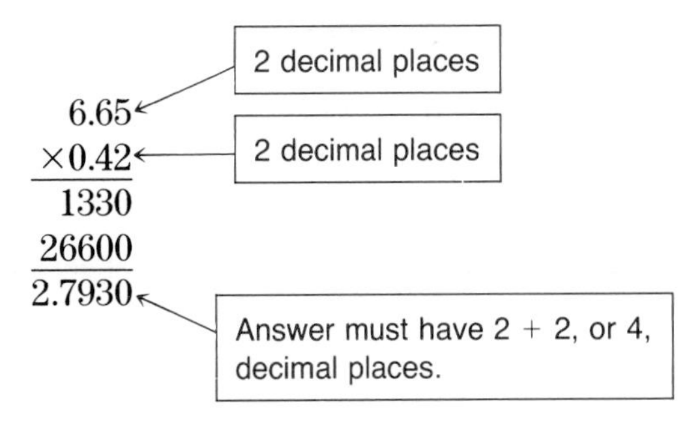

Written Exercises

Add or subtract.

A

1. $0.46 + 0.2$
2. $2.4 + 5.67$
3. $3.95 - 1.02$
4. $4.06 - 1.2$
5. $2.04 + 0.6$
6. $0.7 + 3.002$
7. $6.1 - 0.98$
8. $3.2 + 5.013$
9. $8.4 - 0.635$
10. $7.2 - 3.963$
11. $8.81 + 4.007$
12. $0.4 - 3.567$
13. $6 + 2.9 + 5.63$
14. $8.1 + 0.2 + 3.56$
15. $7.421 + 5 + 8.09$
16. $3.7 + 0.004 + 12$
17. $6.7 + 0.356 + 4$
18. $7.9 + 3.0204 + 10$

Multiply.

19. 2×6.3
20. 4.7×5
21. 8.2×0.1
22. 5.6×10
23. 2.43×10
24. 6.71×100
25. 6.1×100
26. 3.04×0.3
27. 2.04×2.02
28. 7.9×0.04
29. 6.25×0.4
30. 3.71×1.02
31. 0.421×1.006
32. 3.07×0.501
33. 2.006×0.03
34. 4.06×0.031
35. 0.041×0.102
36. 0.04×0.605
37. 1.011×0.204
38. 5.02×3.009
39. 6.7×2.413

40. The Richter Scale measures the power of an earthquake. The Chilean earthquake of 1960 measured 9.5 on the Richter Scale. An earthquake in 1906 in San Francisco measured 7.9 on the Richter Scale. How much greater was the reading on the Richter Scale for the Chilean earthquake?

2 Division with Decimals

In dividing by a decimal, the first step is to make the divisor an integer.

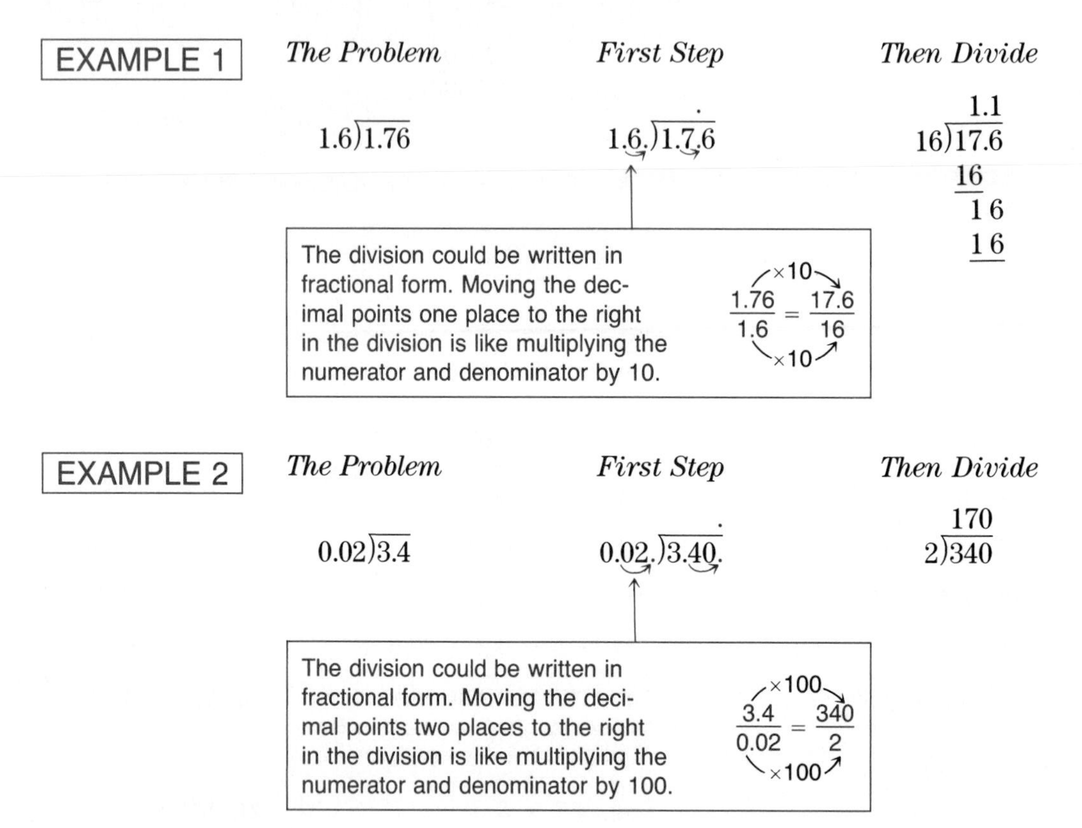

Often divisions do not come out "even." In those cases, you will have to round the answer.

EXAMPLE 3 Find $3.75 \div 0.07$. Round to one decimal place.

$$0.07.\overline{)3.75.00} = 53.57$$

Keep dividing until you have *two* decimal places. Round the answer to one decimal place.

Answer: $3.75 \div 0.07 \approx 53.6$

($\approx$ is approximately equal to)

EXAMPLE 4 Find 14 ÷ 2.3. Round to two decimal places.

$$\begin{array}{r} 6.086 \\ 2.3_{\wedge}\overline{)14.0_{\wedge}000} \end{array}$$

Keep dividing until you have three decimal places. Round the answer to two decimal places.

Answer: $14 \div 2.3 \approx 6.09$

Classroom Practice

Locate the decimal point in the answer.

1. $\begin{array}{r} 59 \\ 0.6\overline{)3.54} \end{array}$
2. $\begin{array}{r} 84 \\ 0.8\overline{)67.2} \end{array}$
3. $\begin{array}{r} 36 \\ 0.72\overline{)2.592} \end{array}$
4. $\begin{array}{r} 164 \\ 0.25\overline{)41} \end{array}$

Round to one decimal place.

5. 0.72
6. 5.63
7. 8.15
8. 0.66
9. 7.89

Round to two decimal places.

10. 0.235
11. 6.477
12. 7.038
13. 5.214
14. 16.739

Written Exercises

Divide.

A

1. $0.4\overline{)2.4}$
2. $0.7\overline{)2.17}$
3. $0.5\overline{)4}$
4. $0.9\overline{)1.008}$
5. $0.08\overline{)3.68}$
6. $2.5\overline{)0.1}$
7. $0.24\overline{)0.264}$
8. $0.125\overline{)5}$
9. $3.2\overline{)0.64}$
10. $0.003\overline{)2.43}$
11. $1.6\overline{)0.896}$
12. $0.8\overline{)1.88}$

Divide. Round to one decimal place.

13. $0.6\overline{)1}$
14. $0.4\overline{)2.7}$
15. $0.5\overline{)6.3}$
16. $1.2\overline{)3.7}$
17. $6.2\overline{)93}$
18. $0.34\overline{)0.65}$
19. $0.07\overline{)40}$
20. $0.51\overline{)4.7}$
21. $0.12\overline{)7.9}$
22. $2.3\overline{)60}$
23. $0.36\overline{)0.92}$
24. $1.8\overline{)8.25}$

Divide. Round to two decimal places.

25. $0.3\overline{)6.74}$
26. $0.9\overline{)0.35}$
27. $0.4\overline{)10}$
28. $1.7\overline{)62}$
29. $0.08\overline{)2.6}$
30. $0.34\overline{)3.1}$
31. $0.26\overline{)62}$
32. $0.05\overline{)84}$
33. $0.75\overline{)40}$
34. $2.3\overline{)38}$
35. $0.06\overline{)76}$
36. $0.54\overline{)160}$

3 Decimals for Fractions

Fractions whose numerators and denominators are integers are called **rational numbers.** Here are some examples of rational numbers.

$$\frac{-2}{4} \qquad \frac{1}{2} \qquad \frac{7}{-5} \qquad \frac{5}{1}$$

Since any integer can be written as a fraction, any integer is a rational number.

To find the decimal form for any rational number, carry out the division shown by the fraction.

EXAMPLE 1 Express $\frac{4}{5}$ as a decimal.

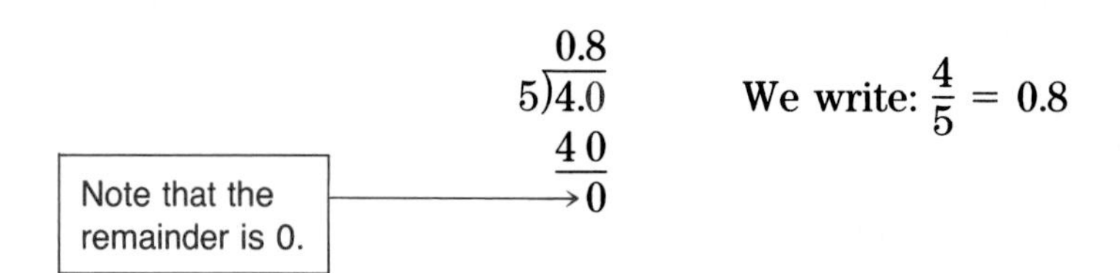

We call 0.8 a **terminating decimal** because the division terminates, or comes out "even."

Sometimes the digits in the decimal are repeating digits.

EXAMPLE 2 Express $\frac{1}{3}$ as a decimal.

```
   0.333
3)1.000
   9
   10
    9
    10
     9
     1
```

We write: $\frac{1}{3} = 0.33\ldots$

We call 0.33 . . . a **repeating decimal.** The three dots show that the 3 repeats.

EXAMPLE 3 Express $\frac{3}{11}$ as a decimal.

$$\begin{array}{r} 0.2727 \\ 11\overline{)3.0000} \\ \underline{2\,2} \\ 80 \\ \underline{77} \\ 30 \\ \underline{22} \\ 80 \\ \underline{77} \\ 3 \end{array}$$

We write: $\frac{3}{11} = 0.2727\ldots$

This is a repeating decimal. The 27 repeats.

Written Exercises

Express the fraction as a decimal. Be sure to use three dots with a repeating decimal.

A

1. $\frac{1}{8}$ **2.** $\frac{5}{8}$ **3.** $\frac{3}{5}$ **4.** $\frac{5}{9}$ **5.** $\frac{3}{8}$

6. $\frac{2}{3}$ **7.** $\frac{2}{7}$ **8.** $\frac{1}{6}$ **9.** $\frac{4}{7}$ **10.** $\frac{4}{9}$

11. $\frac{1}{9}$ **12.** $\frac{7}{9}$ **13.** $\frac{6}{11}$ **14.** $\frac{5}{6}$ **15.** $\frac{2}{5}$

Batting averages are figured as a ratio $\frac{H}{B}$ where H is the number of hits and B is the number of times at bat. Batting averages are given correct to three decimal places. Complete the table.

	Name	H	B	$\frac{H}{B}$
Sample	Ortiz	10	30	0.333
16.	Doggan	15	24	?
17.	Johnson	18	25	?
18.	Fisher	9	19	?
19.	Lykos	17	26	?
20.	Alfonso	20	28	?

4 Equations with Decimals

Some equations contain decimals. The examples below show how to handle them. Recall how to multiply by 10, 100, or 1000.

Multiply by 10.	*Multiply by 100.*	*Multiply by 1000.*
0.6.	2.07.	0.423.

Now use this skill in solving equations.

EXAMPLE 1

$$\begin{aligned} 0.4x + 0.8x &= 3.6 \\ 10(0.4x + 0.8x) &= 10 \times 3.6 \\ 4x + 8x &= 36 \\ 12x &= 36 \\ x &= 3 \end{aligned}$$

It's easier to deal with whole numbers. Multiply both sides by 10.

Check: $0.4x + 0.8x = 3.6$

$$\begin{array}{r|l} 0.4(3) + 0.8(3) & 3.6 \\ 1.2 + 2.4 & \\ 3.6 & \quad \checkmark \end{array}$$

EXAMPLE 2

$$\begin{aligned} 0.7 &= 0.3 + 0.02a \\ 100 \times 0.7 &= 100(0.3 + 0.02a) \\ 70 &= 30 + 2a \\ 40 &= 2a \\ 20 &= a \end{aligned}$$

Multiply both sides by 100.

Check: $0.7 = 0.3 + 0.02a$

$$\begin{array}{r|l} 0.7 & 0.3 + 0.02(20) \\ & 0.3 + 0.4 \\ & 0.7 \quad \checkmark \end{array}$$

EXAMPLE 3

$$\begin{aligned} 0.025x + 0.05x &= 0.3 \\ 1000(0.025x + 0.05x) &= 1000 \times 0.3 \\ 25x + 50x &= 300 \\ 75x &= 300 \\ x &= 4 \end{aligned}$$

Check: $0.025x + 0.05x = 0.3$

$$\begin{array}{r|l} 0.025(4) + 0.05(4) & 0.3 \\ 0.1 + 0.2 & \\ 0.3 & \quad \checkmark \end{array}$$

Classroom Practice

Multiply the expression by 10.

1. $0.4x$
2. $2.3y$
3. $7.4n$
4. $0.3x + 0.5y$
5. $2.5a - 0.4b$
6. $3.7y + 0.6x$
7. $0.8(x + 5)$
8. $2.6(a + b)$

Multiply the expression by 100.

9. $0.04x$
10. $0.35y$
11. $6.02n$
12. $0.04x + 0.02y$
13. $0.31x - 0.12y$
14. $0.05x + 0.5y$
15. $2.3x - 0.06y$
16. $0.7a - 0.43b$

Written Exercises

Solve.

A

1. $0.2x + 0.3x = 25$
2. $0.4a + 0.7 = 55$
3. $0.6x = 2.4$
4. $0.7y = 4.2$
5. $0.10x + 0.12x = 1.9$
6. $0.75x + 0.10x = 85$
7. $0.35x = 0.91$
8. $0.22y = 1.54$
9. $0.2x + 0.04x = 1.2$
10. $0.14a + 0.6a = 5.92$
11. $0.3n - 0.24n = 0.8$
12. $0.05x + 0.5x = 4.4$
13. $0.12n = 0.3 - 0.03n$
14. $0.05d = 2000$
15. $0.04x = 264$
16. $0.03x = 0.15(4 - x)$
17. $90 - x = 0.04(180 - x)$
18. $0.2(x - 3) + 0.4x = 0$
19. $0.2x + 0.3(x + 4) = 37$
20. $0.07x + 0.03(2x) = 390$
21. $0.05n + 0.03(n - 20) = 0$
22. $0.06x + 0.04(10 - x) = 70$

B

23. $0.2x + 0.25(9 - x) = 2$
24. $0.25(16 - n) + n = 0.4(16)$
25. $0.6(15) + s = 0.7(15 + s)$
26. $0.035x + 0.06(600 - x) = 27$
27. $0.4x + 0.24(x - 5) = 0.08$
28. $0.025y - 0.05(20 - 2y) = 0.2$
29. $0.36m - 0.045(m + 1) = 0.711$
30. $2.09z + 1.25(3 - z) = 4.8$

C

31. $0.02(x - 3) + 0.41(x + 1) = 1.21$
32. $0.55(2 - y) - 0.1(y - 3) = 0.75$
33. $3.2x + 0.8(x - 4) = 0.4(x + 10)$
34. $0.5(z + 5) = 1.4(z - 1) + 0.15(10z + 2)$

5 Using Decimal Equations

Many problems involve decimals. In order to solve the problems and their equations, you'll put to work the methods you used in the last section.

EXAMPLE 1

One serving of spinach has about 2.6 mg more iron than a piece of chicken. Together they have about 5.4 mg of iron. How much iron does a serving of spinach have? a piece of chicken?

Let x = milligrams of iron in the chicken.
Then $x + 2.6$ = milligrams of iron in the spinach.

$$\underbrace{\text{chicken}}_{x} + \underbrace{\text{spinach}}_{x + 2.6} = 5.4$$

$$\begin{aligned} x + x + 2.6 &= 5.4 \\ 2x + 2.6 &= 5.4 \\ 10(2x + 2.6) &= 10 \times 5.4 \\ 20x + 26 &= 54 \\ 20x &= 28 \\ x &= 1.4 \end{aligned}$$

Answer: $x = 1.4 \longrightarrow$ 1.4 mg iron in the chicken
$x + 2.6 \longrightarrow$ 4 mg iron in the spinach

Check: $1.4 + 4 = 5.4$ ✓

EXAMPLE 2

A ball player's batting average is 0.330 this year. This is 1.2 times his average last year. What was his average last year?

Let x = batting average last year.
Then $1.2x$ = batting average this year.

$$\begin{aligned} 1.2x &= 0.330 \\ 1000 \times 1.2x &= 1000 \times 0.330 \\ 1200x &= 330 \\ x &= 0.275 \end{aligned}$$

$$1200\overline{)330.000} = 0.275$$

Answer: 0.275 batting average

Check: $1.2 \times 0.275 = 0.330$ ✓

Written Exercises

Solve.

A

1. A number is 0.9 greater than another number.
   The sum of the two numbers is 5.7.
   What are the two numbers?

2. The Falls Trail is 2.4 km longer than the Forks Trail.
   Together the trails are 10.2 km long.
   How long is each trail?

3. Two boards are arranged end to end.
   One board is 2.1 times as long as the other.
   Together they are 133.3 cm long.
   How long is each board?

4. One suitcase is 5.8 kg less than another.
   Together the two suitcases are 37.6 kg.
   How many kilograms is each suitcase?

5. On Friday morning, Charlotte left her home and drove to her office.
   That afternoon, she drove to her tennis lesson, adding an extra
   9.8 km to her trip home.
   Altogether she drove 32.5 km that day.
   How far is it from Charlotte's home to her office?

6. The Mississippi River is 1.24 times as long as the St. Lawrence.
   The Mississippi is about 3777.9 km long.
   How long is the St. Lawrence?

CALCULATOR ACTIVITIES

You can use your calculator to solve equations containing decimals.

Use a calculator to solve each equation.

1. $0.26x = 0.65$
2. $2a = 12.7$
3. $5.6y = 4.2$
4. $0.123x = 0.0246$
5. $\frac{b}{2.9} = 3.7$
6. $\frac{z}{1.325} = -7.1$

SELF-TEST

Vocabulary

rational numbers (p. 352) repeating decimal (p. 352)
terminating decimal (p. 352)

Compute. Round Question 5 correct to two decimal places.

1. $3.6 + 0.05$ 2. $10.735 - 6.25$ 3. 0.6×0.72 *(10-1)*
4. $0.384 \div 0.12$ 5. $4.85 \div 0.72$ *(10-2)*
6. Express $\frac{7}{8}$ as a decimal. *(10-3)*
7. Solve: $0.3x + 0.8x = 5.5$ *(10-4)*
8. The sum of two distances is 7.5 km. *(10-5)*
   One distance is 1.5 km greater than the other.
   Find the two distances.

CAREER NOTES

DEPARTMENT STORES

When you think of positions in a department store do you think only of sales people and store managers? There are many opportunities in other areas of department store work as well.

To provide the merchandise for the store, employees called buyers visit manufacturers and other distributors. Buyers select and buy the goods for one or several departments in the store. The buyers spend much of their time traveling. They often bargain with the distributor for the best price for the merchandise.

Stores must keep a record of the goods they buy, as well as of their sales to consumers. Computer programmers help keep track of the store's inventory and also design programs to handle customer billing and employee payroll. The programmers work with accountants who keep the financial records of the store.

6 Percents

Percents are used so often in everyday life because they are a convenient way to compare two quantities with one another.

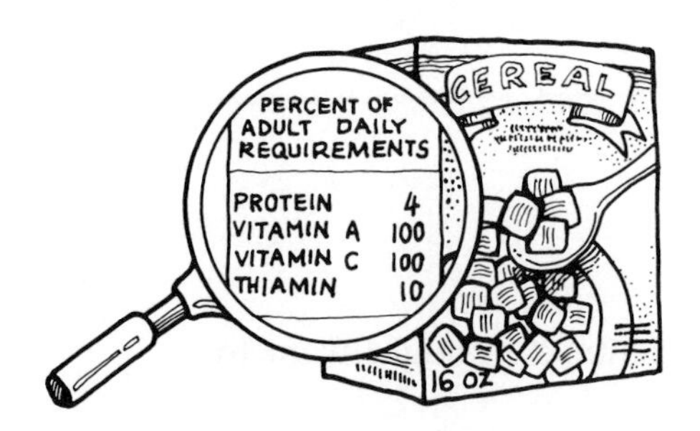

For example, cereal boxes list the percent of the adult daily requirement for several vitamins and minerals in one serving of the cereal. It wouldn't be as easy to compare brands if the percents were listed in fraction form instead.

Percents express the ratio of a number to 100. Fractions and decimals can be rewritten as percents.

Fraction	Fraction with Denominator 100	Decimal	Percent
$\frac{1}{2}$	$\frac{50}{100}$	0.50	50%

Changing Decimals to Percents

Decimals can be written as percents in the following way.

EXAMPLE 1 $0.28 = 28\%$ ← *Think:* 0.28

EXAMPLE 2 $0.275 = 27.5\%$ ← *Think:* 0.275

EXAMPLE 3 $0.04 = 4\%$ ← *Think:* 0.04

Changing Fractions to Percents

Fractions can be written as percents in a couple of ways.

EXAMPLE 4 $\frac{3}{5} = \frac{60}{100}$ *and* $\frac{60}{100} = 60\%$

EXAMPLE 5 $\frac{1}{8} \longrightarrow 8\overline{)1.000}$ with quotient 0.125 *and* $0.125 = 12.5\%$

Changing Percents to Decimals or Fractions

To compute with a percent, you will need to change the percent to a decimal or a fraction. It doesn't matter which one you use, but one might be more convenient than the other.

EXAMPLE 6 $46\% = 0.46$ ← *Think:* 46.%

EXAMPLE 7 $5\% = 0.05$ ← *Think:* 0.05.%

EXAMPLE 8 $25\% = \frac{25}{100} = \frac{1}{4}$

Here's a table of some commonly used percents. You might want to copy it for future use.

Percent	Decimal	Fraction
10%	0.10	$\frac{1}{10}$
20%	0.20	$\frac{1}{5}$
25%	0.25	$\frac{1}{4}$
$33\frac{1}{3}\%$	$0.33\frac{1}{3}$ *or* 0.33 . . .	$\frac{1}{3}$
50%	0.50	$\frac{1}{2}$
$66\frac{2}{3}\%$	$0.66\frac{2}{3}$ *or* 0.66 . . .	$\frac{2}{3}$

Classroom Practice

Express as a percent.

1. 0.42 **2.** 0.30 **3.** 0.05 **4.** 0.6 **5.** 0.7

6. $\frac{1}{2}$ **7.** $\frac{1}{4}$ **8.** $\frac{1}{3}$ **9.** $\frac{3}{8}$ **10.** $\frac{2}{5}$

Express as a decimal.

11. 63% **12.** 59% **13.** 47% **14.** 4% **15.** 7%

Estimate the pieces named.

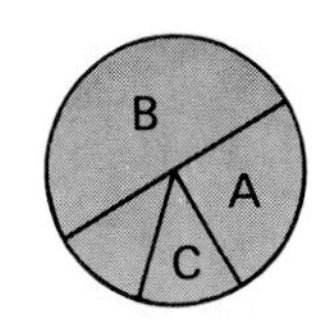

16. In the pie at the right, about what percent of the whole is A?

17. About what percent of the whole is B?
18. About what percent of the whole is C?

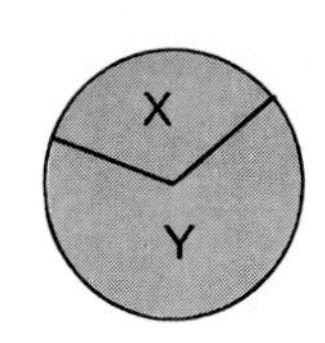

19. About what percent of the whole is X?
20. About what percent of the whole is Y?

Written Exercises

Write as a percent.

A

1. 0.40	2. 0.23	3. 0.65	4. 0.89	5. 0.50
6. 0.14	7. 0.07	8. 0.03	9. 0.09	10. 0.06
11. 0.045	12. 0.036	13. 0.214	14. 0.525	15. 0.125
16. $\frac{2}{5}$	17. $\frac{1}{5}$	18. $\frac{4}{5}$	19. $\frac{3}{4}$	20. $\frac{5}{8}$

Write as a decimal.

21. 64%	22. 86%	23. 95%	24. 30%	25. 20%
26. 56%	27. 67%	28. $66\frac{2}{3}\%$	29. 47%	30. $33\frac{1}{3}\%$
31. 75%	32. 6%	33. 8%	34. 2%	35. 9%
36. 3.5%	37. 6.5%	38. 7.5%	39. 5.5%	40. 5.25%

Write as a fraction.

41. $33\frac{1}{3}\%$	42. $66\frac{2}{3}\%$	43. 25%	44. 75%	45. 50%
46. 20%	47. 10%	48. 40%	49. 60%	50. 80%

Estimate the percent of the whole for each lettered part.

51. A 52. B 53. C 54. D

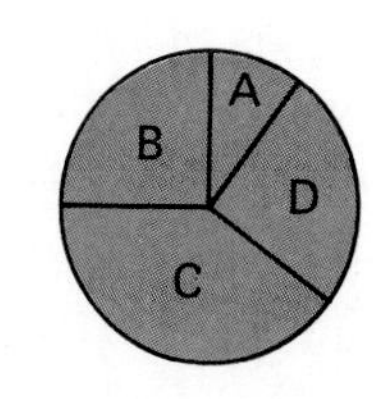

The circle graphs below show the kinds of courses students took at a certain college in the last few years. Use the data in the graphs to complete the table.

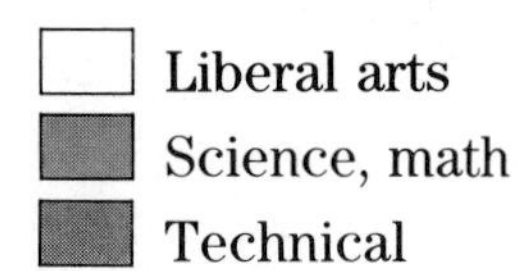

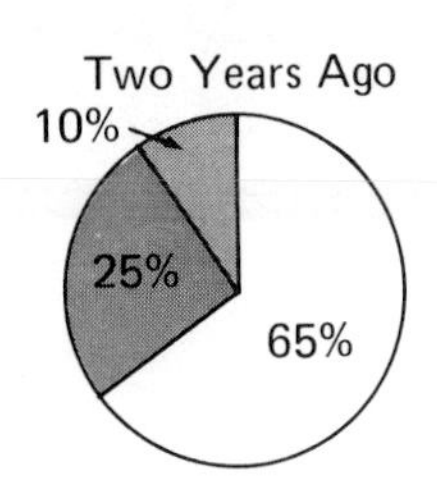

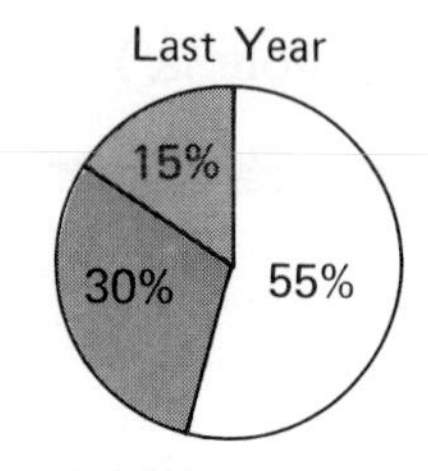

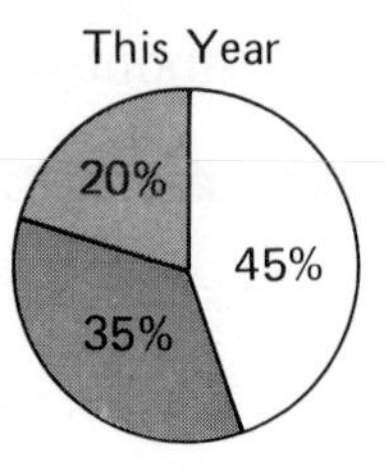

	Courses	Two Years Ago	Last Year	This Year	Next Year (est.)
55.	Liberal arts	65%	?	?	?
56.	Science, math	?	?	?	?
57.	Technical	?	?	?	?

What percent of the whole is the missing piece? Remember, the whole is 100%.

58.

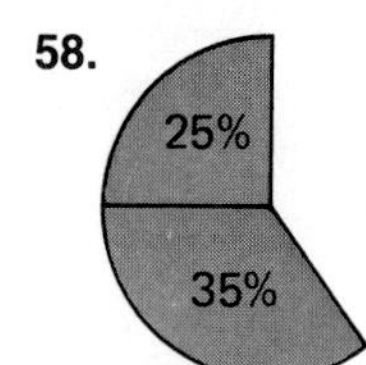

59.

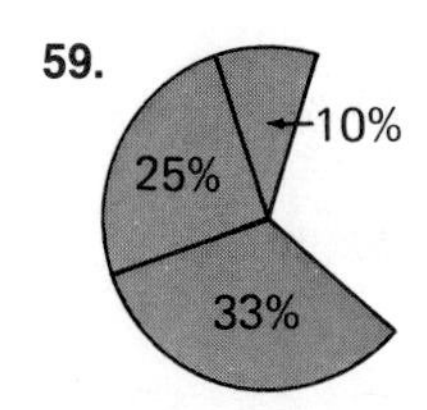

60.

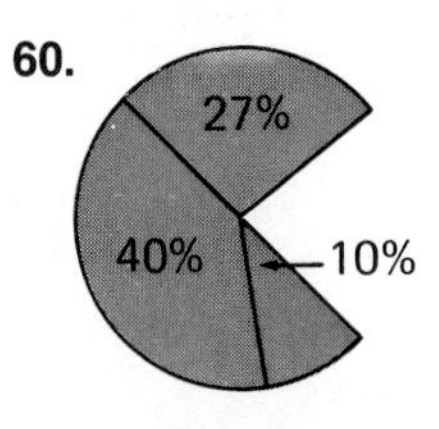

The circle graph at the right shows the ages of cars in the United States.

B **61.** Half of the cars are more than __?__ years old.

62. What percent of the cars are 7 years old or less?

63. Draw a bar graph to represent the information shown in the circle graph.

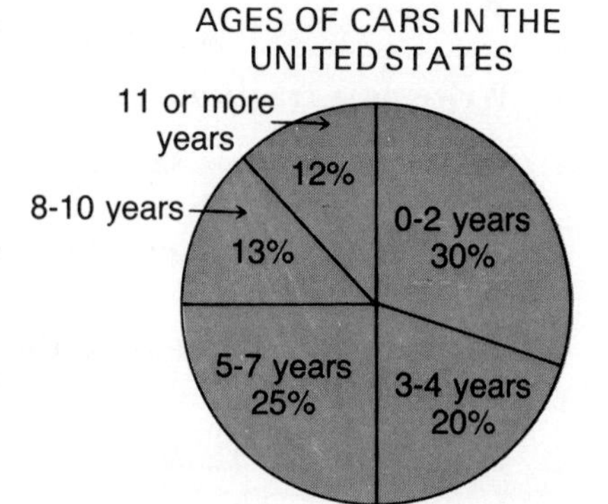

7 Using Percents

Look in any newspaper. You'll probably see some advertisements for sales at local stores. Often the discount (amount you save) is given as a percent of the price the item originally cost.

EXAMPLE 1

Luggage was on sale in one store at a 30% discount. Find the discount on a $56 suitcase.

$$30\% \times \$56 = 0.30 \times \$56$$
$$= \$16.80$$

EXAMPLE 2

During a sale shoes are reduced 20%.
One kind of shoes usually sells for $30.
What is the sale price of the shoes?

Compute discount.

$$20\% \times \$30 = 0.20 \times \$30$$
$$= \$6$$

Compute sale price.

$$\$30 - \$6 = \$24$$

Figuring percents in your head can be a very useful skill. Follow these hints. You can extend these ideas to figure other percents.

	Hints	of $120	of $4.80
To find 10%	Move the decimal point one place to the left.	$12	$.48
To find 20%	Double what you get for 10%.	$24	$.96
To find 25%	Find one fourth (or divide by 4).	$30	$1.20
To find $33\frac{1}{3}\%$	Find one third (or divide by 3).	$40	$1.60
To find 50%	Find one half (or divide by 2).	$60	$2.40
To find $66\frac{2}{3}\%$	Double what you find for $33\frac{1}{3}\%$.	$80	$3.20

Classroom Practice

Figure the work in your head.

1. 50% of 80
2. 10% of 62
3. 25% of 60
4. 10% of 79
5. $33\frac{1}{3}$% of 96
6. $66\frac{2}{3}$% of 45
7. 20% of 28
8. 25% of 76
9. 50% of $2.50
10. 30% of $12
11. 40% of $25
12. 75% of $36

Compute.

13. 36% of 400
14. 24% of 86
15. 5% of 92
16. 1% of 67

Written Exercises

Compute.

A

1. 30% of 120
2. 60% of 46
3. 25% of 18
4. 75% of 50
5. 15% of 56
6. 40% of 65
7. 35% of 80
8. 45% of 25
9. 60% of 150
10. 80% of 210
11. 15% of 250
12. 45% of 500
13. 18% of 100
14. 55% of 200
15. 60% of 480
16. 2% of 80

Complete the tables.

	SALE! 20% off		
	Item	Original price	Discount
17.	shoes	$27	?
18.	shirt	$18	?
19.	jeans	$25	?

	SALE! 15% off		
	Item	Original price	Sale price
20.	shovel	$7	?
21.	rake	$5.60	?
22.	paint	$11.90	?

23. Camping tents are on sale at a $33\frac{1}{3}$% discount.
Find the sale price on a $75 tent.

24. A buyer for a department store buys baskets at $12.40 each.
To sell them in the store, the price is increased 40%.
For how much do those baskets sell in the store?

8 Percents in Equations

Sales taxes are figured as a percent of the cost. Suppose a car costs \$6000 and the sales tax is 7%. The tax is figured this way.

7% of \$6000 is what number?

$$7\% \times \$6000 = n$$
$$0.07 \times 6000 = n$$
$$420 = n$$

The tax is \$420.

You can solve all types of percent problems if you can set up your equation using the pattern shown above.

EXAMPLE 1

Only 40% of the seats in a theater are occupied. The theater has 320 seats. How many seats are occupied?

40% of 320 is what number?

$$40\% \times 320 = x$$
$$0.40 \times 320 = x$$
$$128 = x$$

Answer: 128 seats are occupied.

EXAMPLE 2

In one game, a football quarterback threw 20 passes and completed 12 of them. What percent of the passes were completed?

What percent of 20 is 12?

$$n \times 20 = 12$$
$$n = \frac{12}{20}$$

$$20\overline{)12.00} = 0.60$$

$$n = 0.60$$

Answer: 60% of the passes were completed.

EXAMPLE 3

In a nationwide poll about two shampoos, 2000 people were surveyed. 648 people said they preferred Brand A to Brand B. What percent of the people surveyed preferred Brand A?

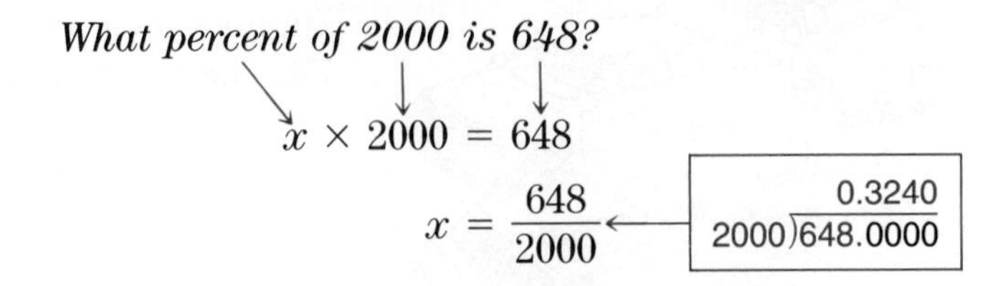

What percent of 2000 is 648?

$$x \times 2000 = 648$$

$$x = \frac{648}{2000}$$

$$2000\overline{)648.0000} = 0.3240$$

Answer: 32.4% of the people surveyed preferred Brand A.

EXAMPLE 4

In the six o'clock news, it was reported that 60% of the registered voters voted in the election. This was 48,000 people. How many registered voters were there?

60% of what number is 48,000?

$$60\% \times x = 48{,}000$$
$$0.60x = 48{,}000$$
$$100 \times 0.60x = 100 \times 48{,}000$$
$$60x = 4{,}800{,}000$$
$$x = 80{,}000$$

Answer: 80,000 registered voters

Classroom Practice

Solve.

1. 20% of 52 is what number?
2. 4% of 96 is what number?
3. What percent of 80 is 20?
4. What percent of 75 is 15?
5. 60% of what number is 90?
6. 30% of what number is 13.5?
7. Candidate Lin received 72% of the 4200 votes in the election. How many votes did she get?
8. James answered 90% of the 60 questions correctly. How many did he answer correctly?
9. A highway is to be 240 km long. So far 180 km are complete. What percent of the road is complete?

Written Exercises

Solve.

A

1. 25% of 88 is what number?
2. 80% of 62 is what number?
3. What percent of 35 is 14?
4. What percent of 56 is 14?
5. 5% of 53 is what number?
6. What percent of 85 is 17?
7. 40% of what number is 22?
8. 65% of what number is 84.5?
9. 95% of 1000 is what number?
10. What percent of 280 is 42?
11. The sales tax rate in one area is 4%.
    How much tax is due on an item costing $3.75?
12. Fifty out of 60 members attended the club meeting.
    What percent is this?
13. About 60% of the human body is water.
    Jerry is 53 kg in all.
    How many kilograms of water is she?
14. About 1 out of 700 blood cells in humans is a white blood cell.
    What percent of the blood cells are white cells?
    (Express your percent correct to one decimal place.)
15. A large company had a blood donation drive.
    68% of the 1200 workers donated blood.
    How many people donated?

16. A book has 420 pages.
    Tom has read 105 pages.
    What percent of the book is this?
17. Consider the same book again.
    Sylvia has read 84 pages.
    What percent of the book is this?

B

18. One day 6% of the students in a school were absent.
    87 students were absent.
    How many students are enrolled in the school?
19. In one vacation spot, hotels add a 4% tax to the room price.
    What is the tax on a room priced at $58 per day?
20. A serving of a lasagna recipe contains 15.6 g of protein.
    This is about 26% of a person's daily requirement.
    How much protein does a person need each day?

21. A person needs about 5000 units of Vitamin A daily.
A cup of tomato soup provides 1200 units.
What percent of the daily need is this?

22. Last year the rent on an apartment was $350 per month.
This year the rent was increased $52.50.
What percent of last year's rent was the increase?

SELF-TEST

Write as a percent.

1. 0.52 2. 0.04 3. $\frac{3}{5}$ 4. $\frac{3}{4}$ 5. $\frac{3}{8}$ *(10-6)*

Write as a decimal.

6. 35% 7. 14% 8. 7% 9. 65% 10. 1.25%

Compute.

11. 30% of 24 12. 15% of 85 *(10-7)*

13. What percent of 35 is 7? 14. 60% of what number is 15? *(10-8)*

15. A bill in a restaurant comes to $5.50.
An 8% meal tax is to be added to the bill.
How much money would the tax be?

16. Thirty out of 45 tickets are sold.
What percent of the tickets are sold?

CALCULATOR ACTIVITIES

Your calculator may have a percent key (%) that you can use in solving equations involving percents. This key will display percents as decimals.

If your calculator does not have a % key, remember that you can divide by 100 to change the percent to a decimal.

Solve.

1. 15% of 30 is what number?
2. 17.5% of 200 is what number?
3. What percent of 20 is 15?
4. What percent of 350 is 7?
5. 20% of what number is 2?
6. 55.5% of what number is 111?

9 Interest

When you put money into a savings account at a bank, you expect to earn **interest** for letting the bank use your money. Here's a formula for computing interest.

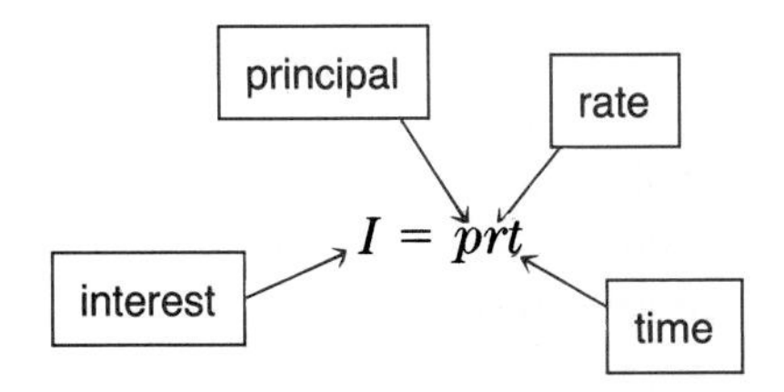

The formula can be used to solve many types of problems.

EXAMPLE 1

Lana put \$450 into a savings account.
In that bank, the money earns $5\frac{1}{2}\%$ interest per year.
How much interest will the money earn in a year?

$$I = prt$$
$$I = 450 \times 5\tfrac{1}{2}\% \times 1$$
$$= 450 \times 0.055 \times 1$$
$$= 24.75$$

$5\frac{1}{2}\% = 5.5\% = 0.055$

Answer: \$24.75 interest

EXAMPLE 2

James put \$200 into a savings account in January.
Six months later he put \$50 into the account.
The interest rate is 6% per year.
How much interest will be earned at the end of the year?

$$I = prt$$
$$I = 200 \times 6\% \times 1$$
$$= 200 \times 0.06 \times 1$$
$$= 12$$

$$I = prt$$
$$I = 50 \times 6\% \times \tfrac{1}{2}$$
$$= 50 \times 0.06 \times \tfrac{1}{2}$$
$$= 1.50$$

The \$50 earns interest for only $\frac{1}{2}$ year.

Answer: \$12 + \$1.50 = \$13.50 ⟶ \$13.50 interest

EXAMPLE 3 How long will it take \$100 to double to \$200 when the interest rate is $6\frac{1}{2}\%$ per year?

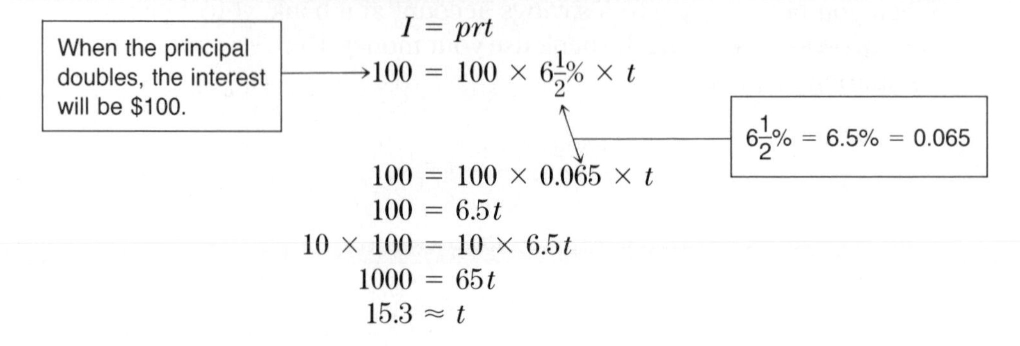

Answer: about 15 years

Written Exercises

Solve.

A

1. How much interest can be earned in one year on \$500 at 8%?
2. How much interest does \$450 earn each year at $5\frac{1}{2}\%$?
3. Ginny invested \$800 at 6.25% interest for 1 year.
   How much interest did the money earn?
4. In January Junior put \$785 into an account.
   Six months later he put \$600 more into the account.
   The account earns 7% interest per year.
   How much interest did the money earn at the end of the year?
5. How long will it take \$100 to double at 6% interest?
6. How long will it take \$500 to double at 6% interest?
7. Kathy's bank offers 5% interest per year.
   The bank across the street pays $6\frac{1}{2}\%$ interest.
   Kathy has \$2500 to put into an account.
   How much more money would be earned in one year at the second bank?
8. Ted wants his \$1000 to earn \$500 interest in 5 years.
   He puts it into an account earning 8.5% interest per year.
   Will the money earn enough interest?
   How much more or less than Ted's goal?

10 Investment

The tables you used to solve motion and cost problems are helpful for investment problems too.

EXAMPLE 1 Mark invested some money at 6% yearly interest and \$800 at 7%. The money earned \$128 interest in a year. How much was invested at 6% interest?

	p	$\times\ r$	$\times\ t =$	I
Amount at 6%	x	0.06	1	$0.06x$
Amount at 7%	800	0.07	1	56
				128 ← total

Multiply each side by 100 to get this step. →

$$0.06x + 56 = 128$$
$$6x + 5600 = 12800$$
$$6x = 7200$$
$$x = 1200$$

Answer: \$1200 at 6% interest

EXAMPLE 2 Della invested \$2000, part at 6% and the rest at 9%. The interest for one year came to \$138. Find how much was invested at each rate.

Let x = principal invested at 6%.
Then $2000 - x$ = principal invested at 9%.

	p	$\times\ r$	$\times\ t =$	I
Amount at 6%	x	0.06	1	$0.06x$
Amount at 9%	$2000 - x$	0.09	1	$0.09(2000 - x)$
				138 ← total

Multiply each side by 100 to get this step. →

$$0.06x + 0.09(2000 - x) = 138$$
$$6x + 9(2000 - x) = 13{,}800$$
$$6x + 18{,}000 - 9x = 13{,}800$$
$$-3x = -4200$$
$$x = 1400$$

Answer: $x = 1400$ ⟶ \$1400 at 6% interest
$2000 - x = 600$ ⟶ \$600 at 9% interest

Written Exercises

A

1. Dr. Hamada invested $1600 at 7% interest.
   She also invested some more money at 8.5% interest.
   At the end of a year her money had earned $197 interest.
   How much money was invested at 8.5%?

	p	$\times\ r$	$\times\ t$	$=\ I$
Amount at 7%	1600	0.07	1	?
Amount at 8.5%	x	?	1	?
				197

2. Troy invested $1200, part at 6% interest, and the rest at 8%.
   The money earned $78 in interest in one year.
   How much money was invested at each rate?

	p	$\times\ r$	$\times\ t$	$=I$
Amount at 6%	x	?	1	?
Amount at 8%	$1200 - x$	?	1	?
				?

3. A conservation group raised $120,000.
   The interest each year, $9200, is used for wildlife protection.
   Part of the $120,000 was invested at 7%, and the rest at 9%.
   How much was invested at each rate?

4. A union invested $100,000 in two unemployment funds.
   One fund paid 7% interest, and the other 6%.
   The interest earned each year was $6,700.
   How much was invested at each rate?

5. An organization invested $9000 in stocks and bonds.
   The stocks pay 6% and the bonds pay 7%.
   The amount earned each year from the two is $575.
   How much is invested in the stocks? in the bonds?

6. The Trans invested $7000 in two small businesses.
   The printing shop pays 8%, while the grocery shop pays 5%.
   The annual income from the two is $431.
   How much is invested in each business?

11 Mixture Problems

Problems about mixtures have many applications in business and industry. Recall the cost formula you used in earlier chapters.

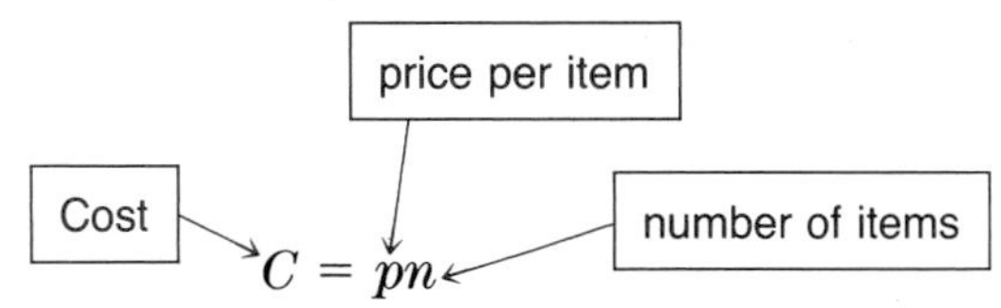

You'll use this formula in Example 1.

EXAMPLE 1

An auto products company mixes two kinds of motor oil to get the blend they sell wholesale at 59¢ a quart. One of the kinds is worth 50¢ a quart, the other, 80¢ a quart. How much of each kind should be mixed to get 1000 quarts of the blend?

Let x = number of quarts of 50¢ oil.
Let $1000 - x$ = number of quarts of 80¢ oil.

	p ×	n =	C
50¢ oil	50	x	$50x$
80¢ oil	80	$1000 - x$	$80(1000 - x)$
Mixture	59	1000	59,000

Cost of 50¢ oil + Cost of 80¢ oil = Cost of mixture

$$50x + 80(1000 - x) = 59{,}000$$
$$50x + 80{,}000 - 80x = 59{,}000$$
$$-30x = -21{,}000$$
$$x = 700$$

Answer: $x = 700 \longrightarrow$ 700 quarts of 50¢ oil
$1000 - x = 300 \longrightarrow$ 300 quarts of 80¢ oil

Check:
$700 \times \$.50 = \$350$
$300 \times \$.80 = \$240$
$\$590$

$1000 \times \$.59 = \$590$ ✓

Labels on bottles or cans often tell you that a mixture contains a certain percent of the substance named. Sometimes mixtures are made to reduce or to increase that percent.

EXAMPLE 2 A salt-water solution contains 6% salt. How much water must be added to 10 liters of solution to get a mixture that is only 4% salt?

Let x = number of liters of water to be added.

	Amount of solution ×	% of salt =	Amount of salt
Original solution	10	6% *or* 0.06	0.6
Water added	x	0%	0
New solution	$10 + x$	4% *or* 0.04	$0.04(10 + x)$

$$\begin{aligned} 0.6 + 0 &= 0.04(10 + x) \\ 0.6 &= 0.04(10 + x) \\ 100 \times 0.6 &= 100 \times 0.04(10 + x) \\ 60 &= 4(10 + x) \\ 60 &= 40 + 4x \\ 20 &= 4x \\ 5 &= x \end{aligned}$$

Answer: 5 liters of water

Check: Original solution: $0.06 \times 10 = 0.6 \longrightarrow 0.6$ L salt
New solution: $0.04 \times 15 = 0.6 \longrightarrow 0.6$ L salt ✓

Written Exercises

A

1. Two kinds of candy are mixed to sell at $.85 per kilogram. One kind sells at $.70 per kilogram, the other at $.95 per kilogram. How much of each kind is used in 100 kilograms of the mixture?

	p ×	n =	C
First kind	70	x	$70x$
Second kind	95	$100 - x$	$95(100 - x)$
Mixture	85	100	8500

2. At its grand opening, a bank gave away 5000 souvenirs.
The bank paid $.50 for each keyring and $.60 for each calendar.
The total cost was $2650.
How many keyrings and calendars did the bank buy?

	$p \times n =$		C
Keyrings	?	x	?
Calendars	?	?	?
			2650

3. At a movie theater, children's tickets cost $2.50, while adults' tickets cost $4.25. The receipts for 600 tickets totaled $1885.
How many children's tickets were sold?

	$p \times n =$		C
Children	?	?	?
Adults	?	?	?
			1885

4. A vegetable oil made of safflower and corn oils sells for $3.05 per liter.
The safflower oil is worth $5.25 per liter; the corn oil, $2.56.
How much safflower oil is used to make 100 liters of the mixture?

	$p \times n = C$		
Safflower oil	?	?	?
Corn oil	?	?	?
			?

5. A grocer mixes two kinds of nuts to sell at $4.00 per pound.
One kind of nuts sells at $3.20 per pound.
The other kind of nuts sells at $4.40 per pound.
How much of each kind is used in 60 pounds of the mixture?
(Make a chart like the one in Exercise 1.)

6. A trucking company has 1000 liters of coolant.
The coolant is 30% antifreeze.
How much water should be added to make the mixture 25% antifreeze?

	Amount of coolant ×	% of antifreeze =	Amount of antifreeze
Old mixture	1000	30%	?
Water	x	0%	?
New mixture	?	25%	?

B **7.** A 1000-kg vat of insect repellent contains 32% active ingredients and 68% inactive ingredients. Chemists plan to concentrate the repellent by boiling away some of the inactive ingredients. The new, concentrated repellent must be 40% active ingredients. How much of the inactive ingredients must be boiled away?

	Amount of repellent ×	% of active ingredients =	Amount of active ingredients
Old repellent	1000	?	?
Boiled-away ingredients	x	0%	?
New repellent	$1000 - x$	?	?

8. The "lead" in pencils is a mixture of clay and graphite.
In a pencil factory, 500 liters of the mixture is 45% clay.
To make softer pencils, the clay is reduced to 40%.
How much graphite must be added to make the new mixture?

	Amount of mixture ×	% of clay =	Amount of clay
Old mixture	500	?	?
Graphite added	x	0%	?
New mixture	?	?	?

C **9.** Marcie has 19 coins in her wallet.
The coins consist of dimes and quarters.
The dimes and quarters together total $4.00.
How many of each kind of coin does she have?

SELF-TEST

Vocabulary

interest (p. 369)

1. How long will it take $400 to double at 6% interest? *(10-9)*
2. Pearl invested $700 at 7% interest for one year.
   How much interest did the money earn?
3. Joe has $700 in one bank paying 5% interest. *(10-10)*
   He has some more money in a bank paying 6%.
   His interest on the two accounts for one year is $65.
   How much money is earning 6% interest?
4. Some dried fruit worth $5 per kilogram is to be mixed with some *(10-11)*
   less expensive dried fruit worth $4 per kilogram.
   The mixture is to be worth $4.50 per kilogram.
   How much of each kind will make 100 kilograms of the mixture?

CALCULATOR ACTIVITIES

Calculators can be used to express fractions as decimals.

To express $\frac{7}{8}$ as a decimal, you can use division and divide 7 by 8.

Enter: 7 ÷ 8 =

If your calculator has a reciprocal key, [1/x], use it to multiply 7 by $\frac{1}{8}$.

Enter: 7 × 8 [1/x] =

Use a calculator to express each fraction as a decimal. Be sure to use three dots with a repeating decimal.

1. $\frac{5}{8}$
2. $\frac{1}{12}$
3. $\frac{8}{15}$
4. $\frac{7}{11}$

If $x = 5$ and $y = 2$, use a calculator to find the value of each expression.

5. $\frac{1}{x} \cdot \frac{1}{y}$
6. $\frac{1}{x} + \frac{1}{y}$
7. $\frac{1}{x - y}$
8. $\frac{1}{x^2 + y}$

PROBLEM SOLVING STRATEGIES

VERBAL EQUATIONS

When you are solving a word problem, you should follow the steps in the problem-solving guide introduced on page 58. If you are having difficulty solving problems, you may want to review the guide before continuing to read this section.

A common cause of solving a word problem incorrectly is writing an equation incorrectly. The following tips may help you avoid making mistakes.

- Read each problem slowly and carefully. Decide which information is unknown and which information is given.
- Look for words that indicate an operation. For example, "a total of" suggests addition. The words "is," "are," and "have" are shown mathematically by an equals sign.
- Before you write an algebraic equation, try writing the equation in words to be certain that you understand the information that the algebraic equation will represent.

Consider the following example.

A collection of 22 stamps consists of some 15¢ stamps and some 35¢ stamps.
The total value of the stamps is $5.90.
How many stamps of each type are in the collection?

If you read the problem carefully, you'll see that the following facts are given:

A. Number of 15¢ stamps + number of 35¢ stamps = 22
B. Value of 15¢ stamps + value of 35¢ stamps = 5.90

The verbal equations can be used to write algebraic equations.

Let x = number of 15¢ stamps.
Let y = number of 35¢ stamps.

$$\begin{aligned} x + \quad y &= 22 \quad \text{(A.)} \\ 0.15x + 0.35y &= 5.90 \quad \text{(B.)} \end{aligned}$$

Solving these two equations, you should find that the collection consists of nine 15¢ stamps and thirteen 35¢ stamps.

EXERCISES

Complete each verbal equation.

1. There are 23 students in the classroom, with 5 more girls than boys.

   A. Number of boys + __?__ = __?__
   B. 5 = __?__ − __?__

2. Catherine invested \$2000, part at 8% interest and the rest at 5.5% interest. The interest earned in one year at 8% was \$52 more than the interest earned at 5.5%.

   A. 2000 = __?__ + __?__
   B. Interest earned at 8% = __?__ + __?__

Complete the verbal equation. Write and solve an algebraic equation.

3. Marla and Danny live 21 km apart.
   At the same time, each leaves home and travels toward the other.
   Marla travels at 8 km/h.
   Danny travels at 6 km/h.
   After how long do they meet?

   Distance Marla travels + __?__ = __?__

4. Mr. Martinez can restock the shelves in his store in 5 hours.
   When his daughter helps him the job takes 3 hours.
   How long would it take his daughter to do the job alone?

   Part of the job done in one hour when they work together = __?__ + __?__

For each exercise, write two verbal equations. Then write and solve two algebraic equations.

5. At a certain fruit stand, a pear costs twice as much as an apple.
   Three apples and 2 pears cost \$2.80.
   Find the cost of each item.

6. Mark is older than Joanne.
   The sum of their ages is 30.
   The difference of their ages is 6.
   How old is each?

SKILLS REVIEW

DECIMALS AND PERCENTS

Find the increase or decrease in price.

1. from \$2.98 to \$3.50
2. from \$6.50 to \$9.00
3. from \$4.25 to \$3.50
4. from \$25 to \$16.80
5. from \$10.35 to \$12.59
6. from \$230 to \$255

Find the percent of increase in price.

SAMPLE from \$3.00 to \$4.50 ⟶ *increase:* \$1.50

What percent of the original price is the increase?

$$
\begin{aligned}
x \times \$3.00 &= \$1.50 \\
3.00x &= 1.50 \\
100 \times 3.00x &= 100 \times 1.50 \\
300x &= 150 \\
x &= 0.5
\end{aligned}
$$

Answer: $0.5 = 0.50 = 50\%$

7. from \$2.00 to \$3.00
8. from \$5.00 to \$8.00
9. from \$4.00 to \$6.00
10. from \$2.00 to \$3.40
11. from \$8.00 to \$9.60
12. from \$6.00 to \$7.80
13. from \$20 to \$25
14. from \$40 to \$50
15. from \$60 to \$84
16. from \$100 to \$150

Find the percent of decrease in price.
Use the guide: *What percent of the original price is the decrease?*

17. from \$20 to \$10
18. from \$18 to \$9
19. from \$30 to \$24
20. from \$50 to \$30
21. from \$80 to \$56
22. from \$24 to \$18
23. from \$200 to \$150
24. from \$125 to \$100

CHAPTER REVIEW

CHAPTER SUMMARY

1. To add or subtract decimals, line up the decimal points and compute as you would with whole numbers.
2. To multiply decimals, compute as you would with whole numbers, ignoring the decimal points. Then put the decimal point in the answer.
3. To divide by a decimal, first make the divisor an integer.
4. To solve equations involving decimals, first multiply by a power of 10 to get rid of the decimals.
5. Percents express the ratio of a number to 100.
6. The formula for interest is

$$\text{Interest} = \text{principal} \times \text{rate} \times \text{time, or } I = prt.$$

REVIEW EXERCISES

Add, subtract, or multiply. *(See pp. 348–349.)*

1. $0.34 + 5.36$
2. $14.55 + 2.605$
3. $0.8 + 50.35$
4. $16 + 50.5 + 0.62$
5. $34.4 - 29.5$
6. $15 - 3.25$
7. $103.6 - 25.52$
8. $5.95 - 3.605$
9. 3×4.08
10. 1.5×0.3
11. 4.06×0.2
12. 0.025×0.6

Divide. Round to one decimal place. *(See pp. 350–351.)*

13. $0.5\overline{)26.2}$
14. $0.05\overline{)50}$
15. $0.02\overline{)42.85}$
16. $1.9\overline{)3.139}$

Write as a decimal. Use three dots with a repeating decimal. *(See pp. 352–353.)*

17. $\frac{3}{8}$
18. $\frac{1}{8}$
19. $\frac{5}{6}$
20. $\frac{1}{11}$
21. $\frac{4}{9}$

Solve for the variable. *(See pp. 354–355.)*

22. $0.3n + 0.9n = 4.8$
23. $1.2x - 0.9x = 33$
24. $0.7y = 4.9$
25. $0.12x = 3.6$
26. $0.12n = 45 - 0.03n$
27. $5.5(2x - 3) = 10x$
28. $0.31x + 0.58 = 0.6x$
29. $4.2z + 0.5(3z) = 57$
30. $2.1(15 + n) = 63$

Solve. *(See pp. 356–357.)*

31. A tire and a generator cost \$109.50 in all.
    The tire cost \$21 less than the generator.
    How much did the tire cost?

32. Daryl grew 8.3 cm in two years.
    He grew 4.5 cm more the second year than the first.
    How much did he grow the first year?

Write as a percent. *(See pp. 359–362.)*

33. 0.04
34. 0.32
35. 0.136
36. 0.025
37. 0.9
38. 0.3
39. $\frac{1}{4}$
40. $\frac{1}{5}$
41. $\frac{1}{8}$
42. $\frac{1}{3}$

Write as a decimal. *(See pp. 359–362.)*

43. 36%
44. 85%
45. 2.5%
46. 3.15%
47. $7\frac{1}{2}\%$

Write as a fraction. *(See pp. 359–362.)*

48. 25%
49. 10%
50. $66\frac{2}{3}\%$
51. 50%
52. 75%

Write an equation and solve. *(See pp. 363–368.)*

53. 40% of 50 is what number?
54. $33\frac{1}{3}\%$ of 15 is what number?
55. What percent of 34 is 17?
56. What percent of 16 is 2?
57. 3% of what number is 3?
58. 32% of what number is 48?

Solve. *(See pp. 369–376.)*

59. Connie invested \$2000, part at 6%, and the rest at 9%.
    Her money earned \$141 in one year.
    How much was invested at each interest rate?

60. Two fund-raising dinners were attended by a total of 550 people.
    People attending one dinner paid \$15 each.
    People attending the other dinner paid \$100 each.
    Together the dinners raised \$21,000.
    How many attended each of the dinners?

61. Ten liters of a salt-water solution has 3% salt.
    The salt concentration is to be reduced to 2% of the solution.
    How much water must be added?

Chapter Test

Add, subtract, or multiply.

1. $9.7 + 5 + 0.468$ 2. $7.913 - 1.85$ 3. 2.05×0.416 *(10-1)*

Divide. If necessary, round to one decimal place.

4. $0.27\overline{)0.837}$ 5. $0.3\overline{)8}$ 6. $0.064\overline{)2.4}$ *(10-2)*

Write as a decimal. Use three dots with a repeating decimal.

7. $\frac{2}{15}$ 8. $\frac{19}{20}$ 9. $\frac{17}{25}$ 10. $\frac{5}{18}$ *(10-3)*

Solve.

11. $0.3x = 1.36 - 0.02x$ 12. $0.4(a + 1) + 0.05a = 2.2$ *(10-4)*

13. A shirt and a hat cost \$13.35 in all. The shirt cost \$3.15 more than the hat. How much did each cost? *(10-5)*

14. Write as a percent. a. 0.04 b. $\frac{7}{10}$ *(10-6)*

15. Write as a decimal. a. 29% b. 4.2%

16. Write as a fraction. a. $33\frac{1}{3}\%$ b. 90%

Compute.

17. 55% of 40 18. 90% of 240 19. 50% of 25 *(10-7)*

20. Find the sale price of a \$15 calculator sold at a 20% discount.

Solve.

21. What percent of 68 is 51? 22. 35% of what number is 4.2? *(10-8)*

23. The cost of a \$28.50 camera increases 10%. Find the new cost.

24. How much interest can be earned in one year on \$300 at 5.5%? *(10-9)*

25. The Robinsons invested \$6000 in two bank accounts. One account paid 6% and the other paid 9%. The money earned \$480 in interest from the two this year. How much was invested at each rate? *(10-10)*

26. At the zoo, a child's ticket costs \$2 and an adult's ticket costs \$3. On Tuesday 438 tickets were sold. Receipts totaled \$1018. How many children's tickets were sold? *(10-11)*

UNIT E **CUMULATIVE REVIEW**

Simplify.

1. $\frac{56xy^2}{24y}$
2. $\frac{68x^3y^2z}{24xyz^4}$
3. $\frac{3z - 6}{4z - 8}$
4. $\frac{5 - b}{b^2 - 25}$

5. A 180-cm board is cut into two pieces. The lengths of the pieces have a ratio of 5 to 1. How long is the longer piece?

6. The sales tax on a $5400 car is $324. At that rate, what is the tax on an $8000 car?

Simplify.

7. $\frac{8}{r^2 - 9} \cdot \frac{3 - r}{16}$
8. $\frac{x^2 - 4}{3x} \div \frac{x + 2}{3x - 6}$
9. $\frac{5}{x} + \frac{7}{3x^2}$

Solve.

10. $\frac{x}{5} + \frac{x}{8} = 13$
11. $\frac{a - 1}{2} + \frac{a + 2}{3} = 1$

12. Pipe A can fill a tank in 16 hours. Pipe B takes 8 hours to fill the tank alone. How long will it take if both pipes are used?

Compute.

13. $3.18 + 5.2$
14. $0.26 + 12.9$
15. $4.25 - 3.125$
16. $0.62 - 5.93$
17. 5.2×6.6
18. 3.22×4.15
19. $3.4\overline{)85}$
20. $7.1\overline{)39.05}$

21. Write as a decimal. a. $\frac{22}{25}$ b. $\frac{8}{9}$ c. 3%

22. Write as a percent. a. 0.65 b. 0.732 c. $\frac{7}{8}$

23. Write as a fraction. a. 85% b. 24% c. 30%

Solve.

24. 24% of 25 is what number?

25. What percent of 80 is 56?

26. The Lis invested $2000, part at 5% and the rest at 7%. They earned $125 in interest at the end of the year. How much was invested at each rate?

27. A vat contains 800 liters of a 40% acid solution. How many liters of water should be added to make a 25% acid solution?

UNIT

F

Here's what you'll learn in this chapter:

To find the positive and negative square roots of a number.

To use a square root table.

To identify rational and irrational numbers.

To simplify square roots.

To solve equations by using square roots.

To multiply and divide with radicals.

To add and subtract expressions with radicals.

Do you see that each column and its shadow form two sides of a right triangle? In this chapter you will learn how to find the length of the third side, the distance from the top of the column to the end of its shadow.

Chapter 11

Squares and Square Roots

1 Square Roots

When you square 3 or -3 you get 9. We say that each of the numbers 3 and -3 is a **square root** of 9.

Read: The positive, or principal, square root of 9 is 3. → $\sqrt{9} = 3$

$-\sqrt{9} = -3$ ← Read: The negative square root of 9 is -3.

The symbol $\sqrt{\ }$ is called the **radical sign.** It means the square root. $\sqrt{9}$ is a **radical number,** or **radical.**

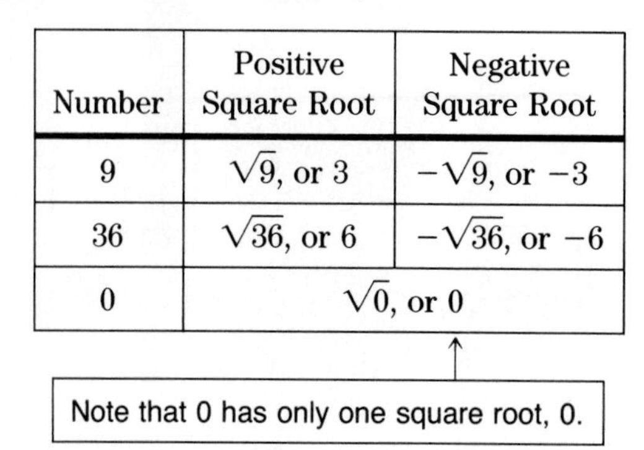

Number	Positive Square Root	Negative Square Root
9	$\sqrt{9}$, or 3	$-\sqrt{9}$, or -3
36	$\sqrt{36}$, or 6	$-\sqrt{36}$, or -6
0	$\sqrt{0}$, or 0	

Note that 0 has only one square root, 0.

Sometimes we want to speak of *both* the positive and negative square roots of a number. In that case, the symbol $\pm$ is used.

Read: positive or negative.

$$\pm\sqrt{25} = \pm 5$$

$\sqrt{9}$, $\sqrt{36}$, $\sqrt{64}$, and $\sqrt{25}$ are easy to find because 9, 36, 64, and 25 are **perfect squares.** Any number that is the square of an integer is called a perfect square.

Classroom Practice

Which numbers are perfect squares?

1. 49 **2.** 50 **3.** 1 **4.** 2 **5.** 0

Find the value.

6. $\sqrt{81}$ **7.** $-\sqrt{81}$ **8.** $\sqrt{100}$ **9.** $-\sqrt{7^2}$ **10.** $\sqrt{13^2}$

Written Exercises

Find the value.

A

1. $\sqrt{4}$
2. $-\sqrt{9}$
3. $\sqrt{25}$
4. $-\sqrt{25}$
5. $\sqrt{16}$
6. $-\sqrt{16}$
7. $-\sqrt{4}$
8. $\sqrt{0}$
9. $\sqrt{1}$
10. $\sqrt{49}$
11. $-\sqrt{49}$
12. $-\sqrt{1}$
13. $-\sqrt{36}$
14. $\sqrt{9}$
15. $\sqrt{64}$
16. $\sqrt{100}$
17. $-\sqrt{64}$
18. $\sqrt{81}$
19. $\sqrt{36}$
20. $-\sqrt{81}$
21. $\sqrt{8^2}$
22. $\sqrt{9^2}$
23. $\sqrt{3^2}$
24. $-\sqrt{5^2}$
25. $-\sqrt{10^2}$

Find the perimeter of a square that has the area shown.

26. 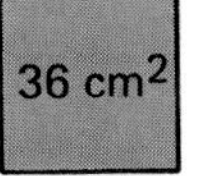

27. 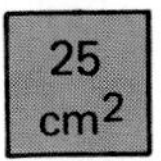

28. 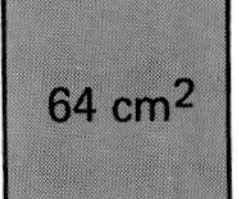

29. Find the perimeter of a square whose area is 16 ft^2.
30. Find the perimeter of a square whose area is 49 in.^2.
31. Find the perimeter of a square whose area is 100 m^2.

B

32. Find the value of $\sqrt{3^2 + 4^2}$. The answer is *not* $3 + 4$.
33. Find the value of $\sqrt{6^2 + 8^2}$. The answer is *not* $6 + 8$.
34. Find the value of $\sqrt{12^2 + 5^2}$.

PUZZLE ◆ PROBLEMS

Can you group four 4's to make whole numbers from 1 to 16? Of course, there might be several ways to write some numbers, but you need to write only one. The first four are done for you.

$$1 = (4 + 4) \div (4 + 4)$$
$$2 = 4 - (4 + 4) \div 4$$
$$3 = (4 + 4 + 4) \div 4$$
$$4 = (\sqrt{4} \cdot 4) \div (4 - \sqrt{4})$$

2 Using a Square Root Table

The table below gives the square roots of integers from 1 to 100. Notice that most of the values in the Square Root column are not integers. These values are not exact, but are rounded to three decimal places. They are *approximate* values.

Number n	Positive Square Root $\sqrt{n}$	Number n	Positive Square Root $\sqrt{n}$	Number n	Positive Square Root $\sqrt{n}$	Number n	Positive Square Root $\sqrt{n}$
1	1	**26**	5.099	**51**	7.141	**76**	8.718
2	1.414	**27**	5.196	**52**	7.211	**77**	8.775
3	1.732	**28**	5.292	**53**	7.280	**78**	8.832
4	2	**29**	5.385	**54**	7.348	**79**	8.888
5	2.236	**30**	5.477	**55**	7.416	**80**	8.944
6	2.449	**31**	5.568	**56**	7.483	**81**	9
7	2.646	**32**	5.657	**57**	7.550	**82**	9.055
8	2.828	**33**	5.745	**58**	7.616	**83**	9.110
9	3	**34**	5.831	**59**	7.681	**84**	9.165
10	3.162	**35**	5.916	**60**	7.746	**85**	9.220
11	3.317	**36**	6	**61**	7.810	**86**	9.274
12	3.464	**37**	6.083	**62**	7.874	**87**	9.327
13	3.606	**38**	6.164	**63**	7.937	**88**	9.381
14	3.742	**39**	6.245	**64**	8	**89**	9.434
15	3.873	**40**	6.325	**65**	8.062	**90**	9.487
16	4	**41**	6.403	**66**	8.124	**91**	9.539
17	4.123	**42**	6.481	**67**	8.185	**92**	9.592
18	4.243	**43**	6.557	**68**	8.246	**93**	9.644
19	4.359	**44**	6.633	**69**	8.307	**94**	9.695
20	4.472	**45**	6.708	**70**	8.367	**95**	9.747
21	4.583	**46**	6.782	**71**	8.426	**96**	9.798
22	4.690	**47**	6.856	**72**	8.485	**97**	9.849
23	4.796	**48**	6.928	**73**	8.544	**98**	9.899
24	4.899	**49**	7	**74**	8.602	**99**	9.950
25	5	**50**	7.071	**75**	8.660	**100**	10

$\sqrt{13} \approx 3.606$

$\sqrt{74} \approx 8.602$

Written Exercises

Use the table to find the value.

A

1. $\sqrt{8}$ 2. $\sqrt{15}$ 3. $\sqrt{40}$ 4. $\sqrt{30}$ 5. $\sqrt{18}$

6. $\sqrt{65}$ 7. $\sqrt{20}$ 8. $\sqrt{51}$ 9. $\sqrt{79}$ 10. $\sqrt{19}$

11. $\sqrt{89}$ 12. $\sqrt{10}$ 13. $\sqrt{35}$ 14. $\sqrt{62}$ 15. $\sqrt{75}$

16. $-\sqrt{53}$ 17. $-\sqrt{38}$ 18. $-\sqrt{29}$ 19. $-\sqrt{50}$ 20. $-\sqrt{74}$

21. $\sqrt{57}$ 22. $\pm\sqrt{81}$ 23. $\pm\sqrt{16}$ 24. $\pm\sqrt{35}$ 25. $\pm\sqrt{67}$

Round to one decimal place.

26. $\sqrt{7}$ 27. $\sqrt{14}$ 28. $\sqrt{10}$ 29. $\sqrt{57}$ 30. $\sqrt{83}$

31. $\sqrt{51}$ 32. $\sqrt{21}$ 33. $\sqrt{19}$ 34. $\sqrt{27}$ 35. $\sqrt{37}$

36. $\sqrt{56}$ 37. $-\sqrt{18}$ 38. $-\sqrt{43}$ 39. $-\sqrt{71}$ 40. $-\sqrt{48}$

41. $\sqrt{32}$ 42. $\sqrt{94}$ 43. $-\sqrt{68}$ 44. $\pm\sqrt{20}$ 45. $\sqrt{82}$

Find the length of a side of the square correct to one decimal place.

B

46. 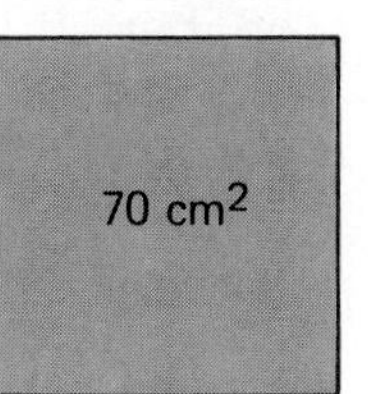

47. 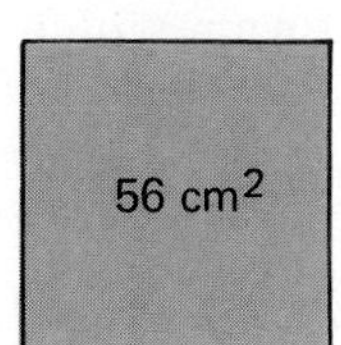

48. 20 cm^2

49. 35 cm^2

Find the perimeter of the square correct to one decimal place.

50. 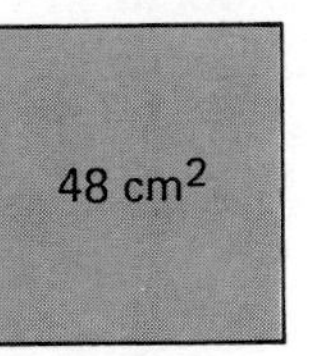

51. 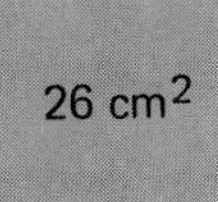

52. 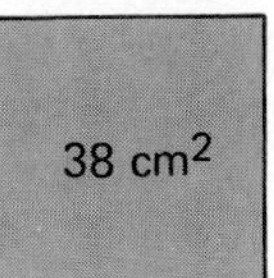

53. 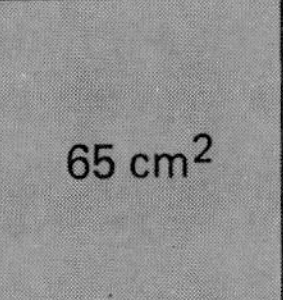

54. 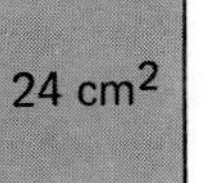

55. 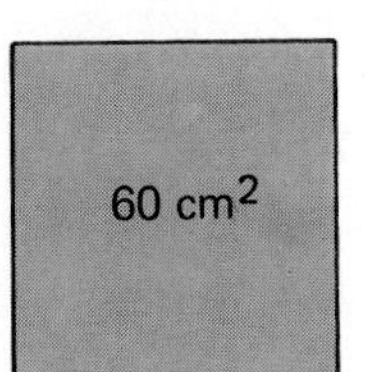

56. 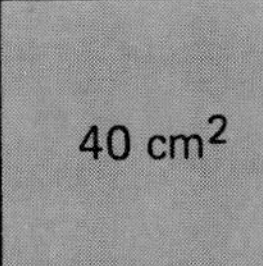

57. 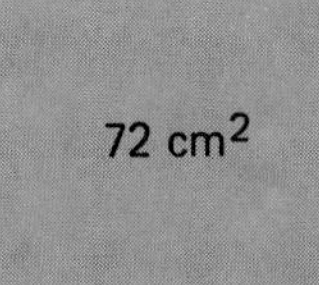

3 Irrational Numbers

All the numbers you've worked with in this course are called real numbers. Real numbers are the positive numbers, negative numbers, and zero. Real numbers can be separated into two kinds of numbers, the rational numbers and the irrational numbers.

These numbers can be expressed as ratios of two integers.

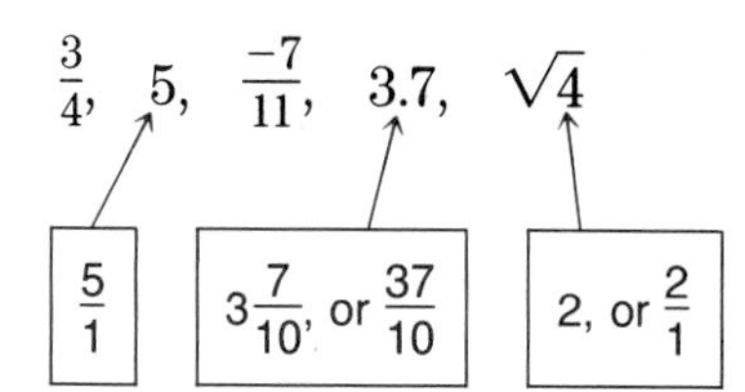

These numbers cannot be expressed as ratios of two integers.

$\sqrt{2}$, $\sqrt{7}$, $\sqrt{11}$, $\sqrt{20}$

Neither 2, 7, 11, nor 20 is a perfect square.

All irrational numbers have decimal forms which go on and on, but never repeat.

$$\sqrt{2} = 1.4142\square\square\square\square\square\square\square\square\square\square\ldots$$
$$\sqrt{7} = 2.6457\square\square\square\square\square\square\square\square\square\square\ldots$$
$$\sqrt{11} = 3.3166\square\square\square\square\square\square\square\square\square\square\ldots$$
$$\sqrt{20} = 4.4721\square\square\square\square\square\square\square\square\square\square\ldots$$

One famous irrational number is π (pi). A common place for it to occur is in formulas like $C = \pi d$ and $A = \pi r^2$ to calculate the circumference and area of a circle. π is the ratio of C to d (circumference to diameter) in any circle.

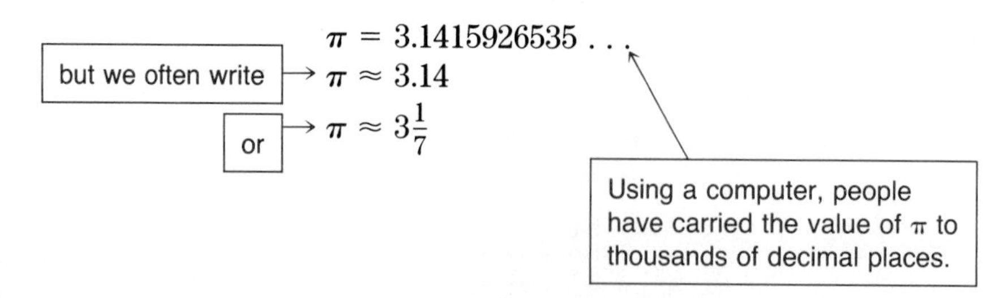

Written Exercises

Is the number rational or is it irrational?

A

1. 4
2. $\frac{1}{5}$
3. $\sqrt{4}$
4. $\frac{1}{8}$
5. -8
6. $\sqrt{7}$
7. $\sqrt{9}$
8. $\frac{3}{4}$
9. $\sqrt{2}$
10. $\frac{1}{3}$
11. $\sqrt{6}$
12. -7
13. $\frac{4}{5}$
14. 11
15. $4\frac{3}{5}$
16. $\sqrt{16}$
17. 6.72
18. 3.715
19. $-\sqrt{25}$
20. 2π
21. $\frac{2}{3}$
22. $-\sqrt{10}$
23. $\frac{5}{8}$
24. -17
25. $-\sqrt{49}$
26. $\sqrt{75}$
27. 4.24
28. $\sqrt{15^2}$
29. 5π
30. $-\sqrt{1}$

Find the approximate area. Use the formula $A = \pi r^2$. Use 3.14 for π.

SAMPLE

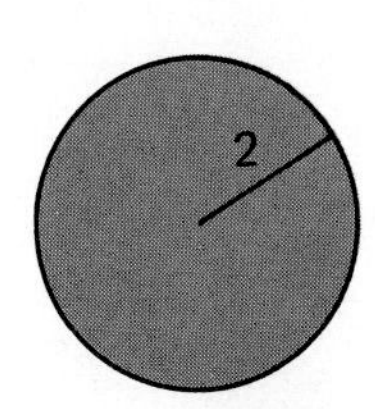

$A = \pi r^2$
$A \approx 3.14 \cdot 2^2$
$A \approx 3.14 \cdot 4$
$A \approx 12.56$

31.
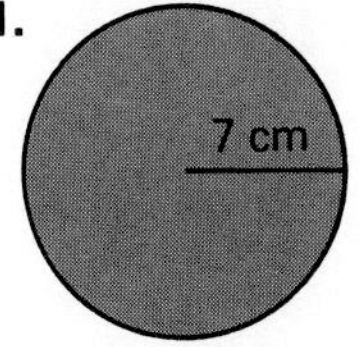

32.
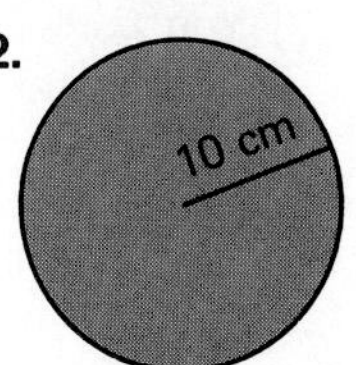

33.
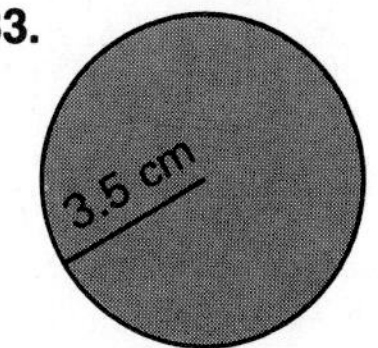

34.
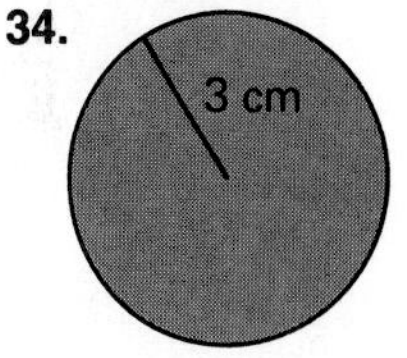

35. If $\sqrt{n}$ is a rational number, n must be a __?__ __?__.

Add or subtract.

B

36. $\frac{2}{3} + \frac{5}{6}$
37. $\frac{4}{5} - \frac{3}{10}$
38. $\frac{4}{7} + \frac{3}{5}$
39. $\frac{2}{9} - \frac{5}{6}$
40. Look at your answers to Exercises 36–39. Do you think that the sum or difference of two rational numbers is always a rational number?
41. Try to find an example that shows that the sum or difference of two irrational numbers is *not* always an irrational number.

4 Simplifying Square Roots

You know → $\sqrt{36} = 6$

but also → $\sqrt{36} = \sqrt{9 \cdot 4}$

$= \sqrt{9} \cdot \sqrt{4}$

$= 3 \cdot 2$

$= 6$

You can see that the following rule has been used.

$$\sqrt{ab} = \sqrt{a} \cdot \sqrt{b}$$

The rule above can be used to find square roots of numbers not found in the table on page 390.

EXAMPLE 1

$$\sqrt{2500} = \sqrt{25 \cdot 100} = \sqrt{25} \cdot \sqrt{100} = 5 \cdot 10 = 50$$

We choose the factors 25 and 100 because they are perfect squares.

EXAMPLE 2

$$\sqrt{125} = \sqrt{5 \cdot 25} = \sqrt{5} \cdot \sqrt{25} = \sqrt{5} \cdot 5 = 5\sqrt{5}$$

You won't always be able to find two factors that are perfect squares.

Product of a rational and an irrational number

Often it is convenient to leave your answer with a radical sign in it, as $5\sqrt{5}$. If your answer expresses a measurement, then you could use the square root table to find $\sqrt{5}$.

$$5\sqrt{5} \text{ cm} \approx 5 \times 2.236 \text{ cm} \approx 11.180 \text{ cm}$$

In working with variables, use the same rules. In this chapter, when you see a variable under a radical sign, assume that it represents a number greater than zero. Consider the following example.

EXAMPLE 3

$$\begin{aligned}\sqrt{8x^2} &= \sqrt{4 \cdot 2 \cdot x^2} \\ &= \sqrt{4} \cdot \sqrt{2} \cdot \sqrt{x^2} \leftarrow \boxed{\sqrt{x^2} = x} \\ &= 2x\sqrt{2}\end{aligned}$$

Written Exercises

Find the value. Leave your answer in simplest radical form if the number is not a perfect square.

A

1. $\sqrt{144}$
2. $\sqrt{243}$
3. $\sqrt{225}$
4. $\sqrt{196}$
5. $\sqrt{1000}$
6. $\sqrt{12}$
7. $\sqrt{2000}$
8. $\sqrt{324}$
9. $\sqrt{128}$
10. $\sqrt{288}$
11. $\sqrt{180}$
12. $\sqrt{338}$
13. $\sqrt{108}$
14. $\sqrt{1764}$
15. $\sqrt{192}$

Find the value. Leave your answer in simplest radical form.

B

16. $\sqrt{16x^2}$
17. $\sqrt{12y^2}$
18. $\sqrt{64x^2y}$
19. $\sqrt{15x^2y^2}$
20. $\sqrt{75a^4}$
21. $\sqrt{36xy^2}$
22. $\sqrt{125x^3}$
23. $\sqrt{3a^2b^2}$

C

24. Show by an example that $\sqrt{x^2} \neq x$ when x is negative.

SELF-TEST

Vocabulary

square root (p. 388)
radical sign (p. 388)
radical (p. 388)
perfect square (p. 388)

Find the value.

1. $\sqrt{49}$ 2. $\sqrt{1}$ 3. $-\sqrt{9}$ 4. $\sqrt{10000}$ *(11-1)*

Use the table on page 390 to find the value.

5. $\sqrt{60}$ 6. $\sqrt{24}$ 7. $\sqrt{52}$ 8. $\pm\sqrt{17}$ *(11-2)*

Is the number rational or irrational?

9. $\sqrt{16}$ 10. 4.25 11. -3 12. $\sqrt{18}$ *(11-3)*

Find the value. Leave your answer in simplest radical form if the number is not a perfect square.

13. $\sqrt{75}$ 14. $\sqrt{90}$ 15. $\sqrt{3600}$ 16. $\sqrt{432}$ *(11-4)*

5 Solving Equations

Some equations can be solved by using square roots.

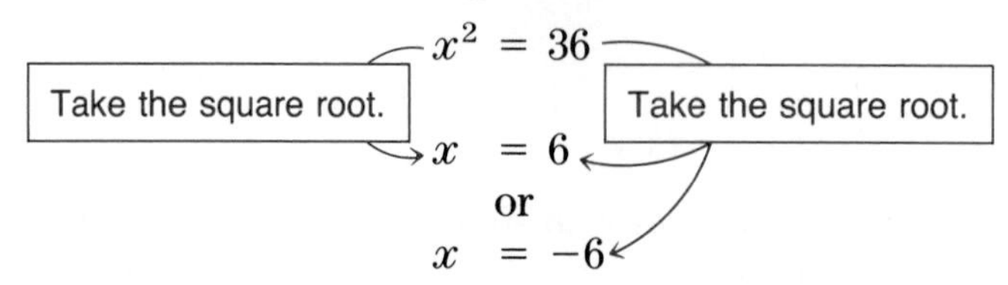

So, the equation $x^2 = 36$ has two solutions, 6 and -6.

Check:

$x^2 = 36$	
6^2	36
36	✓

$x^2 = 36$	
$(-6)^2$	36
36	✓

EXAMPLE 1

$$a^2 = 25$$

Take the square root of both sides.

$$a = \pm 5$$

So, $a = 5$ or $a = -5$.

EXAMPLE 2

$$\frac{x}{9} = \frac{4}{x}$$

Cross multiply first.

$$x^2 = 4 \cdot 9$$
$$x^2 = 36$$
$$x = \pm 6$$

EXAMPLE 3

$$y^2 - 7 = 57$$
$$y^2 = 57 + 7$$
$$y^2 = 64$$

Take the square root of both sides.

$$y = \pm 8$$

EXAMPLE 4

$$x^2 + 3 = 27$$
$$x^2 = 27 - 3$$
$$x^2 = 24$$

Take the square root of both sides.

$$x = \pm\sqrt{24}$$
$$x = \pm\sqrt{4 \cdot 6}$$
$$x = \pm 2\sqrt{6}$$

Written Exercises

Find the two solutions of the equation.

A

1. $x^2 = 64$
2. $x^2 = 81$
3. $m^2 = 16$
4. $n^2 = 100$
5. $y^2 = 9$
6. $n^2 = 49$
7. $n^2 = 121$
8. $x^2 = 144$
9. $\frac{x}{3} = \frac{3}{x}$
10. $\frac{x}{4} = \frac{25}{x}$
11. $\frac{4}{x} = \frac{x}{9}$
12. $\frac{y}{2} = \frac{8}{y}$
13. $y^2 - 1 = 35$
14. $y^2 - 5 = 20$
15. $x^2 + 6 = 31$
16. $n^2 + 4 = 53$
17. $x^2 = 12$
18. $n^2 = 10$
19. $y^2 = 6$
20. $x^2 = 27$
21. $x^2 + 2 = 8$
22. $y^2 - 7 = 20$
23. $5 + y^2 = 12$
24. $8 + y^2 = 30$
25. $x^2 + 5 = 23$
26. $n^2 - 2 = 40$
27. $x^2 - 5 = 11$
28. $7 + a^2 = 9$
29. $x^2 - 1 = 3$
30. $y^2 + 5 = 21$
31. $x^2 - 5 = 31$
32. $a^2 + 1 = 82$

Find the two solutions correct to one decimal place.

33. $n^2 = 40$
34. $x^2 = 56$
35. $y^2 = 38$
36. $x^2 = 91$
37. $x^2 - 1 = 75$
38. $y^2 + 3 = 80$
39. $x^2 + 7 = 52$
40. $n^2 - 4 = 75$

CAREER NOTES

ENVIRONMENTAL SCIENCES

We are becoming more and more aware of the need to protect our environment against pollution, land abuse, and destruction of wildlife. There are many careers in the fields of pollution control and conservation.

Many scientists and environmental engineers study the effects of pollution. For example, air pollution engineers conduct tests to see how pollution affects our atmosphere and weather conditions, and water pollution technicians analyze the chemical and bacterial content of water samples.

Other environmental engineers work with legislators to develop new laws and regulations to control the effects of pollution.

6 Law of Pythagoras

Centuries ago there was a brilliant Greek mathematician named Pythagoras. He is often credited with the discovery of one of the most useful laws in all mathematics. It deals with right angles and right triangles.

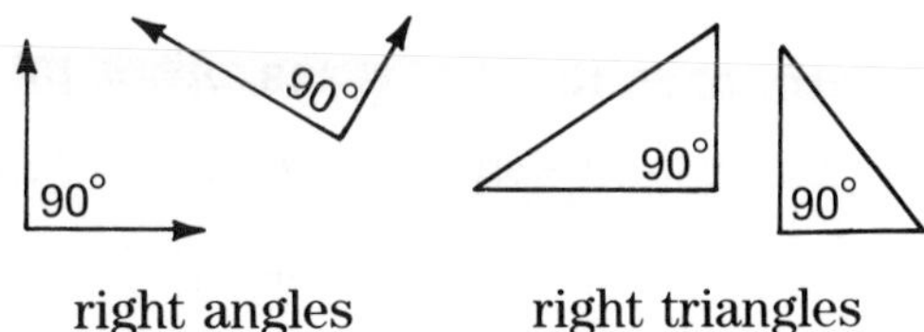

1. Consider this triangle.

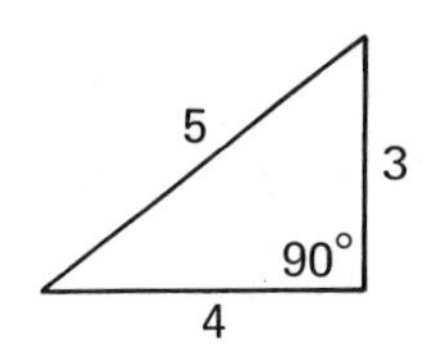

Notice that the side opposite the right angle in the triangle is the longest. This is true in all right triangles. This side is called the **hypotenuse.**

2. Build squares on its sides.

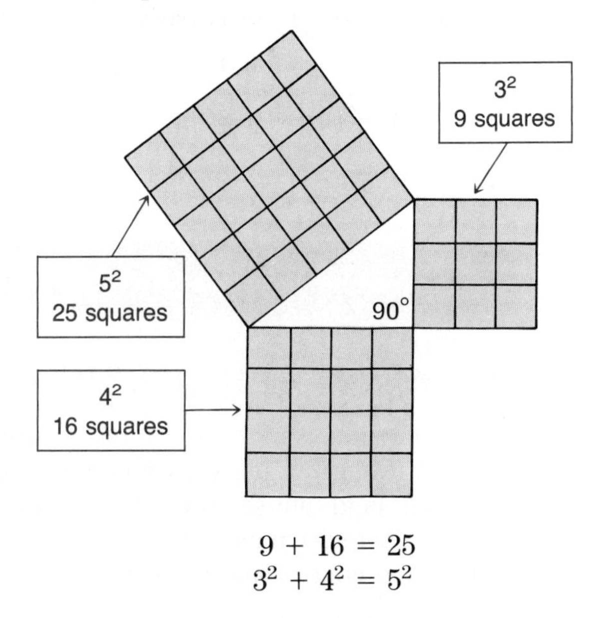

$$9 + 16 = 25$$
$$3^2 + 4^2 = 5^2$$

The relationship among the lengths of the sides of the triangle shown above is true in *all* right triangles.

LAW OF PYTHAGORAS
The square of the length of the hypotenuse of a right triangle equals the sum of the squares of the lengths of the other two sides.

Let's consider what this law of Pythagoras means. Take any right triangle. Call the length of its hypotenuse c. Call the lengths of the other two sides a and b. Then the following formula is true.

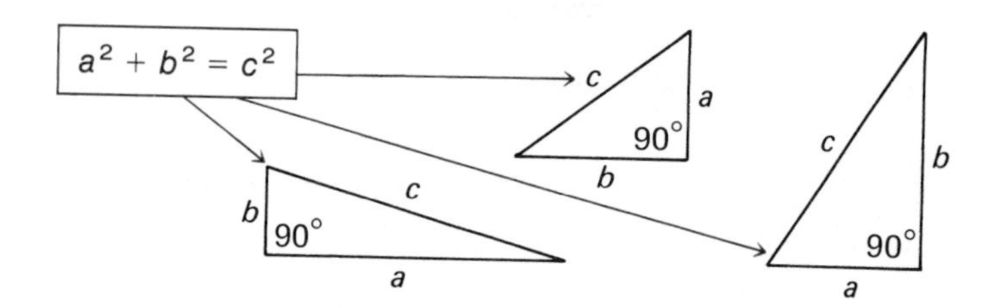

EXAMPLE 1 Find the length of the hypotenuse in this triangle.

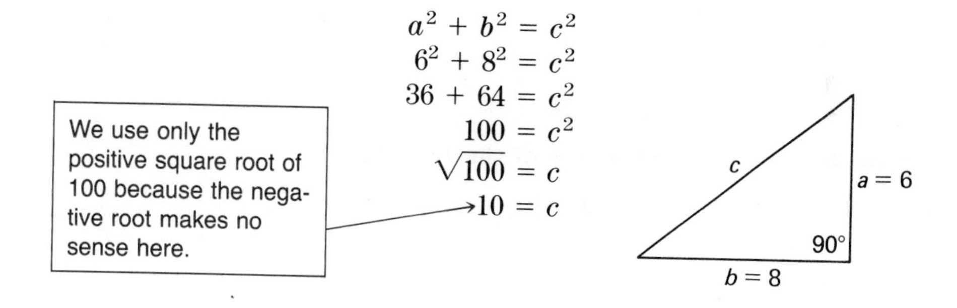

$$
\begin{aligned}
a^2 + b^2 &= c^2 \\
6^2 + 8^2 &= c^2 \\
36 + 64 &= c^2 \\
100 &= c^2 \\
\sqrt{100} &= c \\
10 &= c
\end{aligned}
$$

EXAMPLE 2 Find the length of the hypotenuse in this triangle.

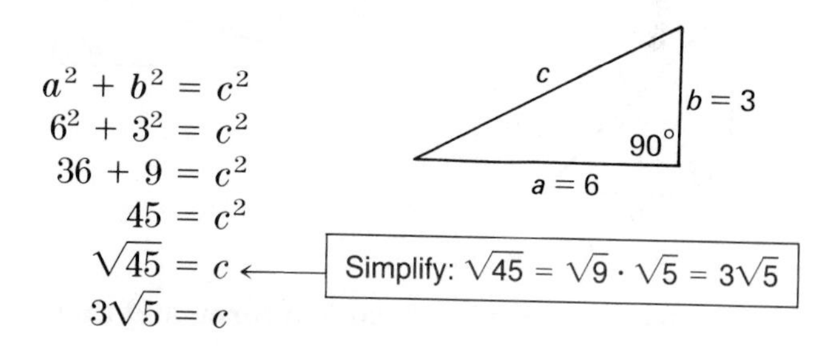

$$
\begin{aligned}
a^2 + b^2 &= c^2 \\
6^2 + 3^2 &= c^2 \\
36 + 9 &= c^2 \\
45 &= c^2 \\
\sqrt{45} &= c \\
3\sqrt{5} &= c
\end{aligned}
$$

EXAMPLE 3 Find b in the following triangle.

$$
\begin{aligned}
a^2 + b^2 &= c^2 \\
5^2 + b^2 &= 10^2 \\
25 + b^2 &= 100 \\
b^2 &= 75 \\
b &= \sqrt{75} \\
b &= 5\sqrt{3}
\end{aligned}
$$

a = 5
c = 10
90°
b

Classroom Practice

Complete.

1. In a right triangle there is always one __?__ angle.
2. The hypotenuse of a right triangle is the side __?__ the right angle.
3. If you know the lengths of __?__ sides of a __?__ triangle, you can use the law of __?__ to find the __?__ side.
4. In the formula $a^2 + b^2 = c^2$, c stands for the length of the __?__ of a __?__ triangle.
5. The law of Pythagoras is true for __?__ right triangles.

Written Exercises

Find the length of the third side of the triangle.

A

1.

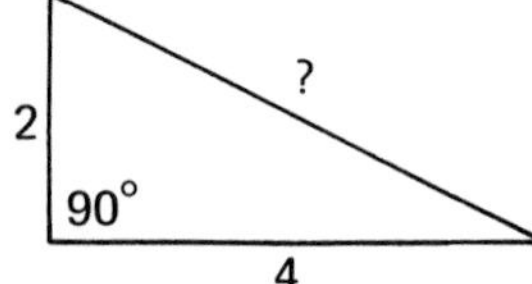

2.

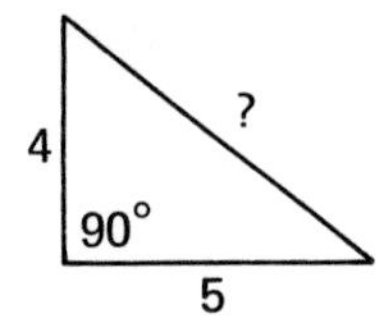

3.

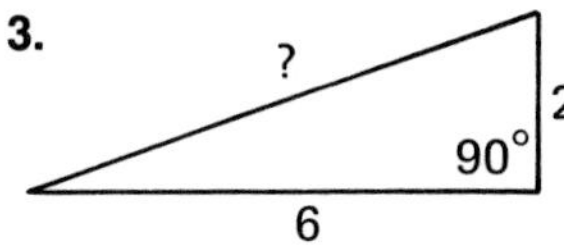

4.

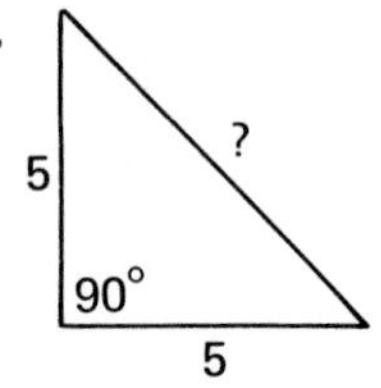

5.

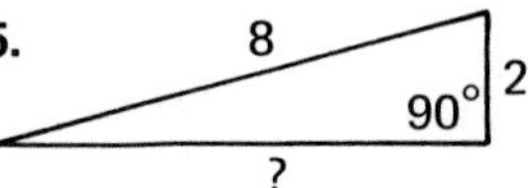

6.

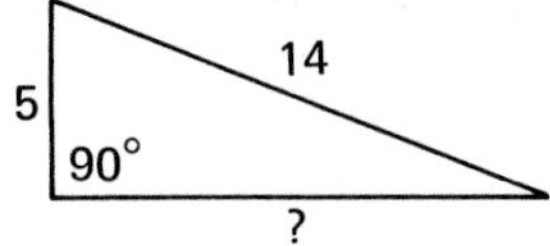

Solve. Give your results in radical form, and then correct to one decimal place.

7. A ladder 6 meters long leans against a wall. Its foot rests on the ground 1 meter from the wall. How high on the wall does the ladder reach?

8. A shelf support must be how long to meet the conditions shown in the picture?

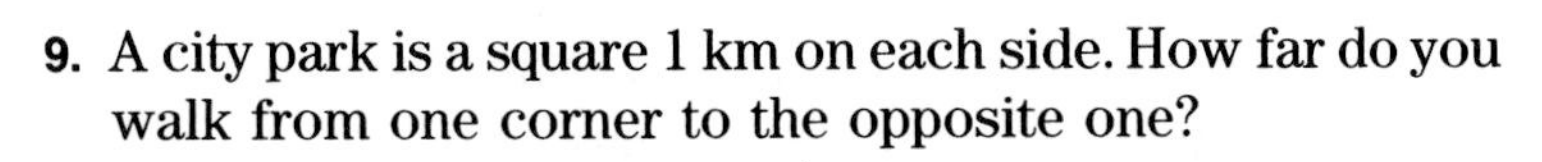

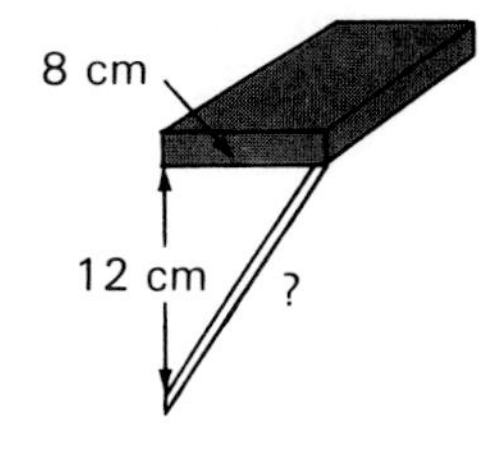

9. A city park is a square 1 km on each side. How far do you walk from one corner to the opposite one?

10. What is the longest line you can draw with a ruler on a piece of paper 20 cm by 30 cm?

11. How long an umbrella can you carry in the bottom of a suitcase 45 cm by 18 cm?

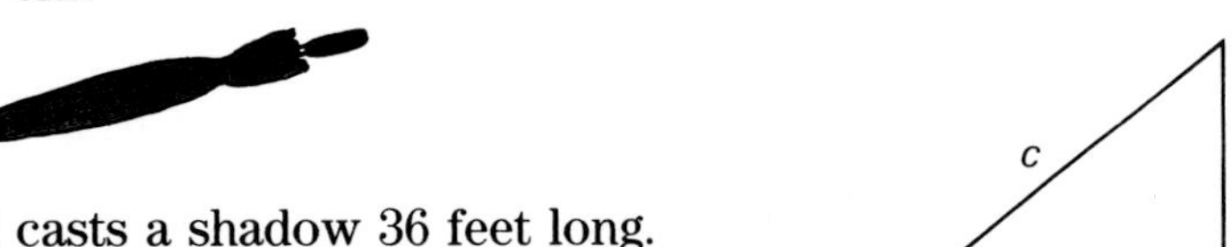

12. A column 27 feet tall casts a shadow 36 feet long. What is the distance, c, from the top of the column to the end of the shadow?

B 13. The drawing at the right shows the diagonal of a cube. Find a formula for the length of the diagonal d. (Hint: first find c, the diagonal of the base.)

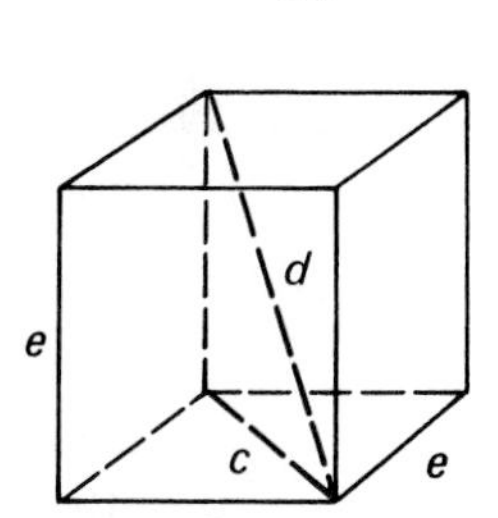

SELF-TEST

Vocabulary

hypotenuse (p. 398) law of Pythagoras (p. 398)

Find the two solutions of the equation. *(11-5)*

1. $x^2 = 49$
2. $y^2 = 100$
3. $\frac{x}{9} = \frac{4}{x}$
4. $x^2 + 2 = 18$
5. $y^2 - 3 = 22$
6. $x^2 + 1 = 10$

Find the length of the third side of the triangle. *(11-6)*

7. 9, ?, 90°, 12

8.

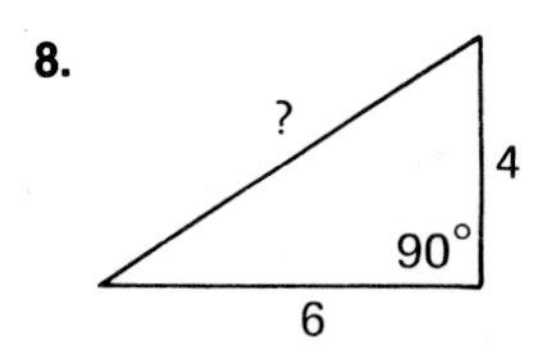

9.

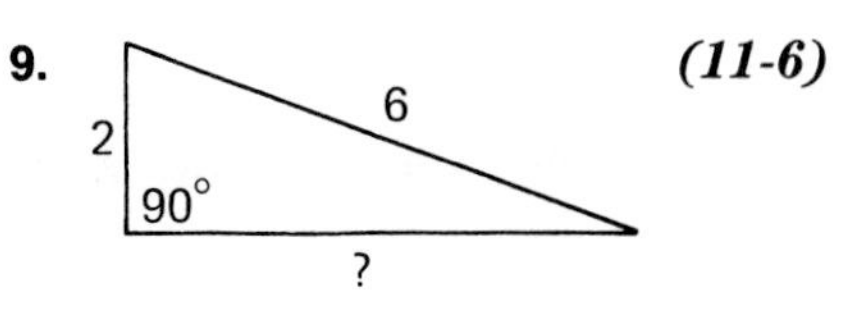

7 Quotients of Square Roots

You may sometimes need to find the square root of a fraction. Here's a method to use when the denominator is a perfect square.

EXAMPLE 1 $\sqrt{\frac{144}{9}} = \frac{\sqrt{144}}{\sqrt{9}} = \frac{12}{3} = 4$

EXAMPLE 2 $\sqrt{\frac{6}{16}} = \frac{\sqrt{6}}{\sqrt{16}} = \frac{\sqrt{6}}{4} = \frac{1}{4}\sqrt{6}$

EXAMPLE 3 $\sqrt{\frac{y^2}{4}} = \frac{\sqrt{y^2}}{\sqrt{4}} = \frac{y}{2}$

EXAMPLE 4 $\sqrt{\frac{3x^2}{16}} = \frac{\sqrt{3x^2}}{\sqrt{16}} = \frac{x\sqrt{3}}{4}$

You can see that the following rule has been used.

$$\sqrt{\frac{a}{b}} = \frac{\sqrt{a}}{\sqrt{b}}$$

Classroom Practice

Find the square root. Results should be in simplest form.

1. $\sqrt{\frac{7}{25}}$
2. $\sqrt{\frac{5}{36}}$
3. $\sqrt{\frac{7}{49}}$
4. $\sqrt{\frac{15}{36}}$
5. $\sqrt{\frac{a^2}{4}}$
6. $\sqrt{\frac{x^2}{9}}$
7. $\sqrt{\frac{2x^2}{25}}$
8. $\sqrt{\frac{4n^2}{49}}$

Find the square root. Express the result correct to two decimal places.

9. $\sqrt{\frac{2}{4}}$
10. $\sqrt{\frac{1}{9}}$
11. $\sqrt{\frac{1}{4}}$
12. $\sqrt{\frac{10}{16}}$

Written Exercises

Find the square root. Results should be in simplest form.

A

1. $\sqrt{\frac{4}{25}}$
2. $\sqrt{\frac{16}{64}}$
3. $\sqrt{\frac{121}{4}}$
4. $\sqrt{\frac{9}{144}}$
5. $\sqrt{\frac{1}{16}}$
6. $\sqrt{\frac{16}{49}}$
7. $\sqrt{\frac{36}{64}}$
8. $\sqrt{\frac{225}{625}}$
9. $\sqrt{\frac{9}{196}}$
10. $\sqrt{\frac{169}{16}}$
11. $\sqrt{\frac{1}{144}}$
12. $\sqrt{\frac{1}{400}}$

Simplify.

13. $\sqrt{\frac{x^2}{36}}$
14. $\sqrt{\frac{y^2}{81}}$
15. $\sqrt{\frac{4y^2}{9}}$
16. $\sqrt{\frac{16x^2}{25}}$
17. $\sqrt{\frac{3x^2}{4}}$
18. $\sqrt{\frac{x^2y^2}{100}}$
19. $\sqrt{\frac{7a^2}{64}}$
20. $\frac{1}{2}\sqrt{\frac{4n^2}{9}}$

Find the square root. Express the result correct to two decimal places.

21. $\sqrt{\frac{3}{4}}$
22. $\sqrt{\frac{10}{16}}$
23. $\sqrt{\frac{12}{9}}$
24. $\sqrt{\frac{22}{49}}$
25. $\sqrt{\frac{20}{25}}$
26. $\sqrt{\frac{50}{4}}$
27. $\sqrt{\frac{21}{9}}$
28. $\sqrt{\frac{15}{36}}$

PUZZLE ◆ PROBLEMS

Many great artists used a special rectangle as a basis for their drawings. It is special because the ratio of the length to the width is $\sqrt{2}$.

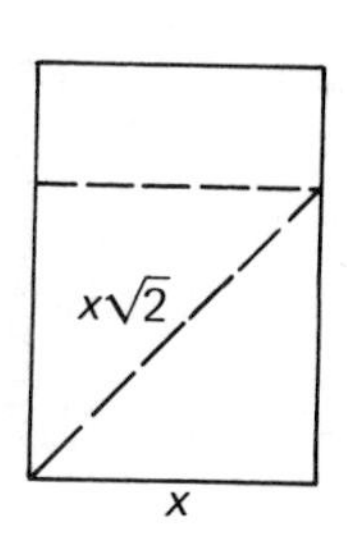

Look at the figure at the right. Do you see why the length of the diagonal of the square is $x\sqrt{2}$?

Now measure the longer side in the large rectangle. Compare its length to the length of the diagonal of the square. What do you notice?

In Germany standard sizes of paper are made with this same ratio of length to width.

8 Multiplication and Division

We can multiply and divide radical expressions. Just turn around the rules you've learned.

$\sqrt{ab} = \sqrt{a} \cdot \sqrt{b}$ *can be written* $\sqrt{a} \cdot \sqrt{b} = \sqrt{ab}$.

$\sqrt{\frac{a}{b}} = \frac{\sqrt{a}}{\sqrt{b}}$ *can be written* $\frac{\sqrt{a}}{\sqrt{b}} = \sqrt{\frac{a}{b}}$.

These two new rules can be helpful when you do not see a perfect square in an expression.

EXAMPLE 1

$$\begin{aligned}\sqrt{2} \cdot \sqrt{8} &= \sqrt{2 \cdot 8} \\ &= \sqrt{16} \\ &= 4\end{aligned}$$

EXAMPLE 2

$$\begin{aligned}3\sqrt{2} \cdot 4\sqrt{3} &= 3 \cdot 4 \cdot \sqrt{2} \cdot \sqrt{3} \\ &= 12\sqrt{6}\end{aligned}$$

EXAMPLE 3

$$\begin{aligned}3\sqrt{a} \cdot 2\sqrt{a} &= 3 \cdot 2 \cdot \sqrt{a} \cdot \sqrt{a} \\ &= 6\sqrt{a^2} \\ &= 6a\end{aligned}$$

Examine divisions with radicals closely too. If you do not see perfect squares in the expressions you are dividing, you can simplify in the following way.

EXAMPLE 4

$$\frac{\sqrt{18}}{\sqrt{2}} = \sqrt{\frac{18}{2}} = \sqrt{9} = 3$$

EXAMPLE 5

$$\frac{\sqrt{12}}{\sqrt{6}} = \sqrt{\frac{12}{6}} = \sqrt{2}$$

EXAMPLE 6

$$\frac{\sqrt{x^3}}{\sqrt{x}} = \sqrt{\frac{x^3}{x}} = \sqrt{x^2} = x$$

Written Exercises

Simplify. Leave no perfect-square factor under the radical sign.

A

1. $\sqrt{3} \cdot \sqrt{3}$
2. $\sqrt{2} \cdot \sqrt{2}$
3. $\sqrt{7} \cdot \sqrt{7}$
4. $\sqrt{5} \cdot \sqrt{5}$
5. $\sqrt{3} \cdot \sqrt{6}$
6. $\sqrt{2} \cdot \sqrt{6}$
7. $\sqrt{2} \cdot \sqrt{8}$
8. $\sqrt{10} \cdot \sqrt{2}$
9. $2 \cdot 2\sqrt{3}$
10. $2\sqrt{2} \cdot 2\sqrt{3}$
11. $4\sqrt{3} \cdot \sqrt{3}$
12. $4\sqrt{5} \cdot 2\sqrt{5}$
13. $\sqrt{18} \cdot 5$
14. $2\sqrt{18} \cdot 3$
15. $\sqrt{2} \cdot \sqrt{18}$
16. $2\sqrt{6} \cdot \sqrt{12}$
17. $\sqrt{12} \cdot \sqrt{3}$
18. $\sqrt{24} \cdot \sqrt{2}$
19. $\sqrt{15} \cdot \sqrt{5}$
20. $2\sqrt{2} \cdot \sqrt{20}$
21. $\sqrt{x} \cdot \sqrt{x}$
22. $\sqrt{a} \cdot \sqrt{a}$
23. $2\sqrt{a} \cdot 4\sqrt{a}$
24. $3x\sqrt{2} \cdot x\sqrt{2}$
25. $4\sqrt{x} \cdot 2\sqrt{x}$
26. $2\sqrt{n} \cdot \sqrt{4n}$
27. $\dfrac{\sqrt{10}}{\sqrt{2}}$
28. $\dfrac{\sqrt{12}}{\sqrt{3}}$
29. $\dfrac{\sqrt{14}}{\sqrt{2}}$
30. $\dfrac{\sqrt{18}}{\sqrt{3}}$
31. $\dfrac{\sqrt{20}}{\sqrt{5}}$
32. $\dfrac{\sqrt{21}}{\sqrt{7}}$
33. $\dfrac{10\sqrt{6}}{\sqrt{3}}$
34. $\dfrac{6\sqrt{10}}{\sqrt{2}}$
35. $\dfrac{8\sqrt{12}}{2\sqrt{3}}$
36. $\dfrac{14\sqrt{14}}{7\sqrt{2}}$
37. $\dfrac{\sqrt{a^3}}{\sqrt{a}}$
38. $\dfrac{2\sqrt{x}}{\sqrt{x}}$
39. $\dfrac{\sqrt{2n^2}}{\sqrt{n}}$
40. $\dfrac{\sqrt{8x}}{\sqrt{2x}}$
41. $\dfrac{\sqrt{27a^2}}{\sqrt{3b^2}}$
42. $\dfrac{\sqrt{3xy^3}}{\sqrt{2xy}}$
43. $\sqrt{\dfrac{1}{9}} \cdot \sqrt{\dfrac{1}{4}}$
44. $\sqrt{\dfrac{2}{3}} \cdot \sqrt{\dfrac{2}{27}}$
45. $\sqrt{\dfrac{2}{3}} \cdot \sqrt{\dfrac{3}{2}}$
46. $\sqrt{\dfrac{4}{5}} \cdot \sqrt{\dfrac{5}{4}}$
47. $\sqrt{\dfrac{10}{4}} \cdot \sqrt{\dfrac{4}{2}}$
48. $\sqrt{\dfrac{12}{5}} \cdot \sqrt{\dfrac{5}{3}}$

B

49. $3\sqrt{x^3} \cdot 2\sqrt{x}$
50. $5\sqrt{n} \cdot \dfrac{1}{5}\sqrt{\dfrac{1}{n}}$
51. $-3x\sqrt{2} \cdot x\sqrt{8}$
52. $\sqrt{\dfrac{1}{7}} \cdot -x\sqrt{14}$
53. $-x^3 \cdot -\sqrt{\dfrac{1}{4x^4}}$
54. $\sqrt{2} \cdot \sqrt{3} \cdot \sqrt{6}$
55. $\sqrt{3x} \cdot \sqrt{2x} \cdot \sqrt{3}$
56. $3x\sqrt{9x^2} \cdot \sqrt{6x^4}$
57. $2a^2\sqrt{3a^2} \cdot 5\sqrt{5a^2}$

C

58. $\sqrt{n} \cdot \dfrac{\sqrt{n^4}}{\sqrt{n}}$
59. $x \cdot \dfrac{-2\sqrt{x^3}}{\sqrt{4x}}$
60. $\dfrac{\sqrt{n^4}}{\sqrt{n^2}} \cdot 3n\sqrt{\dfrac{9n^3}{n}}$

9 Rationalizing the Denominator

Can you find the decimal value of $\frac{\sqrt{2}}{\sqrt{5}}$? You might look in the table of square roots and write $\frac{\sqrt{2}}{\sqrt{5}} \approx \frac{1.414}{2.236}$. Unless you had a calculator, you'd have troublesome division to do. Here is an easier way.

$$\frac{\sqrt{2}}{\sqrt{5}} = \frac{\sqrt{2} \cdot \sqrt{5}}{\sqrt{5} \cdot \sqrt{5}}$$

$$= \frac{\sqrt{10}}{5}$$

$$= \frac{1}{5}\sqrt{10}$$

You can multiply the numerator and denominator by the same number. You choose $\sqrt{5}$ because you'll then have a whole number for the denominator.

Now if you need to find the decimal value, you can find $\sqrt{10}$ in the table and calculate.

$$\frac{1}{5}\sqrt{10} \approx \frac{1}{5} \times 3.162 \approx 0.632$$

When a fraction has a radical in the denominator, you can simplify it by multiplying both numerator and denominator by a number that will make the denominator a whole number. This process is called **rationalizing the denominator.**

EXAMPLE 1

$$\frac{\sqrt{5}}{\sqrt{3}} = \frac{\sqrt{5} \cdot \sqrt{3}}{\sqrt{3} \cdot \sqrt{3}}$$

$$= \frac{\sqrt{15}}{3} \text{ or } \frac{1}{3}\sqrt{15}$$

Either form is correct.

EXAMPLE 2

$$\sqrt{\frac{3}{7}} = \frac{\sqrt{3}}{\sqrt{7}}$$

$$= \frac{\sqrt{3} \cdot \sqrt{7}}{\sqrt{7} \cdot \sqrt{7}}$$

$$= \frac{\sqrt{21}}{7} \text{ or } \frac{1}{7}\sqrt{21}$$

EXAMPLE 3

$$\sqrt{\frac{1}{2}} \cdot \sqrt{3} = \sqrt{\frac{3}{2}}$$
$$= \frac{\sqrt{3}}{\sqrt{2}}$$
$$= \frac{\sqrt{3} \cdot \sqrt{2}}{\sqrt{2} \cdot \sqrt{2}}$$
$$= \frac{\sqrt{6}}{2} \text{ or } \frac{1}{2}\sqrt{6}$$

Keep in mind these suggestions for simplifying radical expressions.

a. Leave no factor under the radical sign if the number is a perfect square.

b. Leave no radical expression in the denominator.

c. Leave no fraction under the radical sign.

Written Exercises

Express in simplest form.

A

1. $\sqrt{\frac{1}{2}}$ **2.** $\sqrt{\frac{3}{2}}$ **3.** $\sqrt{\frac{3}{4}}$ **4.** $\sqrt{\frac{4}{5}}$

5. $\sqrt{\frac{1}{3}}$ **6.** $\sqrt{\frac{3}{5}}$ **7.** $\sqrt{\frac{5}{8}}$ **8.** $\sqrt{\frac{6}{7}}$

9. $\sqrt{\frac{7}{10}}$ **10.** $\sqrt{\frac{5}{6}}$ **11.** $\sqrt{\frac{4}{7}}$ **12.** $\sqrt{\frac{5}{12}}$

13. $\frac{3x}{\sqrt{5}}$ **14.** $\frac{a^2}{\sqrt{a}}$ **15.** $\sqrt{\frac{a^2}{2}}$ **16.** $\sqrt{\frac{x^2}{3}}$

17. $\frac{-x^2}{\sqrt{y}}$ **18.** $\frac{3\sqrt{18}}{\sqrt{2}}$ **19.** $\frac{2\sqrt{8}}{\sqrt{2}}$ **20.** $3\sqrt{\frac{n^2}{5}}$

Find the decimal value correct to two decimal places.

B

21. $5\sqrt{\frac{1}{2}}$ **22.** $4\sqrt{\frac{1}{3}}$ **23.** $6\sqrt{\frac{2}{5}}$ **24.** $2\sqrt{\frac{3}{8}}$

25. $\sqrt{\frac{1}{3}} \cdot \sqrt{\frac{1}{3}}$ **26.** $\sqrt{\frac{1}{3}} \cdot \sqrt{\frac{1}{2}}$ **27.** $\sqrt{\frac{3}{4}} \cdot \sqrt{\frac{1}{2}}$ **28.** $\sqrt{\frac{5}{2}} \cdot \sqrt{\frac{2}{5}}$

10 Addition, Subtraction

Some radicals can be combined. Others cannot. You know that you can combine terms in an expression like $2a + 3a$.

like terms

$$2a + 3a = 5a$$

You can combine terms with radicals, but *only* if you have like terms.

like terms

$$2\sqrt{6} + 3\sqrt{6} = 5\sqrt{6}$$

like terms

$$2\sqrt{7} - 3\sqrt{7} = -\sqrt{7}$$

You cannot combine terms in the expressions below. All three expressions are in their simplest form.

unlike terms

$$\sqrt{2} + \sqrt{3}$$

unlike terms

$$\sqrt{a} + \sqrt{b}$$

unlike terms

$$2\sqrt{x} - 3\sqrt{y}$$

EXAMPLE 1 $3\sqrt{2} + 10\sqrt{2} = 13\sqrt{2}$

EXAMPLE 2 $7\sqrt{3} - \sqrt{3} = 6\sqrt{3}$

Be on the lookout for radicals which you could change in order to make like terms.

EXAMPLE 3

$$\begin{aligned} \sqrt{27} + \sqrt{3} &= \sqrt{9 \cdot 3} + \sqrt{3} \\ &= 3\sqrt{3} + \sqrt{3} \\ &= 4\sqrt{3} \end{aligned}$$

EXAMPLE 4

$$\begin{aligned} 2\sqrt{\frac{1}{2}} + \sqrt{2} &= 2\sqrt{\frac{2}{4}} + \sqrt{2} \\ &= 2 \cdot \frac{1}{2}\sqrt{2} + \sqrt{2} \\ &= \sqrt{2} + \sqrt{2} \\ &= 2\sqrt{2} \end{aligned}$$

Written Exercises

Express in simplest form.

A

1. $2\sqrt{2} + 6\sqrt{2}$
2. $6\sqrt{2} - 4\sqrt{2}$
3. $8\sqrt{5} + 5\sqrt{5}$
4. $-\sqrt{3} + 2\sqrt{3}$
5. $-\sqrt{5} + 5\sqrt{5}$
6. $10\sqrt{7} - \sqrt{7}$
7. $6\sqrt{3} - 5\sqrt{3}$
8. $\sqrt{2} - \sqrt{2}$
9. $\sqrt{18} + \sqrt{2}$
10. $4\sqrt{12} - 2\sqrt{3}$
11. $2\sqrt{8} - \sqrt{2}$
12. $2\sqrt{3} + 3\sqrt{12}$
13. $6\sqrt{3} - 4\sqrt{27}$
14. $\sqrt{2} + \sqrt{3} + 2\sqrt{2}$
15. $\sqrt{5} - \sqrt{3} + 3\sqrt{5}$
16. $\sqrt{6} + \sqrt{3} + 2\sqrt{6}$
17. $4\sqrt{5} - 2\sqrt{2} + 6\sqrt{5}$
18. $\sqrt{2} - 6\sqrt{5} + 5\sqrt{2}$
19. $\sqrt{12} - 4\sqrt{3} + \sqrt{6}$
20. $\sqrt{24} + 5\sqrt{6} - \sqrt{2}$
21. $\sqrt{45} + 2\sqrt{5} - 4\sqrt{7}$

B

22. $\sqrt{2a} + \sqrt{8a}$
23. $3\sqrt{b} + 3\sqrt{b^3}$
24. $\sqrt{49a^2} + 6\sqrt{b} - \sqrt{4b^3}$
25. $\sqrt{27a} - \sqrt{3a} - \sqrt{12a}$
26. $\sqrt{5} - (2\sqrt{5} - \sqrt{125})$
27. $\sqrt{6} - (\sqrt{24} + 6\sqrt{6})$
28. $\sqrt{8} - 2\sqrt{\frac{1}{8}} + 6\sqrt{24}$
29. $\sqrt{5} + 2\sqrt{\frac{1}{5}} + \sqrt{75}$

SELF-TEST

Vocabulary

rationalizing the denominator (p. 406)

Simplify.

1. $\sqrt{\frac{1}{4}}$
2. $\sqrt{\frac{5}{16}}$
3. $\sqrt{\frac{8}{25}}$
4. $\sqrt{\frac{20}{49}}$

(11-7)

5. $\sqrt{2} \cdot 4\sqrt{5}$
6. $\frac{\sqrt{21}}{\sqrt{3}}$
7. $3\sqrt{6} \cdot 4\sqrt{3}$

(11-8)

8. $\sqrt{\frac{7}{12}}$
9. $\sqrt{\frac{3}{10}}$
10. $\sqrt{\frac{5}{27}}$

(11-9)

11. $\sqrt{3} + 9\sqrt{3}$
12. $\sqrt{18} + \sqrt{2}$
13. $10\sqrt{3} - 4\sqrt{27}$

(11-10)

11 Radicals and Binomials (Optional)

You will recall that $(a + b)(a - b) = a^2 - b^2$. You can use the same idea to find a product of binomials with radicals.

EXAMPLE 1

$$\begin{aligned}(2 + \sqrt{3})(2 - \sqrt{3}) &= 2^2 - (\sqrt{3})^2 \\ &= 4 - 3 \\ &= 1\end{aligned}$$

This follows the form $(a + b)(a - b) = a^2 - b^2$.

This idea is often used to rationalize a denominator.

EXAMPLE 2

$$\begin{aligned}\frac{14}{3 - \sqrt{2}} &= \frac{14(3 + \sqrt{2})}{(3 - \sqrt{2})(3 + \sqrt{2})} \\ &= \frac{14(3 + \sqrt{2})}{3^2 - (\sqrt{2})^2} \\ &= \frac{14(3 + \sqrt{2})}{7} \\ &= 2(3 + \sqrt{2}) \\ &= 6 + 2\sqrt{2}\end{aligned}$$

EXAMPLE 3

$$\begin{aligned}\frac{3}{4 + \sqrt{5}} &= \frac{3(4 - \sqrt{5})}{(4 + \sqrt{5})(4 - \sqrt{5})} \\ &= \frac{3(4 - \sqrt{5})}{4^2 - (\sqrt{5})^2} \\ &= \frac{3(4 - \sqrt{5})}{16 - 5} \\ &= \frac{3(4 - \sqrt{5})}{11} \\ &= \frac{12 - 3\sqrt{5}}{11}\end{aligned}$$

Examples 4 and 5 show you how to square a binomial containing a radical.

EXAMPLE 4

$$\begin{aligned}(5 + \sqrt{3})^2 &= (5 + \sqrt{3})(5 + \sqrt{3}) \\ &= 5^2 + 5\sqrt{3} + 5\sqrt{3} + (\sqrt{3})^2 \\ &= 25 + 10\sqrt{3} + 3 \\ &= 28 + 10\sqrt{3}\end{aligned}$$

This is one way you can square a binomial. Just write the two factors and use the FOIL method.

EXAMPLE 5

$$\begin{aligned}(3 - \sqrt{6})^2 &= 3^2 - 2 \cdot 3\sqrt{6} + (\sqrt{6})^2 \\ &= 9 - 6\sqrt{6} + 6 \\ &= 15 - 6\sqrt{6}\end{aligned}$$

Another way to square a binomial is to use the rule

$(a - b)^2 = a^2 - 2ab + b^2$

Written Exercises

Express in simplest form.

A

1. $(1 - \sqrt{2})(1 + \sqrt{2})$
2. $(5 - \sqrt{5})(5 + \sqrt{5})$
3. $(\sqrt{3} - 4)(\sqrt{3} + 4)$
4. $(2 - \sqrt{7})(2 + \sqrt{7})$
5. $(1 - \sqrt{6})(1 + \sqrt{6})$
6. $(7 + \sqrt{5})(7 - \sqrt{5})$
7. $(\sqrt{6} - 2)(\sqrt{6} + 2)$
8. $(\sqrt{8} + 4)(\sqrt{8} - 4)$
9. $(2\sqrt{2} + 1)(2\sqrt{2} - 1)$
10. $(4\sqrt{5} + 2)(4\sqrt{5} - 2)$
11. $(3\sqrt{5} - 1)(3\sqrt{5} + 1)$
12. $(3\sqrt{3} - 2)(3\sqrt{3} + 2)$

Rationalize the denominator.

13. $\dfrac{1}{\sqrt{3} + 1}$
14. $\dfrac{5}{2 + \sqrt{3}}$
15. $\dfrac{2}{\sqrt{2} + 3}$
16. $\dfrac{\sqrt{5}}{2 + \sqrt{6}}$
17. $\dfrac{\sqrt{6}}{1 - \sqrt{6}}$
18. $\dfrac{\sqrt{8}}{4 - \sqrt{2}}$
19. $\dfrac{\sqrt{10}}{3 + \sqrt{6}}$
20. $\dfrac{\sqrt{7}}{\sqrt{3} + 5}$
21. $\dfrac{2\sqrt{7}}{\sqrt{7} - 1}$
22. $\dfrac{4\sqrt{3}}{5 - \sqrt{2}}$
23. $\dfrac{6\sqrt{2}}{2 - \sqrt{5}}$
24. $\dfrac{4\sqrt{3}}{6 - \sqrt{6}}$
25. $\dfrac{\sqrt{3} - 1}{\sqrt{3} + 1}$
26. $\dfrac{1 - \sqrt{3}}{\sqrt{3} + 1}$
27. $\dfrac{2 - \sqrt{7}}{\sqrt{7} - 4}$
28. $\dfrac{2 - 3\sqrt{3}}{3\sqrt{2} + 2}$

Simplify.

29. $(2 - \sqrt{3})^2$
30. $(5 + \sqrt{2})^2$
31. $(1 - \sqrt{5})^2$
32. $(3 + \sqrt{8})^2$
33. $(7 + \sqrt{5})^2$
34. $(6 - \sqrt{6})^2$
35. $(4 - \sqrt{10})^2$
36. $(2\sqrt{2} + 5)^2$
37. $(3\sqrt{5} - 2\sqrt{3})^2$
38. $(4\sqrt{8} - 2\sqrt{7})^2$
39. $(2\sqrt{5} - 3\sqrt{8})^2$
40. $(4\sqrt{2} - 3\sqrt{3})^2$
41. Find the area of a square whose sides measure $7 + 2\sqrt{3}$ each.

CALCULATOR ACTIVITIES

If your calculator has a square-root key, you can use this key to find the square root of any positive number.

Use a calculator to find the value. Round to three decimal places.

1. $\sqrt{120}$
2. $\sqrt{180}$
3. $-\sqrt{400}$
4. $\sqrt{952}$
5. $\pm\sqrt{0.25}$
6. $\sqrt{2.36}$
7. $\sqrt{8524}$
8. $\sqrt{0.0012}$

SKILLS REVIEW

POLYNOMIALS

Add or subtract.

1. $\begin{array}{r} 5x + 3y \\ \underline{+(2x + 9y)} \end{array}$

2. $\begin{array}{r} 7m^2 - 2m + 1 \\ \underline{+(4m^2 - 2m - 8)} \end{array}$

3. $\begin{array}{r} -3y + 5z - 9 \\ \underline{+(3y - 5z - 2)} \end{array}$

4. $\begin{array}{r} 2x - 3 \\ \underline{-(5x - 1)} \end{array}$

5. $\begin{array}{r} 15ab^2 - 3a + 7 \\ \underline{-(9ab^2 + 3a + 4)} \end{array}$

6. $\begin{array}{r} 2m^2n^2 - 3mn - 11 \\ \underline{-(8m^2n^2 - 4mn + 2)} \end{array}$

Multiply. You may wish to review the FOIL method on page 164.

SAMPLE $(x + 1)(x + 8) = x^2 + 9x + 8$

F → x^2; O + I → $9x$; L → 8

7. $(y + 2)(y + 7)$
8. $(k + 5)(k + 4)$
9. $(b + 9)(b + 4)$
10. $(z - 3)(z - 1)$
11. $(t - 8)(t - 3)$
12. $(g - 7)(g + 1)$
13. $(c - 9)(c + 8)$
14. $(m - 11)(m + 3)$
15. $(k - 15)(k + 2)$
16. $(m + 7)(m + 7)$
17. $(y - 10z)(y + 10z)$
18. $(2t - 1)(3t + 1)$

Factor.

19. $6t + 12$
20. $2b^2 + 6b$
21. $3x^2 + 15y$
22. $y^2 + 2y + 1$
23. $t^2 + 9t + 20$
24. $m^2 + 15m + 44$
25. $z^2 - 12z + 32$
26. $b^2 - 11b + 30$
27. $x^2 - 12x + 20$
28. $y^2 - 8y + 16$
29. $y^2 + 10yz + 25z^2$
30. $m^2 - 18m + 81$
31. $a^2 - 4a - 21$
32. $n^2 - n - 42$
33. $t^2 + 7t - 60$
34. $9d^2 - 49$
35. $1 - 36c^2$
36. $25m^2 - 121n^2$

Factor. Look for a common monomial factor first.

37. $2x^2 + 10x + 12$
38. $4z^2 - 8z - 32$
39. $5t^2 - 15t - 90$
40. $28y^2 - 7$
41. $2d^3 - 50d$
42. $p^2q^2 - 9pq^2 + 20q^2$
43. $-x^2 + 16$
44. $3v^2 + 18v - 21$
45. $-m^2 + 16m - 64$

CHAPTER REVIEW

CHAPTER SUMMARY

1. Every positive number n has a positive square root written $\sqrt{n}$ and a negative square root written $-\sqrt{n}$.
2. The symbol $\sqrt{}$ is called a radical sign and a number such as $\sqrt{8}$ is called a radical.
3. The table on page 390 gives the square roots of the integers from 1 to 100 rounded to three decimal places.
4. Irrational numbers are numbers that cannot be expressed as the ratio of two integers.
5. To simplify square roots you can use the rules

$$\sqrt{ab} = \sqrt{a} \cdot \sqrt{b} \quad \text{and} \quad \sqrt{\frac{a}{b}} = \frac{\sqrt{a}}{\sqrt{b}}.$$

6. The law of Pythagoras states:
The square of the length of the hypotenuse of a right triangle equals the sum of the squares of the lengths of the other two sides.
7. To rationalize the denominator of a fraction, multiply it by a number that will make the radical denominator a whole number. You must multiply the numerator by the same number.

REVIEW EXERCISES

Find the value. ***(See pp. 388–389.)***

1. $\sqrt{25}$
2. $-\sqrt{36}$
3. $\sqrt{81}$
4. $-\sqrt{100}$
5. $\sqrt{9}$
6. $\sqrt{8^2}$
7. $-\sqrt{49}$
8. $\sqrt{64}$
9. $\sqrt{12^2}$
10. $-\sqrt{15^2}$

Use the table on page 390 to find the value. Round to one decimal place. ***(See pp. 390–391.)***

11. $\sqrt{19}$
12. $\sqrt{29}$
13. $\sqrt{60}$
14. $\sqrt{42}$
15. $-\sqrt{7}$

Is the number rational or irrational? ***(See pp. 392–393.)***

16. 6
17. $\frac{1}{6}$
18. $\sqrt{6}$
19. 1.6
20. $\sqrt{81}$
21. $\sqrt{9}$
22. $-\sqrt{81}$
23. $4\frac{1}{4}$
24. $6\frac{1}{3}$
25. π

Find the value. Leave your answer in simplest radical form. *(See pp. 394–395.)*

26. $\sqrt{18}$ 27. $\sqrt{32}$ 28. $\sqrt{28}$ 29. $\sqrt{150}$

30. $\sqrt{45}$ 31. $\sqrt{175}$ 32. $\sqrt{140}$ 33. $\sqrt{200}$

Find the two solutions of the equation. *(See pp. 396–397.)*

34. $n^2 = 9$ 35. $x^2 = 81$ 36. $y^2 = 48$ 37. $a^2 - 6 = 43$

Find the length of the third side of the triangle. *(See pp. 398–401.)*

38.

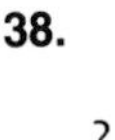

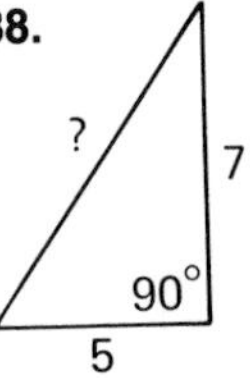

39.

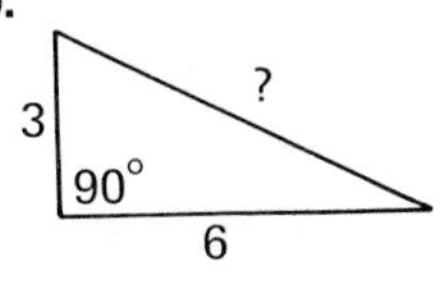

40.

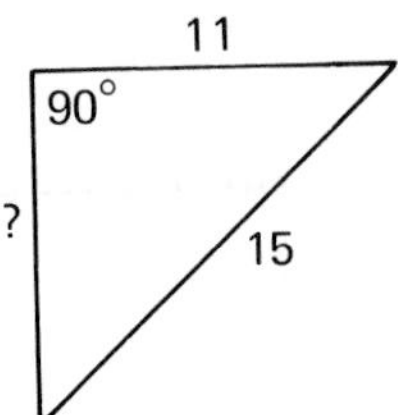

Find the square root. Results should be in simplest form. *(See pp. 402–403.)*

41. $\sqrt{\frac{4}{144}}$ 42. $\sqrt{\frac{121}{64}}$ 43. $\sqrt{\frac{3}{81}}$ 44. $\sqrt{\frac{14x^2}{64}}$ 45. $\sqrt{\frac{x^2y^2}{9}}$

Simplify. Leave no perfect-square factor under the radical sign. *(See pp. 404–407.)*

46. $\sqrt{10} \cdot \sqrt{10}$ 47. $3\sqrt{3} \cdot 2\sqrt{3}$ 48. $2\sqrt{5} \cdot \sqrt{25}$ 49. $3\sqrt{n} \cdot 6\sqrt{n}$

50. $\frac{\sqrt{15}}{\sqrt{3}}$ 51. $\frac{\sqrt{21}}{\sqrt{7}}$ 52. $\frac{\sqrt{2r^2}}{\sqrt{r}}$ 53. $\frac{\sqrt{24x}}{\sqrt{6x}}$ 54. $\sqrt{\frac{1}{5}}$

55. $\sqrt{\frac{7}{8}}$ 56. $\sqrt{\frac{a^2}{5}}$ 57. $\sqrt{\frac{2x}{5}}$ 58. $4\sqrt{\frac{2}{\pi}}$ 59. $\frac{2n}{\sqrt{8}}$

Express in simplest form. *(See pp. 408–409.)*

60. $3\sqrt{3} - \sqrt{3}$ 61. $\sqrt{2} + \sqrt{8} - \sqrt{12}$ 62. $2\sqrt{45} - \sqrt{5} + \sqrt{75}$

63. $5\sqrt{3} + \sqrt{48} - \sqrt{4}$ 64. $3\sqrt{24} + \sqrt{8} - 6\sqrt{20}$ 65. $\sqrt{7} - 3\sqrt{21} + \sqrt{49}$

(Optional) Simplify. *(See pp. 410–411.)*

66. $(\sqrt{5} - 3)(\sqrt{5} + 3)$ 67. $(3\sqrt{2} - 1)(3\sqrt{2} + 1)$ 68. $(4 - 2\sqrt{3})^2$

CHAPTER TEST

Find the value.

1. $\sqrt{25}$ 2. $-\sqrt{100}$ 3. $-\sqrt{6^2}$ 4. $\sqrt{12^2 + 9^2}$ *(11-1)*

Use the table on page 390 to find the value. Round to one decimal place.

5. $\sqrt{28}$ 6. $\sqrt{89}$ 7. $-\sqrt{87}$ 8. $\sqrt{97}$ *(11-2)*

Is the number rational or is it irrational?

9. 0.82 10. $\sqrt{49}$ 11. $\sqrt{3}$ 12. $-\pi$ *(11-3)*

13. A circle has radius 14. Find the approximate area of the circle. Use the formula $A = \pi r^2$. Use 3.14 for π.

Find the value. Leave your answer in simplest radical form.

14. $\sqrt{147}$ 15. $\sqrt{126}$ 16. $\sqrt{180}$ 17. $\sqrt{27x^3y}$ *(11-4)*

Find the two solutions of the equation.

18. $\frac{x}{2} = \frac{32}{x}$ 19. $y^2 = 28$ 20. $1 + n^2 = 4$ *(11-5)*

21. Find the length of the third side of the triangle shown. *(11-6)*

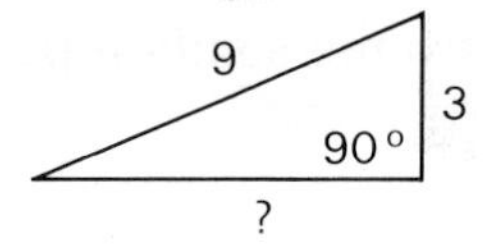

Express in simplest form.

22. $\sqrt{\frac{16}{100}}$ 23. $\sqrt{\frac{144}{169}}$ 24. $\sqrt{\frac{5x^2}{49}}$ *(11-7)*

25. $4\sqrt{9x} \cdot 3\sqrt{x}$ 26. $\frac{8\sqrt{18}}{4\sqrt{8}}$ 27. $\sqrt{\frac{20}{11}} \cdot \sqrt{\frac{11}{5}}$ *(11-8)*

28. $\sqrt{\frac{25}{3}}$ 29. $\sqrt{\frac{n^2}{8}}$ 30. $4\sqrt{\frac{1}{5}}$ *(11-9)*

31. $\sqrt{7} - 7\sqrt{2} + 2\sqrt{7}$ 32. $\sqrt{50a} - \sqrt{32a} + \sqrt{8a}$ *(11-10)*

33. (Optional) Rationalize the denominator: $\frac{3 - \sqrt{5}}{3 + \sqrt{5}}$ *(11-11)*

MIXED REVIEW

1. Two sides of a triangle have equal lengths. Each of the equal sides is 1.6 times as long as the third side of the triangle. The perimeter of the triangle is 18.9 cm. Find the length of each side.

2. Solve by the graphing method: $y = 3x + 5$
   $y - x = 3$

Simplify.

3. $\dfrac{\sqrt{24x^3}}{\sqrt{2x}}$ 4. $(3a - b)^2 - (8a^2 - 5ab)$ 5. $\dfrac{x^2 + 2x - 15}{x^2 - 25} \div (3 - x)$

6. $\dfrac{5}{6x} - \dfrac{3}{10x}$ 7. $\dfrac{2n}{n+3} + \dfrac{6}{n+3}$ 8. $\sqrt{9} - \sqrt{27} + \sqrt{18}$

9. The graph of the equation $2x - y = -4$ is a line having a slope of __?__.

10. Four tires and a heavy-duty battery cost $167. Five tires and two batteries cost $241. Find the cost of each item.

11. Express $\frac{7}{12}$ as a decimal. Round to three decimal places.

12. Find three solution pairs for the inequality $y < 3x - 1$.

Factor.

13. $y^2 + 19y + 48$ 14. $-12a^2 - 12a - 3$ 15. $2x^3y + 4x^2y^2 + 14xy^3$

16. In a trail mix, the ratio of nuts to raisins is 5 to 2. How many kilograms of nuts are needed to make 14 kg of trail mix?

Multiply.

17. $\dfrac{r^2 + 3r}{6} \cdot \dfrac{4r^2}{r+3}$ 18. $(a^3 + 2b)(a^3 - 2b)$ 19. $(m - 4)(3m^2 - 2m - 5)$

Solve.

20. $-7(4 - 3x) = 9x - 52$ 21. $0.04(120 - x) = 0.12x$

22. Sean's August electric bill of $49.60 covered 31 days of service. At that rate, what is the charge for 30 days of service?

23. How long will it take \$800 to double at 8% interest?

24. When you buy 3 containers of yogurt you get one container free.
   a. The number of containers you get free is a function of __?__.
   b. How many free containers do you receive if you buy 7 containers?

25. What percent of 48 is 30?

26. The two shorter sides of a right triangle have the lengths 2 and 3. Find the length of the hypotenuse.

27. A mail carrier needs 6 hours to sort and deliver the mail on her route. A trainee would need 9 hours to do the job alone. How long would it take to do the job together?

Solve.

28. $\frac{x + 5}{3} = \frac{3x - 1}{5}$

29. $1 - 9y = -62$

30. $1.8s = 0.24 + 2.6s$

31. A box is $(x + 2)$ units by $(x - 3)$ units by 4 units. Express its volume as a polynomial.

Simplify.

32. $(-7a^2b^3)(-2ab^2)^2$

33. $\frac{n^2 - 10n + 16}{3n^2 - 6n}$

34. $\frac{-8x^4 + 6x^2}{2x} + x^3$

35. Points $(-1,$ __?__$)$ and $($__?__$, 0)$ are solution pairs for the equation $3x - y = 6$.

36. Haywood is 7 years older than Shelley.
   In 4 years Haywood will be twice as old as Shelley.
   How old is each now?

37. Rita rode her bike for 2 hours at 10 km/h.
   On her return trip she traveled at a speed that was 20% slower.
   How much time did Rita spend on her return trip?

38. Tickets to the symphony cost \$18 in advance and \$20 at the door.
   In all, 573 tickets were sold and \$10,670 was collected.
   How many tickets were sold in advance?

39. Graph the solution of $-2 + x = -7$ on a number line.

40. Compare. Write $<$ or $>$: $(-2)^3$ __?__ $-(-3)^2$

Here's what you'll learn in this chapter:

- To solve quadratic equations by factoring.
- To solve quadratic equations by taking square roots.
- To use the quadratic formula to solve quadratic equations.
- To use quadratic equations to solve word problems.
- To draw the graphs of quadratic functions.

The curves formed by the jets of water from these fireboats are parabolas. In this chapter you will discover that every quadratic function has a parabola as its graph.

Chapter 12

Quadratic Equations

1 Zero Products

Can you solve the puzzle below?

> I'm thinking of two numbers.
> Their product is zero.
> Tell me one of the numbers.

> *If two numbers have a zero product, one of the numbers must be zero.*

The rule above is used to solve some equations.

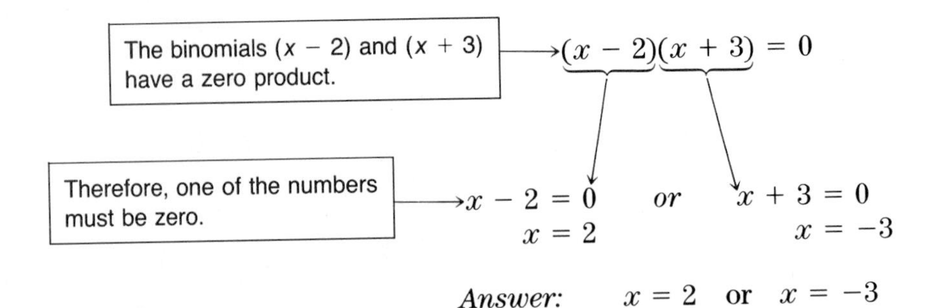

Study these examples.

EXAMPLE 1

$$x(x + 1) = 0$$

$x = 0$ *or* $x + 1 = 0$

$x = -1$

Answer: $x = 0$ or $x = -1$

EXAMPLE 2

$$(n - 5)(2n - 6) = 0$$

$n - 5 = 0$ *or* $2n - 6 = 0$

$n = 5$ $\quad$ $2n = 6$

$n = 3$

Answer: $n = 5$ or $n = 3$

Classroom Practice

Complete.

1. If $x(x + 3) = 0$, then either $x = \underline{\ ?\ }$ or $x + 3 = \underline{\ ?\ }$.
2. If $x(x - 2) = 0$, then either $x = \underline{\ ?\ }$ or $x - 2 = \underline{\ ?\ }$.
3. If $(x + 3)(x - 4) = 0$, either $x + 3 = \underline{\ ?\ }$ or $\underline{\ ?\ } = 0$.

Solve.

4. $x(x - 3) = 0$
5. $a(a + 5) = 0$
6. $3n(n - 2) = 0$
7. $(x - 1)(x - 3) = 0$
8. $(x - 5)(x + 2) = 0$
9. $(2x - 8)(x + 1) = 0$

Written Exercises

Solve.

A

1. $n(n + 3) = 0$
2. $n(n - 3) = 0$
3. $x(x + 9) = 0$
4. $y(y + 6) = 0$
5. $5m(m - 1) = 0$
6. $2m(m - 4) = 0$
7. $2n(n - 9) = 0$
8. $4x(x + 8) = 0$
9. $(y - 3)(y - 1) = 0$
10. $(a + 8)(a + 1) = 0$
11. $(x + 4)(x + 3) = 0$
12. $(b + 1)(b + 6) = 0$
13. $(y + 3)(y + 8) = 0$
14. $(x - 10)(x - 20) = 0$
15. $(n - 14)(n - 2) = 0$
16. $(y - 10)(y - 3) = 0$
17. $(x + 5)(x + 6) = 0$
18. $(n + 20)(n + 40) = 0$
19. $(a - 8)(a + 8) = 0$
20. $(x - 25)(x + 25) = 0$
21. $(y + 12)(y - 12) = 0$
22. $(n - 6)(n + 4) = 0$
23. $(x + 100)(x - 100) = 0$
24. $(x - 4)(x - 75) = 0$

B

25. $(y - 12)(4y + 20) = 0$
26. $(m - 8)(3m - 12) = 0$
27. $(6r - 18)(7r + 49) = 0$
28. $(4n - 12)(6n + 72) = 0$
29. $(12x - 72)(5x + 20) = 0$
30. $x(x + 1)(x + 5) = 0$
31. $n(n - 1)(n + 2) = 0$
32. $(18x - 54)(9x + 18) = 0$
33. $(7x - 28)(15x - 90) = 0$
34. $3y(y - 4)(y + 4) = 0$

2 Solving Quadratic Equations

A **quadratic equation** in x is an equation that has an x^2 term but no term with a higher power of x. Here are three quadratic equations.

$$2x^2 + 6x = 0 \qquad x^2 - 5x + 6 = 0 \qquad x^2 = 5$$

Many quadratic equations can be solved by factoring.

EXAMPLE 1

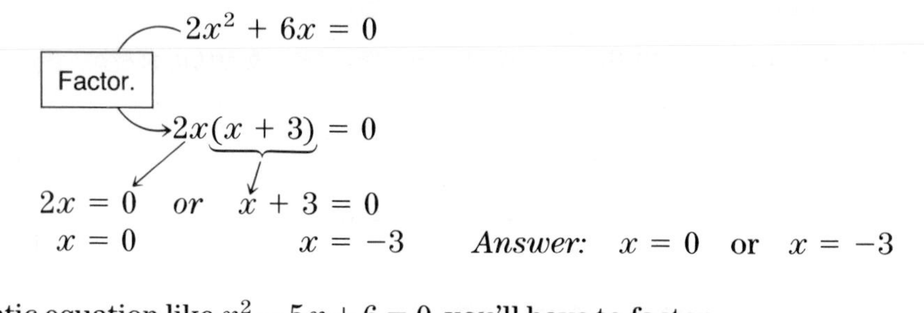

To solve a quadratic equation like $x^2 - 5x + 6 = 0$, you'll have to factor the trinomial $x^2 - 5x + 6$. Recall how to factor.

Factor $x^2 - 5x + 6$.

Step 1: $(\quad)(\quad)$
Step 2: $(x \quad)(x \quad)$
Step 3: $(x\ ?\ )(x\ ?\ )$

Step 4: $(x - 2)(x - 3)$

What two numbers have a product of 6 and a sum of -5?

-2 and -3

Now see how the equation $x^2 - 5x + 6 = 0$ is solved.

EXAMPLE 2

$$x^2 - 5x + 6 = 0$$

Factor.

$$(x - 2)(x - 3) = 0$$

$x - 2 = 0$ *or* $x - 3 = 0$

$x = 2 \qquad x = 3 \qquad$ *Answer:* $x = 2$ or $x = 3$

EXAMPLE 3

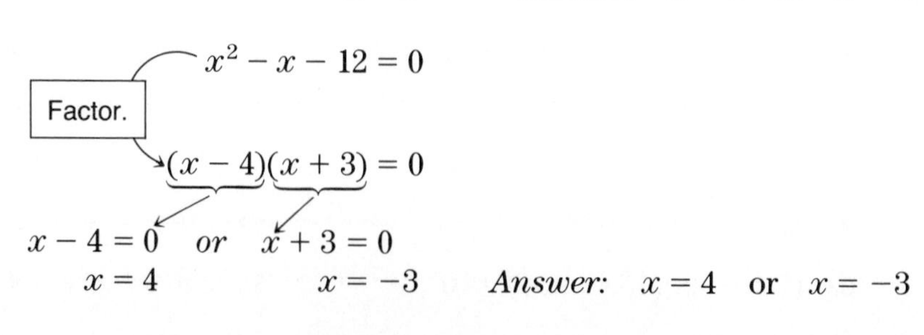

Classroom Practice

1. Factor: $x^2 - 3x$
   Solve: $x^2 - 3x = 0$
2. Factor: $x^2 + 2x$
   Solve: $x^2 + 2x = 0$
3. Factor: $5x^2 - 10x$
   Solve: $5x^2 - 10x = 0$
4. Factor: $3x^2 + 18x$
   Solve: $3x^2 + 18x = 0$
5. Factor: $x^2 - 6x + 8$
   Solve: $x^2 - 6x + 8 = 0$
6. Factor: $x^2 - x - 20$
   Solve: $x^2 - x - 20 = 0$
7. Factor: $x^2 - 8x + 12$
   Solve: $x^2 - 8x + 12 = 0$
8. Factor: $x^2 - 3x - 10$
   Solve: $x^2 - 3x - 10 = 0$

Written Exercises

Factor.

A

1. $x^2 - 2x$
2. $y^2 - 3y$
3. $2x^2 - 4x$
4. $x^2 + 3x + 2$
5. $x^2 + 2x - 3$
6. $x^2 + x - 2$
7. $x^2 + 6x + 8$
8. $x^2 - 7x + 12$
9. $x^2 - 3x - 18$

Solve.

10. $x^2 + x = 0$
11. $x^2 - 3x = 0$
12. $x^2 - 5x = 0$
13. $6x + 2x^2 = 0$
14. $9x - 3x^2 = 0$
15. $16x - 4x^2 = 0$
16. $x^2 - 5x + 4 = 0$
17. $x^2 - 9x + 8 = 0$
18. $x^2 + 4x + 3 = 0$
19. $x^2 - 8x + 7 = 0$
20. $x^2 + 8x + 15 = 0$
21. $x^2 - 10x + 21 = 0$
22. $y^2 + 4y + 4 = 0$
23. $y^2 + 4y - 21 = 0$
24. $y^2 - 7y + 6 = 0$
25. $x^2 - 2x - 8 = 0$
26. $x^2 + x - 6 = 0$
27. $n^2 - 5n - 14 = 0$
28. $a^2 + 4a - 32 = 0$
29. $a^2 + 10a + 25 = 0$
30. $a^2 - 12a + 35 = 0$
31. $x^2 - 11x + 28 = 0$
32. $x^2 - 16 = 0$
33. $x^2 - 9 = 0$
34. $x^2 + 12x + 36 = 0$
35. $x^2 - 13x + 22 = 0$
36. $x^2 - 6x - 27 = 0$
37. $x^2 - 6x - 55 = 0$
38. $x^2 - 22x + 40 = 0$
39. $x^2 - 13x + 40 = 0$
40. $x^2 - 5x - 36 = 0$
41. $a^2 + 6a - 40 = 0$
42. $y^2 + 5y - 24 = 0$
43. $n^2 - 6n - 7 = 0$
44. $m^2 + 20m + 64 = 0$
45. $x^2 + 11x - 80 = 0$

3 Solving Quadratic Equations

Sometimes before you can solve a quadratic equation you must first rewrite the equation so that one side is zero.

EXAMPLE 1

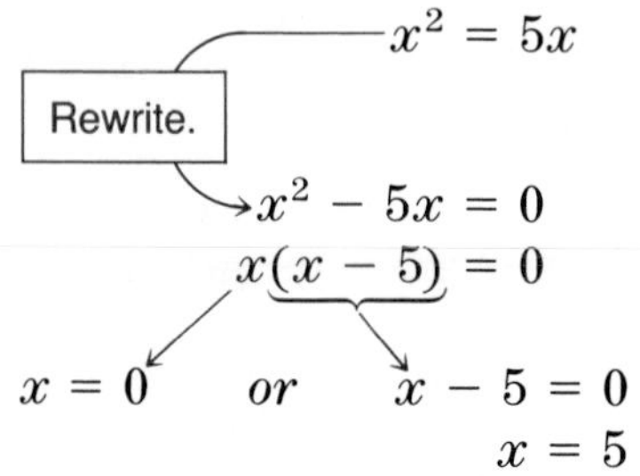

Answer: $x = 0$ or $x = 5$

EXAMPLE 2

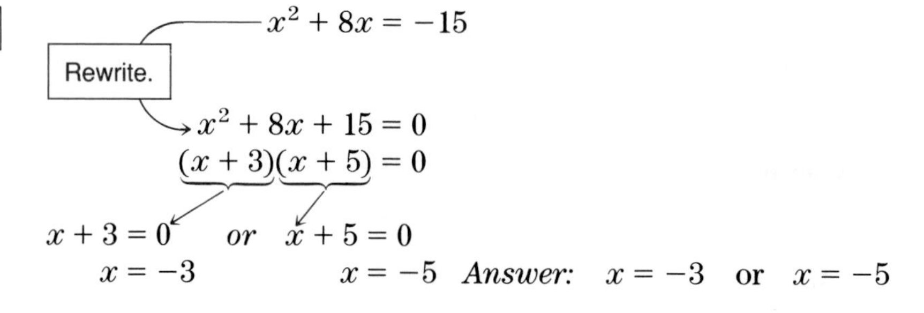

EXAMPLE 3

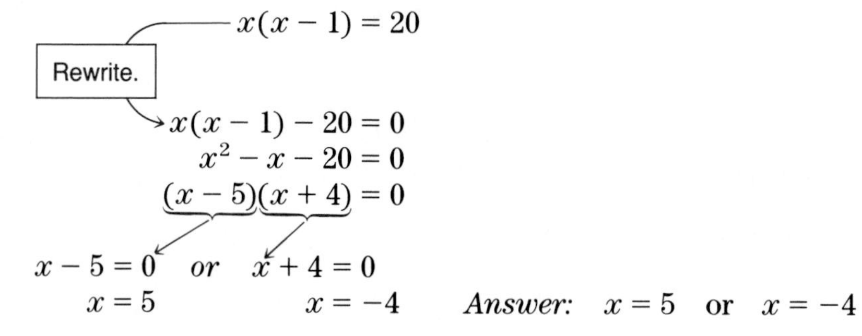

Classroom Practice

Solve.

1. $x^2 = 4x$

2. $x^2 - 4x = 5$

3. $x^2 + 8 = 6x$

4. $6 + x = x^2$

5. $x(x - 1) = 12$

6. $x(x + 2) = 15$

7. $8 = (x + 1)(x - 1)$

8. $\frac{x}{3} = \frac{x + 6}{x}$

9. $\frac{x}{5} = \frac{2x - 5}{x}$

Written Exercises

Solve.

A

1. $x^2 = 3x$
2. $x^2 = -7x$
3. $x^2 = -4x$
4. $x^2 + 5x = 6x$
5. $y^2 + 4y = 32$
6. $x^2 + 5x = 6$
7. $y^2 - 2y = 8$
8. $6 - x = x^2$
9. $6y - 9 = y^2$
10. $x^2 + 22 = 13x$
11. $x^2 + 20 = -12x$
12. $x^2 - 6 = 5x$
13. $y^2 + y = 42$
14. $y^2 = 9y - 20$
15. $x^2 = 7x + 18$
16. $a^2 - 3a = 18$
17. $a^2 = 12a - 35$
18. $x^2 = 5x + 24$
19. $x(x + 4) = 12$
20. $y(y + 3) = 40$
21. $(y - 1)(y + 1) = 48$
22. $(x + 2)(x - 3) = 6$
23. $(n - 2)(n + 5) = 18$
24. $(a - 1)(a + 1) = 15$

B

25. $\frac{x}{2} = \frac{6}{x + 1}$
26. $\frac{x}{4} = \frac{7}{x + 3}$
27. $\frac{x - 1}{3} = \frac{4}{x}$
28. $\frac{x - 2}{x} = \frac{x}{8}$
29. $\frac{y - 3}{5} = \frac{8}{y + 3}$
30. $\frac{4}{x + 3} = \frac{x - 3}{4}$

C

31. $2x^2 + x = 3$
32. $2x^2 - x = 1$
33. $x + 10 = 2x^2$
34. $(2x + 1)(2x - 1) = 0$
35. $(2n + 1)(n + 1) = 28$
36. $\frac{1 + n}{3} = 2n^2$

SELF-TEST

Vocabulary

quadratic equation (p. 422)

Solve.

1. $x(x - 3) = 0$
2. $2n(n + 7) = 0$ *(12-1)*
3. $(y - 6)(y - 3) = 0$
4. $(x + 4)(x - 3) = 0$
5. $x^2 - 3x = 0$
6. $2y^2 - 10y = 0$ *(12-2)*
7. $x^2 + 8x + 15 = 0$
8. $x^2 + 4x - 12 = 0$
9. $x^2 = -8x$
10. $a^2 - 12 = a$ *(12-3)*
11. $y^2 + y = 30$
12. $x^2 = 11x - 28$

4 Using Square Roots

A quadratic equation like $x^2 = 25$ can be solved by factoring or by taking square roots. In Chapter 11 you studied how to solve equations by taking square roots. Let's review these methods here.

SOLVING BY FACTORING

$$x^2 = 25$$
$$x^2 - 25 = 0$$
$$(x - 5)(x + 5) = 0$$
$$x - 5 = 0 \quad or \quad x + 5 = 0$$
$$x = 5 \qquad\qquad x = -5$$

Answer: $x = 5$ or $x = -5$

SOLVING BY TAKING SQUARE ROOTS

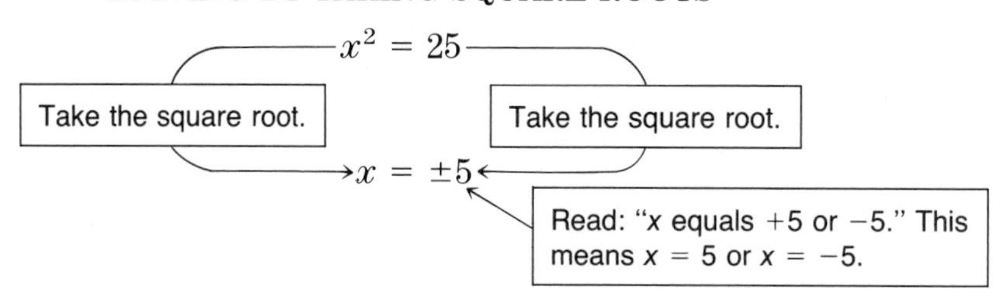

Answer: $x = 5$ or $x = -5$

You'll probably agree that the second method is a little easier. Study these examples of that method.

EXAMPLE 1

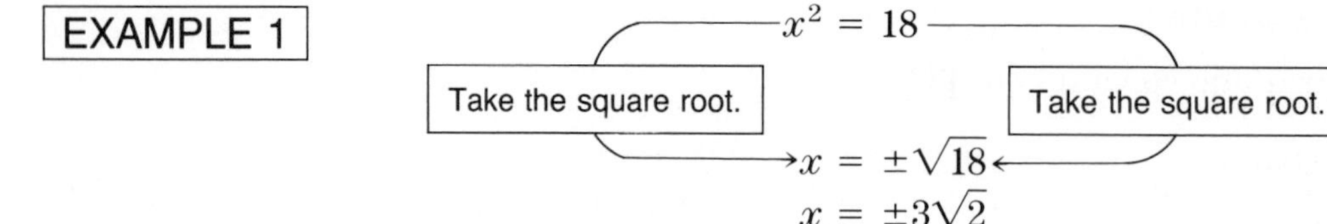

Answer: $x = 3\sqrt{2}$ or $x = -3\sqrt{2}$

EXAMPLE 2

$$2x^2 = 18$$
$$x^2 = 9$$

Take the square root. Take the square root.

$$x = \pm 3$$

Answer: $x = 3$ or $x = -3$

Some quadratic equations have no real-number solutions.

EXAMPLE 3

$$x^2 + 9 = 0$$
$$x^2 = -9$$

It's impossible to find a real-number square root for -9. The square of a real number is never negative.

Answer: no solution

Written Exercises

Solve. If there is no solution, write *no solution*.

A

1. $x^2 = 4$
2. $x^2 = 9$
3. $x^2 = 16$
4. $y^2 = 49$
5. $x^2 = 64$
6. $a^2 = 81$
7. $y^2 = 121$
8. $x^2 = 100$
9. $x^2 = 400$
10. $x^2 = 3$
11. $y^2 = 5$
12. $n^2 = 7$
13. $a^2 = 8$
14. $y^2 = 32$
15. $n^2 = 1$
16. $n^2 = 0$
17. $a^2 = 12$
18. $2x^2 = 18$
19. $3x^2 = 12$
20. $2x^2 = 50$
21. $3y^2 = -6$
22. $2n^2 = -200$
23. $x^2 - 1 = 3$
24. $x^2 - 5 = 4$
25. $x^2 + 1 = 50$
26. $y^2 + 2 = 30$
27. $y^2 - 4 = 20$
28. $n^2 - 2 = 5$
29. $a^2 + 7 = 4$
30. $x^2 + 3 = 2$
31. $n^2 - 5 = 10$
32. $y^2 + 3 = 21$
33. $m^2 - 11 = 49$
34. $x^2 + 8 = 7$
35. $b^2 - 15 = 12$
36. $z^2 - 12 = -4$

B

37. $2x^2 - 1 = 17$
38. $3y^2 + 5 = 50$
39. $2n^2 - 7 = 13$
40. $3n^2 + 9 = 42$
41. $x^2 = \frac{1}{9}$
42. $x^2 = \frac{4}{9}$
43. $\frac{x}{4} = \frac{9}{x}$
44. $\frac{y}{4} = \frac{16}{y}$

Complete the chart in Exercise 45 by using Galileo's law.

Galileo (1564–1642) discovered the following law for falling bodies.

distance it falls $= 5 \times (\text{time})^2$

$$d = 5t^2$$

45.

time in seconds	1	2	4	?	?
distance in meters	?	?	?	125	500

5 Using Square Roots

Equations like $(x + 1)^2 = 25$ can be solved in the same way you solved equations in the last section. All you do is take the square root of both sides of the equation.

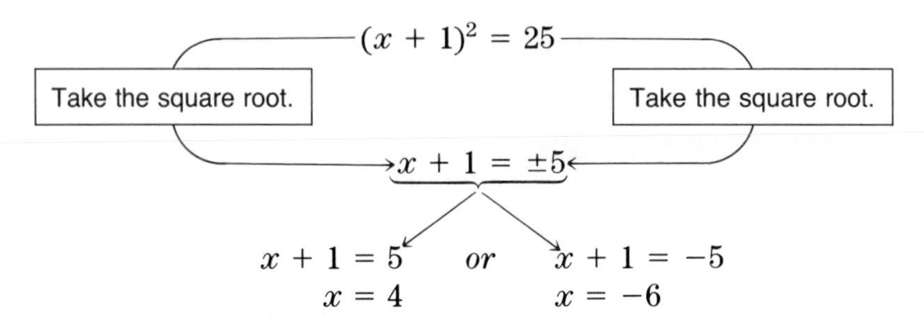

Answer: $x = 4$ or $x = -6$

EXAMPLE 2

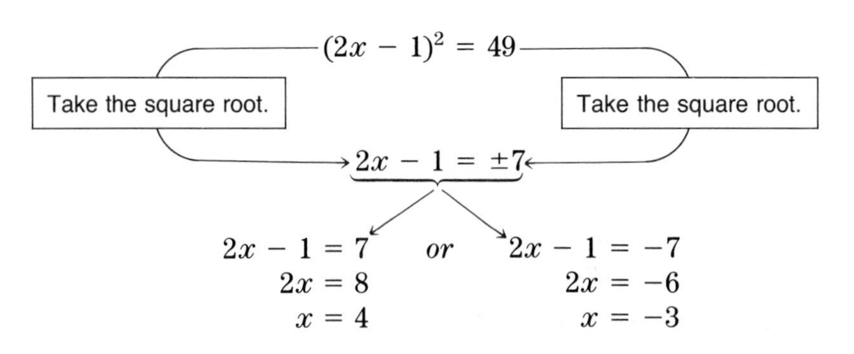

Answer: $x = 4$ or $x = -3$

Classroom Practice

Complete.

1. $(x + 1)^2 = 9$
$x + 1 = \pm$ __?__
$x + 1 = 3$ *or* $x + 1 = -3$
$x =$ __?__ $x =$ __?__

2. $(x - 3)^2 = 16$
__?__ $= \pm 4$
$x - 3 = 4$ *or* $x - 3 = -4$
$x =$ __?__ $x =$ __?__

Solve.

3. $(x - 1)^2 = 4$

4. $(2x + 1)^2 = 25$

5. $(x - 4)^2 = 36$

Written Exercises

Solve.

A

1. $(x - 3)^2 = 36$
2. $(x + 5)^2 = 49$
3. $(x - 4)^2 = 25$
4. $(x + 1)^2 = 16$
5. $(y + 8)^2 = 100$
6. $(y - 2)^2 = 81$
7. $(y - 3)^2 = 64$
8. $(n + 2)^2 = 9$
9. $(a - 7)^2 = 121$
10. $(2x - 1)^2 = 25$
11. $(4n - 2)^2 = 36$
12. $(2y - 2)^2 = 64$
13. $(2x - 3)^2 = 49$
14. $(2x - 1)^2 = 81$
15. $(2x + 3)^2 = 25$
16. $(5p + 5)^2 = 100$
17. $(2p + 3)^2 = 81$
18. $(3x - 6)^2 = 9$
19. $(4a - 8)^2 = 0$
20. $(3x - 3)^2 = 1$
21. $(6n - 6)^2 = 36$
22. $(2m + 5)^2 = 49$

B

23. $(x - 2)^2 = 7$
24. $(x - 4)^2 = 3$
25. $(y + 1)^2 = 2$
26. $(y + 8)^2 = 6$
27. $(a - 5)^2 = 8$
28. $3(x - 1)^2 = 12$
29. $2(x - 7)^2 = 18$
30. $6(a + 1)^2 = 24$
31. $5(n + 2)^2 = 45$
32. $3(a - 1)^2 = 75$
33. $7(x + 3)^2 = 28$
34. $3(y - 6)^2 = 27$
35. $3(2x + 2)^2 = 48$
36. $2(2x - 1)^2 = 50$
37. $2(2y + 4)^2 = 72$
38. $4(a - 7)^2 = 20$

SELF-TEST

Solve.

1. $x^2 = 49$
2. $y^2 = 64$
3. $n^2 = 17$ *(12-4)*
4. $2x^2 = 14$
5. $5x^2 = 30$
6. $m^2 + 5 = 41$
7. $(x - 7)^2 = 49$
8. $(x + 1)^2 = 36$
9. $(x - 3)^2 = 64$ *(12-5)*
10. $(x + 7)^2 = 25$
11. $(2x + 6)^2 = 16$
12. $(2x - 3)^2 = 81$

6 The Quadratic Formula

Not all quadratic equations can be solved by factoring or by taking the square roots. The formula below will help you find the solutions to all quadratic equations.

THE QUADRATIC FORMULA
The solution of an equation $ax^2 + bx + c = 0$ *is either* $$x = \frac{-b + \sqrt{b^2 - 4ac}}{2a} \quad or \quad x = \frac{-b - \sqrt{b^2 - 4ac}}{2a}.$$ *Sometimes these are written in combined form:* $$x = \frac{-b \pm \sqrt{b^2 - 4ac}}{2a}.$$

EXAMPLE 1 Solve $2x^2 + 3x - 9 = 0$.

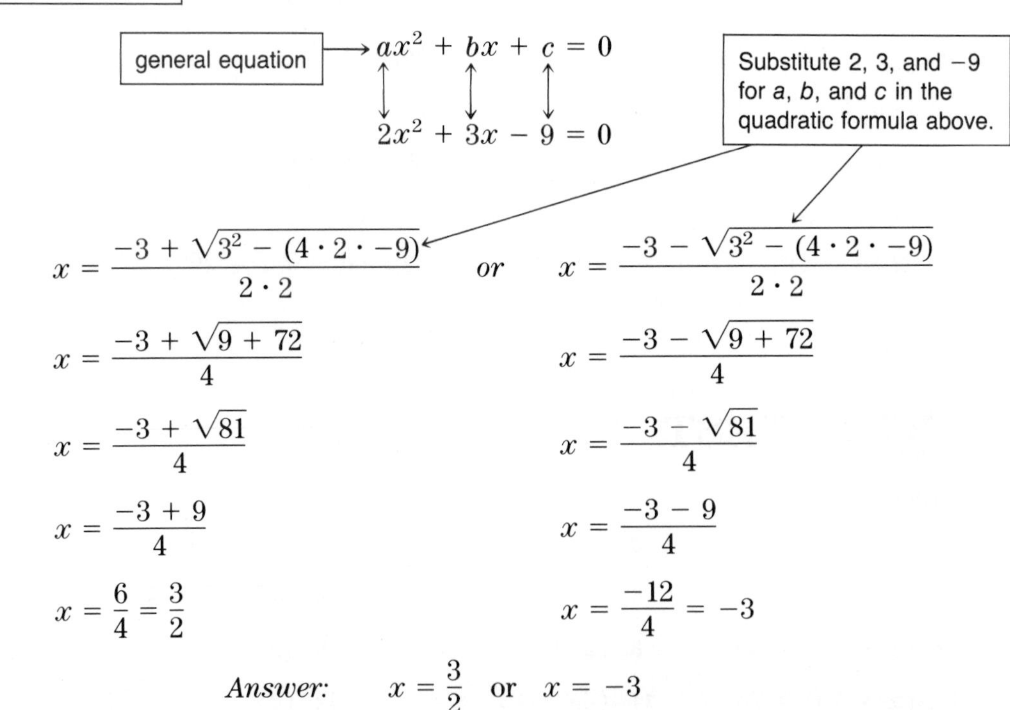

EXAMPLE 2 Solve $x^2 - 6x + 2 = 0$.

$$ax^2 + bx + c = 0$$
$$\updownarrow \quad \updownarrow \quad \updownarrow$$
$$1x^2 - 6x + 2 = 0$$

$a = 1$, $b = -6$, $c = 2$

Once you are comfortable with the ± symbol, use the combined form.

$$x = \frac{-b \pm \sqrt{b^2 - 4ac}}{2a}$$

$$x = \frac{-(-6) \pm \sqrt{(-6)^2 - (4 \cdot 1 \cdot 2)}}{2 \cdot 1}$$

$$x = \frac{6 \pm \sqrt{36 - 8}}{2}$$

$$x = \frac{6 \pm \sqrt{28}}{2}$$

$$x = \frac{6 \pm 2\sqrt{7}}{2}$$

$$x = \frac{2(3 \pm \sqrt{7})}{2}$$

$$x = 3 \pm \sqrt{7}$$

$$x = 3 + \sqrt{7} \quad \textit{or} \quad x = 3 - \sqrt{7}$$

Answer: $x = 3 + \sqrt{7}$ or $x = 3 - \sqrt{7}$

Classroom Practice

Tell what the following expressions mean.

1. $2 \pm \sqrt{5}$ **2.** $1 \pm 3\sqrt{2}$ **3.** $4 \pm 5\sqrt{6}$ **4.** $2 \pm 3\sqrt{2}$

State the values of *a*, *b*, and *c* for each equation.

5. $2x^2 + 3x + 1 = 0$ **6.** $3x^2 - 7x + 2 = 0$ **7.** $x^2 - 3x - 4 = 0$

8. $x^2 - x - 3 = 0$ **9.** $3x^2 - 15 = 0$ **10.** $2y^2 - 3y - 7 = 0$

Complete.

11. If $3x^2 - 5x = 2$,
then $3x^2 - 5x - \underline{\ ?\ } = 0$.
$a = 3$, $b = \underline{\ ?\ }$, and $c = \underline{\ ?\ }$.

12. If $x^2 + 2x = 7$,
then $\underline{\ ?\ } = 0$.
$a = \underline{\ ?\ }$, $b = \underline{\ ?\ }$, and $c = \underline{\ ?\ }$.

13. If $2x^2 = 1 - 3x$, then $\underline{\ ?\ } = 0$.
$a = \underline{\ ?\ }$, $b = \underline{\ ?\ }$, and $c = \underline{\ ?\ }$.

14. If $2x^2 - x = 15$, then $\underline{\ ?\ } = 0$.
$a = \underline{\ ?\ }$, $b = \underline{\ ?\ }$, and $c = \underline{\ ?\ }$.

Written Exercises

Solve by using the quadratic formula. Leave irrational numbers in simplest radical form.

A

1. $2x^2 + 3x + 1 = 0$
2. $4x^2 + 5x + 1 = 0$
3. $3x^2 - 7x + 2 = 0$
4. $2x^2 - 5x - 3 = 0$
5. $4x^2 + 7x - 2 = 0$
6. $x^2 - 5x - 14 = 0$
7. $2x^2 + 7x - 4 = 0$
8. $x^2 + 4x - 21 = 0$
9. $3y^2 - 7y + 2 = 0$
10. $5x^2 + 3x - 2 = 0$
11. $3x^2 - x - 2 = 0$
12. $2x^2 - 2x - 3 = 0$
13. $4x^2 - 4 = 0$
14. $x^2 - 9 = 0$
15. $4x^2 - 1 = 0$
16. $y^2 - y - 1 = 0$
17. $2x^2 + x - 2 = 0$
18. $x^2 - 3x + 1 = 0$
19. $5y^2 + 7y + 2 = 0$
20. $2y^2 + 4y + 1 = 0$
21. $x^2 - 2x - 5 = 0$
22. $2x^2 - 3x = 5$
23. $2x^2 + x = 7$
24. $3x^2 = x + 1$
25. $2x^2 = 5 + x$
26. $3x^2 = 8 - x$
27. $x^2 + 3 = 6x$
28. $2x^2 = 5x - 3$
29. $2x^2 = 7x - 2$
30. $4y^2 = 6y + 2$

Solve by two methods.

a. Factoring
b. Using the quadratic formula

B

31. $x^2 - 5x = 0$
32. $x^2 + 8x + 7 = 0$
33. $x^2 - 4x - 5 = 0$
34. $y^2 + y = 0$
35. $y^2 - 7y + 10 = 0$
36. $s^2 - 6s + 9 = 0$
37. $x^2 - 6x + 8 = 0$
38. $x^2 - 6x - 7 = 0$
39. $x^2 - x - 12 = 0$
40. $x^2 - 4x - 12 = 0$
41. $m^2 + 3m - 10 = 0$
42. $y^2 + y - 20 = 0$
43. $x^2 + 5x = 6$
44. $z^2 - 2z = 15$
45. $x^2 - 8 = 2x$
46. $n^2 - 12 = 11n$
47. $b^2 = -b + 30$
48. $a^2 = 10a - 24$

Mixed Practice Exercises

Solve by factoring.

1. $x^2 - 2x = 0$
2. $y^2 + 3y = 0$
3. $x^2 - 4x = 0$
4. $2x^2 - 4x = 0$
5. $5x^2 - 10x = 0$
6. $2x^2 + x = 0$
7. $x^2 + 3x + 2 = 0$
8. $x^2 - 2x - 3 = 0$
9. $x^2 + 2x - 8 = 0$
10. $x^2 + 10x + 21 = 0$
11. $x^2 - x = 30$
12. $x^2 + 2x = 35$

Solve by taking square roots.

13. $x^2 = 9$
14. $x^2 = 36$
15. $y^2 = 35$
16. $x^2 = 42$
17. $y^2 = 19$
18. $x^2 - 40 = 0$
19. $x^2 - 18 = 0$
20. $(x + 2)^2 = 25$
21. $(x - 1)^2 = 16$
22. $(x + 4)^2 = 49$
23. $(y - 3)^2 = 17$
24. $(y + 2)^2 = 28$

Solve by using the quadratic formula.

25. $x^2 - 2x - 3 = 0$
26. $x^2 + 4x - 5 = 0$
27. $x^2 + 2x - 8 = 0$
28. $2x^2 - 3x + 1 = 0$
29. $3x^2 + 4x + 1 = 0$
30. $x^2 - 5x + 3 = 0$
31. $3x^2 - 2x - 5 = 0$
32. $2x^2 + 5x - 3 = 0$
33. $x^2 - 3x + 2 = 0$
34. $x^2 + 4x - 3 = 0$
35. $2x^2 - x = 4$
36. $2x^2 - 2x = 3$

CONSUMER APPLICATIONS

TIPPING

People in a restaurant who wait on tables usually get tips. When you are the customer, do you know how to figure the tip? The amount you leave depends on you, but people usually leave about a 15% or 20% tip. All the figuring is done in your head.

To figure 15% or 20% of your restaurant check, you may find it easier to first find 10%. All you have to do is round the amount on the check, then move the decimal point one place to the left. Once you know how to figure 10%, then you can find 20% by doubling your result.

Now consider the 15% tip. You can figure the 10% tip, then add to that half of the 10% tip. Here's how it works.

Your check	This is about	about 10%	about $\frac{1}{2}$ of 10%	about 15%
\$9.12	\$9.10	\$.91 *or* \$.90	\$.45	\$1.35 (\$.90 + \$.45 = \$1.35)

Remember, none of your figuring needs to be exact.

7 Problem Solving

In this section you'll use quadratic equations to solve word problems. Your method should be the same as you used earlier in this course.

A GUIDE FOR PROBLEM SOLVING
1. *Read the problem. Read it more than once if you wish.* 2. *Let a variable stand for one of the unknowns. Show the other unknowns in terms of that variable.* 3. *Write an equation and solve it.* 4. *Answer the question.* 5. *Check your work.*

EXAMPLE The length of a rectangle is 2 cm more than the width. The area is 80 cm^2. Find the length and width.

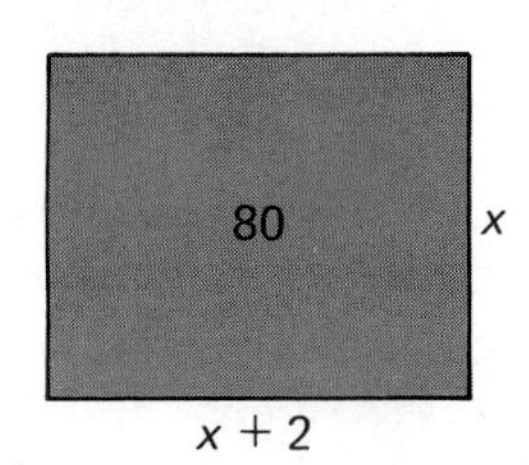

Let x = width
Then $x + 2$ = length

$$\underbrace{\text{width}} \times \underbrace{\text{length}} = \underbrace{\text{Area}}$$

$$x \times (x+2) = 80$$
$$x(x+2) = 80$$
$$x^2 + 2x = 80$$
$$x^2 + 2x - 80 = 0$$
$$(x+10)(x-8) = 0$$

$x + 10 = 0$ *or* $x - 8 = 0$
$x = -10$ $\quad$ $x = 8$

Answer: The width must be a positive number, so the value $x = -10$ is *not* a solution to the problem.

width: $x = 8 \longrightarrow 8$ cm
length: $x + 2 = 10 \longrightarrow 10$ cm

Check: length × width = Area
$10 \times 8 = 80$ ✓

Written Exercises

A

1. The length of a rectangle is 3 cm more than the width. The area is 40 cm^2. Find the length and width.

2. The length of a rectangle is twice its width. The area is 72 cm^2. Find the length and width.

3. One number is 2 more than another. The product of the numbers is 35. Find the numbers.

4. The square of a number is 30 more than the number. The number is positive. Find the number.

5. Shaded area $= 40$
Find x.

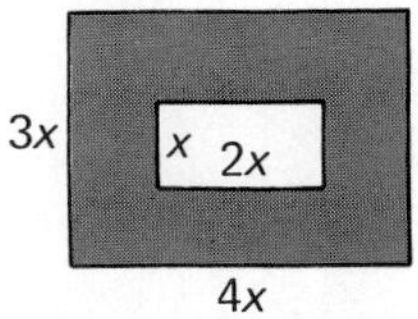

6. Shaded area $= 400$
Find x.

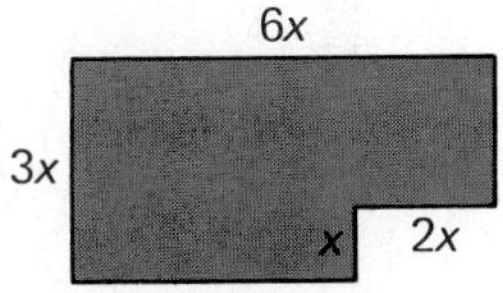

7. Volume $= 60$
Find x.

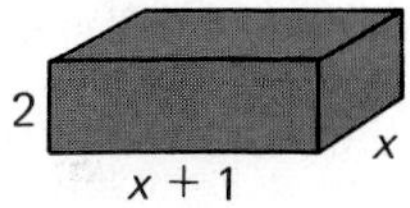

Use the Law of Pythagoras to find *x*.

8.

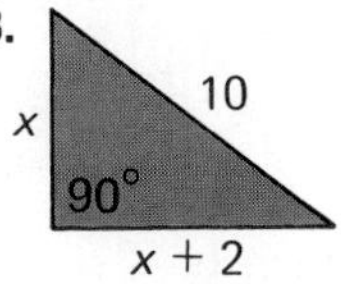

9.

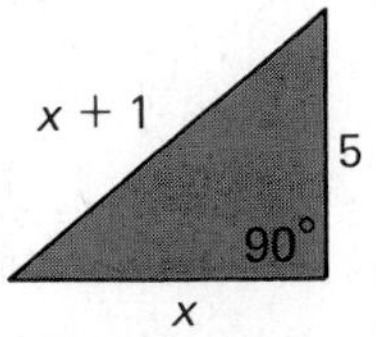

10.

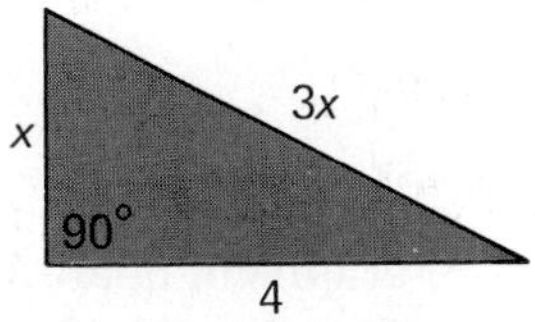

B

11. The perimeter of a rectangle is 32 cm. The area is 48 cm^2. Find the dimensions.

12. The sum of two numbers is 13. The product is 42. Find the numbers.

C

13. Vince bikes 3 km/h faster than Al. It takes Al 1 hour more than Vince to travel 36 km. Find how fast each bikes.

	$D \div r =$		t
Al	36	x	$\frac{36}{x}$
Vince	?	?	$\frac{36}{x+3}$

14. Andrea bikes 1 mph faster than Liane. It takes Liane 1 hour more than Andrea to travel 30 mi. Find the time each girl travels.

	$D \div t =$		r
Andrea	30	x	$\frac{30}{x}$
Liane	30	?	$\frac{30}{x+1}$

8 Quadratic Graphs

The equation $y = x^2 - 1$ shows that y is a **quadratic function** of x, or that y *depends* upon x. This equation has two variables, so its solutions are pairs of numbers.

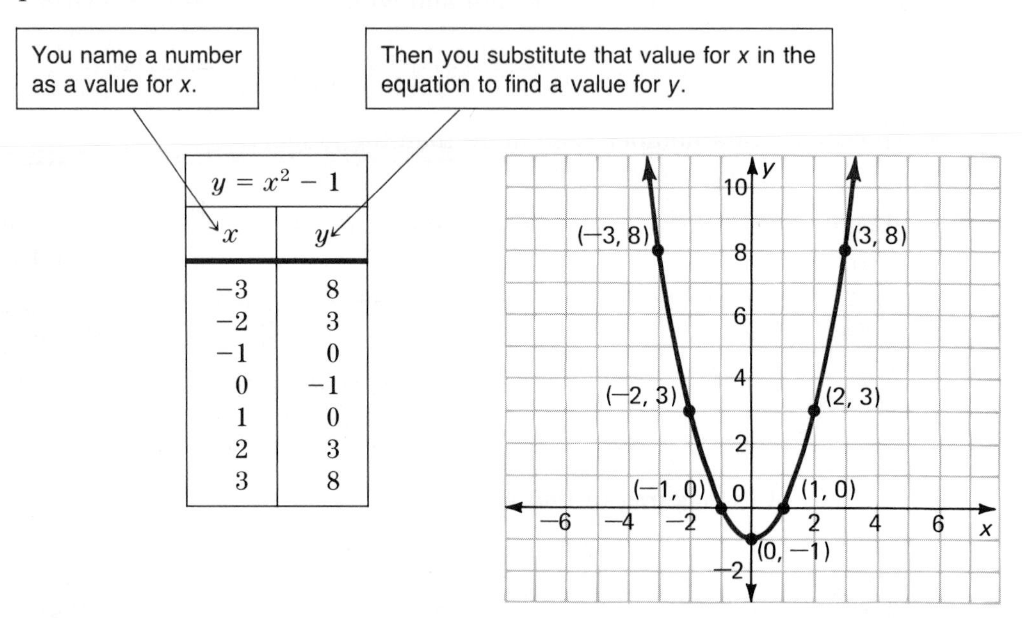

$y = x^2 - 1$	
x	y
−3	8
−2	3
−1	0
0	−1
1	0
2	3
3	8

The graphs of quadratic functions all curve in a similar way. The graph is called a **parabola.** The "tip" of a parabola is called its **vertex.** Study a few more examples of these special graphs.

EXAMPLE 1

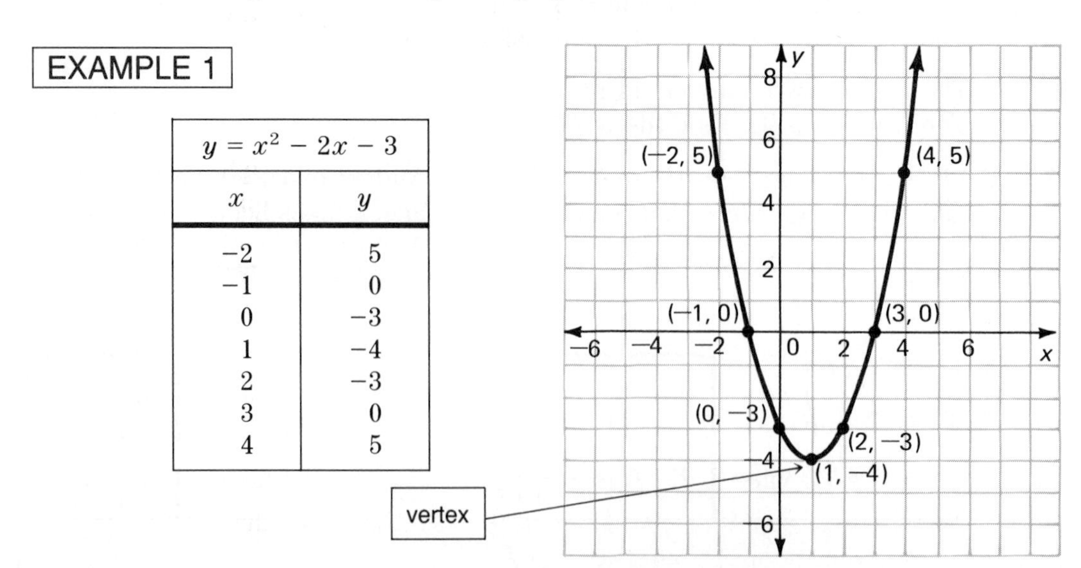

$y = x^2 - 2x - 3$	
x	y
−2	5
−1	0
0	−3
1	−4
2	−3
3	0
4	5

Some graphs of quadratic functions curve upside down.

EXAMPLE 2

$y = -x^2$	
x	y
-3	-9
-2	-4
-1	-1
0	0
1	-1
2	-4
3	-9

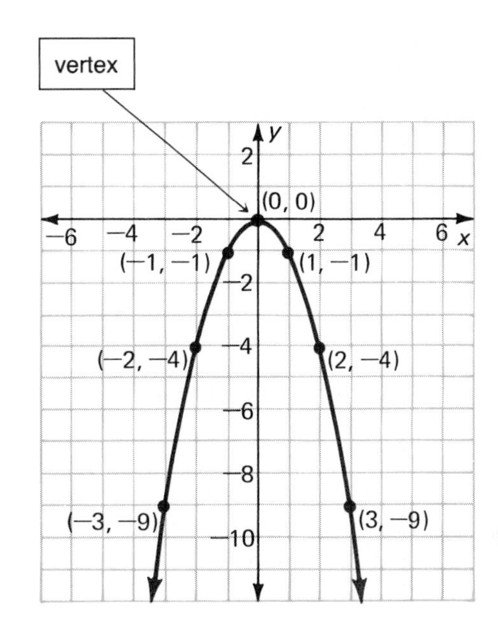

Classroom Practice

Study the graph of $y = x^2 - 1$ shown on the opposite page.

1. What are the coordinates of the vertex of this parabola?
2. How large a value can be substituted for x? How small a value?
3. How many values of x make $y = 3$?
4. How many values of x make $y = -3$?
5. Suppose you are told that the point (2.3, 4.29) is on the graph. Then you also know that the point (−2.3, __?__) is on the graph.

Study the graph of $y = x^2 - 2x - 3$ shown on the opposite page.

6. What is the x-coordinate of the vertex of this parabola?
7. What is the y-coordinate of the vertex?
8. How many values of x make $y = -4$?
9. How many values of x make $y = -3$?
10. Suppose you are told that the point (−1.5, 2.25) is on the graph. Is the point (1.5, 2.25) on the graph?

Written Exercises

For each equation do three things.

a. Make a table of values.

b. Draw the graph.

c. Give the coordinates of the vertex.

A

1. $y = x^2$

2. $y = x^2 + 1$

3. $y = x^2 - 4$

4. $y = 2x^2$

5. $y = -2x^2$

6. $y = \frac{1}{2}x^2$

7. $y = x^2 + 2x$

8. $y = x^2 - 2x$

9. $y = 2x - x^2$

10. $y = x^2 - 2x - 1$

11. $y = x^2 - 4x + 3$

12. $y = x^2 + 4x - 2$

13. $y = x^2 - 2x - 2$

14. $y = x^2 - 6x + 8$

15. $y = x^2 - 6x + 9$

16. $y = x^2 + 6x - 6$

17. $y = (3 - x)(3 + x)$

18. $y = (2 - x)(2 - x)$

19. On the same set of axes, graph both equations:

$$y = x^2 + 2 \quad \text{and} \quad y = -(x^2 + 2).$$

The graph of $y = x^2 - 4x + 2$ is shown at the right.

20. What is the vertex?

21. For what values of x does $y = -1$?

22. How many values of x make $y = 2$?

23. How many values of x make $y = 7$?

24. How many values of x make $y = -2$?

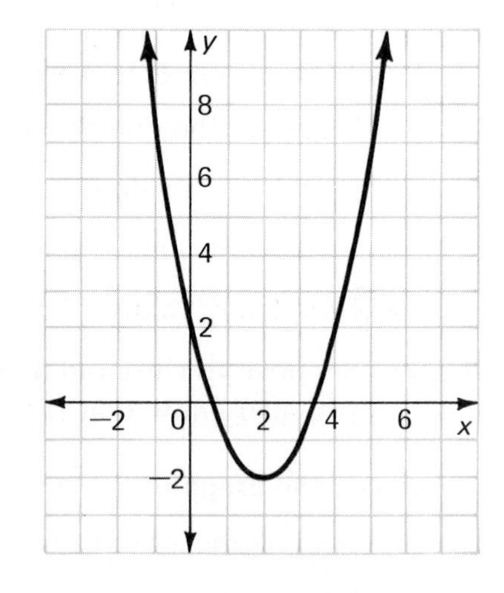

B

25. Use the graph of $y = x^2 - 4x + 2$ shown at the right to draw the graph of $y = -(x^2 - 4x + 2)$.

26. Suppose you are told that (4, 1) is the vertex of the parabola $y = x^2 - 8x + 17$. What are the coordinates of the vertex of the parabola $y = -(x^2 - 8x + 17)$?

The graphs of three equations are shown at the right.

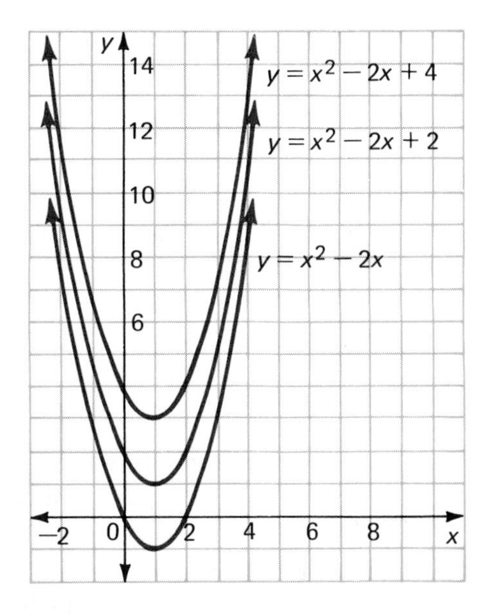

27. Give the vertex of each parabola.

28. What is the vertex of the parabola $y = x^2 - 2x + 6$?

The expression $b^2 - 4ac$ (from the quadratic formula) can tell you whether the graph of a quadratic function crosses the x-axis once, twice, or not at all.

If $b^2 - 4ac > 0 \longleftrightarrow$ graph crosses x-axis 2 times
If $b^2 - 4ac = 0 \longleftrightarrow$ graph crosses x-axis 1 time
If $b^2 - 4ac < 0 \longleftrightarrow$ graph crosses x-axis 0 times

How many times does the graph of the equation cross the x-axis?

C **29.** $y = x^2 - 5x + 2$ **30.** $y = x^2 - 4x + 6$ **31.** $y = x^2 - 10x + 25$

SELF-TEST

Vocabulary

quadratic formula (p. 430)
quadratic function (p. 436)
parabola (p. 436)
vertex of a parabola (p. 436)

Solve by using the quadratic formula. Leave irrational solutions in simplest radical form.

1. $x^2 + 7x - 10 = 0$ **2.** $3x^2 - 2x - 6 = 0$ **3.** $x^2 = 5x - 4$ *(12-6)*

4. The length of a rectangle is 4 cm more than the width. The area is 45 cm^2. Find the length and width. *(12-7)*

5. Draw the graph of $y = x^2 - 2x - 2$. What is the vertex? *(12-8)*

PROBLEM SOLVING STRATEGIES

PROBLEMS WITH NO SOLUTION

Suppose you see the following problem on a college entrance examination. Which answer would you choose?

> Lucille has 9 quarters and dimes worth $1.75.
> How many quarters does she have?
>
> **A.** 5 **B.** 6 **C.** $5\frac{2}{3}$ **D.** No solution
>
> Let q = the number of quarters.
> Then $9 - q$ = the number of dimes.
>
> $10(9 - q) + 25q = 175$

The solution of this equation is $q = 5\frac{2}{3}$, but is $5\frac{2}{3}$ the answer to the problem above? No. Lucille cannot have $5\frac{2}{3}$ quarters. Therefore, not all the pieces of information given in the problem can be true. The correct answer choice is **D.**

Consider another problem that has no solution.

> The sum of two consecutive integers is 52.
> Find the integers.

If the two integers are x and $x + 1$, then the equation would be $x + (x + 1) = 52$. Solving this equation, you should find that $x = 25\frac{1}{2}$, which is not an integer. Therefore, the problem has no solution.

Some problems, like the ones above, have no solution because the given facts contradict each other. Other problems, like the following example, have no solution because not enough information is given.

> Marcus spent one hour longer driving from Arness to Baxter than on his return trip to Arness.
> If he spent a total of 5 hours driving, how fast did he drive each way?

The only information given in the problem is the total time. To solve this problem you would also need information about either the distance between Arness and Baxter or a relationship between the speeds during each part of the trip. Not enough information was given, so the problem has no solution.

EXERCISES

Solve if possible. If the problem cannot be solved, either show that the facts cannot all be true or state which required piece of information is missing.

1. The sum of three consecutive odd integers is 18.
   Find the integers.

2. The Reillys invested equal amounts of money at 6% interest and at 8% interest.
   They earned $280 interest in one year from the investments.
   How much did they invest in all?

3. A solution contains 10% acid.
   How much water must be added to get a solution that is only 5% acid?

4. A jar contains nickels and quarters.
   The number of nickels is one less than five times the number of quarters.
   The nickels are worth 15¢ less than the quarters.
   How many quarters are in the jar?

5. Seth bought 10 pencils and pens.
   A pencil cost 20¢ less than a pen.
   How many of each did he buy?

6. Two fifths of Marcie's stamps are Canadian, one third of them are Mexican, and the remaining 16 are Spanish. How many stamps does she have altogether?

7. Jim can set up the lunch trays in 30 minutes.
   Leo needs 45 minutes to do the job.
   How long would the job take if the two worked together?

8. Elaine and Judy are 18 km apart.
   At the same time they begin to travel toward each other.
   Elaine travels 2 km/h faster than Judy.
   At what time do they meet?

9. Arthur is 2 years older than Rose.
   In three years, the sum of their ages will be 8.
   How old is each now?

10. Student tickets to the school concert cost $2.
    Adult tickets cost $4.
    Three times as many student tickets as adult tickets were sold.
    Ticket sales for the concert totaled $324.
    How many student tickets were sold?

COMPUTER ACTIVITIES

REPEATING DECIMALS

To express a fraction as a decimal, you divide the denominator into the numerator. The result can be a terminating decimal, as in $\frac{3}{4} = 0.75$. Here, all digits after the five are zeros. The result might also be a nonterminating repeating decimal, as in $\frac{23}{99} = 0.232323 \ldots$. Here, a group of digits repeats over and over.

The program below will divide any proper fraction (one with a numerator that is less than the denominator) to give a decimal for the fraction. You can ask for as many digits as you like in the decimal. In line 90 of the program, you will be asked to input the number of digits to be computed.

The part of the program from line 130 through line 190 is a loop. Since I is the final value given in line 130, the computer will work through this loop as many times as you told it to when you gave a value for I (the number of digits to compute).

```
10  REM**CONVERTING FRACTIONS
20  REM**TO DECIMALS
30  PRINT "NUMERATOR";
40  INPUT N
50  PRINT "DENOMINATOR";
60  INPUT D
70  PRINT "HOW MANY DIGITS"
80  PRINT "WOULD YOU LIKE";
90  INPUT I
100  PRINT
110  PRINT "FRACTION: "; N; "/"; D
120  PRINT "DECIMAL: 0.";
130  FOR C = 1 TO I
140  LET N1 = 10 * N
150  LET Q = N1 / D
160  LET Q1 = INT (Q)
170  PRINT Q1;
180  LET N = N1 - D * Q1
190  NEXT C
200  END
```

You may find that a group of repeating digits is very large. In the example $\frac{23}{99} = 0.2323\ldots$, the group of repeating digits contains only two digits, 2 and 3. Other decimals have groups of repeating digits containing 10 or 20 or even 100 digits. In general, the group of repeating digits will contain at most 1 less than the denominator of the fraction. For example, the decimal expression for $\frac{7}{11}$ will contain at most $11 - 1$, or 10, digits before it begins to repeat. The decimal for $\frac{17}{42}$ will contain at most 41 digits before it repeats, and so on.

If you run the program and decide you don't have enough digits, just run it again and ask for more digits.

EXERCISES

1. RUN the program. Write a decimal for the fraction. For repeating decimals, use a bar to show the repeating digits. For example, $0.232323\ldots = 0.\overline{23}$.

a. $\frac{5}{8}$ **b.** $\frac{5}{9}$ **c.** $\frac{9}{13}$ **d.** $\frac{4}{7}$

e. $\frac{13}{20}$ **f.** $\frac{16}{25}$ **g.** $\frac{9}{17}$ **h.** $\frac{11}{18}$

i. $\frac{16}{23}$ **j.** $\frac{48}{125}$ **k.** $\frac{7}{60}$ **l.** $\frac{14}{39}$

2. Find a fraction with at least 10 digits in the repeating part of its decimal.

3. Find a fraction with at least 20 digits in the repeating part of its decimal.

4. Which one of these fractions has a terminating decimal?

$$\frac{5}{6}, \frac{5}{9}, \frac{5}{12}, \frac{5}{16}, \frac{5}{17}$$

5. Which one of these fractions has a repeating decimal?

$$\frac{13}{25}, \frac{13}{32}, \frac{13}{40}, \frac{13}{48}, \frac{13}{50}$$

6. RUN the program to find a fraction between the given fractions.

a. $\frac{8}{15}; \frac{9}{16}$ **b.** $\frac{6}{13}; \frac{16}{31}$ **c.** $\frac{10}{11}; \frac{11}{12}$ **d.** $\frac{6}{7}; \frac{16}{17}$

SKILLS REVIEW

FORMULAS

Find the perimeter.

1.

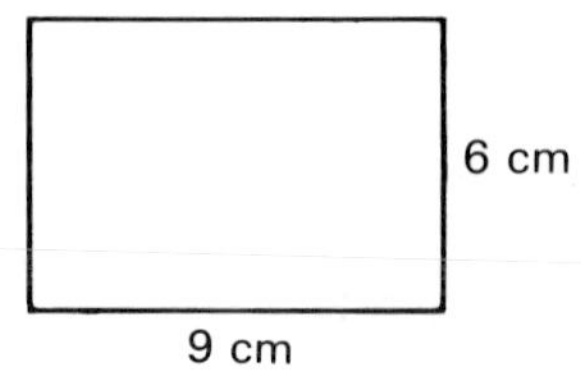

2.

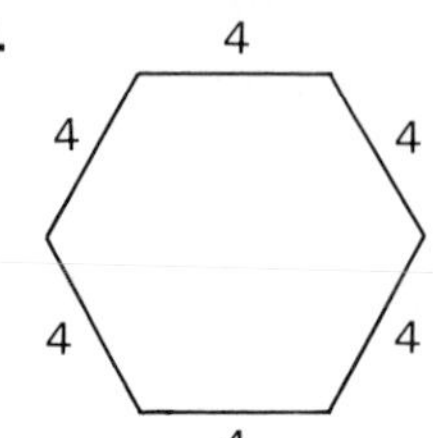

3.

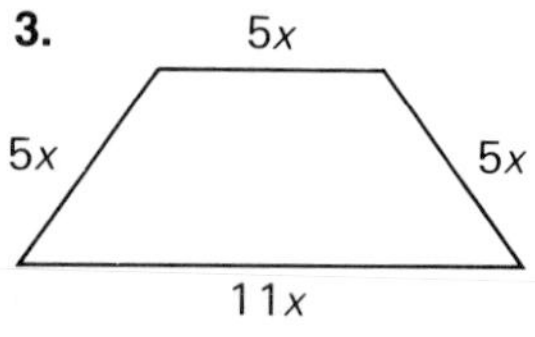

Find the area of the shaded region.

4.

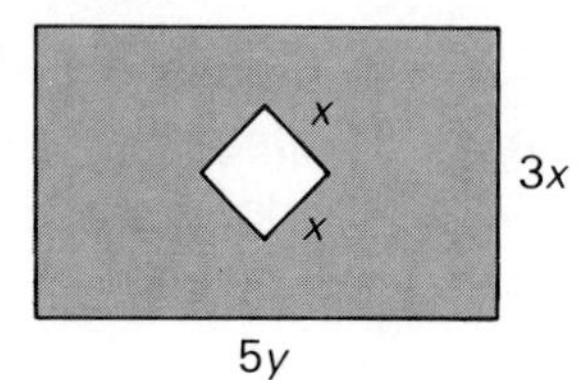

5.

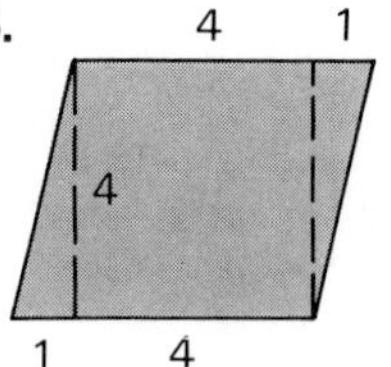

6.

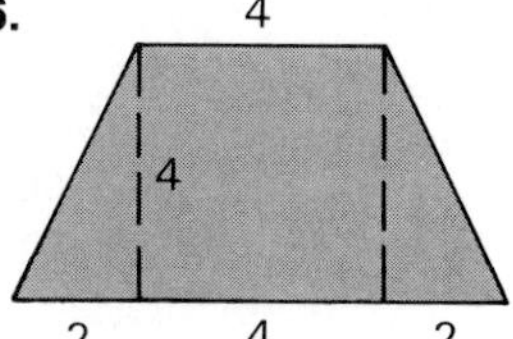

Find the area of the shaded region. Leave your answers in terms of π.

7.

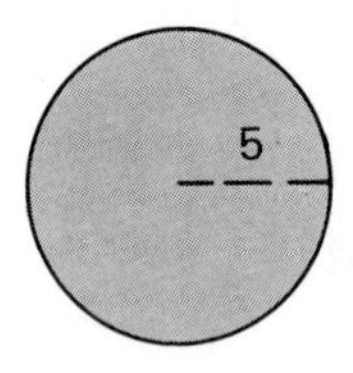

8.

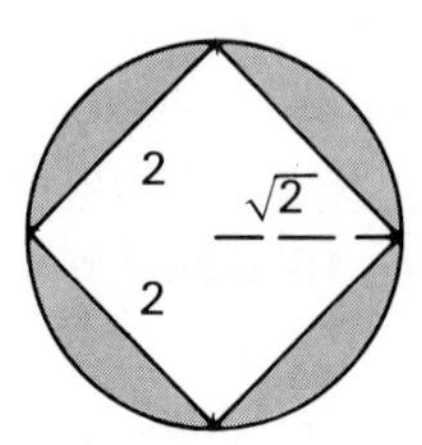

9.

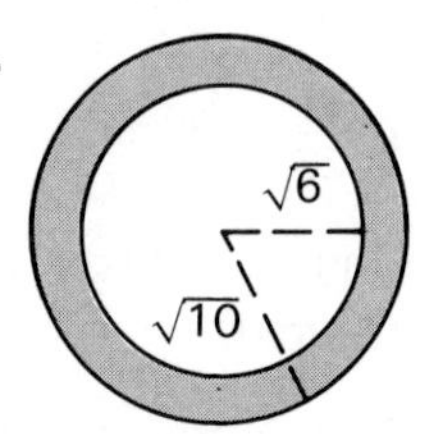

Find the volume.

10.

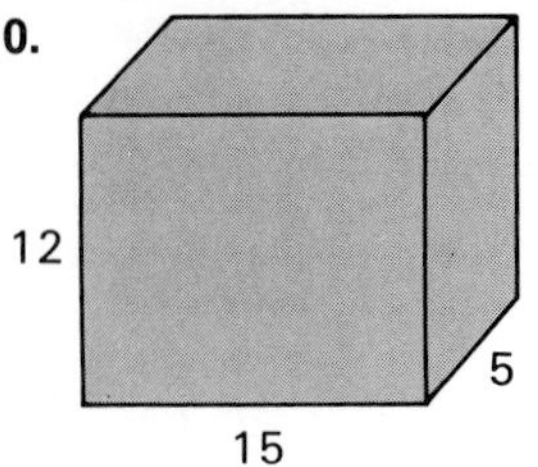

11.

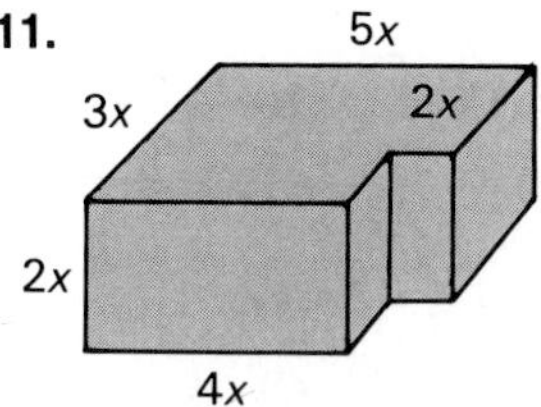

12.

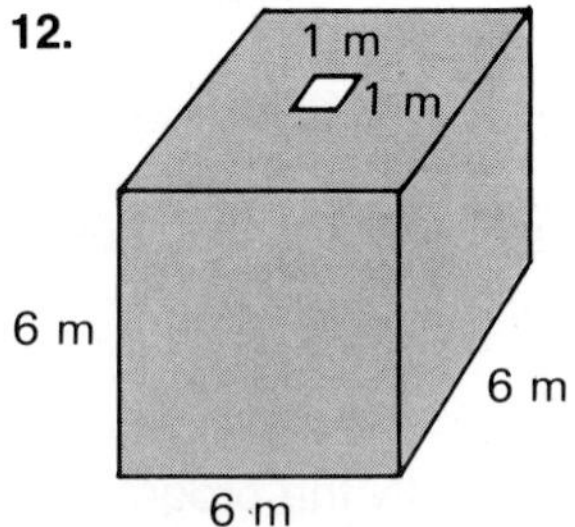

CHAPTER REVIEW

CHAPTER SUMMARY

1. If two numbers have a zero product, one of the numbers must be zero.

2. A quadratic equation in x is an equation which has an x^2 term but no term with a higher power of x.

3. Often the first step in solving a quadratic equation is rewriting the equation so that one side is zero.

4. The Quadratic Formula states:

   The solution of an equation $ax^2 + bx + c = 0$ is either

$$x = \frac{-b + \sqrt{b^2 - 4ac}}{2a} \quad \text{or} \quad x = \frac{-b - \sqrt{b^2 - 4ac}}{2a}.$$

5. An equation like $y = x^2 - 1$ tells us that y is a quadratic function of x. The solutions of the equation are ordered pairs of numbers.

6. The graph of a quadratic function is a curve called a parabola. The "tip" of a parabola is called its vertex.

REVIEW EXERCISES

Solve. *(See pp. 420–421.)*

1. $x(x - 2) = 0$
2. $y(y + 4) = 0$
3. $x(x - 7) = 0$
4. $2a(a - 9) = 0$
5. $4x(x + 6) = 0$
6. $5n(n + 8) = 0$
7. $(x + 1)(x + 5) = 0$
8. $(x - 3)(x + 6) = 0$
9. $(y + 5)(y - 3) = 0$

Solve. *(See pp. 422–423.)*

10. $x^2 - 3x = 0$
11. $n^2 + 4n = 0$
12. $5x + x^2 = 0$
13. $2y^2 - 4y = 0$
14. $x^2 - 8x + 15 = 0$
15. $x^2 - 6x + 8 = 0$
16. $x^2 - x - 20 = 0$
17. $x^2 + 3x - 18 = 0$
18. $x^2 + 2x - 35 = 0$

Solve. *(See pp. 424–425.)*

19. $x^2 = 2x$

20. $x^2 = -3x$

21. $y^2 = -y$

22. $n^2 = -4n$

23. $x^2 + 7x = -10$

24. $x^2 + 6x = -9$

25. $x^2 = -11x - 30$

26. $x^2 = 9x - 14$

27. $x^2 = x + 42$

Solve. *(See pp. 426–427.)*

28. $x^2 = 9$

29. $y^2 = 81$

30. $n^2 = 20$

31. $y^2 = 27$

32. $2x^2 = 50$

33. $3y^2 = 18$

34. $n^2 - 3 = 13$

35. $x^2 + 2 = 38$

36. $y^2 - 6 = 43$

Solve. *(See pp. 428–429.)*

37. $(x - 1)^2 = 16$

38. $(x + 3)^2 = 81$

39. $(x + 4)^2 = 64$

40. $(y - 2)^2 = 36$

41. $(y + 5)^2 = 49$

42. $(a - 7)^2 = 100$

43. $(4x + 4)^2 = 16$

44. $(3x - 9)^2 = 36$

45. $(5x - 15)^2 = 100$

Solve. *(See pp. 430–432.)*

46. $x^2 + 3x - 10 = 0$

47. $2x^2 + 5x - 3 = 0$

48. $3x^2 + 5x + 2 = 0$

49. $4x^2 - 2x - 2 = 0$

50. $2x^2 - x - 3 = 0$

51. $2x^2 + 5x + 3 = 0$

52. $5x^2 + 2x - 2 = 0$

53. $4x^2 + 7x + 2 = 0$

54. $3x^2 - 2x - 4 = 0$

Solve. *(See pp. 434–435.)*

55. The length of a rectangle is 5 cm more than the width. The area is 36 cm^2. Find the length and width.

56. One number is 6 more than another. The sum of their squares is 90. Find the numbers.

Do the following exercises. *(See pp. 436–439.)*

57. Make a table of values for the equation $y = x^2 + 2$. For what values of x does $y = 6$?

58. Make a table of values for the equation $y = x^2 + 3x - 5$. For what values of x does $y = -7$?

Draw the graph of the equation. *(See pp. 436–439.)*

59. $y = x^2 - 3$

60. $y = 2x^2 + 1$

61. $y = x^2 + 2x + 1$

CHAPTER TEST

Solve. If there is no solution, write *no solution*.

1. $3x(x + 2) = 0$
2. $(y - 3)(y + 9) = 0$ *(12-1)*
3. $(2a - 10)(7a - 42) = 0$
4. $r(r + 8)(r - 6) = 0$
5. $n^2 + 10n + 9 = 0$
6. $x^2 - 12x + 36 = 0$ *(12-2)*
7. $y^2 - y - 20 = 0$
8. $a^2 - 1 = 0$
9. $x^2 = 14x - 48$
10. $y^2 + 9y = 22$ *(12-3)*
11. $(n + 4)(n - 2) = -8$
12. $\frac{x + 1}{4} = \frac{3}{x - 3}$
13. $3y^2 = 27$
14. $x^2 = 56$
15. $n^2 - 8 = 2$ *(12-4)*
16. $n^2 + 8 = 2$
17. $5y^2 + 1 = 81$
18. $\frac{8x}{9} = \frac{1}{2x}$
19. $(2x + 5)^2 = 1$
20. $(y - 9)^2 = 4$
21. $(3x - 7)^2 = 0$ *(12-5)*
22. $(n - 3)^2 = 5$
23. $4(a + 5)^2 = 64$
24. $2(3y + 18)^2 = 72$

Solve by using the quadratic formula. Leave irrational numbers in simplest radical form.

25. $x^2 + 6x + 4 = 0$
26. $5x^2 + 3x - 4 = 0$ *(12-6)*
27. $3x^2 + 4x - 4 = 0$
28. $2y^2 - 4y + 1 = 0$
29. Solve **(a)** by factoring and **(b)** by using the quadratic formula: $s^2 + 2s + 1 = 0$.
30. A box is $(x + 3)$ in. long, $(x + 1)$ in. wide, and 4 in. high. If the volume is 192 in.^{3}, find x. *(12-7)*
31. One number is 4 less than another. The sum of their squares is 40. Find the numbers.
32. Make a table of values for the equation $y = x^2 + 3x - 6$. For what values of x does $y = -2$? *(12-8)*

Draw the graph of the equation. Give the coordinates of the vertex.

33. $y = x^2 - 4x + 1$
34. $y = 6x - x^2$
35. Refer to your graph in Exercise 34.
    Complete: Only one value of x makes $y =$ __?__.

UNIT F **CUMULATIVE REVIEW**

Find the value. Use the table on page 390 for Exercises 6–10.

1. $\sqrt{400}$
2. $-\sqrt{100}$
3. $\sqrt{81}$
4. $\sqrt{121}$
5. $-\sqrt{23^2}$
6. $\sqrt{63}$
7. $-\sqrt{17}$
8. $\sqrt{41}$
9. $-\sqrt{22}$
10. $\pm\sqrt{3}$

Find the value. Leave your answer in simplest radical form.

11. $\sqrt{54}$
12. $\sqrt{75}$
13. $\sqrt{432}$
14. $\sqrt{48a^3}$
15. $\sqrt{72x^2y}$

Find the two solutions of the equation.

16. $x^2 = 1$
17. $y^2 - 3 = 13$
18. $x^2 = 18$
19. $b^2 + 17 = 66$
20. The two shorter sides of a right triangle have lengths 4 and 7. Find the length of the hypotenuse.

Express in simplest form.

21. $\sqrt{\frac{9}{16}}$
22. $\sqrt{\frac{4y^2}{121}}$
23. $\sqrt{3a} \cdot \sqrt{12a^3}$
24. $\sqrt{\frac{5}{8}} \cdot \sqrt{\frac{1}{10}}$
25. $\sqrt{\frac{3}{8}}$
26. $2\sqrt{\frac{1}{5}}$
27. $5\sqrt{3} - 3\sqrt{3}$
28. $2\sqrt{18} - 4\sqrt{2}$

Solve.

29. $5y(y + 4) = 0$
30. $(x - 3)(x + 6) = 0$
31. $(3n - 6)(2n + 8) = 0$
32. $t^2 - 7t = 0$
33. $y^2 - 8y + 7 = 0$
34. $x^2 - x - 6 = 0$
35. $x^2 = -5x$
36. $t^2 + 2t = 15$
37. $n^2 = 8n - 16$
38. $z^2 = 11$
39. $m^2 - 3 = 17$
40. $2y^2 + 7 = 19$
41. $(n - 1)^2 = 16$
42. $(2t + 2)^2 = 36$
43. $(a - 4)^2 = 64$

Solve by using the quadratic formula. Leave irrational numbers in simplest radical form.

44. $6x^2 + 5x + 1 = 0$
45. $4x^2 + 4x = 3$
46. $x^2 + x - 1 = 0$

For each equation, make a table of values. Then give the coordinates of the vertex of each parabola.

47. $y = x^2 - 4x + 5$
48. $y = 1 - x^2$
49. $y = x^2 - 2x$

Table of Squares

Number	Square
n	n^2
1	1
2	4
3	9
4	16
5	25
6	36
7	49
8	64
9	81
10	100
11	121
12	144
13	169
14	196
15	225
16	256
17	289
18	324
19	361
20	400
21	441
22	484
23	529
24	576
25	625

Number	Square
n	n^2
26	676
27	729
28	784
29	841
30	900
31	961
32	1024
33	1089
34	1156
35	1225
36	1296
37	1369
38	1444
39	1521
40	1600
41	1681
42	1764
43	1849
44	1936
45	2025
46	2116
47	2209
48	2304
49	2401
50	2500

Number	Square
n	n^2
51	2601
52	2704
53	2809
54	2916
55	3025
56	3136
57	3249
58	3364
59	3481
60	3600
61	3721
62	3844
63	3969
64	4096
65	4225
66	4356
67	4489
68	4624
69	4761
70	4900
71	5041
72	5184
73	5329
74	5476
75	5625

Number	Square
n	n^2
76	5776
77	5929
78	6084
79	6241
80	6400
81	6561
82	6724
83	6889
84	7056
85	7225
86	7396
87	7569
88	7744
89	7921
90	8100
91	8281
92	8464
93	8649
94	8836
95	9025
96	9216
97	9409
98	9604
99	9801
100	10,000

Table of Square Roots

The square roots are given correct to three decimal places.

Number n	Positive Square Root $\sqrt{n}$
1	1
2	1.414
3	1.732
4	2
5	2.236
6	2.449
7	2.646
8	2.828
9	3
10	3.162
11	3.317
12	3.464
13	3.606
14	3.742
15	3.873
16	4
17	4.123
18	4.243
19	4.359
20	4.472
21	4.583
22	4.690
23	4.796
24	4.899
25	5
26	5.099
27	5.196
28	5.292
29	5.385
30	5.477
31	5.568
32	5.657
33	5.745
34	5.831
35	5.916
36	6
37	6.083
38	6.164
39	6.245
40	6.325
41	6.403
42	6.481
43	6.557
44	6.633
45	6.708
46	6.782
47	6.856
48	6.928
49	7
50	7.071
51	7.141
52	7.211
53	7.280
54	7.348
55	7.416
56	7.483
57	7.550
58	7.616
59	7.681
60	7.746
61	7.810
62	7.874
63	7.937
64	8
65	8.062
66	8.124
67	8.185
68	8.246
69	8.307
70	8.367
71	8.426
72	8.485
73	8.544
74	8.602
75	8.660
76	8.718
77	8.775
78	8.832
79	8.888
80	8.944
81	9
82	9.055
83	9.110
84	9.165
85	9.220
86	9.274
87	9.327
88	9.381
89	9.434
90	9.487
91	9.539
92	9.592
93	9.644
94	9.695
95	9.747
96	9.798
97	9.849
98	9.899
99	9.950
100	10

Extra Practice Exercises

CHAPTER 1

For use with pages 4–9

Find the value of the expression.

A

1. $7 \cdot 6$
2. $5 \cdot 8$
3. $9 \cdot 7$
4. $6 \cdot 4$
5. $2 \cdot 3 + 4$
6. $5 \cdot 2 - 6$
7. $(6 - 1) \cdot 3$
8. $(4 + 2) \div 3$
9. $2(3 + 2)$
10. $4(5 + 1)$
11. $8 - (2 \cdot 3)$
12. $10 + (6 \cdot 1)$

If $n = 3$, find the value of the expression.

13. $9n$
14. $9 + n$
15. $9 - n$
16. $9 \div n$

If $x = 10$ and $y = 5$, find the value of the expression.

17. $x + y$
18. $x - y$
19. $x + x + y$
20. $x + 5 - y$

If $y = 2$, find the value of the expression.

21. $6(y + 2)$
22. $6y + 2$
23. $(y - 1) \cdot 4$
24. $2y - 3$

For use with pages 10–13

Simplify.

A

1. $n + n$
2. $n + n + n$
3. $2n + n$
4. $n + 2n$
5. $3x - x$
6. $5y - 2y$
7. $7m + 3m$
8. $10x - x$
9. $4a + 5a$
10. $6x + 3x$
11. $4a + 2a - 3a$
12. $5x + 5x - 3x$
13. $3x + 8x - x$
14. $n + 6n - 7$
15. $a + 2a - b$
16. $3n - n + m$
17. $2x - 5 + 3x$
18. $6n - 6 + 2n$
19. $5m + 2 + 6m - 1$
20. $6 + 7n - 6n + 7$
21. $8 + 6y + 5y - 7$
22. $7 + 3x + 9x + 9$
23. $5x + 3y - y + 11x$
24. $2x + 5y - y + 9x$
25. $4n + r + r - 3n$

For use with pages 14–17

Find the value of the expression.

A

1. 3^2
2. 4^2
3. 2^2
4. 6^2
5. 4^1
6. 2^3
7. 3^3
8. 5^2

Simplify.

9. $y \cdot y \cdot y$
10. $x \cdot x$
11. $a \cdot a \cdot a$
12. $c \cdot c \cdot c \cdot c$
13. $r \cdot s \cdot s$
14. $a \cdot a \cdot b \cdot b$
15. $x \cdot y \cdot y \cdot y$
16. $a \cdot b \cdot b \cdot c \cdot c$
17. $3 \cdot a \cdot a$
18. $2 \cdot c \cdot c$
19. $5 \cdot 2 \cdot n \cdot n$
20. $6 \cdot a \cdot a \cdot b$
21. $2 \cdot (4r)$
22. $6 \cdot 3n$
23. $(4n) \cdot 4$
24. $6n \cdot 5$
25. $(2r)(4r)$
26. $(10x)(5x)$
27. $7y \cdot 8y$
28. $6s \cdot 5s$
29. $(2a)(4b)$
30. $(3x)(4y)$
31. $2 \cdot a \cdot 5 \cdot a$
32. $x \cdot x \cdot 4 \cdot x$

For use with pages 18–19

State the expression without parentheses.

A

1. $3(x + 3)$
2. $7(y + 1)$
3. $5(a - 1)$
4. $3(n + 2)$
5. $8(n + 4)$
6. $6(c - 2)$
7. $4(x + 3)$
8. $a(a + 1)$
9. $x(x - 2)$
10. $y(y + 7)$
11. $n(n + 2)$
12. $c(c + 9)$
13. $6(2n - 8)$
14. $4(5x - 4)$
15. $3(2a - 1)$
16. $5(3x + 2)$
17. $3(2a - 3b)$
18. $2(4x - 2y)$
19. $a(6a + b)$
20. $x(4x - 2y)$

Use the distributive property. Then combine like terms.

21. $2(n + 3) + 1$
22. $4(y + 1) + 3$
23. $3(n + 1) + 8$
24. $4(x + 3) + 2x$
25. $2(x + 4) - 2x$
26. $8(3n + 1) - 2$

B

27. $4(2a + 3) - 7a + 1$
28. $3(5x + 7) + 2x - 20$
29. $2(3y + 5) - 5y + 4$
30. $5(2 + 3x) - 4x - 7$
31. $6(a + 1) + 4(a + 2)$
32. $7(n + 3) + 2(n - 6)$

For use with pages 20–23

Find the value if possible. If not, write *impossible*.

A

1. $6 \cdot 0$
2. $5 \cdot 1$
3. $8 \cdot 0$
4. $\frac{0}{4}$
5. $\frac{0}{7}$
6. $\frac{9}{0}$
7. $6 \div 1$
8. $\frac{8}{0}$
9. $\frac{5 \cdot 0}{3}$
10. $\frac{7}{1}$
11. $\frac{7 + 1}{0}$
12. $12 \div 0$

Let $x = 1$. Find the value of the expression.

13. $7(x - 1)$
14. $x \div x$
15. $3x \div x$
16. $\frac{4}{4}x$

Let $a = 6$. Find the value of the expression.

17. $(a - 1) \div 5$
18. $(2a - 2) \div 10$
19. $\frac{a + 2}{8}$
20. $\frac{a}{6} \cdot 1$

Simplify.

21. $3 + 2x - 3$
22. $7 + 4y - 7$
23. $4a + 5 + 2a - 5$
24. $2(x + 4) - 8$
25. $\frac{6}{6} \cdot n$
26. $\frac{9}{9} \cdot c$
27. $\frac{x}{3} \cdot 3$
28. $\frac{2n}{2}$

For use with pages 25–28

Tell which of the numbers shown in color is a solution.

A

1. $n + 3 = 4$ 2, 1, or 0
2. $x - 2 = 6$ 7, 8, or 9
3. $5 + y = 8$ 3, 2, or 1
4. $a - 4 = 8$ 14, 13, or 12
5. $6 - a = 4$ 4, 3, or 2
6. $n - 5 = 10$ 14, 15, or 16

Tell which of the numbers shown in color are solutions.

7. $n < 4$ 1, 2, 4, 5
8. $x > 6$ 6, 7, 9, 10
9. $y < 17$ 15, 16, 19, 20
10. $a > 9$ 6, 8, 10, 12
11. $x + 1 > 3$ 1, 3, 4, 10
12. $2c > 3$ 4, 2, 1, 9

CHAPTER 2

For use with pages 38–41

Solve.

A

1. $x - 2 = 4$ **2.** $y - 4 = 2$ **3.** $n - 3 = 3$ **4.** $t - 4 = 4$

5. $y - 4 = 7$ **6.** $x - 3 = 6$ **7.** $n - 7 = 6$ **8.** $m - 3 = 8$

9. $n + 2 = 6$ **10.** $x + 1 = 7$ **11.** $x + 7 = 8$ **12.** $y + 6 = 10$

13. $x + 5 = 20$ **14.** $y + 6 = 14$ **15.** $n + 9 = 15$ **16.** $m + 12 = 18$

Solve.

17. $y - 5 = 20$ **18.** $x - 6 = 13$ **19.** $x + 6 = 19$ **20.** $n + 7 = 14$

21. $n + 9 = 17$ **22.** $n - 5 = 16$ **23.** $x + 14 = 21$ **24.** $r - 11 = 22$

25. $t + 16 = 22$ **26.** $n + 18 = 25$ **27.** $x - 6 = 18$ **28.** $30 = n - 5$

29. $15 = y + 8$ **30.** $x - 4 = 0$ **31.** $y + 3 = 27$ **32.** $16 = x - 14$

For use with pages 42–45

Solve.

A

1. $2x = 12$ **2.** $3y = 12$ **3.** $4n = 12$ **4.** $5n = 25$

5. $3x = 15$ **6.** $6x = 18$ **7.** $7y = 28$ **8.** $9x = 63$

9. $\frac{x}{3} = 4$ **10.** $\frac{n}{2} = 9$ **11.** $\frac{a}{4} = 3$ **12.** $\frac{x}{2} = 6$

Solve.

13. $9y = 54$ **14.** $7x = 21$ **15.** $\frac{n}{3} = 8$ **16.** $\frac{a}{6} = 7$

17. $4x = 36$ **18.** $8x = 72$ **19.** $7y = 42$ **20.** $9a = 81$

21. $\frac{x}{7} = 5$ **22.** $9 = \frac{a}{4}$ **23.** $\frac{n}{8} = 8$ **24.** $9x = 54$

25. $15x = 150$ **26.** $\frac{m}{6} = 5$ **27.** $7a = 196$ **28.** $\frac{x}{12} = 10$

For use with pages 46–49

Solve.

A

1. $2x + 4 = 6$
2. $4y - 3 = 9$
3. $3n - 7 = 11$
4. $5n - 3 = 22$
5. $5y + 5 = 15$
6. $6x + 18 = 36$
7. $4n + 5 = 45$
8. $2a - 6 = 20$
9. $3a + 6 = 36$
10. $7n - 8 = 41$
11. $3x - 8 = 13$
12. $8x - 5 = 35$
13. $7a + 14 = 21$
14. $2m + 1 = 13$
15. $4n + 3 = 51$
16. $9n - 40 = 32$
17. $6x - 12 = 30$
18. $4a + 2 = 46$
19. $7x + 12 = 75$
20. $5b - 13 = 17$

For use with pages 52–53

Solve.

A

1. $2x + 3x = 15$
2. $4a - 2a = 10$
3. $6y + 3y = 18$
4. $2x + 3x - 5 = 30$
5. $5c - 2c + 8 = 20$
6. $9a + a - 2 = 28$
7. $2n - 5 + 8n = 45$
8. $7x + 3 - 2x = 18$
9. $4n + 7 + 3n = 21$
10. $15n - 6 - 7n = 34$
11. $8y - 3 + y = 24$
12. $10a + 6 - 7a = 30$
13. $28 = y - 4 + 3y$
14. $40 = 13x - 6x - 2$
15. $56 = 7a - 16 + 2a$
16. $46 = 2x + 5x - 3$
17. $19 = 8a + 7 - 6a$
18. $21 = 6x + 3x - 4 - 4x$

For use with pages 58–61

A

1. Three times a number plus 5 times the number is 64.
   What is the number?
2. Cynthia's father is 26 years older than Cynthia.
   Together their ages total 42.
   How old is Cynthia?
3. Amy has twice as much money as Brian.
   Together they have $15.
   How much does Amy have?
4. Our class has 90 students.
   There are 4 more boys than girls.
   How many boys are there?

For use with pages 62–63

Solve.

A

1. $2n = 10 + n$
2. $4x = x + 15$
3. $8x = 2x + 6$
4. $16 - 2y = 6y$
5. $5x - 20 = 3x$
6. $10x - 14 = 3x$
7. $7x - 12 = 4x$
8. $3y + 9 = 4y$
9. $14 - 2x = 5x$
10. $6y - 2 = 4y + 8$
11. $3x + 1 = x + 7$
12. $5a - 4 = 8 - a$
13. $2n + 11 = 7 + 6n$
14. $5x - 8 = 10 + 2x$
15. $7c - 4 = 2c + 6$

For use with pages 64–65

Solve.

A

1. $2(n - 4) = 16$
2. $7(a - 3) = 4a$
3. $5(z + 2) = 10z$
4. $6(y - 3) = 0$
5. $8(x - 2) = 0$
6. $5(m + 2) = 25$
7. $4(x + 3) = 6x$
8. $6(y + 3) = 18$
9. $3(2 - x) = 3x$
10. $4(x - 2) = 3x + 2$
11. $3(x + 1) = 2x + 10$
12. $2(5 - x) = 3x$

B

13. $2(c + 3) = 3(c - 1)$
14. $4(r - 2) = 2(r + 1)$
15. $4(x - 4) = 2(x - 1)$

For use with pages 66–67

Solve.

A

1. Al has 3 times as much money as Bob.
   Bob has \$10 more than Ed.
   Together the three boys have \$65.
   How much money does each have?
2. Jo, May, and Sara save dimes.
   Jo has twice as many dimes as May.
   Sara has 4 more dimes than May.
   Together the three have 100 dimes.
   How many dimes has Jo?

B

3. Nadine is 11 years older than Rob.
   Next year Nadine will be twice as old as Rob.
   How old is Nadine?

CHAPTER 3

For use with pages 78–82

Compare the numbers. Write > or <.

A

1. $-2 \underline{\ ?\ } -5$
2. $3 \underline{\ ?\ } -2$
3. $0 \underline{\ ?\ } -4$
4. $-6 \underline{\ ?\ } -1$
5. $6 \underline{\ ?\ } -7$
6. $3 \underline{\ ?\ } 0$
7. $2 \underline{\ ?\ } -2$
8. $-7 \underline{\ ?\ } -3$

Graph the solutions on the number line.

9. $n + 1 = 4$
10. $y - 3 = 7$
11. $4 + x = 6$
12. $a - 4 = 3$
13. $x < 3$
14. $x > -1$
15. $y < -2$
16. $x > 3$

For use with pages 83–87

Add.

A

1. $2 + (-6)$
2. $6 + (-2)$
3. $-8 + 8$
4. $-3 + 5$
5. $-4 + (-1)$
6. $-2 + (-8)$
7. $3 + (-1) + (-8)$
8. $-4 + 8 + (-4)$
9. $5 + 0 + (-7)$

Subtract.

10. $4 - (-8)$
11. $7 - 11$
12. $-9 - (-2)$
13. $-6 - 4$
14. $18 - 4$
15. $3 - (-9)$
16. $-4 - 7$
17. $8 - (-5)$
18. $-3 - 10$
19. $-2 - (-5)$
20. $10 - 15$
21. $6 - (-3)$

For use with pages 89–90

Simplify.

A

1. $2n - 5n$
2. $-a + 6a$
3. $-4b - 6b$
4. $x + 9x$
5. $-2y - 4y$
6. $x^2 - 3x^2$
7. $-5a + 6a + 1$
8. $7x - (-x) + y$
9. $-4y + 3 + 4y$
10. $15 - (-3x) + x$
11. $-7x - (-8x) + 2$
12. $6y + 2x - (-6y)$
13. $x^2 - y + 2x^2$
14. $8a - (-2a) + a^2$
15. $2n - n^2 - (-3n)$

For use with pages 91–94

Simplify.

A

1. $12 \cdot 4$
2. $-3(6)$
3. $-8(-7)$
4. $3(-9)$
5. 5^2
6. $(-5)^2$
7. $-(5)^2$
8. $-(-4)^2$
9. $3n(-7)$
10. $-2y(-8x)$
11. $2(2a - b)$
12. $-5(x + y)$
13. $-(m - 4)$
14. $-(8 - a)$
15. $x - (x + 2)$
16. $2y - (y - 3)$

For use with pages 95–96

Divide.

A

1. $72 \div -8$
2. $14 \div -14$
3. $-63 \div -9$
4. $-54 \div 6$
5. $56 \div -8$
6. $-24 \div -6$
7. $15 \div -5$
8. $-30 \div -3$
9. $\frac{40}{-5}$
10. $\frac{-18}{-2}$
11. $\frac{-16}{4}$
12. $\frac{81}{9}$

For use with pages 97–98

Solve for the variable. Check.

A

1. $n + 3 = 6$
2. $n - 3 = 7$
3. $x + 9 = 6$
4. $a + 4 = -4$
5. $b - 2 = -2$
6. $15 = y - 12$
7. $3x = 15$
8. $-3x = -15$
9. $6x = -24$
10. $33 = -11y$
11. $5n - n = -16$
12. $9y - 11y = 20$
13. $5n = 50 - 45$
14. $3n + 60 = 2n$
15. $-3n + 2 = -4n$
16. $3x - 9 = 12x$
17. $5n - 4 = 3n + 8$
18. $14x - 4 = 13x$
19. $-4x + 1 = -x + 19$
20. $6x - 14 = 2x - 2$
21. $-7y - 3 = y + 29$

B

22. $x - (4 - x) = 20$
23. $5n - (n + 3) = 25$
24. $11x - (x - 20) = 100$
25. $4 - (10 - x) = 34$
26. $(w - 4) - 3 = 11$
27. $-10(v + 4) - 20 = 10v$

CHAPTER 4

For use with pages 112–114

Find the perimeter.

A

1. 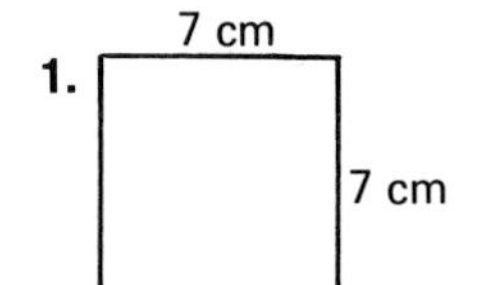

2. 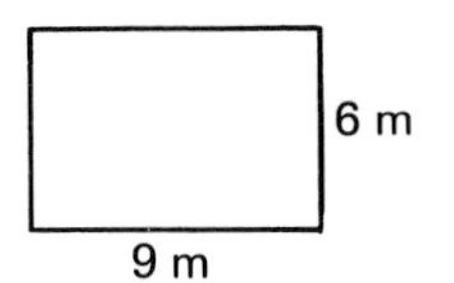

3. 4. 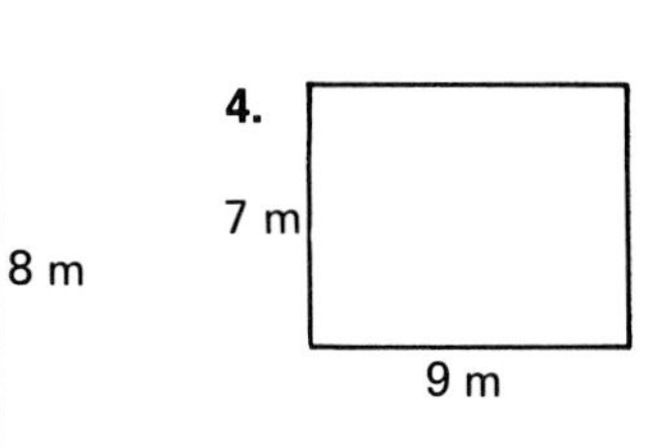

5. 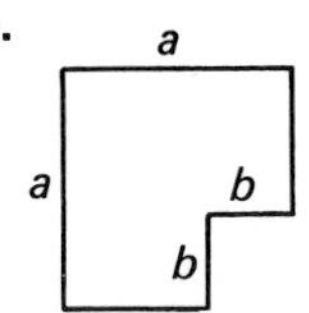

6. 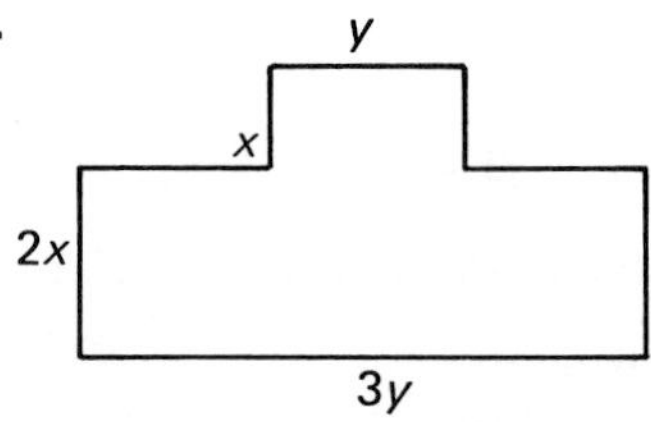

7. 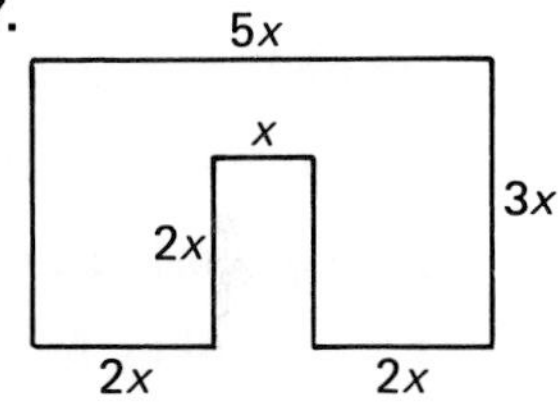

8. The length of a rectangle is 4 in. longer than the width. The perimeter of the rectangle is 20 in. Find its length and its width.

9. The perimeter of the triangle shown is 23. Find the value of x.

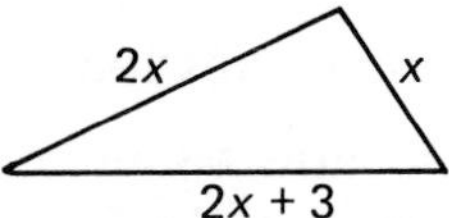

10. A circle has a diameter of 4 in. What is its circumference? (Use the formula $C = 3.14d$.)

For use with pages 115–118

Find the area.

A

1. 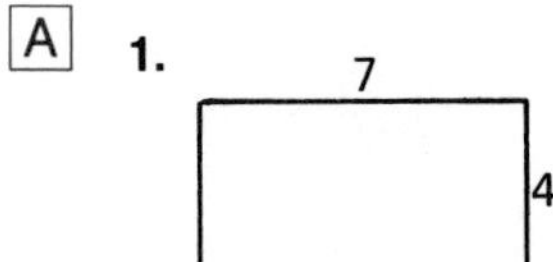

2. 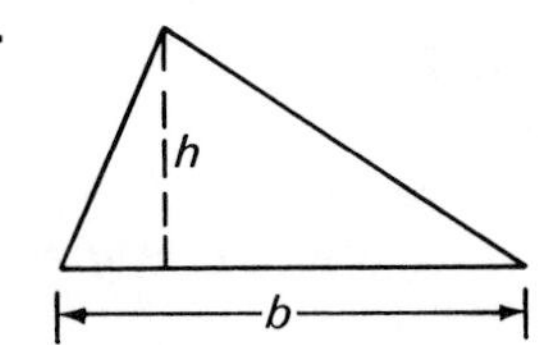

3. 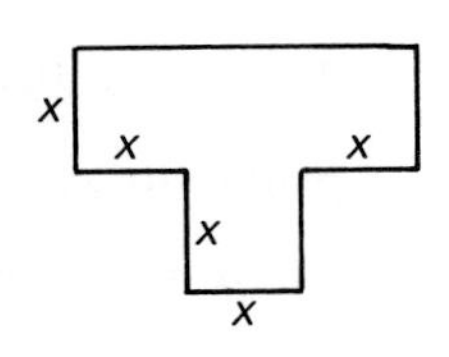

4. A rectangular lot measures 30 m by 35 m. Find its value at $12.50 per square meter.

5. Washable wallpaper costs $3.95 per square meter. How much would it cost to paper a kitchen wall 3 m high and 4 m long?

For use with pages 119–122

A 1. State the formula for finding the volume of a prism.

2. The area of the base of a rectangular box is 20 cm^2. Its height is 3 cm. What is its volume?

3. Find the volume of a cube 6 cm on a side.

4. Find the volume of a box that measures $6x$ units by $4x$ units by $2x$ units.

For use with pages 124–133

Complete.

A 1. You walk K hours at h km/h. $D =$ __?__ km.

2. You work x hours at \$7 per hour. You earn __?__.

3. John bikes from Midland to Newton at 20 km/h. He returns by bus at an average of 80 km/h. The trip going takes 3 hours more than returning. The distance from Midland to Newton is __?__.

	rate ×	time =	Distance
Bike	20	$x + 3$	?
Ride	80	x	?

4. Di earns \$8 an hour and so does her sister Jo. If Di works 3 hours more per week and together they earn \$664 per week, how many hours per week does Jo work?

	pay per hour ×	number of hours =	pay per week
Di	8	?	?
Jo	?	?	?
			?

For use with pages 134–137

Solve for x.

A 1. $y + x = n$

2. $2x - y = 4$

3. $4y - x = 12$

4. $4x = y$

5. $nx = y$

6. $kx = y$

7. $10y + x = 15$

8. $3y - x = 9$

9. $3x - y = 0$

10. $5x + 5 = 15y$

11. $x - y = k$

12. $-x + y = -2$

CHAPTER 5

For use with pages 152–155

Add.

A

1. $\begin{array}{r} n + 3 \\ \underline{3n + 1} \end{array}$

2. $\begin{array}{r} 7y^2 + 3 \\ \underline{8y^2 + 9} \end{array}$

3. $\begin{array}{r} a^2 + b^2 \\ \underline{a^2 - b^2} \end{array}$

4. $\begin{array}{r} 3x + 7y \\ \underline{-5x + 8y} \end{array}$

5. $\begin{array}{r} -4xy - z \\ \underline{4xy + 6z} \end{array}$

6. $\begin{array}{r} a^2 + ab \\ \underline{3a^2 + ab} \end{array}$

7. $\begin{array}{r} 4x - 6y \\ \underline{-7x + 4y} \end{array}$

8. $\begin{array}{r} 9x^2 + 4y^2 \\ \underline{6x^2 - 8y^2} \end{array}$

9. $(5a - b) + (8a + 4b)$

10. $(x + 2y + 9) + (-3x - y + 7)$

11. $(x^2 - 3x + 4) + (2x^2 + 3x - 4)$

12. $(k^2 + 1) + (-3k^2 + 2k - 1)$

Subtract.

13. $\begin{array}{r} 3a - b \\ \underline{-(2a - b)} \end{array}$

14. $\begin{array}{r} -6n + 5 \\ \underline{-(-7n + 2)} \end{array}$

15. $\begin{array}{r} y^3 + 4 \\ \underline{-(y^3 - 4)} \end{array}$

16. $\begin{array}{r} 3x + y \\ \underline{-(7x - y)} \end{array}$

17. $\begin{array}{r} 2n^2 + 2n \\ \underline{-(4n^2 + 2n)} \end{array}$

18. $\begin{array}{r} 5x - 3 \\ \underline{-(3x + 4)} \end{array}$

19. $\begin{array}{r} 3a - 7 \\ \underline{-(7a + 1)} \end{array}$

20. $\begin{array}{r} 6x^2y^2 + 3 \\ \underline{-(x^2y^2 - 3)} \end{array}$

21. $(2x + 1) - (x - 4)$

22. $(3x - 6y) - (2x - 5y)$

23. $(6a + 2b) - (a - b)$

24. $(7m - 2n) - (m + 3n)$

B

25. Jon walked $2n$ km to the corner, $(n + 1)$ km to the store, and $(2n + 4)$ km back home. How far did he walk?

26. May hiked $(5n + 4)$ km. Paulo hiked $(4n - 1)$ km. May hiked how much farther than Paulo?

For use with pages 156–159

Multiply.

A

1. $y \cdot y^3$

2. $(3x^2)(7x^2)$

3. $(-2n)(-n^2)$

4. $(-4x^2)(7x^3)$

5. $x(-xy)$

6. $(4x^2y)(3xy^2)$

7. $n^5 \cdot n^2$

8. $x^4 \cdot x^3$

Simplify.

9. $(n^3)^4$
10. $(x^2)^5$
11. $(y^5)^2$
12. $(x^4)^4$
13. $(ab)^2$
14. $(xy)^2$
15. $(mn)^4$
16. $(2a)^2$
17. $(-xy)^3$
18. $(3a^2)^2$
19. $(-3x^2)^3$
20. $5(-ab^2)^5$

B

21. $-(ab^2)(ab)^2$
22. $(-xy)^2(-xy)$
23. $(4rs)^3(2s)^3$
24. $(3mn)^3(-3mn)$
25. $(2x^2y^2)(2x^3y^3)^2$
26. $-(2ab)^4(4a^4b^4)$

For use with pages 160–163

Multiply.

A

1. $3(n + 5)$
2. $-4(x - y)$
3. $6(a^2 + b^2)$
4. $-2(a - 3b)$
5. $a(x - y)$
6. $-3x(2x + 3y)$
7. $-3(2a + 4b + 6c)$
8. $-1(5x + 2y + z)$
9. $4x(x^2y + xy^2 + 3y^3)$
10. $5a(a^3 + a^2 - a + 1)$
11. $-x(3x^3 - 5x^2 + x - 1)$
12. $m^2(m^2 + 2mn + n)$
13. $(5 - a)(4 - a)$
14. $(n + 5)(n - 3)$
15. $(x + 4)(x + 8)$
16. $(3a - 1)(a + 1)$
17. $(5x - 2)(5x - 4)$
18. $(n + 6)(n - 9)$

B

19. $(a - 1)(a^2 + 2a + 1)$
20. $(x - y)(x^2 - 2xy - y^2)$
21. $(m - n)(m^2 - n^2)$
22. $(x + y)(x^2 + 2xy + y^2)$
23. $(a^2 - 1)(a^2 + 1)$
24. $(n + 2)(n^2 + 4n + 4)$

For use with pages 164–169

Multiply. You may use the FOIL method.

A

1. $(n + 3)(n + 4)$
2. $(x + 6)(x - 3)$
3. $(m - 5)(m - 2)$
4. $(3x - 2)(2x + 3)$
5. $(5y - 1)(5y + 1)$
6. $(7a + 3)(3a - 1)$
7. $(2x + y)(4x - y)$
8. $(2a + b)(a - 2b)$
9. $(5x - y)(3x - y)$
10. A rectangle measures $(3n + 1)$ cm by $(2n + 5)$ cm. What is its area?

Express as a trinomial.

11. $(n + 2)^2$ 12. $(y - 5)^2$ 13. $(n + 7)^2$ 14. $(x - 10)^2$

15. $(x + 3y)^2$ 16. $(a - 2b)^2$ 17. $(a - 10b)^2$ 18. $(3m + 2n)^2$

19. $-1(n + 2)^2$ 20. $-2(4x + y)^2$ 21. $4(2r + s)^2$ 22. $-3(2x - 3y)^2$

Find the area. Use $A = \frac{1}{2}bh$ or $A = lw$.

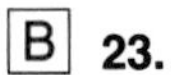

23. 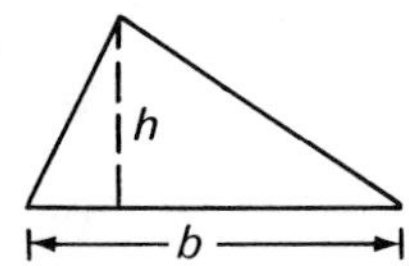

$b = 8n + 1$

$h = 4n$

24. 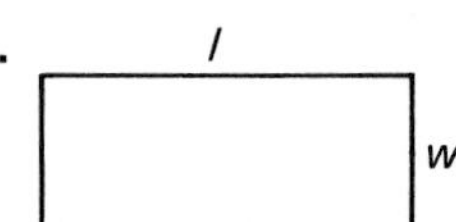

$l = 3n + 1$

$w = 6n$

For use with pages 170–171, 174–175

Divide.

1. $\frac{3x}{x}$ 2. $\frac{a^8}{a^2}$ 3. $\frac{5x}{5}$ 4. $\frac{4ab}{4b}$

5. $\frac{6x^3}{2x}$ 6. $\frac{-12a^5}{2a^3}$ 7. $\frac{63n^4}{-7n^2}$ 8. $\frac{-49a^2b^2}{-7ab}$

9. $\frac{4x - 8y}{2}$ 10. $-\frac{6a - 6b}{6}$ 11. $\frac{x^2y^2 - 3xy}{xy}$ 12. $\frac{15m + 30mn}{5m}$

13. $\frac{20n^3 - 10n^2 - 5n}{5n}$ 14. $\frac{x^3 + 3x^4 - 5x^5}{-x^2}$ 15. $\frac{18a^3b^3 - 12a^2b^2 + 6ab}{6ab}$

16. $\frac{8x^2y + 12xy^2 - 16y^3}{4y}$ 17. $\frac{a^3 - 6a^5 + 8a^7}{-a^3}$ 18. $\frac{11x^4y^5 + 9x^3y^4 + 3x^3y^3}{x^2y^2}$

Simplify.

19. $\frac{12xy}{4x} - \frac{10y^3}{5y^2}$ 20. $\frac{-21ab}{7b} + \frac{4a^2}{a}$ 21. $\frac{3x^3y^5}{xy^3} - \frac{x^4y^5}{x^2y^3}$

Divide.

22. $\frac{9x^4y^4 - 12x^3y^3 + 21x^2y^2 - 3xy}{-3xy}$ 23. $\frac{8a^5b^5 + 6a^4b^6 - a^3b^7 + 3a^2b^8}{a^2b^3}$

CHAPTER 6

For use with pages 186–189

Write each number as a product of prime factors.

A

1. 20
2. 40
3. 250
4. 110
5. 140
6. 60
7. 150
8. 200
9. 180
10. 630

Find the greatest common factor of each pair.

11. 10, 25
12. 32, 28
13. 26, 39
14. 63, 14
15. 15, 90
16. 40, 48
17. 70, 105
18. 55, 242

Factor.

19. $14 - 2x$
20. $20n - 4$
21. $12 - 36x$
22. $2x^2 + 6x$
23. $y^2 - 3y$
24. $14x^2 + 7x$
25. $10xy - y^2$
26. $11n^2 - 44n$
27. $6ab^2 + 24a^2b$
28. $3n^2 - 9n + 12$
29. $4x^2 + 12xy + 24y^2$
30. $7a + 14b + 70c$

B

31. $5 - 25n + 75n^2 + 125n^3$
32. $6xy + 42x^2y^2 - 66x^3y^3 + 72x^4y^4$
33. $a^4b - a^3b^2 + a^2b^3 - ab^4$
34. $12a^2b - 36ab^2 + 48b^2$
35. $100x^4y^4 - 75x^3y^3 + 50x^2y^2 + 25xy$
36. $15m^2n^4 + 45m^3n^5 - 60m^4n^6$

For use with pages 192–197

Factor.

A

1. $y^2 + 7y + 12$
2. $n^2 + 10n + 24$
3. $n^2 + 16n + 60$
4. $x^2 + 14x + 33$
5. $x^2 + 13x + 36$
6. $n^2 + 16n + 55$
7. $b^2 + 19b + 90$
8. $m^2 + 15m + 36$
9. $n^2 + 13n + 12$
10. $a^2 + 14a + 24$
11. $n^2 + 19n + 34$
12. $m^2 + 17m + 16$
13. $x^2 + 14x + 13$
14. $y^2 + 10y + 16$
15. $x^2 + 16x + 28$

16. $b^2 - 10b + 16$
17. $m^2 - 9m + 20$
18. $n^2 - 8n + 15$
19. $y^2 - 14y + 45$
20. $n^2 - 10n + 9$
21. $y^2 - 17y + 60$
22. $x^2 - 19x + 90$
23. $n^2 - 18n + 80$
24. $y^2 - 15y + 44$

B

25. $x^2 + 42x + 80$
26. $x^2 + 29x + 100$
27. $x^2 - 101x + 100$
28. $y^2 - 34y + 93$
29. $m^2 - 32m + 60$
30. $y^2 - 22y + 72$
31. $x^2 + 20x + 64$
32. $n^2 - 40n + 144$
33. $n^2 + 21n + 68$

For use with pages 198–203

Factor. Many are trinomial squares.

A

1. $y^2 - 6y + 9$
2. $x^2 + 16x + 64$
3. $n^2 - 14n + 49$
4. $36 - 12x + x^2$
5. $n^2 + 20n + 100$
6. $81 - 18x + x^2$
7. $n^2 + 5n + 4$
8. $x^2 - 13x + 42$
9. $y^2 - 22y + 121$
10. $25 - 10x + x^2$
11. $m^2n^2 - 4mn + 4$
12. $b^2 - 10b + 25$

Factor. Watch the signs.

13. $y^2 - 7y - 8$
14. $n^2 - 2n - 15$
15. $y^2 + 6y - 40$
16. $n^2 - 8n - 20$
17. $m^2 - 5m - 50$
18. $x^2 - 7x - 60$
19. $n^2 - 8n - 9$
20. $n^2 + 13n - 14$
21. $n^2 - 3n - 4$

For use with pages 204–207

Multiply at sight.

A

1. $(x + 4)(x - 4)$
2. $(m - n)(m + n)$
3. $(1 - 5x)(1 + 5x)$
4. $(3y + 1)(3y - 1)$
5. $(2x - y)(2x + y)$
6. $(a + 2b)(a - 2b)$
7. $(5x + 2y)(5x - 2y)$
8. $(8m - n)(8m + n)$
9. $(4n + 3)(4n - 3)$
10. $(1 - 6a)(1 + 6a)$
11. $(9 - 2b)(9 + 2b)$
12. $(8x - 3)(8x + 3)$

Factor.

13. $x^2 - 25$
14. $n^2 - 36$
15. $x^2 - 100$
16. $4a^2 - 9b^2$
17. $9x^2 - 36y^2$
18. $m^2 - 49$
19. $16a^2 - 1$
20. $64 - 4a^2$
21. $16s^2 - 4t^2$
22. $25a^2 - b^2$
23. $121 - 9x^2$
24. $9b^2 - 4c^2$

Multiply.

B

25. $(n^2 + 1)(n^2 - 1)$
26. $(x^2 - 5)(x^2 + 5)$
27. $(y^2 + 8)(y^2 - 8)$

Factor.

28. $-1 + x^2$
29. $-9 + 16y^2$
30. $-b^2 + a^2$
31. $-64 + a^2$
32. $-4m^2 + k^2$
33. $-16b^2 + 81a^2$
34. $m^2 - 144$
35. $n^2 - 625$
36. $a^2 - 169$
37. $400 - r^2$
38. $x^2y^2 - 16$
39. $m^2n^2 - 25p^2$

For use with pages 208–209

Factor.

A

1. $2x^2 - 50$
2. $3y^2 - 12y + 12$
3. $4n^2 + 20n + 16$
4. $8 + 4n - 4n^2$
5. $4n^2 + 12n - 16$
6. $4x^2 - 64$
7. $10x^2 - 40$
8. $a^3 - 16a$
9. $3n^2 - 9n - 12$
10. $5n^2 - 40n - 100$
11. $12x^2 + 36x + 24$
12. $3x^2 - 24x + 45$
13. $3n^2 + 12n + 12$
14. $4n^2 + 28n + 40$
15. $5a^2 - 5b^2$
16. $12x^2 + 36x + 24$
17. $4x^2 - 24x + 36$
18. $7x^2 + 42x + 63$
19. $4n^2 - 8n - 96$
20. $4x^2 - 36$
21. $x - 3x^2$
22. $7x^2 - 56x + 49$
23. $x^2y^2 - y^3$
24. $4x^2 + 48$

B

25. $10a^2 - 10ab - 60b^2$
26. $3x^2 - 6x - 189$
27. $-4b^2 + 9a^2$
28. $-6 - 6x^4$
29. $4r^2 - 24rs + 32s^2$
30. $9b^2 + 36b - 45$
31. $5x^2 - 55x - 300$
32. $24n^2 + 216n - 864$
33. $x^4 - y^4$

CHAPTER 7

For use with pages 224–227

The table below shows the average lifespan of several animals.

Monkey, 7 years	Fox, 8 years	Bear, 31 years
Bat, 6 years	Dog, 16 years	Elephant, 47 years

A

1. Make a bar graph using the information in the table.

2. Use your graph to answer:
   a. A bear lives how many years longer than a fox?
   b. An elephant lives about how many times as long as a monkey?

For use with pages 233–236

Plot the following points on graph paper.

A

1. $(3, -2)$
2. $(1, -2)$
3. $(1, 3)$
4. $(-3, -3)$

B

5. Plot the following points on graph paper. Connect them in order.

$$(0, -3), (4, 6), (8, -3)$$

For use with pages 237–240

Complete the tables to show four solutions for each equation.

A

1. $y = x + 4$

x	y
0	?
1	?
−1	?
2	?

2. $y = -x - 1$

x	y
0	?
−1	?
2	?
3	?

3. $y = 3x + 2$

x	y
0	?
1	?
3	?
4	?

4. $y = -2x + 3$

x	y
0	?
1	?
−1	?
2	?

Solve the equation for *y*.

5. $x + y = 7$
6. $x + y = -7$
7. $y - x = 3$
8. $4x + y = 8$
9. $x - y = 5$
10. $2x - y = 9$
11. $10x + y = -2$
12. $5x - y = 0$

Find three solution pairs for the equation.

13. $y = x - 3$
14. $y + x = 5$
15. $y + 2x = 1$
16. $y + 4x = 7$
17. $2x + y = 0$
18. $4x + y = 12$
19. $y - 2x = 0$
20. $y - 5x = 10$

Guess an equation to go with the table.

B

21.

x	y
0	−5
1	−4
3	−2

22.

x	y
0	0
1	2
−1	−2

23.

x	y
0	−1
1	2
2	5

24.

x	y
0	3
1	7
2	11

For use with pages 241–244

Draw the graph of the equation.

A

1. $y = 5x$
2. $y = x + 5$
3. $y = 3x - 3$
4. $x + y = 7$
5. $y - 2x = 6$
6. $2x + y = 0$
7. $y = -2x$
8. $y = 3x - 1$
9. $y = 2x - 4$

What is the slope of the line whose equation is given?

10. $y = 2x$
11. $y = -3x$
12. $y = -x + 4$
13. $y = \frac{1}{2}x$

For use with pages 245–248

If y is a function of x, state an equation relating x and y.

A

1.

x	1	2	3
y	3	6	9

2.

x	1	−1	0
y	3	1	2

The distance a car goes in 3 hours is a function of its rate.

3. Complete the table.

Rate (km/h)	Distance (km)
10	30
30	?
40	?

4. Draw a graph based on the table. Show the rate along the x-axis and the distance along the y-axis as on page 247.

CHAPTER 8

For use with pages 262–265

Solve by the graphing method.

A

1. $y = -x$
 $y = x + 2$

2. $y = 2x$
 $y = 3x - 3$

3. $x + y = 3$
 $-2x + y = -3$

4. $3x + 1 = y$
 $2x - 1 = y$

B

5. $y = x - 3$
 $y = 3x + 2$

6. $2x + y = 3$
 $x - y = 7$

7. $y = 3x - 4$
 $y = \frac{1}{2} - 3x$

8. $x + y = 0$
 $3x - y = 6$

For use with pages 266–269

Find the slope of the line whose equation is given.

A

1. $y = 3x$

2. $y = 2x + 1$

3. $y = -3x - 1$

4. $y = \frac{1}{2}x + 5$

Do the equations share *one* solution pair, *no* solution pair, or *all* solution pairs?

5. $y = 3x$
 $y = 3x - 2$

6. $x + y = 6$
 $x - y = 2$

7. $x + y = 0$
 $3x + 3y = 0$

8. $x + y = 3$
 $x + y = 12$

For use with pages 270–275

Solve by the substitution method.

A

1. $y = 3x$
 $x + y = 12$

2. $y = 4x$
 $y - x = 6$

3. $x - 2y = -4$
 $3x + y = 2$

4. $3x - y = 3$
 $x + y = 5$

Solve by the addition or subtraction method.

5. $x + y = 4$
 $x - y = 6$

6. $2x + 2y = 5$
 $2x + 3y = 4$

7. $3x - 2y = 0$
 $x - 2y = 4$

8. $x - 2y = 18$
 $5x + 2y = 0$

Solve by the method easiest for you.

9. $3x - 7 = y$
 $4x + 2 = y$

10. $y = 4x$
 $3x - y = 0$

11. $x = y$
 $3x + 5y = 8$

12. $x - 4y = 3$
 $x - 6y = 1$

Solve by the method easiest for you.

B

13. $x - 5y = 0$
    $x - 4y = 9$

14. $3x + 4y = 6$
    $x + 4y = 7$

15. $5x - y = 5$
    $15x + y = 10$

16. $4x - 12y = 3$
    $4x + 10y = -2$

For use with pages 276–283

Solve.

A

1. $x + 2y = 5$
   $2x + 3y = 4$

2. $2x - y = 4$
   $3x + 2y = 6$

3. $x - 4y = 2$
   $3x - 10y = 4$

4. $4x - 5y = 9$
   $x + y = 0$

5. $3x - y = 6$
   $x + 2y = 2$

6. $2x + 3y = 1$
   $x + y = 4$

7. $x - 3y = 4$
   $3x - y = 4$

8. $3x - 5y = 2$
   $x - 2y = 5$

Use two variables in solving these problems.

9. The sum of two numbers is 15.
   One number is twice the other.
   What are the numbers?

10. Two hats and 3 ties cost $88.50.
    One hat and 2 ties cost $51.00.
    What does each item cost?

11. Two bats and 1 ball cost $29.
    One bat and 2 balls cost $22.
    What does a bat cost?

12. The sum of two numbers is 15.
    One number plus twice the other is 25.
    Find the numbers.

13. Al's score is 6 more than Maya's.
    Their scores total 16.
    Find their scores.

14. One egg and 2 glasses of juice cost $2.30.
    Two eggs and 1 juice cost $3.40.
    What does *one* egg plus *one* juice cost?

Solve.

B

15. $2x + 3y = 0$
    $3x + 4y = 6$

16. $4x - 3y = 1$
    $3x - 2y = 0$

17. $2x + 5y = 0$
    $5x - 2y = 0$

18. $3x - 4y = 5$
    $2x - 5y = 1$

19. $2x - 4y = 4$
    $3x - 5y = 1$

20. $4x - 3y = -2$
    $3x + 4y = 1$

21. Tom's club bought some pencils at 25¢ each and some pens at 75¢ each. The total number of pens plus pencils was 25 and the total cost was $6.75. How many of each did the club buy?

CHAPTER 9

For use with pages 298–303

Simplify.

A

1. $\frac{4}{16}$
2. $\frac{9}{15}$
3. $\frac{3}{12}$
4. $\frac{4}{20}$
5. $\frac{6}{18}$
6. $\frac{5}{25a}$
7. $-\frac{2x}{14x}$
8. $\frac{21r}{7r^2}$
9. $\frac{5y}{-20xy^2}$
10. $\frac{2mn^2}{10m^2n}$
11. $\frac{-4xy^2}{12x^2}$
12. $\frac{5x^3y^3}{30xy}$
13. $-\frac{40n^2s}{5n^3s^2}$
14. $\frac{6d^2e^2}{24d^4e^4}$
15. $\frac{24x^4y^3}{12x^3y^2}$
16. $\frac{n+1}{2n+2}$
17. $\frac{3x+3y}{10x+10y}$
18. $\frac{x-y}{x^2-y^2}$
19. $\frac{4a-4b}{12a-12b}$

Write the expression with a factor −1.

20. $1 - n$
21. x
22. $6 - 4y$
23. $-n^2 + z^2$

Simplify.

24. $\frac{m-n}{n-m}$
25. $\frac{d-z}{z-d}$
26. $\frac{3x-3y}{y-x}$
27. $\frac{x^2-y^2}{y-x}$
28. $\frac{4n^2-4}{1-n}$
29. $\frac{x-3}{9-x^2}$
30. $\frac{4n-8}{48-12n^2}$
31. $\frac{1-n}{3n^2-3}$

B

32. $\frac{x^2+2x+1}{x^2+6x+5}$
33. $\frac{x^2-7x+12}{x^2-2x-8}$
34. $\frac{x^2-3x-10}{x^2-6x-16}$

For use with pages 304–313

Write the ratios in simplest form.

A

1. 18 points to 12 points
2. 4 days to 40 days
3. \$6 to \$24
4. 9 students to 36 students
5. There are 21 students in a math class.
   The ratio of girls to boys is 3 to 4.
   How many boys in the class? How many girls?

Solve the proportion.

6. $\frac{2}{5} = \frac{x}{20}$
7. $\frac{3}{9} = \frac{4}{x}$
8. $\frac{5}{a} = \frac{10}{100}$
9. $\frac{3}{1} = \frac{n}{8}$
10. $\frac{x}{3 + x} = \frac{2}{3}$
11. $\frac{2n - 3}{n + 1} = \frac{3}{4}$
12. $\frac{x + 2}{x - 2} = \frac{1}{5}$
13. $\frac{2y + 6}{y} = \frac{4}{5}$
14. Five apples cost 70¢. How much will 15 apples cost?
15. You can walk 4 km in 2 hours. How many kilometers can you walk in 3 hours?

Give the ratio in simplest form. Use the same units in writing the ratio.

B

16. 2 weeks to 21 days
17. 50¢ to $2.50
18. 25 cm to 6 m
19. 15 seconds to 3 minutes
20. A poll showed that TV audiences favored three programs in the ratio 6 to 5 to 2. Out of a viewing audience of 260,000, how many might be watching each program?

For use with pages 314–324

Simplify.

A

1. $\frac{1}{2} \cdot \frac{1}{8}$
2. $\frac{4}{7} \cdot \frac{21}{2}$
3. $\frac{a}{b} \cdot \frac{2a}{b}$
4. $\frac{x^2 - y^2}{2x + 2y} \cdot \frac{6}{5}$
5. $\frac{9}{10} \div \frac{3}{20}$
6. $\frac{3x - 3}{2x} \div \frac{3}{4x}$
7. $\frac{x^2 - 9}{3} \div (x + 3)$
8. $\frac{14}{x - 1} \div \frac{7}{x^2 - 1}$
9. $\frac{5n}{3} - \frac{9n}{3}$
10. $\frac{12}{10x} - \frac{4}{10x}$
11. $\frac{3n}{n - 1} - \frac{1}{n - 1}$
12. $\frac{a - 2}{a + b} - \frac{a - 4}{a + b}$
13. $\frac{3}{4} + \frac{1}{2}$
14. $\frac{6}{5} + \frac{1}{4}$
15. $\frac{x}{2} - \frac{x}{5}$
16. $\frac{2n}{3} - \frac{n}{2}$
17. $\frac{1}{x} + \frac{3}{y}$
18. $\frac{2}{a} - \frac{3}{a^2}$
19. $\frac{9}{ab^2} - \frac{1}{a^2b}$
20. $\frac{3}{2xy} - \frac{2}{xy}$

B

21. $\frac{x^2 - y^2}{x^2 - 64} \cdot \frac{x - 8}{x + y}$
22. $\frac{x - 3}{1 - x} \cdot \frac{x - 1}{x^2 - 9}$
23. $\frac{3a - 6}{5a} \cdot \frac{5a^3 - 15a^2}{a - 2}$

For use with pages 327–332

Solve.

A

1. $\frac{x}{5} - \frac{x}{2} = 30$
2. $\frac{x}{6} - \frac{x}{5} = 3$
3. $\frac{x}{3} = \frac{1}{6} + \frac{1}{2}$
4. $\frac{1}{4} - \frac{1}{6} = \frac{3}{n}$
5. $\frac{2x}{3} - \frac{x}{5} = \frac{14}{15}$
6. $\frac{2}{n} + \frac{3}{4n} = \frac{1}{8}$
7. $\frac{n}{3} + 10 = \frac{4n}{2}$
8. $\frac{x}{2} + \frac{3x}{8} = 7$
9. $\frac{3n}{4} + 5 = \frac{n}{3}$

10. Frank can shovel snow from the driveway in 2 hours. His younger brother Mark would need 3 hours. If they work together, how long will it take?

	Frank	Mark	Together
Hours needed	2	3	h
Part done in one hour	?	?	?

11. Mary would need 4 hours to paper a room. Her assistant, Brenda, would need 6 hours. How long would it take them together?

	Mary	Brenda	Together
Hours needed	?	?	?
Part done in one hour	?	?	?

Solve.

B

12. $\frac{x - 1}{2} + \frac{x + 2}{4} = 9$
13. $\frac{12 - x}{3} + \frac{3 + x}{2} = \frac{1}{2}$
14. $\frac{n + 5}{2} - \frac{n - 5}{6} = \frac{5}{8}$

15. A gardener can do his weekly job in 5 hours. One week he had a helper, and they finished in 3 hours. How long would the helper take working alone?

CHAPTER 10

For use with pages 348–353

Add or subtract.

A

1. $0.25 + 0.36$
2. $0.05 + 9.12$
3. $6.25 - 0.95$
4. $9.14 - 8.39$
5. $0.3 + 0.205$
6. $6.27 - 4.39$
7. $5.1 - 3.46$
8. $2.41 + 1.995$

Multiply.

9. 5×9.5
10. 9.3×0.1
11. 6.25×10
12. 100×4.5
13. 0.59×6.03
14. 0.78×1.35
15. 0.45×9.5
16. 0.7×6.703

Divide. Round to one decimal place.

17. $3\overline{)3.7}$
18. $0.03\overline{)3.74}$
19. $0.9\overline{)6.17}$
20. $1.5\overline{)10.06}$
21. $6.2\overline{)3.06}$
22. $7.8\overline{)9.31}$
23. $0.16\overline{)8.45}$
24. $0.55\overline{)6.95}$

Express the fraction as a decimal. Use three dots with a repeating decimal.

25. $\frac{2}{5}$
26. $\frac{1}{6}$
27. $\frac{1}{7}$
28. $\frac{5}{6}$
29. $\frac{5}{7}$
30. $\frac{7}{8}$
31. $\frac{1}{12}$
32. $\frac{1}{3}$
33. $\frac{2}{3}$
34. $\frac{1}{4}$

For use with pages 354–357

Solve. Round answers to the nearest tenth.

A

1. $0.20x = 0.1x + 1$
2. $0.5 - 0.5n = 0.2n - 3$
3. $0.06x = 360$
4. $0.05y - 5 = 3y$
5. $0.07x + 25 = 0.124$
6. $0.01(5 - n) = 0.04(6)$
7. The sum of two numbers is 50. The product of 0.75 and the smaller is the same as the product of 0.5 and the greater. What are the numbers?

B

8. $0.83n + 0.63(45 - n) = 0.71 - 45$
9. $0.07(8000 + n) = 600 + 0.05n$
10. A rectangular garden is 1.4 times as long as it is wide. Its perimeter is 19.6 meters. What is the length of the garden?

For use with pages 359–364

Complete the table.

	Percent	Decimal	Fraction
A 1.	5%	?	?
2.	?	?	$\frac{1}{10}$
3.	?	?	$\frac{4}{5}$
4.	?	0.125	?
5.	9%	?	?
6.	50%	?	?
7.	?	0.06	?

Compute.

8. 20% of $5.80 **9.** 10% of $6.30 **10.** 25% of $5 **11.** 33% of $9.30

Complete the table.

	On Sale, 50% off!		
	Item	Original price	Discount
12.	Skates	$45	?
13.	Cap	$6.20	?
14.	Gloves	$22	?

	On Sale, $33\frac{1}{3}$% off!		
	Item	Original price	Sale price
15.	Chair	$27	?
16.	Table	$63	?
17.	TV	$360	?

For use with pages 365–368

Solve.

A **1.** What percent of 20 is 4?

2. 10% of $50 is what amount?

3. 35% of 1000 is what number?

4. 5% of what number is 45?

5. 60% of what number is 36?

6. What percent of 300 is 60?

7. The meals tax in one state is 8%. What is the tax on a bill of $12.25?

8. At one college, 26% of the students own cars. This means that __?__ students out of every 1000 own cars.

B 9. A security system installed on one car model costs $180. What percent is this of the $7200 selling price?

For use with pages 369–372

A 1. Find the interest earned in one year on $6500 at 6%.

2. How long will it take $2000 to double at 5% interest?

3. How much must Cora invest at 5% to earn $300 interest in a year?

4. A college invested $10,000, part at 5% and part at 7%. If it receives $620 in interest for a year, how much was invested at each rate?

B 5. The Ramseys owe $200. On part of it they must pay 9% interest and on the rest 18%. At the end of the year they must pay $22.50 in interest. How much do they owe at each rate?

For use with pages 373–376

A 1. The Sports Club took in $4125 at its annual show. Student tickets were $1.50; all others $2.25. If 2000 people bought tickets, how many were student tickets?

	Number	× unit price	= receipts
Student	n	?	?
Others	$2000 - n$	?	?
			$4125

2. A solution of salt and water contains 4% salt. How much water should be added to 15 liters of the solution to reduce the salt content to 2%?

	Amount of solution	× % salt	= Amount salt
Original solution	15	4%	0.04×15
Water added	x	0	?
New solution	$15 + x$	2%	?

CHAPTER 11

For use with pages 388–393

Find the value. Use the table on page 390 where necessary.

A

1. $\sqrt{49}$ 2. $\sqrt{100}$ 3. $\sqrt{25}$ 4. $-\sqrt{16}$ 5. $\sqrt{1}$

6. $-\sqrt{64}$ 7. $\sqrt{0}$ 8. $\sqrt{4^2}$ 9. $\sqrt{6^2}$ 10. $\sqrt{8^2}$

11. $-\sqrt{12}$ 12. $-\sqrt{7}$ 13. $-\sqrt{3}$ 14. $\pm\sqrt{66}$ 15. $-\sqrt{76}$

Round the value to one decimal place.

16. $\sqrt{13}$ 17. $\sqrt{98}$ 18. $-\sqrt{52}$ 19. $\sqrt{63}$ 20. $\sqrt{70}$

Is the number rational or irrational?

21. $\sqrt{16}$ 22. $\frac{1}{2}$ 23. $\sqrt{2}$ 24. 10 25. π

B

26. The area of a square is 60 cm^2.
Find the length of one side correct to two decimal places.

For use with pages 394–395

Find the value. Leave your answers in simplest radical form.

A

1. $\sqrt{500}$ 2. $\sqrt{320}$ 3. $\sqrt{6400}$ 4. $\sqrt{405}$

5. $\sqrt{147}$ 6. $\sqrt{112}$ 7. $\sqrt{729}$ 8. $\sqrt{120}$

B

9. $\sqrt{81x^2}$ 10. $\sqrt{49x^4}$ 11. $\sqrt{5x^4y^2}$ 12. $\sqrt{10a^2b^3}$

For use with pages 396–397

Find the two solutions to the equation.

A

1. $x^2 = 49$ 2. $n^2 = 25$ 3. $y^2 = 36$

4. $\frac{n}{4} = \frac{4}{n}$ 5. $x^2 = 28$ 6. $x^2 + 1 = 65$

Find the two solutions correct to one decimal place.

B

7. $x^2 = 18$ 8. $y^2 = 24$ 9. $x^2 - 5 = 15$

For use with pages 398–401

Solve. Give your results correct to one decimal place.

A

1. The lengths of the sides of a right triangle are 4 cm and 6 cm. Find the length of the hypotenuse.

2. The hypotenuse of a right triangle is 15 m long. Find the length of one side if the length of the other is 10 m.

3. The hypotenuse of a right triangle is 9 cm long. Find the length of one side if the length of the other is 5 cm.

B

4. A guy wire is attached to a tower 10 meters tall. It is to be fastened at a point on the ground 5 meters from the foot of the tower. How long must the wire be?

For use with pages 402–407

Simplify. Leave results in simplest radical form.

A

1. $\sqrt{\frac{1}{9}}$ 2. $\sqrt{\frac{4}{49}}$ 3. $\sqrt{\frac{9}{100}}$ 4. $\sqrt{\frac{n^2}{4}}$ 5. $\sqrt{\frac{x^2}{25}}$ 6. $\sqrt{\frac{4n^2}{36}}$

7. $\sqrt{6} \cdot \sqrt{6}$ 8. $\sqrt{8} \cdot \sqrt{2}$ 9. $\sqrt{4} \cdot 2\sqrt{9}$

10. $\sqrt{n} \cdot \sqrt{4n}$ 11. $5\sqrt{3x} \cdot 3\sqrt{12x}$ 12. $y\sqrt{6} \cdot 8\sqrt{2y^2}$

B

13. $\sqrt{3} \cdot \sqrt{5} \cdot \sqrt{15}$ 14. $-x\sqrt{5} \cdot x\sqrt{75}$ 15. $\sqrt{\frac{1}{2}} \cdot \sqrt{8x^4}$

Find the decimal value correct to two decimal places.

16. $5\sqrt{\frac{2}{3}}$ 17. $\sqrt{\frac{5}{8}}$ 18. $\sqrt{\frac{1}{2}} \cdot \sqrt{\frac{1}{3}}$ 19. $\sqrt{\frac{1}{6}} \cdot \sqrt{\frac{5}{12}}$

For use with pages 408–409

Express in simplest radical form.

A

1. $3\sqrt{3} - 9\sqrt{3}$ 2. $10\sqrt{2} - \sqrt{2}$ 3. $\sqrt{8} + \sqrt{2}$

4. $-\sqrt{6} + 3\sqrt{6}$ 5. $4\sqrt{18} + \sqrt{50}$ 6. $6\sqrt{5} - \sqrt{3} + \sqrt{5}$

B

7. $\sqrt{3n} - \sqrt{27n}$ 8. $\sqrt{n} + 4\sqrt{n^3}$ 9. $\sqrt{18} - \sqrt{32}$ 10. $\sqrt{3} - (\sqrt{12} + \sqrt{45})$

CHAPTER 12

For use with pages 420–425

Solve.

A

1. $n(n + 5) = 0$
2. $x(x - 10) = 0$
3. $m(m - 2) = 0$
4. $(y + 2)(y + 3) = 0$
5. $(n - 5)(n - 2) = 0$
6. $(x + 3)(x - 4) = 0$
7. $n^2 + n = 0$
8. $x^2 - 4x = 0$
9. $y^2 - 9y = 0$
10. $x^2 + 2x - 15 = 0$
11. $n^2 + 11n + 24 = 0$
12. $y^2 + 3y - 18 = 0$
13. $n^2 = 3n - 2$
14. $x^2 + x = 12$
15. $y^2 - 10 = -3y$

B

16. $(4x - 5)(x + 1) = 0$
17. $(3n - 18)(4n - 4) = 0$
18. $\frac{x}{3} = \frac{5}{x - 2}$

For use with pages 426–429

Solve by using square roots. If there is no solution, write "no solution."

A

1. $x^2 = 0$
2. $2x^2 = 72$
3. $4y^2 = -8$
4. $x^2 + 64 = 100$
5. $(x - 2)^2 = 36$
6. $(2x - 1)^2 = 49$
7. $(x - 4)^2 = 2$
8. $n^2 = 1$
9. $(3n + 1)^2 = 8$

B

10. $2y^2 + 3 = 19$
11. $3n^2 - 9 = 15$
12. $3n^2 + 11 = 38$

For use with pages 430–433

Solve by using the quadratic formula. Leave irrational numbers in simplest radical form.

A

1. $a^2 - 5a + 6 = 0$
2. $n^2 + 5n - 3 = 0$
3. $x^2 - 5x + 1 = 0$
4. $x^2 - 2x - 3 = 0$
5. $x^2 + x - 20 = 0$
6. $3y^2 + 13y + 4 = 0$
7. $x^2 - x - 5 = 0$
8. $2x^2 - 11x + 12 = 0$
9. $2x^2 - x - 4 = 0$
10. $x^2 - 16 = 0$
11. $4n^2 - 3n - 1 = 0$
12. $x^2 - 6x - 2 = 0$

Solve by two methods:

a. Using the quadratic formula

b. Factoring

B **13.** $x^2 - 5x - 14 = 0$ **14.** $n^2 - n = 0$ **15.** $y^2 - 8y + 16 = 0$

16. $x^2 - 6x + 8 = 0$ **17.** $3x^2 - 9x + 6 = 0$ **18.** $5x^2 - 10x - 75 = 0$

For use with pages 434–435

A **1.** One number is 5 more than another number. Their product is 14. Find the numbers.

2. One number is 3 less than another number. Their product is 28. What are the numbers?

3. Find two consecutive numbers whose product is 210.

4. The width of a rectangle is 2 cm less than its length. Its area is 24 cm^2. Find its dimensions.

B **5.** The sum of the squares of three consecutive numbers is 29. Find the numbers.

6. The sum of the length and width of a rectangle is 12 cm. Its area is 32 cm^2. Find its dimensions.

For use with pages 436–439

For each equation do these three things:

a. Make a table of values.

b. Draw the graph.

c. Give the coordinates of the vertex.

A **1.** $y = x^2 - 1$ **2.** $y = x^2 - 4x$ **3.** $y = -x^2$

4. $y = \frac{1}{4}x^2$ **5.** $y = x^2 - 2x + 1$ **6.** $y = x^2 - 2x - 3$

7. $y = x^2 + 4x + 2$ **8.** $y = (2 - x)(2 + x)$ **9.** $y = x^2 + 4x - 3$

Cumulative Review for Chapters 1–3

BASIC SKILLS

Find the value of the expression.

1. $(5 + 3) \cdot 8$
2. $6 - (3 \cdot 1)$
3. $9 + (6 - 3)$
4. 6^2
5. 5^1
6. 3^3
7. $5 \cdot 0$
8. $5 + 0$
9. $\frac{5}{5}$
10. $\frac{5}{1}$
11. $5 \cdot 1$
12. $(5 - 5)(2 \cdot 5)$
13. $-1 + (-4)$
14. $-1(-4)$
15. $-1 - (-4)$
16. $(-4)^2$
17. $-(4)^2$
18. $-(-4)^2$

If $a = 1$, $b = 2$, and $c = 3$, find the value of the expression.

19. $a + b$
20. $-a + b$
21. $-ab$
22. $c \div (-a)$
23. $2(a + c)$
24. $2a + c$
25. $4(a - b)$
26. $4a - b$
27. $-abc$

Simplify.

28. $3x + 4x$
29. $10b - b$
30. $2y - 4y$
31. $4x + 5x - 2$
32. $3a - 4a - 7a$
33. $2a + 4 - 2a$
34. $3a - 5a + c - c$
35. $3 \cdot a \cdot a \cdot b \cdot b$
36. $3(x - 1) + 2$
37. $4 \cdot \frac{x}{4}$
38. $3x - (-2x)$
39. $-(a - 2)$
40. $(x + y)(-2)$
41. $4(x - 3) - x + 15$
42. $-(x^2 + x - 2)$

EQUATIONS AND INEQUALITIES

1. Is $x + 2 = 7$ true if $x = -5$?
2. Is -1 a solution of $x - 1 = -1$?

3. a. Which of the numbers 1, 2, 3, 4 are solutions of $x > 3$?
   b. Are there other solutions of $x > 3$?
   c. Graph the solution set of $x > 3$ on the number line.

Solve.

4. $x - 2 = 4$
5. $y + 6 = 5$
6. $x + 3 = 3$
7. $9 = x + 3$
8. $3x = 18$
9. $-3x = 18$
10. $\frac{x}{2} = 9$
11. $3x - 1 = 20$
12. $5x + 25 = 0$
13. $3 + 3y - y = 21$
14. $7x = 25 + 2x$
15. $3n - n = -16$
16. $2(x - 4) = -4$
17. $5 - (3 - y) = 10$
18. $4 - (2x - 1) = x - 1$
19. $-(y - 4) + 5 = 9y - 1$
20. $\frac{x}{9} = -9$
21. $\frac{x}{-3} = -3$

Write >, <, or = in place of the ?.

22. -2 __?__ -4
23. $3 - (-4)$ __?__ $-3(-4)$
24. $6 - (-2)$ __?__ $(-2)(-4)$
25. Can you give a value for $\frac{0}{6}$? for $\frac{6}{0}$? State the value if possible.

PROBLEMS

1. A number is doubled. Then 4 is added. The sum is 80. What is the number?
2. Find two consecutive whole numbers whose sum is 65.
3. Mary has twice as many dimes as Don. Together they have 105 dimes. How many dimes does each have?
4. South High School played 32 games. The team won three times as many as they lost. How many did they win?
5. Tom says "If I triple the number I'm thinking of and then subtract 5, I get 25." What is the number?

Cumulative Review for Chapters 1–6

BASIC SKILLS

Simplify.

1. $15 \cdot \frac{0}{15x}$

2. $-(-3)^2$

3. $-4a + a - 7a$

4. $2(x - 8) + 2x$

5. $10x^2 - x^2$

6. $(-a)^2 + 4a + 1$

7. $(x - 2y) + (3x + 2y)$

8. $(x^2 - 2x + 1) - (x^2 + 2x - 1)$

Multiply.

9. $4a \cdot a^3$

10. $(-3xy)(4x^2y^2)$

11. $(-3ab)^3$

12. $2x(5 - y)$

13. $(x + 6)(x - 1)$

14. $(n - 4)(n + 4)$

15. $(y + 2)(y - 3)$

Express as a trinomial.

16. $(x - 3)^2$

17. $(x + 4)^2$

18. $(2x - 3)^2$

19. $(3x - 2y)^2$

Divide.

20. $\frac{6n^4}{-n^2}$

21. $\frac{8n^3}{2n}$

22. $\frac{-24a^2b^2}{-12ab^2}$

23. $\frac{-72xy^2}{8y^2}$

24. $\frac{x^2 - 3x^3 + 5x^4}{-x^2}$

25. $\frac{5a^2 - a^3 - a^4}{a}$

26. $\frac{5xy - 15x^2y^2}{-5x}$

Factor.

27. $3xy - 6x^2y^2$

28. $x^2 + 5x + 6$

29. $y^2 + 7y + 6$

30. $x^2 + 4x + 4$

31. $n^2 + 9n - 22$

32. $x^2 - 9y^2$

EQUATIONS AND FORMULAS

Solve.

1. $x - 5 = 20$

2. $40 = 20 - n$

3. $4a + 1 = 13$

Solve.

4. $\frac{n}{4} = -4$

5. $2(x - 1) = 14$

6. $3a = 25 - 2a$

7. $12x + 8 - 10x = 16$

8. $5x + 30 = 3x$

9. $-10 = -3x + 2$

10. $3x - (2x + 2) = 5x + 6$

11. $(2a + 1) - (5a - 1) = -1$

Solve for x.

12. $6x = 24$

13. $6x = y$

14. $kx = y$

Solve for y.

15. $y + 7 = 10$

16. $y + x = 2$

17. $4x - y = 0$

18. Solve $P = 8x$ for x.

19. Solve $D = rt$ for r.

PROBLEMS

1. Find the perimeter of a 4-cm by 4-cm square.

2. Find the area of a rectangle with length $2x$ and width $4x$.

3. A triangle has an 8-cm base and a 4-cm height. Find its area.

4. Write an expression for the volume of a rectangular box whose dimensions are $4x$, $3x$, and $2x$.

5. Find two consecutive numbers whose sum is 25.

6. Wendy bought some hooks at 12¢ each. She used two and sold the rest at 20¢ each. Her profit was $1.20. How many hooks did she buy?

	price	× number	= Cost
Sell	?	?	?
Buy	?	?	?
			120

Cumulative Review for Chapters 1–9

BASIC SKILLS

Let $x = 3$. Find the value if possible. If not, write *impossible*.

1. $\frac{x}{0}$
2. $\frac{x - 5}{x - 5}$
3. $\frac{x - 3}{x - 3}$
4. $\frac{x}{3}$
5. $\frac{0}{x}$

Simplify.

6. $3n - 15n$
7. $x(3xy)(-3y)$
8. $5(2a - 3b) + 15b$
9. $(2ab^2)^2$
10. $(-4x^2y^2)^2$
11. $-10(-3y^2)^3$
12. $(3x - 4y) - (9x - 4y)$
13. $\frac{4x^2y}{-4y}$
14. $\frac{x}{3} - \frac{2x}{6}$

Multiply.

15. $(x + 4)(x + 2)$
16. $(n - 5)(n + 5)$
17. $(2x - 1)(4x + 3)$
18. $(2x + 5)^2$
19. $x(x - 2)(x + 3)$
20. $(2a + b)(3a - 2b)$

Factor.

21. $2x^2 - 4y^2$
22. $3a^2 - 3b^2$
23. $n^2 - 3n + 2$
24. $m^2 - 5m - 14$
25. $x^2 + 8x + 16$
26. $x^2 - 6x - 27$
27. $4y^2 - 9$
28. $n^2 - 2n - 8$
29. $x^2 + x - 6$

EQUATIONS AND GRAPHS

Solve for x.

1. $4(x + 1) = 16$
2. $5x - 15 = 2x$
3. $x + 8 = -3x$
4. $x + 3x + 2 = 18$
5. $6x - 1 = 3x + 11$
6. $-x + 9 = 4x + 4$

Solve for y.

7. $3 + y = x$
8. $2x + y = 4$
9. $5x - y = 10$
10. $x - y = 4$
11. $y - 2x = 8$
12. $5x + y = 0$

Draw the graph of the equation.

13. $y = x - 1$
14. $x + y = 2$
15. $y + 2x = 4$

Give the slope of the line without drawing the graph.

16. $y = -x - 9$
17. $y = \frac{1}{2}x + 2$
18. $y = 2x + 7$

Solve by graphing.

19. $\begin{cases} y - x = 0 \\ y + x = 6 \end{cases}$
20. $\begin{cases} x + y = 2 \\ x + 2 = y \end{cases}$
21. $\begin{cases} x + 1 = -y \\ y - x = 3 \end{cases}$

Solve by substitution.

22. $\begin{cases} y = x - 1 \\ x + y = 3 \end{cases}$
23. $\begin{cases} x = 2y \\ x + y = 9 \end{cases}$
24. $\begin{cases} x = y + 2 \\ y = 3x - 4 \end{cases}$

Solve by addition or subtraction.

25. $\begin{cases} 5x - y = 1 \\ 3x + y = 15 \end{cases}$
26. $\begin{cases} 2x - 3y = 1 \\ 3x - 2y = 4 \end{cases}$
27. $\begin{cases} 3x + y = 4 \\ x + 2y = 3 \end{cases}$

PROBLEMS

1. You can do a typing job in 2 hours. If Jean works alone, it takes 4 hours. How long would it take if you work together?

	You	Jean	Together
Hours needed	2	4	n
Part done in one hour	?	?	?

2. *Use two variables for this problem.* The sum of two numbers is 75. Their difference is 25. What are the numbers?

Cumulative Review for Chapters 1–12

BASIC SKILLS

Simplify.

1. $(3x - 4) - (5x + 6)$
2. $(n + 9) + (4n - 18)$
3. $(a^2 + 3a + 1) - (a^2 - 2a)$
4. $(n^2 - 3) - (n + 8)$
5. $(5x + y) - (6x + y)$
6. $3(n + 2) + 4(n - 2)$
7. $\frac{b^2 - 4}{b + 2}$
8. $\frac{x^2 - 4}{8} \cdot \frac{4}{x - 2}$
9. $\frac{3x^2 - 3}{2} \div \frac{x + 1}{16}$

Multiply.

10. $(3n - 1)(n + 6)$
11. $(n - 4)(n + 4)$
12. $(n + 3)(n + 3)$
13. $(5n - 2)(2n + 5)$
14. $(4x - 1)(x - 4)$
15. $(2x + 7)(7 - x)$

Factor completely.

16. $5n^2 + 25n + 100$
17. $3n^2 + 9n + 6$
18. $2n^2 - 8$
19. $x^2 - 7x + 6$
20. $x^2 - 5x - 6$
21. $x^2 - 5x - 14$

Find the square root. Leave your answer in simplest radical form.

22. $\sqrt{\frac{2}{16}}$
23. $\sqrt{\frac{4y^2}{9}}$
24. $\sqrt{\frac{1}{200}}$
25. $\frac{1}{4}\sqrt{\frac{16}{x^2}}$

Simplify.

26. $\sqrt{3} \cdot \sqrt{12}$
27. $3\sqrt{2} \cdot \sqrt{2x^2}$
28. $\sqrt{\frac{1}{4}} \cdot \sqrt{\frac{1}{3}}$
29. $\frac{\sqrt{2}}{\sqrt{5}}$
30. $\sqrt{8} - \sqrt{2}$
31. $3\sqrt{3} - 2\sqrt{12}$
32. $2\sqrt{27} - \sqrt{3}$
33. $4\sqrt{2} - \sqrt{18}$

EQUATIONS AND INEQUALITIES

Solve.

1. $3x - 1 = 4x - 9$
2. $2(2x - 1) = 6$
3. $2n + 3 - 3n + 5 = 0$

4. On the number line, graph the solution of $y < 2$.

Solve. (There will be two solutions to some equations.)

5. $\frac{2}{x} = \frac{x}{8}$ 6. $\frac{n}{3} - \frac{n}{5} = 16$ 7. $\frac{x}{x + 2} = \frac{4}{3}$

8. $a^2 - 3 = 13$ 9. $y^2 = 50$ 10. $0.3x + 0.9x = 3.6$

11. $0.4x = 4.8$ 12. $(n + 4)(n - 6) = 0$ 13. $0.04(100 - x) = 12 - 0.2x$

14. $(x + 2)(x - 2) = 0$ 15. $x(x - 4) = 0$ 16. $(3x - 6)(2x + 14) = 0$

17. Solve for y: a. $3y = 24$ b. $ay = 24$ c. $ay = c$

Solve for x and y.

18. $\begin{cases} x + y = 4 \\ x - y = 2 \end{cases}$ 19. $\begin{cases} x + 3y = 4 \\ 2x - 5y = -3 \end{cases}$ 20. $\begin{cases} 4x - 3y = 1 \\ 3x + 4y = 7 \end{cases}$

Solve by factoring.

21. $n^2 - 7n + 6 = 0$ 22. $y^2 - 2y - 3 = 0$ 23. $x^2 - 16 = 0$

Solve by the quadratic formula.

24. $2n^2 + 3n + 1 = 0$ 25. $3x^2 - 2x - 3 = 0$ 26. $2n^2 - 3n = 5$

PROBLEMS

1. A wholesaler sells nuts at 50¢ a kilogram. She mixes one kind worth 65¢ a kilogram with another worth 40¢. How much of each kind does she use to make 100 kg of the mixture?

	p	$\times\ n$	$= C$
65¢ Kind	65	?	?
40¢ Kind	?	?	?
Mixture	?	100	?

2. A lot is shaped like a right triangle. What is the length of the longest side of the lot if the other sides measure 30 m and 40 m?

Cumulative Review of Word Problems

Note: Students who have difficulty solving any of these problems should turn to the page number in parentheses for help.

1. I'm thinking of a number. If I add 7 to it and multiply the sum by 4, I get 68. What is the number? (p. 53)

2. The Eagles played 27 games and won 5 more games than they lost. How many games did they win? (p. 58)

3. Gregory is 6 years older than Lara. In two years the sum of their ages will be 30. How old is Gregory now? (p. 66)

4. After Armando and Pat bike in opposite directions for $2\frac{1}{2}$ hours, they are 80 km apart. If Pat's rate is 2 km/h more than Armando's rate, find how far each travels. (p. 126)

5. Sarah bought some erasers at 12 cents each and sold all but 2 for 20 cents each. If her profit was $1.52, how many erasers did she buy? (p. 130)

6. Harry has $3 more than Carol, and Carol has twice as much as Leon. Together they have $108. How much does each person have? (p. 59)

7. Five times a number exceeds twice the number by 36. What is the number? (p. 63)

8. A woman invests $6000, part at 6% interest and the rest at 5% interest. Her annual interest is $335. How much is invested at 6% interest? (p. 371)

9. If I use 5 L of gasoline to go 80 km, how many liters will I need to go 112 km? (p. 310)

10. The ratio of Carl's money to Rita's money is 4:5. Together they have $279. How much does Rita have? (p. 306)

11. Sean jogs from his home to the stadium at 10 km/h. His trip home takes exactly one hour less because he drives a car traveling at 60 km/h. How far is it from Sean's home to the stadium? (p. 126)

12. Juanita earns \$1.00 per hour more than Kelly. After working for 8 hours last Saturday, their combined pay was \$112. How much does each earn per hour? (p. 130)

13. It takes Simon 3 hours to mow the lawn and it takes Roger 2 hours. How long would it take them if they worked together? (p. 329)

14. In order to make a certain kind of concrete, you must mix sand, gravel, and cement in the ratio of 3 to 4 to 2. How much sand will be needed to make 45 cubic meters of concrete? (p. 306)

15. 120% of 55 is what number? (p. 365)

16. What percent of 75 is 30? (p. 365)

17. 30% of what number is 72? (p. 365)

18. A motorboat went 32 km upstream in 4 hours. The return trip took 2 hours. Find the rate of the current. (p. 284)

19. The sum of 2 numbers is 73, and their difference is 11. Find the numbers. (p. 277)

20. I am thinking of two numbers whose difference is 32. Twice the smaller number is 12 more than the larger number. What are the numbers? (p. 277)

21. The length of a rectangle is 5 cm more than twice its width. The perimeter is 46 cm. Find the length and width. (p. 279)

22. Find 3 consecutive integers whose sum is 363. (p. 60)

23. Two shirts and 3 pairs of pants cost \$96. Four shirts and 2 pairs of pants cost \$88. How much does each item cost? (p. 282)

24. Marcia is 4 years younger than Emile. She is also twice as old as Ken. The sum of their ages is 64. How old is Marcia? (pp. 66 and 278)

25. The length of a rectangle is 5 cm more than its width. The area is 150 cm^2. Find the length and width. (p. 434)

26. The square of a positive number is 20 more than 8 times the number. Find the number. (p. 435)

27. A specialty food store mixes Brazil nuts and cashews to get mixed nuts, which will sell for \$5.00 per kilogram. If the Brazil nuts cost \$4.50 per kilogram and the cashews cost \$6.00 per kilogram, how many kilograms of each should be combined to give 30 kilograms of mixed nuts? (p. 373)

28. Tom buys several notebooks for \$2.00 each. He sells some of them for \$3.00 each, but then has to lower his price to \$2.40 to sell the remaining 20. His profit is \$18.00. How many notebooks did he buy? (p. 130)

29. Hilary needs 5 hours to rake her lawn. But when Sylvia helps her, they can do the job in 3 hours. How long would it take Sylvia to rake the lawn by herself? (p. 329)

30. A baseball team buys 12 dozen game balls and 15 dozen practice balls for \$1440. For the same amount of money they could have bought 8 dozen game balls and 20 dozen practice balls. Find the cost of a dozen game balls. (p. 282)

31. A sport coat usually priced at \$48 is on sale at a 20% discount. If the sales tax is 5%, how much do you pay for the coat? (p. 363)

32. A pump can fill a 2700 L tank in 36 minutes. How long will it take to fill an 1800 L tank? (p. 312)

33. The perimeter of a rectangle is 40 cm, and the area is 64 cm^2. Find the dimensions of the rectangle. (p. 434)

34. A saltwater solution contains 25% salt. How much water should be added to 12 L of solution to get a mixture which is only 20% salt? (p. 374)

35. I am thinking of two consecutive even integers. The sum of their squares is 340. What are the integers? (p. 435)

Glossary

A

absolute value (p. 99) $|-5| = 5$; $|5| = 5$; $|0| = 0$. The symbol $| \, |$ means "absolute value of."

approximate values (p. 390) Values that are not exact.
In $\frac{1}{3} \approx 0.33$, the 0.33 is an approximate value.

area (p. 115) The amount of surface of a figure. Area is measured in square units.

associative property of addition (p. 12) $(a + b) + c = a + (b + c)$. It tells us that the way terms are grouped makes no difference.

associative property of multiplication (p. 16) $(ab)c = a(bc)$. It tells us that the way factors are grouped makes no difference.

average (p. 104) The sum of n numbers divided by n. If $n = 3$ and the numbers are 11, 20, and 35, the average is $\frac{11 + 20 + 35}{3} = \frac{66}{3} = 22$.

B

bar graph (p. 224) A statistical graph in which bars of different lengths stand for quantities.

BASIC (p. 102) A computer language.

binomial (p. 152) A polynomial with two terms. $2x + 3y$ is a binomial.

boundary line (p. 251) Graph of the equality part of certain inequalities.

broken line graph (p. 228) A statistical graph in which points are joined by straight lines.

C

circle graph (pp. 361, 362) A circular region that shows the relation of parts to a whole.

circumference (p. 112) The perimeter of a circle.

coefficient (pp. 10, 14) The number 5 is the (numerical) coefficient of x in the expression $5x$.

common factor (p. 186) A number that is a factor of two or more numbers. 2 is a common factor of 4 and 14.

common monomial factor (p. 188) A monomial that is a common factor of each of the terms of a polynomial.

commutative property of addition (p. 12) The property $a + b = b + a$. It tells us that the order of terms makes no difference.

commutative property of multiplication (p. 16) $ab = ba$. It tells us that the order of factors makes no difference.

consecutive numbers (p. 60) Numbers that follow each other in order.

coordinates (p. 233) The numbers in an ordered pair representing a point; in $(3, -2)$, 3 and -2 are coordinates of the point $(3, -2)$.

cubed (p. 14) The third power of x, namely x^3, may be read "x cubed."

D

denominator (p. 298) In the fraction $\frac{1}{x}$, x is called the denominator.

diameter (p. 112) A line through the center of a circle and with endpoints on the circle.

distributive property (p. 18) $a(b + c) = ab + ac$ and $a(b - c) = ab - ac$.

E

equation (p. 25) Any number sentence containing the = sign. $6 + 4 = 10$ and $x + 2 = 7$ are equations.

equivalent expressions (p. 102) Two expressions are equivalent if they represent the same number for all values of the variable.

evaluate (p. 16) To find the value of.

exponent (p. 14) In a power, the number of times the base appears as a factor. In 5^3, the number 5 is the base and ^{3} is the exponent.

F

factor (pp. 14, 188) In a multiplication, each number multiplied is a factor of the product. In $2 \cdot 4$, 2 and 4 are factors of the product 8. In $2y$, 2 and y are factors of $2y$.

formula (p. 112) A rule for finding a desired value. $P = 4s$ is a formula for the perimeter P of a square when the length of one side s is known.

function (p. 245) A quantity that depends on a second quantity is a function of it.

G

graph of an equation (p. 241) The line representing all the solutions of an equation.

graph of an inequality (p. 251) Represented by a half-plane.

graph of a solution (p. 80) The point or points on the number line paired with the solution of an equation or inequality.

greatest common factor (p. 187) The greatest number that is a common factor of two or more numbers. 9 is the greatest common factor of 18 and 81.

H

horizontal axis (p. 233) Horizontal number line on which one can locate the first number in an ordered pair.

hypotenuse (p. 398) Side opposite the right angle in any right triangle.

I

inductive reasoning (p. 212) Reaching a conclusion based on specific cases.

inequality (p. 27) A number sentence stating that two numbers are not equal. $a > 6, b < 2, 6 \neq 3$ are all inequalities.
integers (p. 78) The positive and negative whole numbers and zero.

$$\ldots -3, -2, -1, 0, 1, 2, 3 \ldots$$

interest (p. 369) Money paid for the use of money.
intersection (p. 262) The point where two lines cross.
irrational numbers (p. 392) Numbers that cannot be expressed as the ratios of two integers. $\sqrt{2}$ and $\sqrt{7}$ are irrational numbers; 2 and 7 are not perfect squares. The decimal form of an irrational number never repeats and never ends.

L

law of Pythagoras (p. 398) In any right triangle, the square of the length of the hypotenuse equals the sum of the squares of the lengths of the other two sides.
like terms (p. 10) $2x$ and $5x$ are like terms. The variable part of each term is the same.
liter (p. 121) A metric measure of volume or capacity.

M

monomial (p. 152) A polynomial of one term. $5x$ is a monomial.

N

negative numbers (p. 78) All numbers less than zero.
number line (p. 78) A line on which numbers may be pictured.
numerator (p. 298) In the fraction $\frac{2}{3}$, 2 is called the numerator.

O

opposite of a number (p. 85) The opposite of 8 is -8; the opposite of -4 is 4.
ordered pair (pp. 233, 237) A pair of numbers in which their order is important. The point (2, 5) is not the same as the point (5, 2).
origin (p. 233) Point (0, 0); intersection of horizontal axis and vertical axis.

P

parabola (p. 436) The graph of a quadratic function.
parallel lines (p. 266) Lines (in a plane) which never intersect.
parentheses (p. 8) The marks ( ) used to group terms.
percent (p. 359) The ratio of a number to 100. A number like 25% is called a percent. It expresses the ratio of 25 to 100. $25\% = \frac{25}{100}$.
perfect square (p. 388) Any number which is the square of an integer.
perimeter (p. 112) The distance around a figure.

pi (π) (pp. 190, 392) The Greek letter standing for the ratio of the circumference of a circle to its diameter, approximately 3.14 or $\frac{22}{7}$.

pictograph (p. 224) A kind of bar graph in which rows of symbols are used instead of bars.

plotting a point (p. 234) Locating it by its coordinates.

point on a graph (p. 233) Located by an ordered pair of numbers.

polynomial (p. 152) An algebraic expression of one or more terms connected by + or − signs. $3x + y$ and $x^2 + 2xy + y^2$ are polynomials.

positive numbers (p. 78) All numbers greater than zero.

power (p. 14) A number such as x^3 used to show a product in which all the factors are the same. The number of factors is shown by the exponent.

prime factors (p. 186) Factors that are prime numbers.

prime number (p. 186) A number with no factors except itself and 1. For example, 2 and 5 are prime numbers.

prism (p. 119) A 3-dimensional figure with two parallel bases. The bases are the same size and shape. An ordinary box is an example of a prism.

proportion (p. 308) A statement that two ratios are equal. $\frac{4}{6} = \frac{2}{3}$ is a proportion.

Q

quadratic equation (p. 422) If an equation in x has an x^2 term but no term with a higher power of x, it is called a quadratic equation. $x^2 + 6x + 5 = 0$ is a quadratic equation but $x^3 + x^2 + 1 = 0$ is not.

quadratic formula (p. 430) The formula $x = \frac{-b \pm \sqrt{b^2 - 4ac}}{2a}$ may be used to solve an equation of the form $ax^2 + bx + c = 0$.

quadratic function (p. 436) The equation $y = x^2 - 1$ is an example of such a function. y depends on the value of x.

R

radical (p. 388) A number in which the radical sign $\sqrt{}$ appears. $\sqrt{5}$ is a radical number, or radical.

radius of a circle (p. 190) The distance from the center to the circle.

ratio (p. 304) The ratio of two numbers is their quotient. The ratio of 6 to 4 is $\frac{6}{4}$, or $\frac{3}{2}$.

rational numbers (pp. 352, 392) Numbers that can be expressed as ratios of two integers. $\frac{1}{2}$, $\frac{2}{1}$, and $\frac{-5}{7}$ are rational numbers.

rationalizing the denominator (p. 406) When a fraction has a radical in the denominator, it can be simplified by multiplying both numerator and denominator by a number that will make the denominator a whole number. This is called "rationalizing the denominator." For example:

$$\frac{1}{\sqrt{3}} = \frac{1 \cdot \sqrt{3}}{\sqrt{3} \cdot \sqrt{3}} = \frac{\sqrt{3}}{3}$$

real numbers (p. 392) The positive numbers, negative numbers, and zero.

reciprocals (p. 316) $\frac{3}{4}$ and $\frac{4}{3}$ are reciprocals. Their product is 1.

repeating decimal (p. 352) When a fraction is changed to decimal form, sometimes a digit or group of digits repeats without end. Such a decimal is called a repeating decimal.

In $\frac{2}{3} = 0.666\ldots$, the 6 repeats indefinitely.

In $\frac{3}{11} = 0.2727\ldots$, the 27 repeats indefinitely.

right angle (p. 398) An angle of 90°.

right triangle (p. 398) A triangle with a right angle.

root of an equation (p. 25) A value of the variable which makes the equation true. 4 is a root of $2 + x = 6$.

S

simplify (p. 10) To replace a number or expression by the simplest equivalent one.

slope of a line (pp. 243, 266) The ratio of rise to run; the coefficient of x when an equation is solved for y. In the equation $y = \frac{1}{2}x + 2$, $\frac{1}{2}$ is the slope.

solution of an equation (p. 25) A value of the variable which makes the equation true. 2 is a solution of $x + 2 = 4$.

solution of two equations (p. 262) The ordered pair which is the solution of both equations. (1, 3) is the solution of $y = 3x$ and $y = -x + 4$.

solution pair (p. 237) A pair of numbers which is a solution of an equation in two variables.

square root (p. 388) Each of the numbers 5 and −5 is a square root of 25. 5 is the positive, or principal, square root of 25. −5 is the negative square root of 25. $\sqrt{25}$ means the positive square root. $-\sqrt{25}$ means the negative square root. $\pm\sqrt{25}$ means both roots.

squared (p. 14) The second power of x, namely x^2, may be read "x squared."

T

term (p. 10) In the expression $2a + 3b - 4c$ the terms are $2a$, $3b$, and $4c$.

terminating decimal (p. 352) When a fraction is changed to its decimal form, the decimal terminates if the division "comes out even." Such a decimal is called a terminating decimal.

In $\frac{3}{4} = 0.75$, the 0.75 is a terminating decimal.

trinomial (p. 152) A polynomial with three terms. $a^2 + 2ab - 4b^2$ is a trinomial.

trinomial square (pp. 168, 198) The square of a binomial; when the two factors of a trinomial are equal it is a trinomial square. $(x + 2)^2 = x^2 + 4x + 4$ (a trinomial square).

U

unlike terms (p. 10) $6x + 3y$ contains unlike terms. Their variable parts (x and y) are not the same.

V

value of an expression (p. 5) The value when numbers are substituted for letters in the expression.
variable (p. 5) A letter used to represent a number.
variable expression (p. 5) An expression containing a variable.
vertex of a parabola (p. 436) Highest or lowest point.
vertical axis (p. 233) Vertical number line on which one can locate the second number in an ordered pair.
volume (p. 119) The amount of space in a 3-dimensional figure. Volume is measured in cubic units.

X

***x*-axis** (p. 233) Horizontal axis.

Y

***y*-axis** (p. 233) Vertical axis.

Index

K

L

M

N

O

S

CREDITS

Book design and production: TEXTART, Inc.
Cover design: Sylvia Steiner
Illustrations: Eva Burg Vagreti
Technical art: Network Graphics and ANCO / Boston

PHOTOGRAPHY

Chapter openers
pages 2–3: Georg Gerster / Photo Researchers
pages 36–37: Peter Menzel / Stock Boston
pages 76–77: Jen & Des Bartlett / Photo Researchers
pages 110–111: Daniel Brody / Stock Boston
pages 150–151: Kachaturian / International Stock Photo
pages 184–185: Cary Wozinsky / Stock Boston
pages 222–223: Courtesy of Santa Fe Industries, Inc.
pages 260–261: Susie Fitzhugh / Stock Boston
pages 296–297: Stephen F. Rosenthal / Cambridge Seven Assoc., Inc.
pages 346–347: Georg Gerster / Photo Researchers
pages 386–387: George Holton / Photo Researchers
pages 418–419: Steve Elmore

Other photographs
page 54: Christopher Morrow / Stock Boston
page 56: Dean Abramson / Stock Boston
page 58: Andrew D. Bernstein / Focus On Sports
page 61: Philip Jon Bailey
page 78: L.L.T. Rhodes / Taurus
page 87: Robert Hauser / Photo Researchers
page 88: Michael Grecco / Stock Boston
page 112: David Madison / duomo
page 134: Focus On Sports
page 138: Tim Carlson / Stock Boston
page 195: Ken Lax
page 278: Hilary M. Wallace
page 312: Courtesy of American Airlines, Inc.
page 318: Robert A. Isaacs / Photo Researchers
page 336: Jeffry W. Meyers / Stock Boston
page 358: Michael Weisbrot / Stock Boston
page 397: Ted Spiegel / Black Star

Answers to Self-Tests

CHAPTER 1

Page 13 **1.** 20 **2.** 8 **3.** 42 **4.** 2 **5.** 11 **6.** 8 **7.** 0 **8.** 7 **9.** $2xy$ **10.** not possible **11.** $8y$ **12.** $9w - 5$ **13.** $2a - 5$ **14.** $5m - 3$ **15.** $3x + 2y$ **16.** $6k + 5j$

Page 24 **1.** r^3s^2 **2.** $6m^2np$ **3.** $20bc^2$ **4.** $30x$ **5.** $28a^2$ **6.** $8e^3$ **7.** $3x - 6$ **8.** $3y^2 + y$ **9.** $8a - 12b$ **10.** $18d + 30$ **11.** $2a$ **12.** m **13.** m **14.** a

Page 28 **1.** 13 **2.** 7 **3.** 7 **4.** 7 **5.** 3, 5, 21 **6.** 4, 5 **7.** $<$ **8.** $<$ **9.** $>$ **10.** $<$

CHAPTER 2

Page 45 **1.** 22 **2.** 20 **3.** 10 **4.** 23 **5.** 13 **6.** 11 **7.** 12 **8.** 10 **9.** 9 **10.** 7 **11.** 8 **12.** 3 **13.** 12 **14.** 11 **15.** 15 **16.** 42 **17.** 36 **18.** 30

Page 49 **1.** 2 **2.** 3 **3.** 4 **4.** 6 **5.** 7 **6.** 6 **7.** 9 **8.** 11 **9.** 4 **10.** 7 **11.** 20 **12.** 5

Page 53 **1.** 3 **2.** 1 **3.** 6 **4.** 14 **5.** $3x + 5x = 40$; 5

Page 61 **1.** $m + 3$ **2.** $x + 3$ **3.** \$69 **4.** 108

Page 67 **1.** 12 **2.** 1 **3.** 4 **4.** 5 **5.** 5 **6.** 9 **7.** 5 **8.** 7 **9.** Jo has \$25. Lou has \$50. Sherry has \$14. **10.** Vic has \$30. Al has \$60. Connie has \$35.

CHAPTER 3

Page 82 **1.** $>$ **2.** $<$ **3.** $<$ **4.** $<$ **5.** $<$ **6.** $>$ **7.** $<$ **8.** $>$ **9.** $>$ **10.** $<$ **11.** $<$ **12.** $>$

13. 2 3 4 5 6 7 8

14. −8 −7 −6 −5 −4 −3 −2

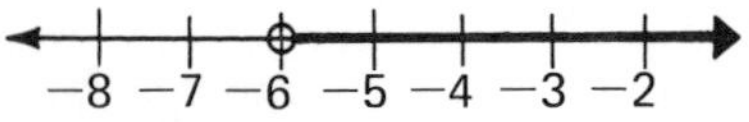

15. −6 −5 −4 −3 −2 −1 0

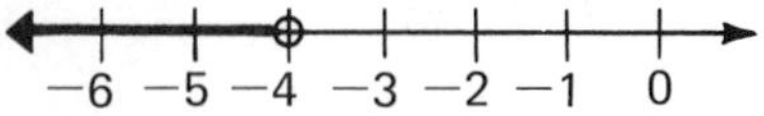

16. 0 1 2 3 4 5 6

17. −1 0 1 2 3 4 5

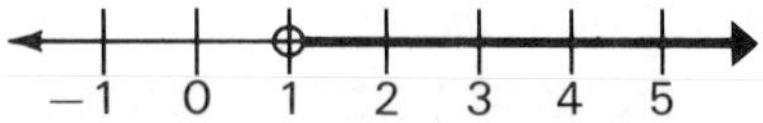

18. −6 −5 −4 −3 −2 −1 0

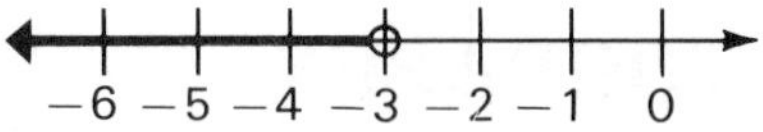

19. −9 −8 −7 −6 −5 −4 −3

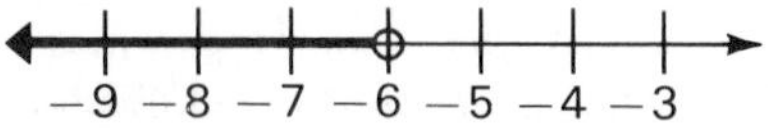

20. 0 1 2 3 4 5 6

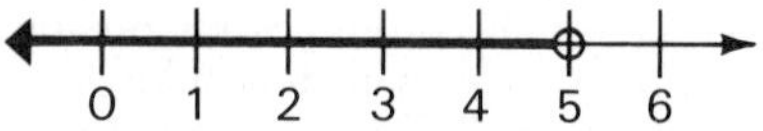

21. −6 −5 −4 −3 −2 −1 0

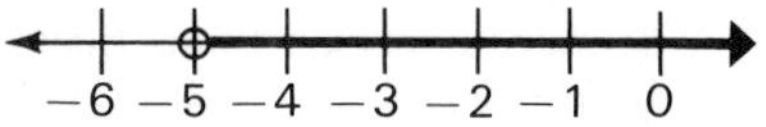

Page 90 **1.** 6 **2.** -6 **3.** -1 **4.** -2 **5.** 2 **6.** -11 **7.** 12 **8.** 2 **9.** x **10.** n **11.** $-11c$ **12.** $-2m^2$

Page 98 **1.** -30 **2.** 14 **3.** 49 **4.** $-27x$ **5.** $6ab$ **6.** $-4y + 20$ **7.** $-x - 6$ **8.** $-a - 5$ **9.** $2y + 7$ **10.** $-x + 3$ **11.** 3 **12.** -3 **13.** 3 **14.** -7 **15.** -2 **16.** -3 **17.** -2

CHAPTER 4

Page 122 **1.** 20 **2.** $b + 2c$ **3.** 14 **4.** $\frac{1}{2}ab$ **5.** 648 cm^3

Page 133 **1.** km **2.** xy **3.** st **4.** $c - d$ **5.** 80 km/h **6.** Louisa worked 35 hours and Bill worked 41 hours.

Page 137 **1. a.** 5 **b.** $\frac{y}{9}$ **c.** $\frac{y}{a}$ **2. a.** 18 **b.** $y - 2$ **c.** $y - k$ **3.** $x + 10$ **4.** $8 - 2x$ **5.** $2x - 8$ **6.** $2x + 8$ **7.** $9 - 3x$ **8.** $\frac{12}{x}$ **9.** -8 **10.** $z - x$

CHAPTER 5

Page 155 **1.** $7x + 5$ **2.** $2m + 4$ **3.** $b^3 + b^2 + 3b + 1$ **4.** $a^2 - b + 1$ **5.** $-2x + 6$ **6.** $5a - 20$ **7.** $-3b^2 + b - 9$ **8.** $5m - 1$

Page 161 **1.** x^4 **2.** $28x^4$ **3.** $-2m^3n^4$ **4.** m^{32} **5.** a^3b^3 **6.** $49x^4y^2$ **7.** $xy + 7x$ **8.** $m^2 - mn$ **9.** $-a^3 + ab$ **10.** $-6x + 8y$ **11.** $20x^2 + 5x^3$ **12.** $3m^5 - m^4 + 2m^3$

Page 169 **1.** $x^2 + 6x + 8$ **2.** $a^2 - 6a + 5$ **3.** $a^3 - 3a^2 + 3a - 1$ **4.** $m^2 - 4m - 5$ **5.** $4a^2 - 8ab - 21b^2$ **6.** $6m^2 + 17m + 5$ **7.** $a^2 + 8a + 16$ **8.** $n^2 - 6n + 9$ **9.** $36x^2 + 12xy + y^2$ **10.** $16a^2 - 56ab + 49b^2$

Page 175 **1.** 2 **2.** $2a^3$ **3.** $2n$ **4.** $-3y$ **5.** $3m - 4n$ **6.** $2a + 8$ **7.** $xy - y$ **8.** $a - 2b$

CHAPTER 6

Page 191 **1.** 8 **2.** 8 **3.** $3x$ **4.** $2mn$ **5.** $5xy^2(3x + 5y)$ **6.** $9mn^2(2 - m)$ **7.** $8a^2(b^3 + 7c^3)$ **8.** $2x^2(4 - \pi)$

Page 203 **1.** $(x + 9)(x + 2)$ **2.** $(y + 12)(y + 2)$ **3.** $(m + 13)(m + 2)$ **4.** $(a - 4)(a - 2)$ **5.** $(a - 3)(a - 9)$ **6.** $(m - 14)(m - 4)$ **7.** $(m + 8)^2$ **8.** $(x - 6)^2$ **9.** $(x + 10)^2$ **10.** $(x + 12)(x - 4)$ **11.** $(y + 7)(y - 6)$ **12.** $(b + 27)(b - 3)$

Page 209 **1.** $m^2 - 169$ **2.** $9 - 16x^2$ **3.** $25a^2 - 49$ **4.** $(a - 8)(a + 8)$ **5.** $(m - 6n)(m + 6n)$ **6.** $(4x - 3y)(4x + 3y)$ **7.** $3(x + 3)(x + 2)$ **8.** $2(y - 5)(y + 3)$ **9.** $2(m - 2n)(m + 2n)$

CHAPTER 7

Page 232 **1.** Brand C **2.** 19 **3.** 20 **4.** 1 P.M. **5.** 2 A.M. **6.** 7°C

Page 236 **1.** (0, −2) **2.** (−2, 1) **3.** (−2, 0) **4.** (−1, −2) **5.** (2, −2) **6.** (−2, −2) **7.** H **8.** J **9.** I **10.** L **11.** K **12.** G

Page 249 Note: Answers to Ex. 1–3 may vary. **1.** (0, 0), (1, 5), (2, 10) **2.** (0, −6), (1, −3), (2, 0) **3.** (0, 1), (1, 5), (2, 9)

4.

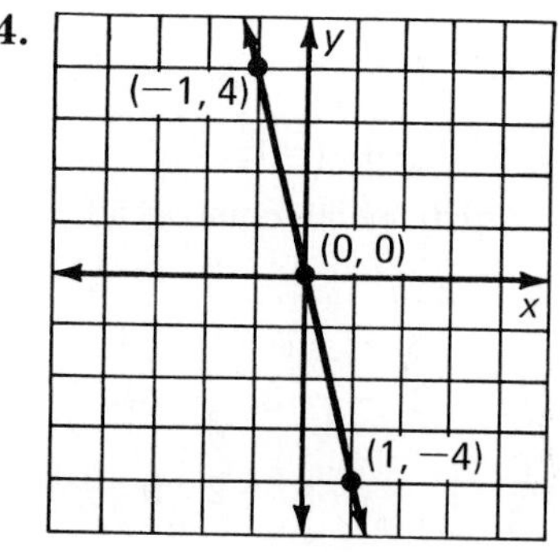

5.

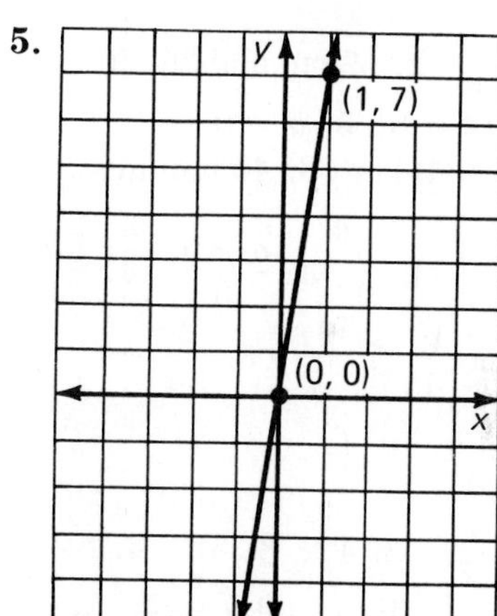

6.

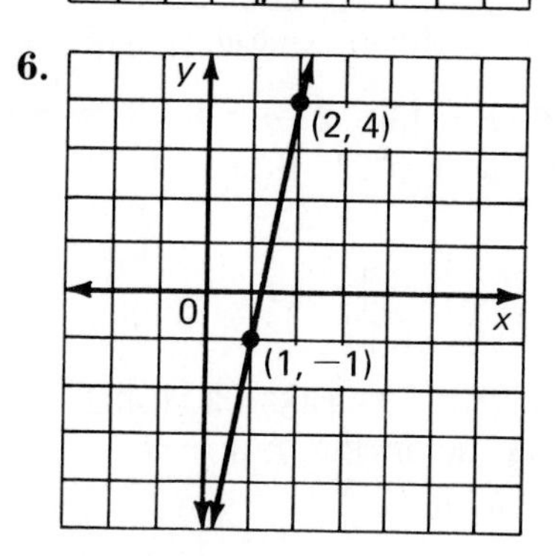

7. 9 **8.** −3 **9.** −1 **10.** $y = 7; x = 4$ **11.** $2x + 1$

CHAPTER 8

Page 269 **1.** (3, 9) **2.** about $\left(1\frac{2}{3}, 4\frac{2}{3}\right)$ **3.** (1, −1) **4.** 3 **5.** 3 **6.** −1 **7.** no **8.** one **9.** all

Page 275 **1.** (2, 3) **2.** (3, 1) **3.** (−6, −8) **4.** (2, 4) **5.** (1, −1) **6.** (2, −1) **7.** (0, 2) **8.** (−2, 5) **9.** (3, −1)

Page 283 **1.** 21 and 4 **2.** Yvette: 15; Lauren: 18 **3.** (2, 0) **4.** $\left(-1\frac{1}{5}, 1\frac{1}{5}\right)$ **5.** (−2, 1) **6.** brush: $1.80; comb: $.50

CHAPTER 9

Page 303 **1.** $\frac{8}{7x}$ **2.** $-\frac{1}{3}$ **3.** $\frac{4}{3s}$ **4.** 5 **5.** $3 - m$ **6.** $\frac{2}{3}$ **7.** −1 **8.** −2 **9.** −2

Page 313 **1.** $\frac{3}{4}$ **2.** 39 cm and 26 cm **3.** $x = 28$ **4.** $x = 8$ **5.** $a = 3$ **6.** $x = -8$ **7.** 100 km **8.** 40 minutes

Page 320 **1.** $\frac{3}{2}$ **2.** $\frac{3}{2}$ **3.** $\frac{a+1}{3}$ **4.** $\frac{5m}{3}$ **5.** $\frac{b}{3a^2}$ **6.** $\frac{x-1}{x^2+1}$ **7.** $\frac{9x}{7}$ **8.** $\frac{2}{9a}$ **9.** $\frac{2x}{y}$

Page 324 **1.** 28 **2.** $4x$ **3.** $3n$ **4.** $5x$ **5.** $\frac{8}{15}$ **6.** $\frac{8x}{21}$ **7.** $\frac{31}{36a}$ **8.** $\frac{9m-4}{3m^2}$

Page 333 **1.** $x = 132$ **2.** $a = 6$ **3.** $x = 36$ **4.** $n = 5$ **5.** $x = 22$ **6.** $x = 14$ **7.** $1\frac{7}{8}$ hours

CHAPTER 10

Page 358 **1.** 3.65 **2.** 4.485 **3.** 0.432 **4.** 3.2 **5.** 6.74 **6.** 0.875 **7.** $x = 5$ **8.** 3 km and 4.5 km

Page 368 **1.** 52% **2.** 4% **3.** 60% **4.** 75% **5.** $37\frac{1}{2}\%$ **6.** 0.35 **7.** 0.14 **8.** 0.07 **9.** 0.65 **10.** 0.0125 **11.** 7.2 **12.** 12.75 **13.** 20% **14.** 25 **15.** 44¢ **16.** $66\frac{2}{3}\%$

Page 377 **1.** about 17 years **2.** $49 **3.** $500 **4.** 50 kilograms of each kind

CHAPTER 11

Page 395 **1.** 7 **2.** 1 **3.** −3 **4.** 100 **5.** 7.746 **6.** 4.899 **7.** 7.211 **8.** ±4.123 **9.** rational **10.** rational **11.** rational **12.** irrational **13.** $5\sqrt{3}$ **14.** $3\sqrt{10}$ **15.** 60 **16.** $12\sqrt{3}$

Page 401 **1.** ±7 **2.** ±10 **3.** ±6 **4.** ±4 **5.** ±5 **6.** ±3 **7.** 15 **8.** $2\sqrt{13}$ **9.** $4\sqrt{2}$

Page 409 **1.** $\frac{1}{2}$ **2.** $\frac{\sqrt{5}}{4}$ **3.** $\frac{2\sqrt{2}}{5}$ **4.** $\frac{2\sqrt{5}}{7}$ **5.** $4\sqrt{10}$ **6.** $\sqrt{7}$ **7.** $36\sqrt{2}$ **8.** $\frac{\sqrt{21}}{6}$ **9.** $\frac{\sqrt{30}}{10}$ **10.** $\frac{\sqrt{15}}{9}$ **11.** $10\sqrt{3}$ **12.** $4\sqrt{2}$ **13.** $-2\sqrt{3}$

CHAPTER 12

Page 425 **1.** 0 or 3 **2.** 0 or −7 **3.** 6 or 3 **4.** −4 or 3 **5.** 0 or 3 **6.** 0 or 5 **7.** −3 or −5 **8.** −6 or 2 **9.** 0 or −8 **10.** 4 or −3 **11.** −6 or 5 **12.** 7 or 4

Page 429 **1.** ±7 **2.** ±8 **3.** $\pm\sqrt{17}$ **4.** $\pm\sqrt{7}$ **5.** $\pm\sqrt{6}$ **6.** ±6 **7.** 0 or 14 **8.** 5 or −7 **9.** 11 or −5 **10.** −2 or −12 **11.** −1 or −5 **12.** 6 or −3

Page 439

1. $\frac{-7 \pm \sqrt{89}}{2}$

2. $\frac{1 \pm \sqrt{19}}{3}$

3. 4 or 1

4. length: 9 cm; width: 5 cm

5. See right; vertex: (1, −3)

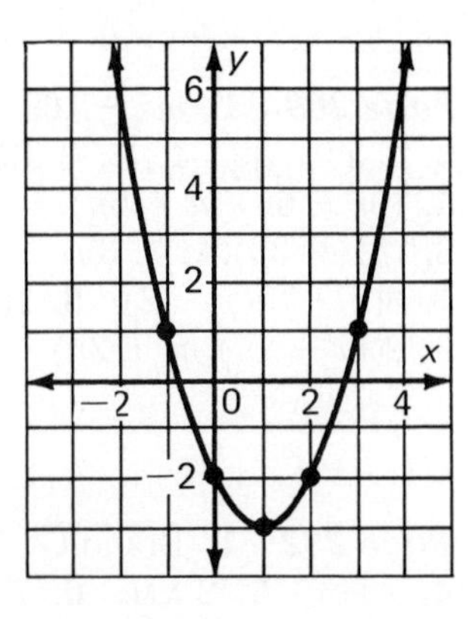

Answers to Selected Exercises

CHAPTER 1

Written Exercises, pages 6–7 **1.** 1 **3.** 10 **5.** 3 **7.** 10 **9.** 16 **11.** 0 **13.** 9 **15.** 24 **17.** 24 **19.** 4 **21.** 17 **23.** 10 **25.** 10 **27.** 10 **29.** 26 **31.** 96 **33.** $x + 7$ **35.** $2n$ **37.** ba **39.** Subtract 4. **41.** Divide by 8. **43.** t

Written Exercises, page 9 **1.** 24 **3.** 36 **5.** 2 **7.** 17 **9.** 7 **11.** 31 **13.** 14 **15.** 18 **17.** 9 **19.** 20 and 14 **21.** 40 and 68 **23.** 12 and 18 **25.** 6, 13 **27.** $2x, 2x + 7$ **29.** 11, 22

Written Exercises, page 11 **1.** $6x$ **3.** $10a$ **5.** $3y$ **7.** $7a + 6$ **9.** $12a$ **11.** $4 + 11y$ **13.** $6s - 2$ **15.** $3 + 9y$ **17.** $4a - 4$ **19.** $4 + 4x$ **21.** $5xy + 7$ **23.** $8d$ **25.** $5y + 9x$ **27.** $5ab + 5b$ **29.** $2xy + 2x$ **31.** Not possible **33.** Not possible **35.** $5ar$ **37.** Not possible

Written Exercises, page 13 **1.** 46 **3.** 143 **5.** 1571 **7.** $8n + 6$ **9.** $7c + 4$ **11.** $31 + 6a$ **13.** $10m + 11$ **15.** $ab + 10$ **17.** $4s + 5$ **19.** $4k + 4n$ **21.** $5n - 9b$ **23.** $7a + 6b$ **25.** $8a + 5b + 4$ **27.** $7d - 7a$ **29.** $3a + 4b$

Written Exercises, page 15 **1.** 9 **3.** 4 **5.** 100 **7.** y^4 **9.** x^2y^3 **11.** rs^3 **13.** t^2m^2 **15.** f^3g^2h **17.** de^2f^2 **19.** $7m^3$ **21.** $2a^2m^2$ **23.** $5n^3p$ **25.** $6a^2b$

Puzzle Problems, page 15

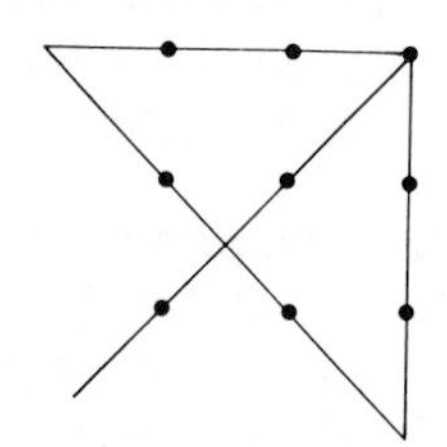

Written Exercises, page 17 **1.** 2900 **3.** 3300 **5.** 730 **7.** $6x$ **9.** $15a$ **11.** $2b^2$ **13.** $18y$ **15.** $20n$ **17.** $72f$ **19.** $4a^2$ **21.** $8ab$ **23.** $16x^2$ **25.** $35a^3$ **27.** $10b^3$ **29.** $6x$ **31.** $2x^2$ **33.** Multiply by 4.

Written Exercises, page 19 **1.** Yes **3.** Yes **5.** $2x + 6$ **7.** $4x + 8$ **9.** $7a + 21$ **11.** $9x + 36$ **13.** $3x - 6$ **15.** $c^2 - 3c$ **17.** $14x + 7$ **19.** $15x - 10$ **21.** $6x + 8y$ **23.** $3x^2 - xy$ **25.** $2x + 13$ **27.** $7x + 15$ **29.** $36 + 5y$ **31.** $8x + 23$ **33.** $8 + 32x$ **35.** $20b + 4$ **37.** $7m + 9$ **39.** $x^2 + xy + 2$

Written Exercises, page 21 **1.** 0 **3.** 35 **5.** 0 **7.** Impossible **9.** 0 **11.** 0 **13.** $8x$ **15.** $6c$ **17.** $15a$ **19.** x **21.** x **23.** x **25.** Add 7. **27.** Subtract 9. **29.** Add 16. **31.** Add 11.

Written Exercises, page 23 **1.** 1 **3.** 1 **5.** 1 **7.** 3 **9.** c **11.** 4 **13.** x **15.** a **17.** x **19.** x **21.** Divide by 6. **23.** x **25.** Multiply by 7. **27.** 4

Calculator Activities, page 24 **1.** 2 **3.** 32 **5.** 21

Written Exercises, page 26 **1.** 3 **3.** 12 **5.** 2 **7.** 4 **9.** 3 **11.** 5 **13.** 5 **15.** Answers may vary. **17.** Answers may vary. **19.** 8 **21.** 6 **23.** 8 **25.** Any number **27.** 40 **29.** Any number **31.** 2 **33.** 0

Puzzle Problems, page 26 The lion tamer must be bald; since Geraldine and Amos have hair, Christopher is the lion tamer. Then, by clue 3, the ringmaster is Geraldine. Thus, Amos is the elephant trainer.

Written Exercises, page 28 **1.** $<$ **3.** $>$ **5.** $>$ **7.** $>$ **9.** 4 **11.** 1, 3, 7 **13.** 20 **15.** 6, 2 **17.** 4, 6 **19.** 10, 11 **21.** 2, 5 **23.** 4, 5, 6 **25.** $<$ **27.** $=$ **29.** $<$

Reading Algebra, page 29 **1.** Answers may vary. For example, fifteen minus eight; subtract eight from fifteen; the difference of fifteen and eight; eight less than fifteen **3.** Answers may vary. For example, seventeen plus four; add four to seventeen; the sum of seventeen and four; four more than seventeen **5.** 30 − 5; 25 **7.** 3^2; 9 **9.** 80 + 8; 88

Skills Review, page 30 **1.** 56 **3.** 993 **5.** 229 **7.** 712 **9.** 822 **11.** 34 **13.** 77 **15.** 535 **17.** 354 **19.** 217 **21.** 288 **23.** 1825 **25.** 2040 **27.** 43,680 **29.** 75,705 **31.** 17 **33.** 68 R2 **35.** 74 R1 **37.** 237 R5 **39.** 104 R13 **41.** 221 R14

Chapter Review Exercises, pages 31–32 **1.** 11 **2.** 16 **3.** 2 **4.** 21 **5.** 26 **6.** 21 **7.** 15 **8.** 38 **9.** 14 **10.** 21 **11.** 21 **12.** 1 **13.** 7 **14.** 10 **15.** $7x$ **16.** $2a$ **17.** $7b + 2$ **18.** $9c$ **19.** $6c + 3$ **20.** $6y + 12$ **21.** $8x + 3$ **22.** $9a - 4$ **23.** $10a + b$ **24.** 16 **25.** 243 **26.** 125 **27.** 27 **28.** 25 **29.** 64 **30.** 49 **31.** 1 **32.** $10x$ **33.** $12a$ **34.** $16b$ **35.** $4x^3$ **36.** $6b^3$ **37.** a^3b^2 **38.** $6x^2$ **39.** $12y^2$ **40.** $6n^3$ **41.** $2x + 10$ **42.** $6a - 21$ **43.** $20x - 10y$ **44.** $15y - 20$ **45.** $8c - 12d$ **46.** $18x + 14y$ **47.** $3a - 18b$ **48.** $24x + 18$ **49.** 0 **50.** 0 **51.** Impossible **52.** 0 **53.** Impossible **54.** 1 **55.** 0 **56.** Impossible **57.** $8x$ **58.** 6 **59.** a **60.** 0 **61.** $6x$ **62.** $5x$ **63.** $10s$ **64.** 1 **65.** $5n$ **66.** 4 **67.** 9 **68.** 3 **69.** 1 **70.** 8 **71.** 3

Mixed Review, pages 34–35

Arithmetic Review **1.** 43 **3.** 1140 **5.** 94 **7.** 333 **9.** 490 **11.** 7489 **13.** 75.9 **15.** a **17.** g **19.** b **21.** 36 **23.** 0.0129 **25.** 8.009 **27.** 0.27408 **29.** 427.5

Chapter 1 Review **1.** $35x$ **3.** $10x^3y$ **5.** $24xy^3$ **7.** x **9.** Subtract 1. **11.** x **13.** 14 **15.** 3 **17.** 0 **19.** 54 **21.** 11 **23.** 513 **25.** Impossible **27.** $<$ **29.** no **31.** $7x + 16$ **33.** 2 **35.** 0, 2

CHAPTER 2

Written Exercises, page 39 **1.** 16 **3.** 10 **5.** 12 **7.** 13 **9.** 24 **11.** 19 **13.** 15 **15.** 9 **17.** 9 **19.** 30 **21.** 25 **23.** 8 **25.** 13 **27.** 5 **29.** 35 **31.** 51 **33.** 10 **35.** 6 **37.** 11 **39.** 21 **41.** 15 **43.** 9 **45.** 15 **47.** 25

Written Exercises, page 41 **1.** 10 **3.** 8 **5.** 4 **7.** 12 **9.** 14 **11.** 15 **13.** 17 **15.** 12 **17.** 5 **19.** 17 **21.** 12 **23.** 13 **25.** 10 **27.** 13 **29.** 15 **31.** 8 **33.** 7 **35.** 4 **37.** 0 **39.** 24

Mixed Practice Exercises, page 41 **1.** 9 **3.** 18 **5.** 11 **7.** 18 **9.** 6 **11.** 5 **13.** 6 **15.** 17 **17.** 19 **19.** 29 **21.** 33 **23.** 30 **25.** 4 **27.** 17 **29.** 27 **31.** 46 **33.** 60 **35.** 18

Written Exercises, page 43 **1.** 4 **3.** 7 **5.** 12 **7.** 5 **9.** 10 **11.** 3 **13.** 3 **15.** 7 **17.** 9 **19.** 12 **21.** 13 **23.** 14 **25.** 180 **27.** 25 **29.** 18 **31.** 44 **33.** 54 **35.** 12 **37.** 15 **39.** 56 **41.** 48 **43.** 72

Mixed Practice Exercises, pages 44–45 **1.** 11 **3.** 5 **5.** 4 **7.** 7 **9.** 20 **11.** 6 **13.** 23 **15.** 14 **17.** 6 **19.** 11 **21.** 21 **23.** 21 **25.** 2 **27.** 21 **29.** 10 **31.** 5 **33.** 14 **35.** 27 **37.** 51 **39.** 18 **41.** 30 **43.** 5 **45.** 3 **47.** 21 **49.** 20 **51.** 13 **53.** 19 **55.** 21 **57.** 19 **59.** 28 **61.** 6

Puzzle Problems, page 45 If Melba and Fran are both wearing black hats, then Brent would know he must be wearing red (since there are only 2 black hats). Since Brent does not know, either Melba or Fran (or both) must be wearing red. If Melba is wearing a black hat, Fran would thus know she must be wearing red. Since Fran does not know, Melba must be wearing a red hat.

Written Exercises, pages 47–49 **1.** $5x = 10$; $x = 2$ **3.** $3b = 3$; $b = 1$ **5.** $2a = 18$; $a = 9$ **7.** $4c = 12$; $c = 3$ **9.** $6n = 18$; $n = 3$ **11.** 3 **13.** 3 **15.** 4 **17.** 2 **19.** 5 **21.** 5 **23.** 6 **25.** 5 **27.** 10 **29.** 5 **31.** 2

33. 5 **35.** 3 **37.** 3 **39.** 7 **41.** 6 **43.** 3 **45.** 5 **47.** 7 **49.** 6 **51.** 19 **53.** 4 **55.** $2n + 8 = 32; n = 12$ **57.** $2n + 4 = 16;$ $n = 6$ **59.** $4n + 5 = 17; n = 3$ **61.** $2n - 15 = 25; n = 20$

Written Exercises, page 51 **1.** Arithmetic: 8, 40, 44, 880, 800; Algebra: n, $5n$, $5n + 4$, $100n + 80$, $100n$ **3.** 9 **5.** 12 **7.** If you use algebra to solve this problem, you will notice that the next to the last step is $n + 20$. It doesn't matter what number you begin with. Subtracting the original number, n, from $n + 20$ will always obtain 20.

Written Exercises, pages 52–53 **1.** 6 **3.** 4 **5.** 8 **7.** 8 **9.** 5 **11.** 1 **13.** 2 **15.** 2 **17.** 2 **19.** 5 **21.** 5 **23.** 7 **25.** $4x + 3x = 63; 9$ **27.** $2x + 5 = 73; 34$ **29.** $2x + 9 = 25; 8$

Written Exercises, pages 55–57 **1.** $n + 5$ **3.** $j + 2$ **5.** $b + 11$ **7.** $3x$ **9.** $4x$ **11.** $3x$ **13.** $x + 1$ **15.** $r - 1$ **17.** $m - 1$ **19.** $x + 2$; $(x + 2) + 4$, or $x + 6$ **21.** $k + 3$; $k - 1$; $(k + 3) - 1$, or $k + 2$ **23.** $x + 2$; $x + 4$

Written Exercises, pages 59–61 **1.** 24 computers **3.** 790 students **5.** 21 cars **7.** sheep, 95 kg; hog, 190 kg **9.** 85 games **11.** 37, 38 **13.** 130 km **15.** Sue, \$75; Liz, \$150; Wendy, \$225

Written Exercises, page 63 **1.** 14 **3.** 4 **5.** 3 **7.** 5 **9.** 2 **11.** 4 **13.** 1 **15.** 2 **17.** 1 **19.** 1 **21.** 2 **23.** 2 **25.** 7 **27.** 3 **29.** 2 **31.** 9 **33.** 1 **35.** 10 **37.** $7n = 3n + 12; 3$ **39.** $2n + 24 = 8n; 4$ **41.** $n + 10 = 6n; 2$

Written Exercises, page 65 **1.** $6x - 15$ **3.** $16x - 40$ **5.** $48y + 24$ **7.** $24 + 48b$ **9.** 2 **11.** 3 **13.** 4 **15.** 10 **17.** 2 **19.** 4 **21.** 3 **23.** 0 **25.** 3 **27.** 5 **29.** 0 **31.** 2 **33.** 3 **35.** 6 **37.** 4 **39.** 21 **41.** 6 **43.** a **45.** b **47.** a **49.** b **51.** c

Written Exercises, pages 66–67 **1.** pear, 100 calories; melon, 120 calories **3.** 18 and 25 **5.** Mona, \$53; Paul, \$61; Tran, \$122 **7.** 20 years old **9.** 41 years

Problem Solving Strategies, page 69 **1. a.** 20 min **b.** 20 min; 5 min **c.** 5 min; 10 min **3. a.** $2p$; $2p + 4$ **b.** number of oranges = 20; $2p + 4 = 20$ **c.** 8 pears and 16 apples **d.** 44 pieces of fruit

Skills Review, page 70 **1.** $<$ **3.** $>$ **5.** $>$ **7.** $>$ **9.** $<$ **11.** $<$ **13–20.** See below. **21.** $0, \frac{1}{2}, 2$ **23.** $\frac{1}{7}, 7, 17$ **25.** $\frac{2}{5}, 1, \frac{6}{5}$ **27.** $\frac{1}{4}, \frac{1}{2}, \frac{3}{4}$ **29.** $\frac{5}{6}, \frac{5}{3}, \frac{5}{1}$ **31.** $2\frac{1}{5}, 2\frac{1}{4}, 2\frac{1}{3}$

Chapter Review Exercises, pages 71–72 **1.** 18 **2.** 15 **3.** 8 **4.** 7 **5.** 8 **6.** 5 **7.** 14 **8.** 39 **9.** 28 **10.** 4 **11.** 21 **12.** 7 **13.** 2 **14.** 12 **15.** 34 **16.** 12 **17.** 5 **18.** 13 **19.** $2n + 7 = 25; n = 9$ **20.** $9y - 3y = 42; y = 7$ **21.** 2 **22.** 2 **23.** 7 **24.** $x + 5$ **25.** $a - 3$ **26.** Phil, \$11; Marta, \$33 **27.** 32 games **28.** 460 girls **29.** 167 cm **30.** Bert, \$11; Cindy, \$22; Mona, \$30 **31.** 5 **32.** 12 **33.** 5 **34.** 4 **35.** 2 **36.** 3 **37.** 5 **38.** 6 **39.** 7 **40.** 4 **41.** 3 **42.** 6

Cumulative Review, page 74 **1.** 10 **3.** 3 **5.** 6 **7.** 125 **9.** $17t$ **11.** $6x + 5y$ **13.** $6r^2$ **15.** $3xy - 4y^2$ **17.** $2y$ **19.** 7 **21.** 2 **23.** 19 **25.** 147 **27.** 3 **29.** 1 **31.** $4n - 17 = 11; n = 7$ **33.** Jon, \$19; Miyoshi, \$37

CHAPTER 3

Written Exercises, page 79 **1.** +25 **3.** −10 **5.** −200 **7.** $<$ **9.** $>$ **11.** $>$ **13.** $<$ **15.** $<$ **17.** $>$ **19.** −274 **21.** Answer will vary with location.

Page 70, Exercises 13–20

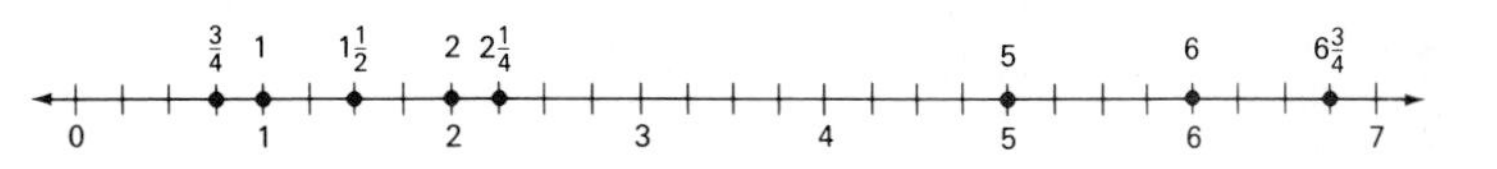

Written Exercises, pages 81–82
1. $-2 < 3$ or $3 > -2$ **3.** $-1 > -6$ or $-6 < -1$ **5.** $5 > -4$ or $-4 < 5$ **7.** $-1 > -7$ or $-7 < -1$ **9.** $8 > -8$ or $-8 < 8$ **11–17.** Check to be sure you have graphed each solution correctly.
11. 6 **13.** 3 **15.** 8 **17.** 10

19.

21.

23.

25.

27.

29.

31. $-4 < 0 < 4$ **33.** $-5 < -3 < 0$ **35.** $-4 < -2 < 6$

37.

39.

41.

43.

Puzzle Problems, page 82 Answers may vary.

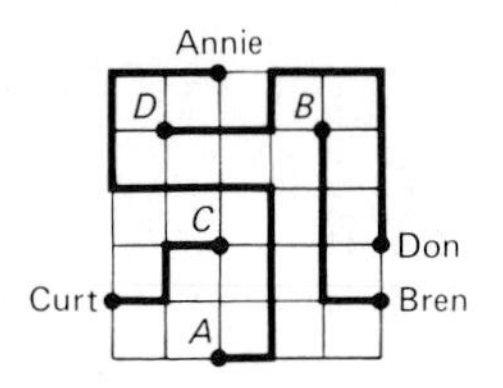

Written Exercises, page 84 **1.** −5 **3.** −1 **5.** −8 **7.** −8 **9.** 3 **11.** 2 **13.** −6 **15.** −3 **17.** −6 **19.** 0 **21.** 0 **23.** −6 **25.** −9 **27.** −5 **29.** −3 **31.** −1 **33.** 10

Written Exercises, page 87 **1.** −10, 5 **3.** −7, −23 **5.** 2, 6 **7.** −1 **9.** −3 **11.** −7 **13.** −13 **15.** −12 **17.** 8 **19.** 8 **21.** 13 **23.** −1 **25.** −6 **27.** −1 **29.** −2 **31.** 8 **33.** −3 **35.** −7 **37.** −5 **39.** 34° **41.** −6 **43.** 53 **45.** −30

Mixed Practice Exercises, page 88
1. −1 **3.** −5 **5.** 9 **7.** −7 **9.** 4 **11.** 8 **13.** 0 **15.** −6 **17.** −1 **19.** 0 **21.** 15 **23.** −9 **25.** 11 **27.** −4 **29.** 11 **31.** 10

Written Exercises, page 90 **1.** $4x$ **3.** $10x$ **5.** $-y$ **7.** $2x^2$ **9.** $6c^2$ **11.** $-a$ **13.** $6x$ **15.** $-10a$ **17.** $-3n^2$ **19.** $-9y + 10$ **21.** 1 **23.** x **25.** x **27.** $-2x^2 + y$ **29.** $7x$ **31.** $10 - 7r$ **33.** $(5n - 2)$ km

Written Exercises, page 92 **1.** 32 **3.** 120 **5.** −14 **7.** 20 **9.** −24 **11.** −6 **13.** 30 **15.** −7 **17.** −12 **19.** −4 **21.** 4 **23.** −28 **25.** −16 **27.** −25 **29.** 25 **31.** 36 **33.** 16 **35.** −100 **37.** −25 **39.** 100 **41.** −36 **43.** −100 **45.** 0 **47.** −48 **49.** 25 **51.** 16

Written Exercises, page 94 **1.** $2x$ **3.** $-6x$ **5.** 0 **7.** $-12x$ **9.** $-6ab$ **11.** $6xy$ **13.** $-8xy$ **15.** $-14ab$ **17.** $36 - 4x$ **19.** $2a - 2b$ **21.** $-6x + x^2$ **23.** $-3a - 3b$ **25.** $-a - 4$ **27.** $-y + 1$ **29.** $x - 5$ **31.** $-7 + a$ **33.** −3 **35.** $x - 4$ **37.** $7a + 4$ **39.** $7x - 3$ **41.** 13 **43.** −7 **45.** 1 **47.** $28x - 4 - 24x + 6 = 4 + 3x - 2$; $4x + 2 = 3x + 2$; $x = 0$

Written Exercises, page 96 **1.** −1 **3.** −8 **5.** −2 **7.** 2 **9.** −7 **11.** 5 **13.** −7 **15.** 3 **17.** −6 **19.** −2 **21.** 7 **23.** 3 **25.** 5

Calculator Activities, page 96 **1.** 3 **3.** −12 **5.** −136 **7.** −4

Written Exercises, page 98 **1.** −6 **3.** −24 **5.** 1 **7.** −3 **9.** −5 **11.** −1 **13.** 9 **15.** 4 **17.** 3 **19.** −6 **21.** −5 **23.** 1 **25.** 4 **27.** −2 **29.** −14 **31.** 5 **33.** 2 **35.** 1

Written Exercises, page 100 **1.** false **3.** true **5.** true **7.** false **9.** false **11.** false **13.** $<$ **15.** $>$ **17.** $<$ **19.** $>$ **21.** $>$ **23.** -2 **25.** 11 **27.** -1 **29.** 4 or -4 **31.** 6 or -6 **33.** 14 or -14 **35.** 3 or -3 **37.** 7 or -7 **39.** 30 or -30

Puzzle Problems, page 100 Fill the large pitcher. Pour water from it to fill the small pitcher. Empty the small pitcher. Pour the 2 L from the large pitcher into it. Fill the large pitcher again. Pour water from it into the small pitcher until the small one is full. The large pitcher now has 4 L.

Reading Algebra, page 101 **1.** positive and negative numbers **3.** to the left of zero **5.** negative three

Computer Activities, page 103 **1.** yes **3. a.** $1 - x$ **b.** $(-2)(-x)$ **c.** $-8x - 5$ **d.** $x - 3$ **4. a.** 0, 2 **b.** 2 **c.** 2 **d.** 5, -5 **e.** 0, -4 **f.** $x \leq 0$

Skills Review, page 104 **1.** about 80 **3.** about 110 **5.** about 90 **7.** about 120 **9.** about 200 **11.** about 190 **13.** about 120 **15.** about 80 **17.** about \$1.60 **19.** about \$1.40 **21.** about \$3.60 **23.** about \$2.70 **25.** Not reasonable **27.** Not reasonable **29.** Not reasonable **31.** Not reasonable **33.** Not reasonable **35.** Reasonable **37.** 13 **39.** 50 **41.** 9 **43.** 14

Chapter Review Exercises, pages 105–106 **1.** $<$ **2.** $>$ **3.** $>$ **4.** $>$ **5.** $>$ **6.** $<$ **7.** $>$ **8.** $<$

9. $x = 2$
−2 −1 0 1 2 3 4

10. $x = 1$
−2 −1 0 1 2 3 4

11. $x = -6$
−7 −6 −5 −4 −3 −2 −1 0 1

12.
−3 −2 −1 0 1 2 3

13.
−3 −2 −1 0 1 2 3

14.
−6 −5 −4 −3 −2 −1 0

15. -13 **16.** -11 **17.** -3 **18.** 0 **19.** 1 **20.** -3 **21.** -11 **22.** -4 **23.** -6 **24.** 4 **25.** -11 **26.** -4 **27.** 18 **28.** -4 **29.** 4 **30.** -1 **31.** -13 **32.** 23 **33.** -12 **34.** 13 **35.** -5 **36.** -4 **37.** 1 **38.** 15 **39.** $-3x$ **40.** $-3x - 10$ **41.** -15 **42.** $-h + 3$ **43.** $2x - y$ **44.** $-3b + 3$ **45.** $8b^2$ **46.** $7a - 10$ **47.** $-6x + 3y$ **48.** -24 **49.** -54 **50.** 24 **51.** 27 **52.** -72 **53.** -36 **54.** $y^2 - 5y$ **55.** $x^2 - 4x$ **56.** $-2x - 6$ **57.** $-5x + 20$ **58.** $-3a + 3b$ **59.** $-6m + 6n$ **60.** $-x^2$ **61.** $-b^2$ **62.** $-y + 7$ **63.** $-3a - b$ **64.** $x + 9$ **65.** $-x + 3 - y$ **66.** -27 **67.** -9 **68.** 1 **69.** -4 **70.** 3 **71.** -10 **72.** 4 **73.** -4 **74.** -8 **75.** -5 **76.** 12 **77.** -2 **78.** $x = -5$ **79.** $x = -8$ **80.** $x = 8$ **81.** $x = 9$ **82.** $x = -4$ **83.** $x = -3$ **84.** $x = 2$ **85.** $x = 5$ **86.** $x = -3$ **87.** -5 **88.** 0 **89.** 3

Mixed Review, pages 108–109 **1.** 16 **3.** $2x$ **5.** $-2 + a$ **7.** $x = 16$ **9.** $a = 3$ **11.** $n = 1$ **13.** 2 **15.** 4 **17.** $>$ **19.** 37 **21.** 100 **23.** $7a^2 - 5ab$ **25.** $x = -7$ **27.** $y = 4$ **29.** 5 **31.** -3 **33.** 18 **35.** -7 **37.** -12 **39.** orange juice, 120 calories; whole milk, 150 calories **41.** Algebra: n, $3n$, $3n - 6$, $12n - 24$, $12n - 16$, $6n - 8$; Original number: 7 **43.** $w + 8$ **45.** $+9$ **47.** -5

CHAPTER 4

Written Exercises, pages 113–114 **1.** 22 **3.** $9x$ **5.** $3x + 3$ **7.** $12a$ **9.** $14x$ **11.** width, 10 cm; length, 15 cm **13.** 12.56 cm **15.** 56.52 cm **17.** 9.42 m **19. a.** 175.84 cm **b.** 17,584 cm

Written Exercises, pages 117–118 **1.** 49 cm^2 **3.** 16 m^2 **5.** 12 cm^2 **7.** 16 cm^2 **9. a.** $6x^2$ **b.** $5x^2$ **11. a.** $12xy$ **b.** $10xy$ **13.** $7xy$ **15.** $25y^2$ **17.** $4x^2$ **19.** centimeters **21.** \$1920 **23.** $s^2 + \frac{1}{2}sh$ **25.** $41bh$

Written Exercises, pages 121–122
1. 180 cm^3 **3.** 3000 cm^3 or 3 L **5.** $24x^3$ **7.** 48 **9.** 20 **11.** 40,000 cm^3 **13.** 40,000 grams **15.** more **17.** $16x^3$

Written Exercises, page 125 **1.** qk km **3.** $2t$ dollars **5.** $6t$ **7.** $6t + 3f$ **9.** $6z$ dollars **11.** $(s - b)$ dollars **13.** $(2u + 3w)$ dollars **15.** $(6d + s)$ cents

Written Exercises, pages 128–129
1. 2 hours **3.** Kim, 18 km/h; Lucia, 22 km/h **5.** 75 km **7.** 12 km/h

Written Exercises, pages 131–133 **1.** 36 bagels **3.** 6 pens **5.** Maureen, \$11/h; Karl, \$10/h **7.** 308 advance tickets; 154 at door

Written Exercises, pages 136–137
1. a. 12 **b.** $x + 7$ **c.** $x + t$ **3.** 3 **5.** 8 **7.** $2x + 5$ **9.** 2 **11. a.** 3 **b.** $\frac{y}{8}$ **c.** $\frac{y}{a}$ **13.** 8 **15.** $4 - 2y$ **17.** $6 - 4y$ **19.** $y - 5$ **21.** $w = \frac{A}{8}$ **23.** $t = \frac{D}{r}$ **25.** $b = N - a$ **27.** $d = \frac{W}{f}$ **29.** $H = AT$ **31.** 2 **33.** 128 km/h **35.** 12,000 km/h

Puzzle Problems, page 137

25¢	10¢	5¢	1¢
1	0	0	2
0	2	1	2
0	2	0	7
0	1	3	2
0	1	2	7
0	1	1	12
0	1	0	17
0	0	5	2
0	0	4	7
0	0	3	12
0	0	2	17
0	0	1	22
0	0	0	27

Written Exercises, page 140 **1.** clockwise **3.** *A:* 16 teeth; *B:* 8 teeth; 4 turns/second **5.** 12 m **7.** About 750 kilograms

Reading Algebra, page 141 **1.** They walked in opposite directions; Ben, 4 km/h; Christina, 5 km/h **3.** Let n = number of hours each walked; Ben, $4n$ km; Christina, $5n$ km **5.** Yes; no; you must add 3 hours to 1:15 P.M. to get a time of 4:15 P.M.

Problem Solving Strategies, pages 142–143

1. c.

First box	Second box	Third box
2	2	1
3	1	1
1	2	2
1	3	1
1	1	3
2	1	2

3. 22 students

Skills Review, page 144 **1.** b **3.** b **5.** b **7.** c **9.** −3 **11.** 9 **13.** 65 **15.** 0 **17.** −43 **19.** 32 **21.** −4 **23.** 3

Chapter Review Exercises, pages 145–146 **1.** $4x + 6$ **2.** $2y + 5x$ **3.** $16x$ **4.** $12x^2$ **5.** $5x^2$ **6.** $10xy$ **7.** \$132 **8.** 40 **9.** $30x^3$ **10.** 30,000; 30 **11.** $90h$ km **12.** st dollars **13.** 93 km/h **14.** Marsha, 7 km/h; Cammie, 5 km/h **15.** 29 monitors **16. a.** $x = 3$ **b.** $x = \frac{21}{k}$ **c.** $x = \frac{a}{k}$ **17. a.** $x = 7 - y$ **b.** $x = 7 + y$ **c.** $x = -7 + y$ **18.** $t = 24$ **19.** $f = \frac{W}{d}$

Cumulative Review, page 148
1. > **3.** >
5. $x = 7$ 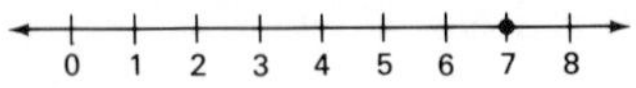

7. 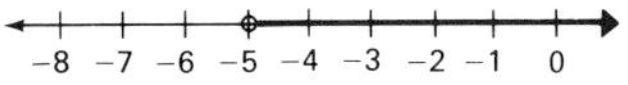

9. 0 **11.** −2 **13.** 4 **15.** 12 **17.** 140 **19.** 45 **21.** $-21ay$ **23.** $-2t + 5$ **25.** −3 **27.** 6 **29.** −6 **31.** −2 **33.** 16 cm^2 **35.** $2w + t$ **37.** 68 pies

CHAPTER 5

Written Exercises, page 153 **1.** $3y + 11$ **3.** $2m^2$ **5.** $5ab + 2b^2$ **7.** $2x^2 + xy + 2y^2$ **9.** $-5y^2 + 1$ **11.** $12n + 2$ **13.** $10m + 12n + 15$ **15.** $-4m + 8$ **17.** $10x^2 - 20x + 2$ **19.** $5x^2 + 2x - 1$ **21.** $k^3 + k^2 - 2k + 6$ **23.** $x^3 + 6x^2 - 3$ **25.** $a^2 - ab - b^2$

Written Exercises, page 155 **1.** $x + 4y$ **3.** $-2x^2 + 2y^2$ **5.** $2x^2 + 4$ **7.** $-4xy$ **9.** $x^2y + 4xy^2 + 2$ **11.** $b + 2$ **13.** $4m - 18$ **15.** $4y + 14$ **17.** $-3x^2 + 5x - 6$ **19.** $-a^2 + 5ab - 10c$ **21.** $2y^2 + 5y - 18$ **23.** $n = 10$ **25.** $n = 80$

Written Exercises, page 157 **1.** a^3 **3.** $3x^3$ **5.** n^7 **7.** $4x^3$ **9.** $-5c^5$ **11.** $-6x^7$ **13.** $12x^6$ **15.** $-5a^2b^2$ **17.** $-3cd^4$ **19.** $20x^3y^3$ **21.** $10r^4s^3$ **23.** $3x^5yz^2$ **25.** $-5a^4b^3$ **27.** $-2x^4y^5$ **29.** $-4m^6n^3p^4$

Puzzle Problems, page 157 The word "nothing" has two different meanings here: in the first sentence, the absence of something; in the second, no thing. Thus, logic cannot be used to conclude the third sentence.

Written Exercises, page 159 **1.** x^6 **3.** b^{12} **5.** c^{15} **7.** $4x^2$ **9.** a^4b^4 **11.** $36a^2x^2$ **13.** $25a^2$ **15.** $10x^2$ **17.** $9a^4$ **19.** $25x^{10}$ **21.** $4a^4b^2$ **23.** x^6y^6 **25.** $3x^4y^2$ **27.** $-16n^4$ **29.** $-x^3y^5$ **31.** $72m^3n^5$ **33.** $-r^{11}s^{12}$

Written Exercises, page 161 **1.** $2x + 8$ **3.** $4x + 4y$ **5.** $-6n - 12m$ **7.** $a^2 - ab$ **9.** $-ca - cb$ **11.** $-15x^2 - 10xy$ **13.** $4a^2 + 8ab + 12a$ **15.** $-2x - y - z$ **17.** $-4y^4 + 8y^2 - 4y$ **19.** $-x^3 - 2x^4$ **21.** $-y^5 + 2y^4 - 4y^3$ **23.** $(30n + 10)$ cents **25.** $n = -11$ **27.** $y = 72$ **29.** $a = 1$ **31.** $n = 2$ **33.** $a = 3$ **35.** $x = 6$

Written Exercises, page 163
1. $x^2 + 3x + 2$ **3.** $n^2 + 10n + 24$ **5.** $x^2 + 8x + 7$ **7.** $x^2 + 2xy + y^2$ **9.** $x^2 + xy + 4x + 4y$ **11.** $y^2 - 7y + 12$ **13.** $x^2 + x - 2$ **15.** $y^2 - y - 20$ **17.** $16x^2 + 8x + 1$ **19.** $a^2 - b^2$ **21.** $2a^2 - a + 2ab - b$ **23.** $36x^2 - 60x + 25$ **25.** $n^3 + 3n^2 + 3n + 1$ **27.** $x^3 - 6x^2 + 12x - 8$ **29.** $m^3 - 4m^2 + 8m - 8$ **31.** $a^3 + 3a^2b + 3ab^2 + b^3$ **33.** $n^4 + 4n^3 + 6n^2 + 4n + 1$ **35.** $10a^4 - 3a^3b + 21a^2b^2 + 3ab^3 + 4b^4$ **37.** $n^5 - 4n^4 + 4n^3 - n^2 + n - 1$

Written Exercises, pages 166–167
1. $x^2 + 7x + 12$ **3.** $x^2 + 8x + 12$ **5.** $x^2 - 3x + 2$ **7.** $x^2 - 7x + 12$ **9.** $y^2 - 3y - 4$ **11.** $n^2 + n - 2$ **13.** $y^2 + 4y - 12$ **15.** $2x^2 + 9x + 4$ **17.** $a^2 - b^2$ **19.** $36a^2 - 49$ **21.** $9x^2 - 16$ **23.** $6x^2 + 13x - 5$ **25.** $21x^2 + 13xy + 2y^2$ **27.** $4x^2 - 23xy + 15y^2$ **29.** $24x^2 + 8xy - 2y^2$ **31.** $x^2 + 10x + 24$ **33.** $2a^2 + 5a - 12$ **35.** $(6y^2 + 11y + 3)$ cm^2 **37.** $\frac{1}{2}(2n^2 + 5n - 3)$ **39.** $6x^2$

Written Exercises, page 169
1. $x^2 + 8x + 16$ **3.** $r^2 + 4r + 4$ **5.** $x^2 + 6x + 9$ **7.** $x^2 - 4x + 4$ **9.** $x^2 - 8x + 16$ **11.** $y^2 - 14y + 49$ **13.** $x^2 + 18x + 81$ **15.** $m^2 + 14m + 49$ **17.** $n^2 - 10n + 25$ **19.** $9a^2 + 12a + 4$ **21.** $16x^2 - 16x + 4$ **23.** $9x^2 + 6xy + y^2$ **25.** $100a^2 - 20ab + b^2$ **27.** $16x^2 - 24xy + 9y^2$

29. $2a^2 + 4a + 2$ **31.** $3x^2 - 12x + 12$ **33.** $-2x^2 + 8x - 8$ **35.** $8x^2 + 8xy + 2y^2$ **37.** $(a + b)^2 = a^2 + ab + ab + b^2$

Written Exercises, page 171
1. 1 **3.** a **5.** $-5x$ **7.** c^5 **9.** $7n$ **11.** -6 **13.** $-4rs$ **15.** $3x^2y$ **17.** $-7x^2$ **19.** $-7b$ **21.** $2cd^2$ **23.** $-7ab$ **25.** $25mn$ **27.** $5a^4b$ **29.** $12xy^2$ **31.** $2x$ **33.** $4a$ **35.** $4ab$

Puzzle Problems, page 171

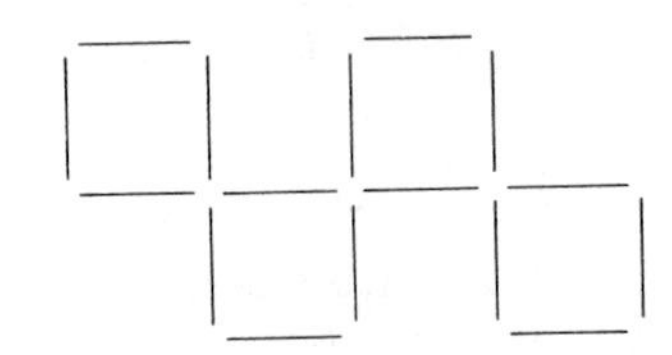

Written Exercises, page 173
1. a^3 **3.** $\frac{1}{x^4}$ **5.** $9x^3$ **7.** $\frac{-2}{n^3}$ **9.** $-3a$ **11.** $-4x^3y^2$ **13.** $2x^3$ **15.** $\frac{6m^2}{n}$ **17.** $-4x^5y$ **19.** $3x^8y^3z$ **21.** $\frac{-7a^4}{c^2}$ **23.** $3r^3t^2$ **25.** $-6x$ **27.** $4abc^2$ **29.** $-7a^2b$ **31.** $27a^3c^4$ **33.** $\frac{8b}{a}$ **35.** $\frac{-z}{x^2}$

Puzzle Problems, page 173 Since the number doubles every minute, and the basket is full after 1 hour, the basket was half full in 1 hour − 1 minute, or 59 minutes.

Written Exercises, page 175
1. $a + 3b$ **3.** $2x - y$ **5.** $4x - 2y$ **7.** $8m + 9n$ **9.** $a + 9$ **11.** $b^2 - ab$ **13.** $a^2 + ab + b^2$ **15.** $3mn^2 - 1 + 2m$ **17.** $x - 3x^2y + 2$ **19.** $2rs + 3s^2 - 6r^2$ **21.** $-4a^2b + a^2b^2 - 2a^3$ **23.** $-2 - 6n$ **25.** $3x^2 - 8xy + 9y^2$ **27.** $-3xb + 4x^3 - 6b^2$

Written Exercises, page 177
1. $x + 2$ **3.** $y + 4$ **5.** $x + 2$ **7.** $x - 4$ **9.** $a - 5$ **11.** $x - 6$ **13.** $x - 2$

Calculator Activities, page 177 **1.** 256 **3.** −4608 **5.** 221,184

Skills Review, page 178 **1.** Let n = the number. $2n + 7 = 11$ **3.** Let x = the number of tickets Clarisse sold. $x + x + 82 = 286$ **5.** Let n = the larger number. $2n = 5(n - 8) + 4$ **7.** Let r = André's rate. $\frac{1}{2}r + \frac{1}{2}(r + 12) = 13$

Chapter Review Exercises, pages 179–180 **1.** $2x - 2$ **2.** $3x^2 - 6$ **3.** $7n + 2$ **4.** $5a^2 - 1$ **5.** $3n + 8$ **6.** $11x - 8y$ **7.** $4n^2$ **8.** $4x + 7$ **9.** $6a^2 + 8b^2$ **10.** $6 + 8x^2$ **11.** 5 **12.** $y - 2$ **13.** $2x - 12y$ **14.** $4m^2 + 3n$ **15.** $4m - 5n$ **16.** $-a - 2b$ **17.** $7x - 2y$ **18.** x **19.** $-7x$ **20.** $7a + 6b$ **21.** a^6 **22.** $2a^3$ **23.** a^4 **24.** $-30a^3$ **25.** $28x^4$ **26.** $24x^3$ **27.** $-56y^5$ **28.** $18n^5$ **29.** a^6 **30.** $-a^6$ **31.** $-a^6$ **32.** $-27x^6$ **33.** m^4n^8 **34.** $27m^3n^6$ **35.** $-x^6y^9$ **36.** $9x^2y^8$ **37.** $30 + 5n$ **38.** $42x + 21$ **39.** $a^2 + 4a$ **40.** $n^2 - 5n + 6$ **41.** $12n^2 - 7n + 1$ **42.** $6x^2 - 5xy - y^2$ **43.** $2r^2 + 16r + 32$ **44.** $10a^2 - 31a - 14$ **45.** $3x^2 - 2xy - y^2$ **46.** $n^2 - 9n + 18$ **47.** $14x^2 - 3x - 2$ **48.** $30n^2 + 16n + 2$ **49.** $1 + 2n - 8n^2$ **50.** $3y^2 + 13y + 4$ **51.** $x^2 - 12x + 35$ **52.** $n^2 + 8n + 16$ **53.** $n^2 - 8n + 16$ **54.** $9m^2 - 12mn + 4n^2$ **55.** $16a^2 + 24ab + 9b^2$ **56.** $3n$ **57.** $10x^2$ **58.** $-10x$ **59.** $-3y$ **60.** $-6ab^3$ **61.** $5a^3b$ **62.** $11n^2$ **63.** $-12y$ **64.** $6n + 1$ **65.** $m + 3$ **66.** $2x - 1$ **67.** $4ab - 3$ **68.** $2a^2 - a + 1$ **69.** $3x + 4y + 5$ **70.** $\frac{1}{y^4}$ **71.** $\frac{3x}{y}$ **72.** $-\frac{2m^3}{n^5}$ **73.** $\frac{3a^2}{c^2}$ **74.** $x - 5$ **75.** $y + 7$ **76.** $m + 3$

Mixed Review, pages 182–183
1. $-x^2 + 2x + 13$ **3.** $x = -6$ **5.** $n = 1$ **7.** -12 **9.** 17 and 18 **11.** $y = -7 + 5x$ **13.** $-8a^2b^3 + 6ab^2 - 7b$

15. 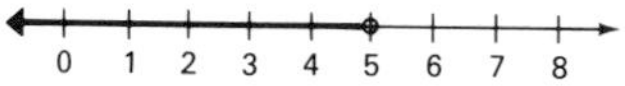

17. $25a^2 - 10a + 1$ **19.** $a = 4$ **21.** $-12x^4y^6$ **23.** $x^3 - x^2 - 10x + 12$ **25.** $-7 < 0 < 2$ **27.** $(x^3 + 2x^2)$ cubic units **29.** $45m^7n^6$ **31.** 15 cm **33.** 0, 1, 2, 3 **35.** $>$ **37.** $a = 2$ **39.** $h = 9$

CHAPTER 6

Written Exercises, page 187

1. $2 \cdot 3 \cdot 7$ **3.** $2 \cdot 2 \cdot 3 \cdot 3$ **5.** $2 \cdot 5 \cdot 31$ **7.** $2 \cdot 2 \cdot 2 \cdot 3 \cdot 5$ **9.** $2 \cdot 2 \cdot 2 \cdot 2 \cdot 3 \cdot 13$ **11.** 3 **13.** 7 **15.** 1 **17.** 11 **19.** 16 **21.** 12 **23.** 14 **25.** 25 **27.** 4 **29.** 8 **31.** 4 **33.** 20 **35.** 6 **37.** 100

Written Exercises, page 189

1. $3(3 + x)$ **3.** $2(x - 5)$ **5.** $x(3x - 1)$ **7.** $2x(x - 3)$ **9.** $7n(4n - 1)$ **11.** $4x(x - 2)$ **13.** $5mn(5 - mn)$ **15.** $9x^2(1 - 3y)$ **17.** $3(x^2 - 2x + 7)$ **19.** $11(5y^2 + 2y + 4)$ **21.** $4(a^2 + 3ab - 4b^2)$ **23.** $6(x^2 + x + 4xy + 7)$ **25.** $25(-2a^2 + b^2 + 3ab)$ **27.** $8xy(7x^2y^2 - 9xy - 8)$ **29.** $(x + 2)(a - b)$ **31.** $(n + y)(m - x)$ **33.** $(2 - a)(a + b)$ **35.** $(x^2 - 4)(y + 2)$

Written Exercises, page 191

1. $(4 - \pi)r^2$ **3.** $(a^2 - b^2)\pi$ **5.** 72 **7.** $x^2(1 - \pi) + 5(x + 3)$

Written Exercises, pages 194–195

1. 1, 5 **3.** 2, 4 **5.** 3, 4 **7.** 3, 5 **9.** 3, 6 **11.** 3, 7 **13.** 3, 8 **15.** 6, 6 **17.** $(a + 6)(a + 1)$ **19.** $(x + 6)(x + 3)$ **21.** $(r + 3)(r + 7)$ **23.** $(n + 1)(n + 16)$ **25.** $(a + 1)(a + 4)$ **27.** $(x + 1)(x + 10)$ **29.** $(n + 4)(n + 2)$ **31.** $(x + 7)(x + 2)$ **33.** $(y + 1)(y + 11)$ **35.** $(b + 3)(b + 9)$ **37.** $(x + 4)(x + 8)$ **39.** $(x + 3)(x + 10)$ **41.** $(a + 2)(a + 10)$ **43.** $(x + 25)(x + 2)$ **45.** $(m + 25)(m + 3)$ **47.** $(a + 21)(a + 3)$

Written Exercises, page 197

1. $-3, -2$ **3.** $-6, -3$ **5.** $-6, -4$ **7.** $(x - 5)(x - 2)$ **9.** $(x - 3)(x - 2)$ **11.** $(y - 5)(y - 1)$ **13.** $(x - 3)(x - 8)$ **15.** $(n + 6)(n + 3)$ **17.** $(n - 4)(n - 1)$ **19.** $(x + 4)(x + 2)$ **21.** $(y - 7)(y - 4)$ **23.** $(n + 5)(n + 5)$ **25.** $(y - 6)(y - 5)$ **27.** $(x - 7)(x - 7)$ **29.** $(m - 5)(m - 8)$ **31.** $(z - 2)(z - 16)$ **33.** $(y - 3)(y - 12)$ **35.** $(d - 5)(d - 9)$ **37.** $(x - 2)(x - 50)$

Written Exercises, page 199

1. yes **3.** yes **5.** no **7.** $(a + 5)^2$ **9.** $(x + 4)^2$ **11.** $(x - 2)^2$ **13.** $(y + 7)^2$ **15.** $(n + 9)^2$ **17.** $(x - 9)(x - 10)$ **19.** $(a - b)^2$ **21.** $(1 - 10x)^2$

Written Exercises, page 202

1. $-5, 2$ **3.** $-4, 1$ **5.** $-2, 1$ **7.** $-10, 3$ **9.** $7, -4$ **11.** $14, -1$ **13.** $(x + 7)(x - 3)$ **15.** $(x + 9)(x - 2)$ **17.** $(b - 3)(b + 4)$ **19.** $(n - 6)(n + 3)$ **21.** $(x - 5)(x + 4)$ **23.** $(y - 1)(y + 15)$ **25.** $(b + 6)(b - 4)$ **27.** $(b - 10)(b + 3)$ **29.** $(x + 14)(x - 2)$ **31.** $(y + 8)(y - 4)$ **33.** $(y + 9)(y - 3)$ **35.** $(m - 4)(m + 16)$ **37.** $(n - 9)(n + 7)$ **39.** $(y - 6)(y + 7)$ **41.** $(y + 7)(y - 8)$ **43.** $(c - 4)(c + 20)$ **45.** $(a - 4)(a + 25)$

Mixed Practice Exercises, page 203

1. $(x + 9)(x + 1)$ **3.** $(n - 6)(n - 2)$ **5.** $(y - 3)(y + 4)$ **7.** $(y - 7)(y + 3)$ **9.** $(x + 3)(x + 7)$ **11.** $(b - 1)(b + 5)$ **13.** $(b - 7)(b - 1)$ **15.** $(y - 4)(y + 5)$ **17.** $(x + 7)(x + 5)$ **19.** $(n - 7)(n - 8)$ **21.** $(x + 9)(x + 6)$ **23.** $(n + 9)(n + 5)$ **25.** $(n - 10)(n + 5)$ **27.** $(y + 26)(y - 2)$ **29.** $(a + 4)(a - 11)$ **31.** $(c - 6)(c + 10)$ **33.** $(m - 4)(m - 10)$ **35.** $4x - 4$ **37.** $4y - 4$

Written Exercises, page 205

1. $n^2 - 49$ **3.** $a^2 - 100$ **5.** $m^2 - n^2$ **7.** $x^2 - y^2$ **9.** $1 - 4x^2$ **11.** $9x^2 - 4$ **13.** $25y^2 - 4$ **15.** $81m^2 - n^2$ **17.** $25x^2 - 36$ **19.** $25x^2 - 9y^2$ **21.** $100x^2 - 25y^2$ **23.** $a^4 - 4$ **25.** 399 **27.** 2496 **29.** 896 **31.** 8091 **33.** 864 **35.** 3575 **37.** 6384 **39.** 2496

Written Exercises, pages 206–207
1. yes **3.** no **5.** yes **7.** no
9. $(x - 4)(x + 4)$ **11.** $(n - 3)(n + 3)$
13. $(a - 3b)(a + 3b)$
15. $(x - 2y)(x + 2y)$
17. $(b - 8c)(b + 8c)$
19. $(2a - 5b)(2a + 5b)$
21. $(x - 11)(x + 11)$
23. $(8x - 3y)(8x + 3y)$
25. $(2 + x)(2 - x)$ **27.** $(8 - x)(8 + x)$
29. $(x - y)(x + y)$
31. $(2x^3 - 3)(2x^3 + 3)$
33. $(12 - 11x)(12 + 11x)$
35. $(6bc - 2a)(6bc + 2a)$
37. $(y^2 + 9)(y - 3)(y + 3)$
39. $(4a^2 + 1)(2a + 1)(2a - 1)$
41. $(16 + x^2y^2)(4 + xy)(4 - xy)$
43. $(x^4y^2 + 1)(x^2y + 1)(x^2y - 1)$

45.
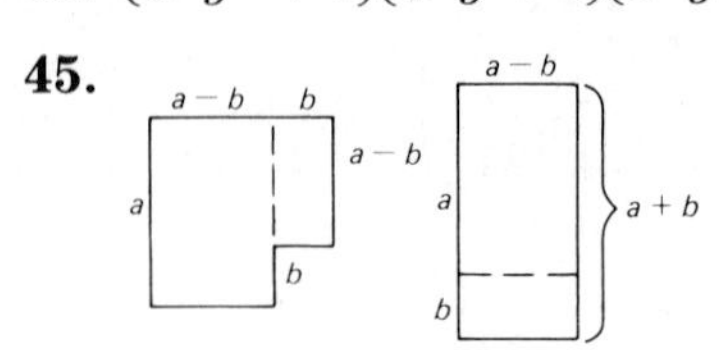

Puzzle Problems, page 207 No, it is not possible.

Written Exercises, page 209
1. $3(y + 4)(y + 2)$ **3.** $2(y - 5)(y - 3)$
5. $5(a - 6)(a + 2)$ **7.** $4(x - 3)(x + 5)$
9. $2(x - 4)^2$ **11.** $8(x - 2)(x + 2)$
13. $4(x - 3)(x + 3)$
15. $6(x - 2y)(x + 2y)$
17. $-1(x - 3)(x - 1)$
19. $2(x - 9)(x + 9)$
21. $-4(a - 2)(a + 1)$
23. $2(a - 10b)(a + 10b)$
25. $3(3 + 2z)(3 - 2z)$
27. $25(a - 2)(a + 4)$
29. $a^2(b + 6)^2$

Written Exercises, page 211
1. $(3x + 1)(x + 1)$ **3.** $(2a + 3)(a + 3)$
5. $(2b + 3)(b + 1)$ **7.** $(3a - 5)(a - 1)$
9. $(3x - 4)(x - 2)$ **11.** $(3y + 1)(y - 2)$
13. $(2x - 1)(x + 3)$ **15.** $(2x + 3)(x - 3)$
17. $(4a + 9)(a - 1)$ **19.** $(3b + 5)(2b - 5)$
21. $(4a - 5)(a + 1)$ **23.** $(2x - 5)(2x + 1)$
25. $(4r + 3)(2r - 3)$ **27.** $(4x + 3)(2x - 1)$
29. $2(2c + 1)(c + 1)$ **31.** $3(3y + 2)(y + 1)$
33. $3(2x + 1)(x + 2)$ **35.** $(5y - 2)(2y - 5)$
37. $(3x - 11)(2x + 11)$ **39.** $(3x + 4)(4x - 3)$
41. $(5n - 1)(3n + 4)$ **43.** $(5x + 1)(5x - 2)$
45. $(2c - 3)(5c - 4)$ **47.** $(3z + 2)(6z + 5)$
49. $(2x + 3)^2$ **51.** $(2b + 3)(9b - 7)$
53. $(4g - 3)(4g - 9)$ **55.** $(4x - 3)(6x + 5)$
57. $(3d - 1)(16d + 3)$ **59.** $(3x + 2)(12x + 5)$

Puzzle Problems, page 211 There is a built-in contradiction. Consider: (1) *The barber shaves himself.* Then (as the barber) he *cannot* shave himself. (2) *The barber does not shave himself.* Since he does not shave himself, then (as the barber) he *must* shave himself.

Problem Solving Strategies, pages 212–213 **1. a.** 7; 9; 3; 1; 7; 9; 3; 1; 7 **b.** 9 **3. a.** $1 + 4 + 9 + 16 = 30$; $1 + 4 + 9 + 16 + 25 = 55$ **b.** 204 squares

Computer Activities, pages 214–215
1. a. \$735 **b.** \$20,669 **c.** \$7765 **3. a.** 6 years **b.** 7 years **c.** 9 years **d.** 12 years **5. a.** Total values: \$105, \$110, \$116, \$122, \$128, \$134, \$141, \$148, \$155, \$163 **b.** Total values: \$110, \$121, \$133, \$146, \$161, \$177, \$195, \$214, \$236, \$259 **7.** Answers may vary.

Skills Review, page 216
1. $\frac{1}{2}$ **3.** $\frac{1}{4}$ **5.** $\frac{1}{2}$ **7.** $\frac{4}{5}$ **9.** $\frac{4}{3}$ **11.** $\frac{2}{3}$
13. $\frac{3}{4}$ **15.** $\frac{2}{3}$ **17.** $\frac{4}{7}$ **19.** $\frac{5}{7}$ **21.** $\frac{1}{2}$
23. $\frac{1}{2}$ **25.** $\frac{10}{13}$ **27.** $\frac{2}{5}$ **29.** $\frac{5}{9}$ **31.** $\frac{1}{2}$
33. $\frac{7}{12}$ **35.** $\frac{3}{13}$ **37.** $\frac{1}{2}$ **39.** $\frac{1}{4}$ **41.** $\frac{1}{3}$

Chapter Review Exercises, pages 217–218 **1.** 3 **2.** 4 **3.** 3
4. 8 **5.** 8 **6.** 3 **7.** $3y$ **8.** $8n$
9. x^2 **10.** $6x$ **11.** 4 **12.** $5x^2$
13. $2(n^2 + 2)$ **14.** $3(x^2 - 3)$

15. $ab(b + a)$ **16.** $7x(x - 3)$
17. $2y^2(y + 4)$ **18.** $3(xy - 9)$
19. $2x^2(1 - 4x)$ **20.** $x(x + 4)$
21. $6(6x + 1)$ **22.** $(n + 3)^2$
23. $(x + 1)(x + 7)$ **24.** $(x - 5)^2$
25. $(x - 2)(x - 7)$ **26.** $(n + 4)(n + 5)$
27. $(n + 2)(n + 10)$ **28.** $(y - 1)(y - 8)$
29. $(y - 5)(y - 6)$ **30.** $(n - 2)(n - 11)$
31. $(y + 2)(y + 15)$ **32.** $(n + 5)(n + 7)$
33. $(y + 4)(y + 8)$ **34.** $(y - 5)(y - 8)$
35. $(n - 2)(n - 16)$ **36.** $(y + 3)(y + 6)$
37. $(y + 2)(y + 9)$ **38.** $(x + 5)(x + 6)$
39. $(h - 1)(h - 23)$ **40.** $(n + 4)(n + 16)$
41. $(b - 5)(b - 20)$ **42.** $(x + 3)(x + 9)$
43. $(x - 3)(x + 6)$ **44.** $(n + 2)(n - 9)$
45. $(y + 4)(y - 6)$ **46.** $(m - 2)(m + 8)$
47. $(x + 3)(x - 8)$ **48.** $(y + 2)(y - 12)$
49. $(x + 3)(x - 4)$ **50.** $(y - 6)^2$
51. $(n + 4)(n - 5)$ **52.** $(a - 2)(a + 5)$
53. $(a + 1)(a - 10)$ **54.** $(b - 1)(b + 6)$
55. $x^2 - 4$ **56.** $y^2 - 16$ **57.** $n^2 - 25$
58. $1 - 4a^2$ **59.** $4y^2 - 9$ **60.** $16n^2 - 1$
61. $x^2 - y^2$ **62.** $x^2y^2 - 1$ **63.** $16x^2 - 9y^2$
64. $(a + 1)(a - 1)$ **65.** $(a + b)(a - b)$
66. $(x + 2y)(x - 2y)$ **67.** $(5 + n)(5 - n)$
68. $(n + 5)(n - 5)$ **69.** $(2x + y)(2x - y)$
70. $(3n+ 1)(3n - 1)$ **71.** $(4a + b)(4a - b)$
72. $(xy + 1)(xy - 1)$ **73.** $4(x + 2)(x - 2)$
74. $3(a + 1)^2$ **75.** $2(n - 2)(n + 5)$
76. $2(y + 4)(y - 6)$ **77.** $3(x + 3)(x - 4)$
78. $x(x + y)(x - y)$ **79.** $2(n - 3)(n + 7)$
80. $4(a + 1)(a - 6)$ **81.** $a(b + 2c)(b - 2c)$
82. $(2x + 1)(x + 1)$ **83.** $(3y - 1)(y + 2)$
84. $(5z + 2)(z - 1)$ **85.** $(2t + 3)(2t + 1)$
86. $(5m + 1)(2m - 3)$ **87.** $(5k + 6)(2k + 1)$

Cumulative Review, page 220

1. $7x - y$ **3.** $-2m - 10n$ **5.** $x - 3$
7. $3t - 2$ **9.** $-3y^7$ **11.** $2x^3y - 3x^2y^2 + xy^3$
13. $y^2 - 5y + 4$ **15.** $(6x^2 - x - 12)$ cm^2
17. $13m^2$ **19.** $4y^4z^2$ **21.** $2x^4 + 3x^2 - 5$
23. 3 **25.** $3x$ **27.** $7(m^2 - 3)$
29. $(x + 2)(x + 8)$ **31.** $(t - 3)(t - 15)$
33. $(1 + 3n)^2$ **35.** $(x + 1)(x - 9)$
37. $(3 + t)(3 - t)$ **39.** $2(x + 2)(x - 10)$
41. $5(2 + a)(2 - a)$ **43.** $-2(y + 5)(y + 15)$

CHAPTER 7

Written Exercises, pages 226–227

1. July, August **3.** 2500

5.

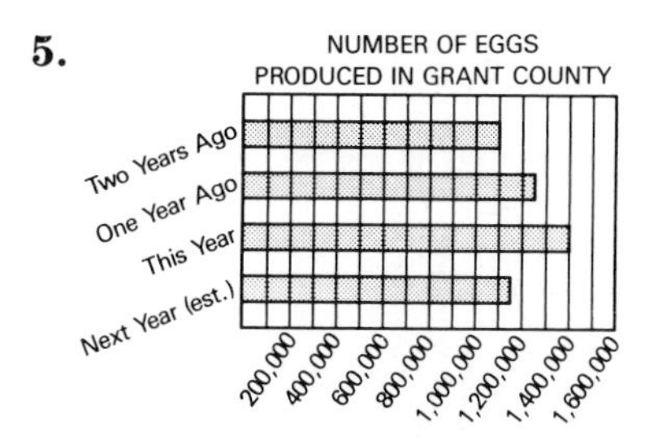

7. Swimming; about 23
9. Answers will vary.

11.

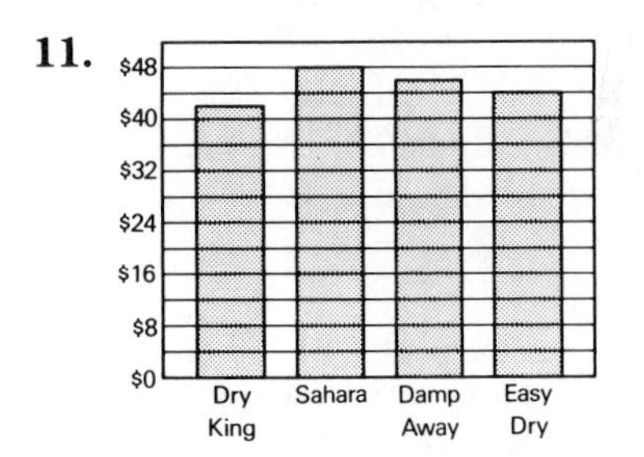

Written Exercises, pages 229–231

1. 0.9 **3.** 1992 **5.** Answers may vary considerably; about 451 million **7.** about 70
9. 40 **11.** 10; about 70 **13.** about 46°F
15. May **17.** 1905 **19.** about 370 million
21. $2500 **23.** 2 years ago **25.** 5¢

27.

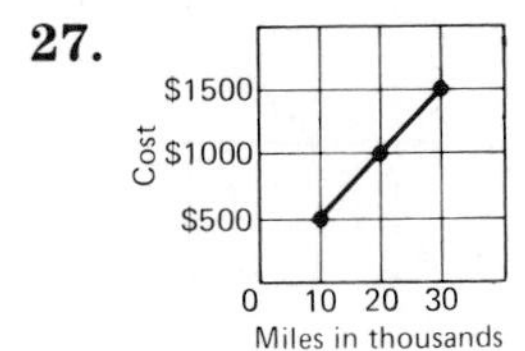

29. 4¢ for 1 mile; 8¢ for 2 miles

Written Exercises, pages 235–236

1. $(-5, 4)$ **3.** $(-3, -5)$ **5.** $(0, 4)$
7. $(2, -4)$ **9.** $(-2, 3)$ **11.** M **13.** R
15. N **17.** Q **19.** 2 **21.** $(2, 4)$ **23.** $(5, 1)$

25.

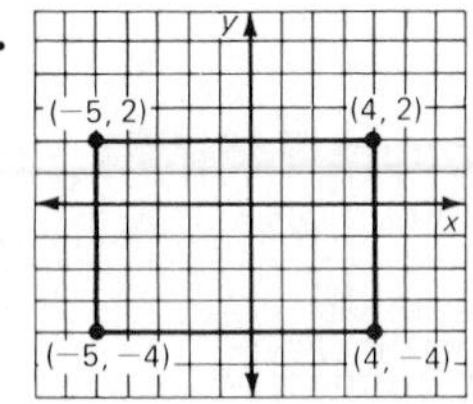

27.

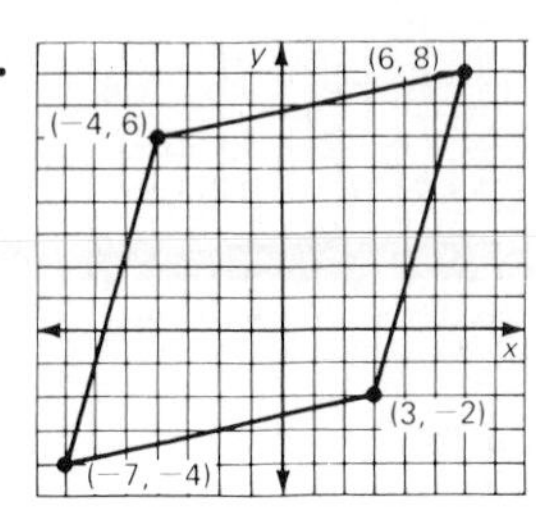

Written Exercises, pages 239–240 **1.** a, c **3.** c **5.** b **7.** 6; 10; 14; 16 **9.** 3; 5; −1; 8 **11.** 2; 1; 0; 4 **13.** 6; 3; 9; 0 **15.** $y = 9 - x$ **17.** $y = 6 + x$ **19.** $y = 5 - 4x$ **21.** $y = 9 - 4x$ **23.** $y = 9x - 7$ **25.** $2x - 5 = y$
Note: Answers to Exercises 27–37 may vary.
27. (1, 0), (2, 1), (−1, −2) **29.** (1, −1), (2, −2), (−1, 1) **31.** (0, 6), (1, 7), (−1, 5) **33.** (0, 4), (2, 8), (−2, 0) **35.** (0, 7), (1, 11), (−1, 3) **37.** (0, 2), (1, 0), (−1, 4) **39.** $y = 4x$ **41.** $y = 4x + 1$ **43.** a, b **45.** a, b

Written Exercises, page 242

1.

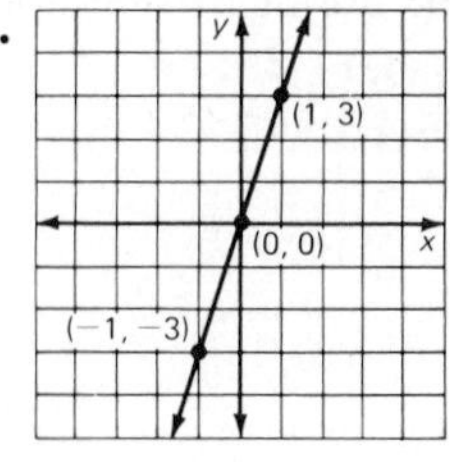

3.

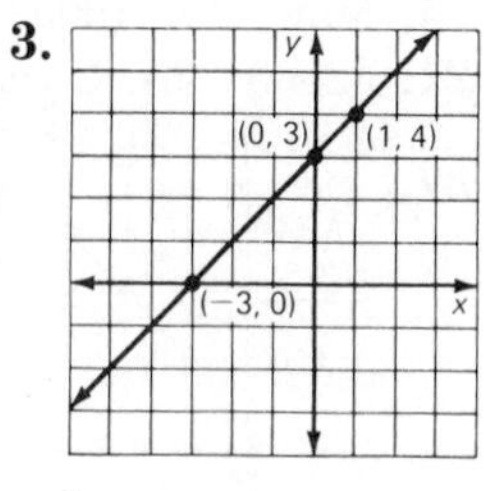

5.

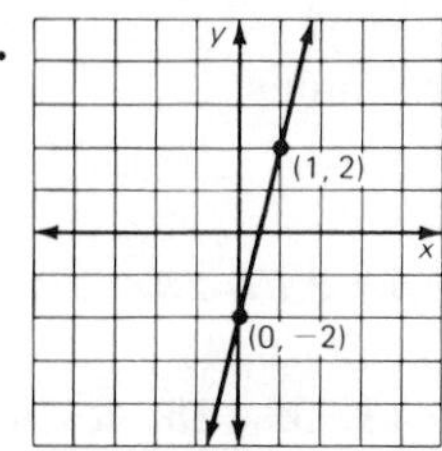

7.

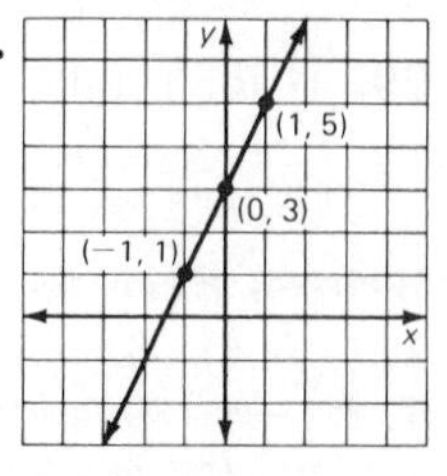

9.

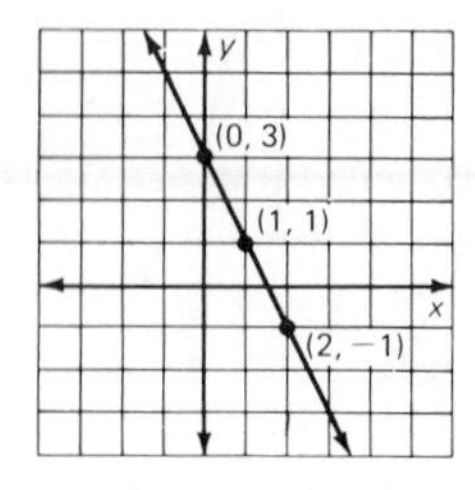

11.

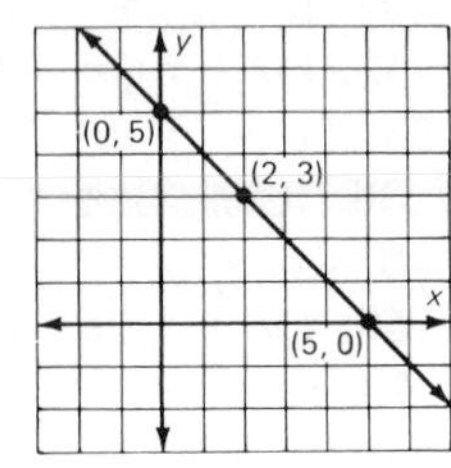

13.

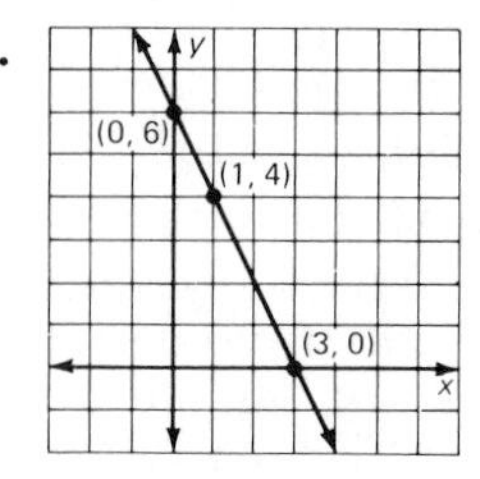

15.

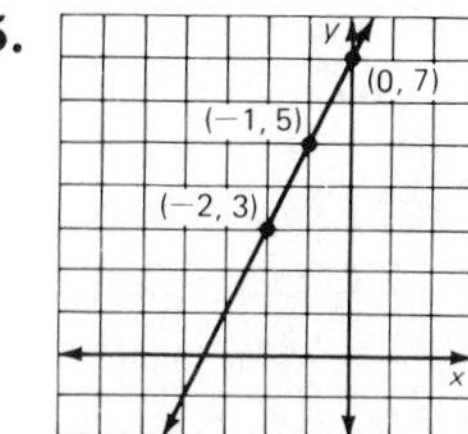

17.

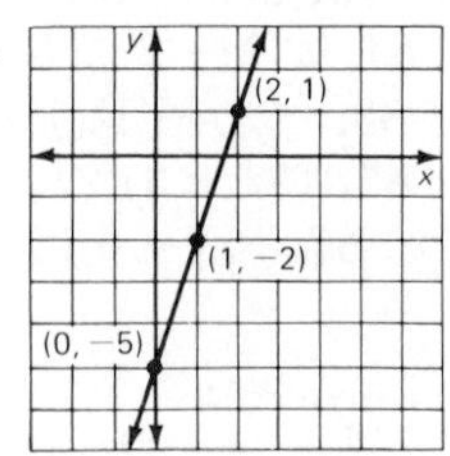

19.

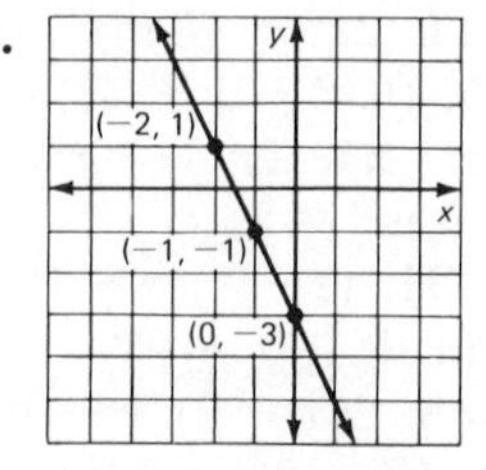

21.

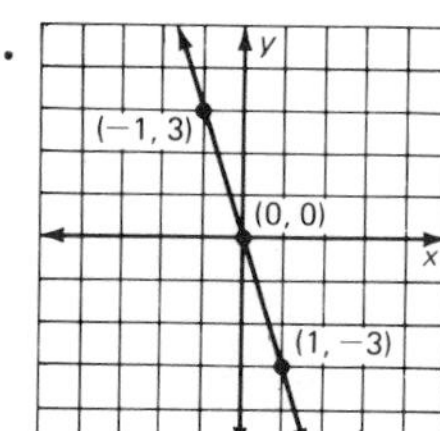

23. $y = -x + 2$

Written Exercises, page 244 **1.** $\frac{2}{3}$ **3.** $\frac{3}{2}$ **5.** $-\frac{1}{3}$ **7.** 3 **9.** 2 **11.** −1 **13.** −3 **15.** $\frac{1}{3}$ **17.** $-\frac{1}{5}$

19. Answers may vary.

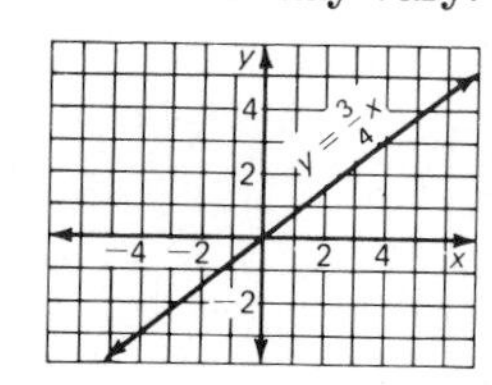

21. 0 **23.** 0 **25.** The rise varies. For example, the rise from (1, 0) to (1, 2) is 2. **27.** No

Written Exercises, pages 247–248
1. $y = -4$; $x = 5$ **3.** $x = -9$

5.

n	1	2	3	12	10
C	48	96	144	576	480

7. 80, 120, 160 **9.** −17°C **11.** 40 **13.** about 15°C **15.** about 4°C

Calculator Activities, page 249 **1.** a, c **3.** a, b **5.** b, d

Written Exercises, page 252

1.

3.

5.

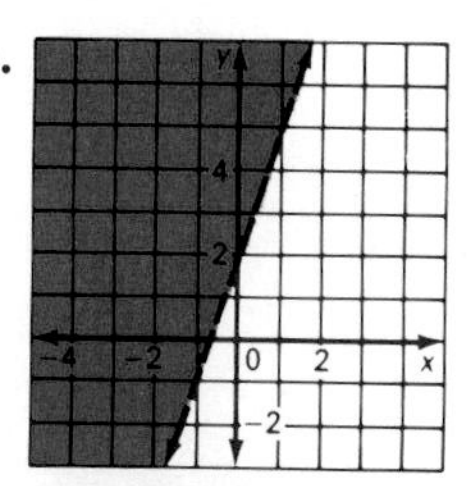

7.

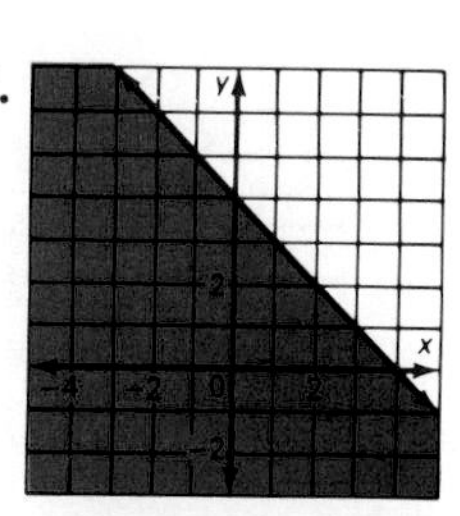

9.

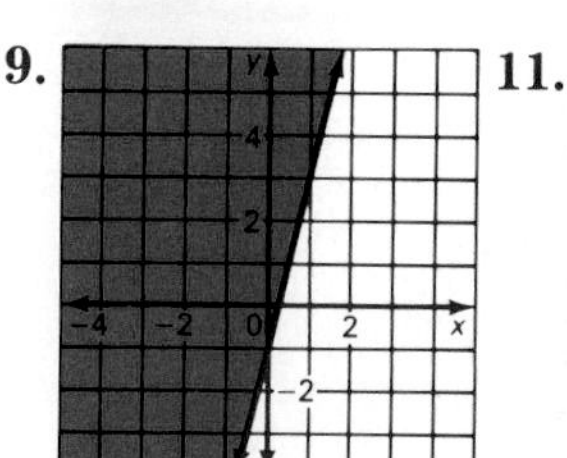

11.

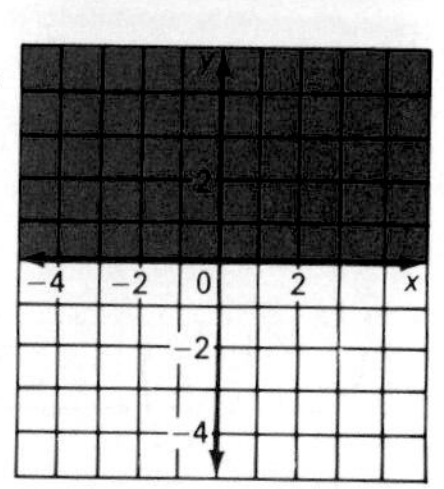

Reading Algebra, page 253 **1.** exponent **3.** trinomial **5.** coordinates

Skills Review, page 254

1. $\frac{3}{12}$ **3.** $\frac{3}{9}$ **5.** $\frac{3}{18}$ **7.** $\frac{18}{48}$ **9.** $\frac{20}{35}$ **11.** $\frac{4}{12}$, $\frac{3}{12}$ **13.** $\frac{2}{8}$, $\frac{1}{8}$ **15.** $\frac{5}{20}$, $\frac{4}{20}$ **17.** $\frac{3}{18}$, $\frac{8}{18}$ **19.** $\frac{6}{14}$, $\frac{1}{14}$ **21.** $\frac{21}{28}$, $\frac{12}{28}$ **23.** $\frac{21}{30}$, $\frac{20}{30}$ **25.** $\frac{32}{72}$, $\frac{27}{72}$ **27.** $\frac{11}{30}$ **29.** $\frac{4}{9}$ **31.** $\frac{1}{6}$ **33.** $\frac{1}{20}$ **35.** $\frac{4}{21}$ **37.** $\frac{41}{42}$ **39.** $\frac{43}{35}$ **41.** $\frac{5}{8}$ **43.** $\frac{16}{33}$ **45.** $\frac{1}{24}$

Chapter Review Exercises, pages 255–256 **1.** about 7.5 cm **2.** January **3.** July **4.** 8 cm **5.** about 117 million **6.** 1967 **7.** 0, −4, −8, 8 **8.** 3, 7, 11, 15 **9.** 10, 9, 8, 11 **10.** 5, 2, −1, 11

11.

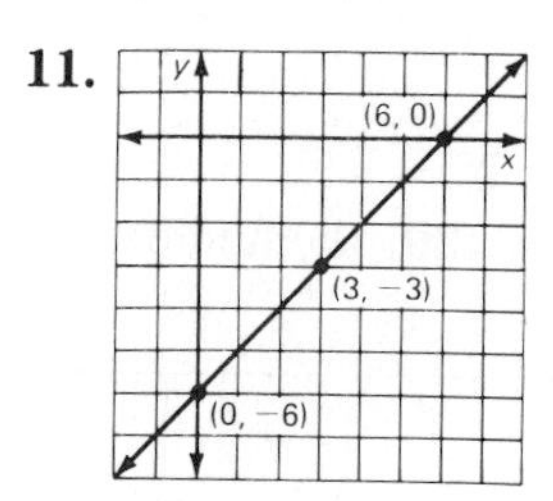

12.

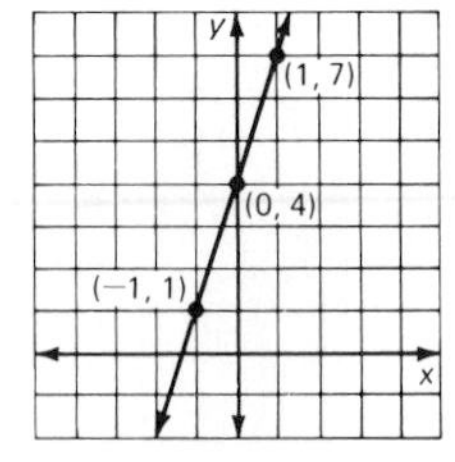

13.

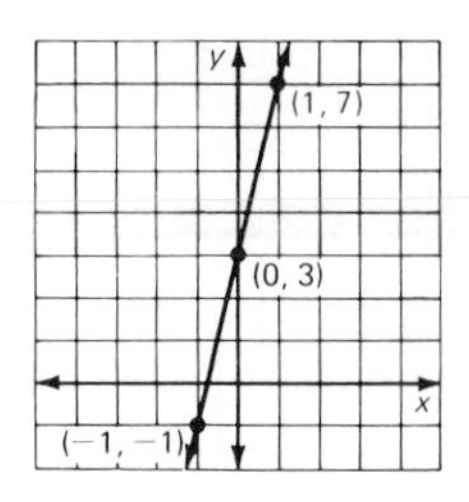

14.

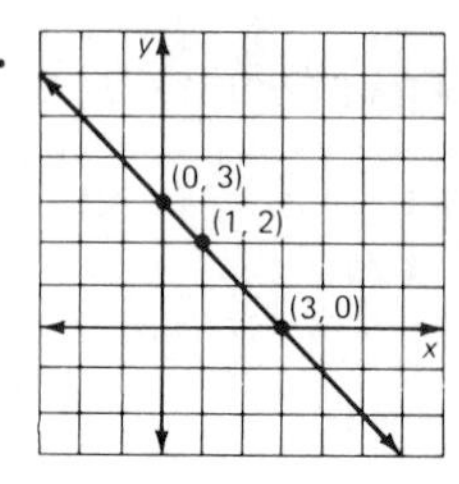

15.

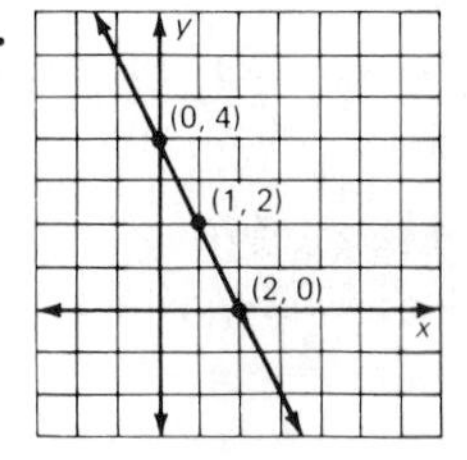

16.

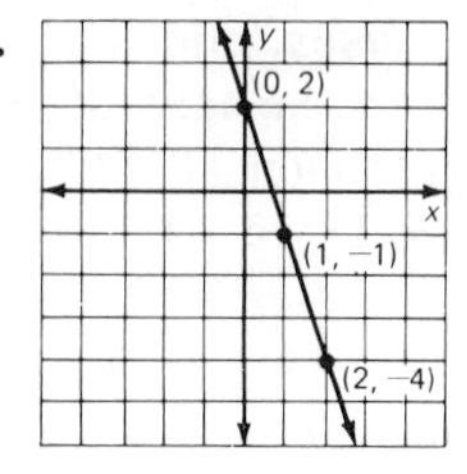

17. 3 **18.** −1 **19.** $\frac{3}{4}$ **20.** the number of yards purchased

21.

n	1	2	3	5	8
C	$4.20	$8.40	$12.60	$21.00	$33.60

22. $4.20n$ **23.** the number of bumper stickers sold

24.

n	10	15	20	50	110
A	$5	$7.50	$10	$25	$55

25.

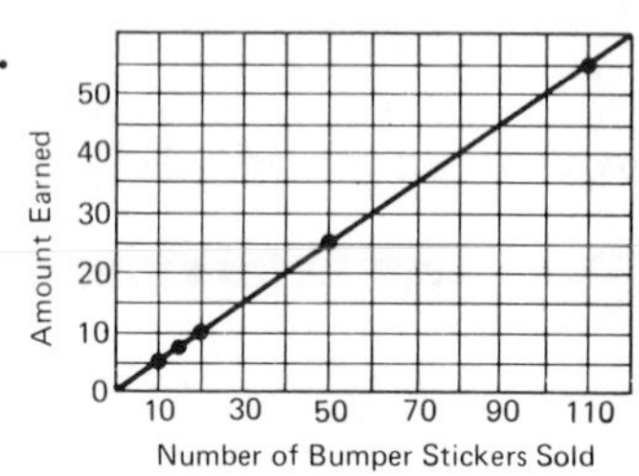

26.

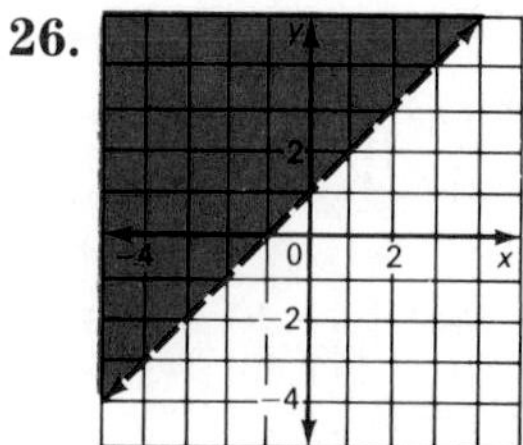

27.

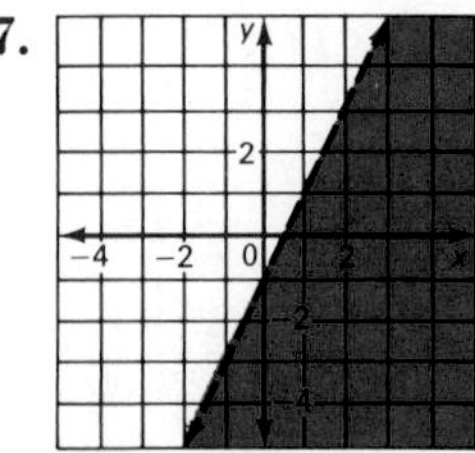

28.

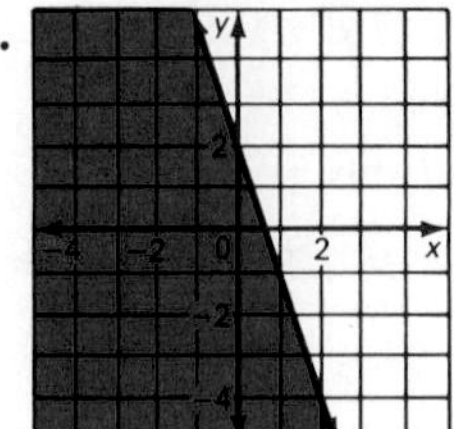

Mixed Review, pages 258–259

1. Answers may vary. See graph at right.
3. 50 km **5.** $-18r^4s^5$
7. $n^3 - 8$ **9.** 1980

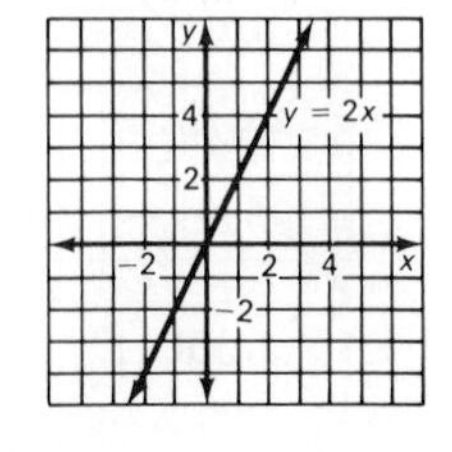

11.

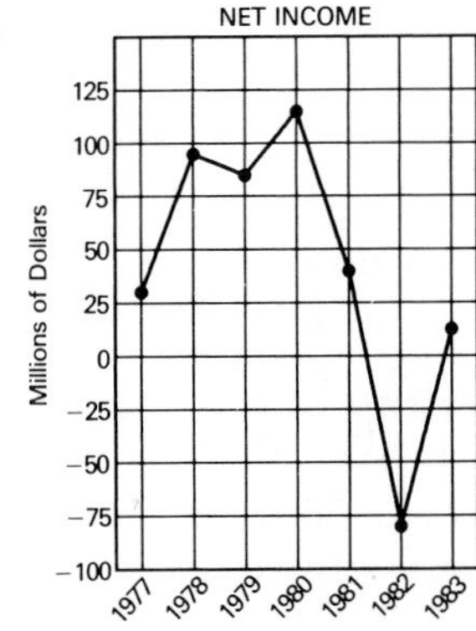

13. a. $6n + 14$ **b.** $2n^2 + 11n + 12$
15. $(m - 6)(m - 8)$
17. $7x(x^2 + 3xy + 4)$
19. $(x - 4y)^2$ **21. a.** 42 **b.** 84

23. −5 −4 −3 −2 −1 0 1

25. $-27x^2 + 18xy - 3y^2$

27.

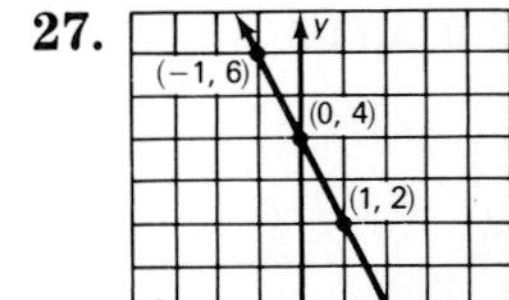

29. $-4x^{14}y^6$ **31.** $3a^2 - 4ab - b^2$
33. 146¢, or $1.46 **35.** $x = -2$

CHAPTER 8

Written Exercises, pages 264–265
1. (1, 2) **3.** (2, −1) **5.** Answers may vary. (0, 12), (6, 6) **7.** Answers may vary. (0, 1), (1, −1) **9.** (1, 2) **11.** (−1, −2) **13.** (−3, 0) **15.** (−2, 1) **17.** (−1, −1)
19. (2, −1) **21.** about $\left(-1\frac{2}{5}, 4\frac{4}{5}\right)$
23. about $\left(-3\frac{1}{3}, 1\frac{1}{3}\right)$
25. about $\left(2, -2\frac{1}{2}\right)$

27.

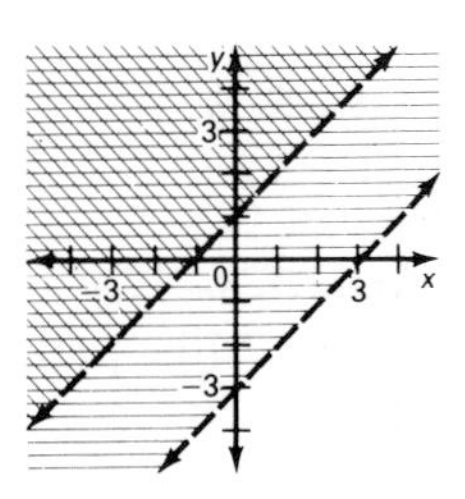

29.

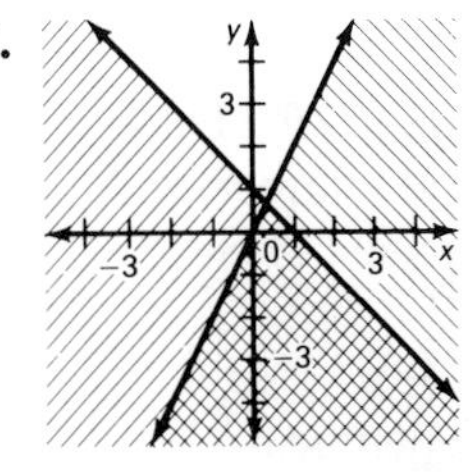

Written Exercises, pages 268–269 **1.** 4 **3.** 1 **5.** 3 **7.** 4 **9.** 2 **11.** −7 **13.** no **15.** one **17.** one **19.** one **21.** all **23.** all **25.** no **27.** one **29. a.** no **b.** yes **c.** no **31. a.** yes **b.** no **c.** no

Written Exercises, page 271 **1.** (3, 6) **3.** (5, 2) **5.** (9, 13) **7.** (1, 4) **9.** (3, −1) **11.** (5, 2) **13.** (10, 2) **15.** (3, 2) **17.** $A = 2h^2$ **19.** $V = 6h^3$ **21.** $x = 19$; $y = 31$; $z = 12$

Written Exercises, pages 273–274
1. (6, −4) **3.** (1, −4) **5.** (2, 1) **7.** (4, 2) **9.** (−2, 2) **11.** (3, −2) **13.** (2, 6) **15.** (−1, 3) **17.** (7, 2) **19.** (−2, 4) **21.** (3, −2) **23.** (2, −6) **25.** (2, −1) **27.** (1, −1) **29.** (−1, −1) **31.** (3, 5)
33. $\left(-\frac{1}{3}, \frac{2}{3}\right)$ **35.** $\left(\frac{1}{4}, -\frac{1}{2}\right)$
37. $\left(2, -\frac{1}{3}\right)$

Puzzle Problems, page 274 Pass the center of one cord under the string circling the other person's wrist, over his hand, then back under the string again.

Mixed Practice Exercises, page 275
1. about $\left(1\frac{1}{3}, 1\frac{2}{3}\right)$ **3.** (−2, −1) **5.** (9, 3) **7.** (−20, 11) **9.** (4, 3) **11.** (3, 4) **13.** (3, 5)

15. (3, 3) **17.** $\left(5, \frac{2}{3}\right)$ **19.** (−1, 5)
21. (2, −2) **23.** (6, −8)

Written Exercises, pages 278–279
1. 24, 11 **3.** 31, 22 **5.** coat, $135; pants, $45 **7.** 1986, 84 cm; 1985, 66 cm **9.** 18 cm, 6 cm **11.** Sam, 22 books; Norm, 34 books **13.** 7 g

Puzzle Problems, page 279 Let b = the number of balls, c = the number of cups, and p = the number of pennies. Then $b + c = 12p$ and $4b + 2p = c$; $b = 12p - c$, so $4(12p - c) + 2p = c$; $48p - 4c + 2p = c$; $50p = 5c$; $10p = c$. 10 pennies

Written Exercises, page 281 **1.** (6, 3) **3.** (−2, 3) **5.** (2, 4) **7.** (−1, −1) **9.** (3, 0) **11.** (2, −3) **13.** (−2, 6) **15.** (3, 2) **17.** (−3, −12) **19.** (1, 2) **21.** (1, 4)

Written Exercises, pages 282–283
1. plate, $1.35; glass, $1.00 **3.** adult, $3; student, $2 **5.** tuna, $1.15; rice, $1.40
7. 5 expensive balls (x), 15 cheaper balls (y)

Written Exercises, page 285 **1.** 5 km/h **3.** 100 km/h **5.** 483 km/h

Reading Algebra, page 287 **1.** Answers may vary. At a pet show, 34 adult's tickets and 85 child's tickets were sold. An adult's ticket cost twice as much as a child's ticket. If a total of $153 in ticket sales was collected, what was the cost of each kind of ticket?

Problem Solving Strategies, page 289
1. If the store bought 60 coats, the profit would be (60 − 10)($95) − (60)($70) = $4750 − $4200 = $550 ≠ $500. **3.** If the cup cost $1.75, then the saucer cost $1.75 − $1.20 = $.55. $1.75 + $.55 = $2.30 ≠ $2.50.
5. 37 tapes **7.** 8 cars

Skills Review, page 290 **1.** $\frac{2}{15}$ **3.** $\frac{5}{12}$ **5.** $\frac{3}{10}$ **7.** $\frac{8}{15}$ **9.** $\frac{2}{5}$ **11.** $\frac{4}{21}$ **13.** $\frac{1}{3}$ **15.** $\frac{1}{6}$ **17.** $\frac{2}{3}$ **19.** $\frac{8}{15}$ **21.** $\frac{8}{5}$ **23.** $\frac{15}{8}$ **25.** $\frac{4}{5}$ **27.** $\frac{5}{8}$ **29.** $\frac{7}{3}$ **31.** $\frac{7}{6}$

Chapter Review Exercises, pages 291–292 **1.** (2, 4) **2.** (1, 2) **3.** (−2, 1) **4.** (−1, 3) **5.** (2, −4) **6.** (0, 1) **7.** 9 **8.** −1 **9.** 1 **10.** −2 **11.** $\frac{1}{2}$ **12.** $\frac{2}{3}$ **13.** no pair **14.** one **15.** all pairs **16.** one **17.** no pair **18.** one **19.** (3, 9) **20.** (−2, 8) **21.** (1, 3) **22.** (2, 10) **23.** (1, 2) **24.** (−5, 25) **25.** (2, −4) **26.** (1, 4) **27.** (8, 2) **28.** (−4, 1) **29.** (−6, 6) **30.** (1, 1) **31.** (3, 2) **32.** (5, −1) **33.** (5, 2) **34.** (3, 2) **35.** (3, 2) **36.** (−1, 1) **37.** (−2, −3) **38.** (1, 1) **39.** (1, 6) **40.** 20, 7 **41.** 17, 3 **42.** saw, $12; hammer, $15 **43.** adult, $4.00; student, $2.50 **44.** 15 km/h

Cumulative Review, page 294
1. 37°C **3.** between 2 A.M. and 3 A.M.
5. $y = 3x - 2$

x	−2	−1	0	1	2
y	−8	−5	−2	1	4

7. $40n$ **9.** −2 **11.** (1, 3) **13.** (1, −1)
15. 24, 32

CHAPTER 9

Written Exercises, page 299
1. $\frac{2}{3}$ **3.** $\frac{1}{6a}$ **5.** $-\frac{1}{2}$ **7.** $\frac{1}{2}$ **9.** $-\frac{4x}{3}$ **11.** $\frac{4}{3r}$ **13.** $\frac{ab}{3}$ **15.** $-\frac{a^2}{5}$ **17.** $\frac{2}{a}$ **19.** $\frac{2a}{x}$ **21.** $\frac{2x^2}{5y}$ **23.** $\frac{5m}{4}$ **25.** $\frac{pq}{8r}$ **27.** $-2yz$ **29.** $\frac{2x}{7w}$ **31.** $-\frac{5b^2}{ac}$ **33.** $\frac{t^2}{12rs^2}$ **35.** $-\frac{11c^3}{ab}$ **37.** $-\frac{17x}{3yz}$
39. 0 **41.** 2 **43.** 2

Written Exercises, page 301
1. $\frac{1}{3}$ **3.** $\frac{3}{7}$ **5.** x **7.** $\frac{1}{3y - 2}$ **9.** $\frac{2}{c - 3}$
11. $\frac{2}{x - 5}$ **13.** $\frac{n - 3}{n}$ **15.** $\frac{2x}{y}$
17. $\frac{5}{k + 2}$ **19.** $\frac{6}{u - v}$ **21.** $\frac{4x}{y + 2x}$
23. $\frac{x + 3}{x - 3}$ **25.** $x - 5$ **27.** $\frac{x}{2x + y}$
29. $\frac{x - 3}{x + 3}$ **31.** $\frac{y - 5}{y - 3}$ **33.** $\frac{z - 10}{z + 6}$

Written Exercises, page 303
1. $-1(2x)$ **3.** $-1(a + b)$
5. $-1(3y - 2)$ **7.** $-1(8m - 3)$
9. $-1(v - 25)$ **11.** $-1(4 + y^2)$ **13.** -1
15. -1 **17.** -1 **19.** -1 **21.** -2
23. -1 **25.** $-a - 1$ **27.** $-\frac{1}{r + 1}$
29. $-\frac{1}{c + 2}$ **31.** $-x - 5$ **33.** $-\frac{4}{t + 3}$
35. $-\frac{p + 3}{2}$ **37.** $-\frac{r - 4}{3}$

Written Exercises, page 305 **1.** $\frac{5}{8}$ **3.** $\frac{2}{3}$
5. $\frac{3}{5}$ **7.** $\frac{2}{1}$ **9.** $\frac{3}{2}$ **11.** $\frac{3}{2}$ **13.** $\frac{1}{7}$ **15.** $\frac{1}{5}$
17. $\frac{3}{1}$

Written Exercises, page 307 **1.** 20 boys
3. 48 tails **5.** 25 m **7.** $22,200 **9.** 420 mL of vinegar and 315 mL of oil **11.** 21,000 votes, 17,500 votes, and 17,500 votes

Written Exercises, page 309
1. $x = 9$ **3.** $x = 12$ **5.** $x = 2$ **7.** $x = 8$
9. $a = 2$ **11.** $x = 10$ **13.** $x = 8$
15. $a = 5$ **17.** $x = 2$ **19.** $a = -4$
21. $n = 5$ **23.** $y = -5$ **25.** $x = 3$
27. $m = -1$ **29.** $y = -1$ **31.** $x = 12$
33. $c = 7$

Written Exercises, pages 311–313
1. $1.55 **3.** $6.90 **5.** 7 hours **7.** 8 g
9. 21 cm **11.** 8 km **13.** 880 cm **15.** $495
17. $18,000

Puzzle Problems, page 313 There are 19 different ways as shown:

Written Exercises, page 315
1. $\frac{1}{12}$ **3.** $\frac{8}{21}$ **5.** $\frac{y^2}{2}$ **7.** $\frac{2}{3}$ **9.** $\frac{2y}{5}$
11. $\frac{1}{2y^2}$ **13.** $6r$ **15.** $\frac{5a^2}{4b}$ **17.** $\frac{x - 1}{2}$
19. $\frac{x - 4}{2}$ **21.** $5a - 5$ **23.** $\frac{a}{ab - b}$
25. $\frac{a + b}{2}$ **27.** $\frac{r}{2}$ **29.** $-\frac{1}{3x + 6}$
31. $\frac{x + 2}{2}$ **33.** $\frac{2x - 10}{x - 1}$ **35.** $\frac{x + 2}{x + 4}$
37. $-7n - 7$ **39.** $\frac{m - 2}{m + 2}$

Written Exercises, pages 317–318
1. $\frac{5}{2}$ **3.** $\frac{6}{5}$ **5.** $\frac{x^2}{12}$ **7.** $\frac{1}{ab}$ **9.** $\frac{21m}{8}$
11. $6b$ **13.** $\frac{ab}{8}$ **15.** 2 **17.** $\frac{a^2 + 1}{a}$
19. $\frac{2}{y + 3}$ **21.** 1 **23.** $\frac{x + 4}{2x}$ **25.** $\frac{a - 3}{a + 3}$
27. $\frac{3y + 6}{2y^2 - 10y + 12}$ **29.** $\frac{y - 3}{2}$ **31.** $d - 5$
33. $\frac{y + 2}{y - 6}$

Written Exercises, page 320

1. $\frac{7}{5}$ **3.** $\frac{9}{2a}$ **5.** $\frac{2n}{3}$ **7.** $\frac{x}{2}$ **9.** $-y$

11. $-x$ **13.** $\frac{x+1}{y}$ **15.** $\frac{x+1}{3x}$

17. $\frac{5}{a+b}$ **19.** $\frac{2}{y-2}$ **21.** 2 **23.** 1

25. $\frac{a+b}{ab}$ **27.** $\frac{-2c+4}{7c}$

Written Exercises, page 322 **1.** 8 **3.** 16 **5.** $4x$ **7.** $3y$ **9.** $6x$ **11.** 9 **13.** $6a$ **15.** $4y$ **17.** $15y$ **19.** $8m$ **21.** $5x$ **23.** $2b$

Puzzle Problems, page 322 The man borrowed one horse from a neighbor so that he then had 18 horses. The ratios $\frac{1}{2}$, $\frac{1}{3}$, and $\frac{1}{9}$ can be expressed as $\frac{9}{18}$, $\frac{6}{18}$, and $\frac{2}{18}$ respectively. Floyd then received 9 horses, Denise 6 horses, Harriet 2 horses, and the borrowed horse was returned to the neighbor.

Written Exercises, page 324

1. $\frac{5}{6}$ **3.** $\frac{3y}{8}$ **5.** $\frac{7x}{6}$ **7.** $\frac{7x}{6}$ **9.** $\frac{5}{2a}$

11. $-\frac{13}{6y}$ **13.** $\frac{13x}{12}$ **15.** $\frac{11}{2x}$ **17.** $\frac{23}{4n}$

19. $\frac{7x}{24}$ **21.** $\frac{x-15}{5x}$ **23.** $\frac{8+x}{2x}$

25. $\frac{4y+1}{6xy}$ **27.** $\frac{2y+3x}{xy}$ **29.** $\frac{x}{4}$

31. $\frac{4x-1}{4}$ **33.** $\frac{21x-10}{6x}$

Written Exercises, page 326

1. $\frac{x+7}{8}$ **3.** $\frac{12t-7}{12}$ **5.** $\frac{11x-15}{12}$

7. $\frac{a-3}{6}$ **9.** $-\frac{7x+9}{12}$ **11.** $\frac{3x+1}{4}$

13. $\frac{x+8}{3}$ **15.** $\frac{4y+20}{3}$ **17.** $\frac{5x+5}{3}$

19. $\frac{5r}{16}$ **21.** 0

Puzzle Problems, page 326 Let $x =$ his age at death. Then you can write the equation that follows:

$\frac{1}{6}x + \frac{1}{12}x + \frac{1}{7}x + 5 + \frac{1}{2}x + 4 = x$

When you solve the equation you find that $x = 84$. Diophantus lived to be 84 years old.

Written Exercises, page 328 **1.** $x = 30$ **3.** $x = 12$ **5.** $x = 2$ **7.** $n = 10$ **9.** $y = -60$ **11.** $a = -9$ **13.** $n = 2$ **15.** $x = 3$ **17.** $x = -18$ **19.** $x = -7$ **21.** $x = 1$ **23.** $x = -5$

Written Exercises, pages 330–332

1. 6 hours **3.** $1\frac{1}{3}$ hours **5.** $4\frac{4}{9}$ hours

7. $3\frac{1}{13}$ hours **9.** 12 hours **11.** Dana, 12 hours; Jack, 6 hours

Calculator Activities, page 333 **1.** 6.5 **3.** 3 **5.** 7

Written Exercises, page 335

1. $\frac{8k+5}{k(k+1)}$ **3.** $\frac{-3(x+2)}{x(x-3)}$

5. $\frac{-8a}{(a+4)(a-4)}$ **7.** $\frac{-y^2+3y-12}{(y+3)(y-2)}$

9. $\frac{x^2+10x-21}{(x+3)(x-3)}$ **11.** $\frac{b(3b+11)}{(b+6)(b-1)}$

13. $\frac{7x-7y-x^2-xy}{(x+y)(x-y)}$

15. $\frac{4a-4b+2a^2+2ab}{(a+b)(a-b)}$

17. $\frac{8+5c+5d}{(c+d)^2}$ **19.** $\frac{1+3x+3y}{(x+y)(x-y)}$

21. $\frac{3(a-5)}{(a+2)(a-2)}$ **23.** $x = \frac{1}{5}$

25. $m = 4$

Skills Review, page 339 **1.** $<$ **3.** $<$ **5.** $<$ **7.** $<$ **9.** 6.7 **11.** 9.3 **13.** 4.3 **15.** 9.21 **17.** 0.08

Chapter Review Exercises, pages 340–341

1. $\frac{1}{3}$ **2.** $\frac{1}{5x}$ **3.** $\frac{x}{2}$ **4.** $\frac{4a}{b}$ **5.** $2y^2$

6. $a-5$ **7.** $\frac{1}{3}$ **8.** $n+2$ **9.** $\frac{3+x}{3-x}$

10. -1 **11.** -1 **12.** $-\frac{1}{2}$ **13.** $-\frac{1}{a + 7}$
14. 6 cans **15.** 18 players **16.** $x = 4$
17. $n = 4$ **18.** $x = 4$ **19.** $b = -1$
20. $5.40 **21.** $1.30 **22.** $\frac{1}{21}$ **23.** $\frac{3}{x}$
24. $\frac{2m^2}{3}$ **25.** $\frac{1}{15x - 15}$ **26.** 9 **27.** 3
28. $\frac{1}{5ab}$ **29.** $\frac{2m}{n}$ **30.** $\frac{x - 2}{5}$
31. $\frac{a + b}{4a - 4b}$ **32.** $\frac{3}{5}$ **33.** $2x$
34. $\frac{n - 4}{2}$ **35.** $\frac{15}{a + 1}$ **36.** $\frac{3a}{a + 2}$
37. $\frac{8}{15}$ **38.** $\frac{2n}{15}$ **39.** $\frac{11x}{14}$ **40.** $\frac{5a}{12}$
41. $\frac{4 + x}{4x}$ **42.** $\frac{x - 9}{3x}$ **43.** $\frac{3b - 8}{6b^2}$
44. $\frac{y - x}{xy}$ **45.** $\frac{x + 6}{6}$ **46.** $\frac{3y - 7}{10}$
47. $\frac{7z + 1}{8}$ **48.** $x = 2$ **49.** $n = 12$
50. $x = -\frac{1}{4}$ **51.** $n = 7$ **52.** $2\frac{8}{11}$ hours
53. $\frac{5t - 6}{t(t - 2)}$ **54.** $\frac{5x - 2}{x(x - 1)}$
55. $\frac{z(3z - 5)}{(z + 1)(z - 1)}$

Mixed Review, pages 344–345 **1.** $(2, -6)$
3. $(0, 4)$ **5.** $\frac{z + 7}{z - 1}$ **7.** Answers may vary.
$(0, 4), (1, 1), (2, -2)$ **9.** $26.25 **11.** $(x - 9)(x - 7)$ **13.** $>$ **15.** $x = 15$ **17.** $x = 4$
19. 24 **21.** $y^4 - 5y^3 + y^2 - 10y + 9$

23.

n	40	37	35.5	22.5
W	400	370	355	225

25. $\frac{2b}{9a}$ **27.** $\frac{4a^2 + 5b^3}{2ab}$ **29.** $-\frac{2}{a}$
31. $2a^3 + a^2b - ab^2$ **33.** $15y^2 + 19y - 56$
35. $6\frac{2}{3}$ hours **37.** 10:00

CHAPTER 10

Written Exercises, page 349 **1.** 0.66 **3.** 2.93 **5.** 2.64 **7.** 5.12 **9.** 7.765 **11.** 12.817 **13.** 14.53 **15.** 20.511 **17.** 11.056 **19.** 12.6 **21.** 0.82 **23.** 24.3 **25.** 610 **27.** 4.1208 **29.** 2.5 **31.** 0.423526 **33.** 0.06018 **35.** 0.004182 **37.** 0.206244 **39.** 16.1671

Written Exercises, page 351 **1.** 6 **3.** 8 **5.** 46 **7.** 1.1 **9.** 0.2 **11.** 0.56 **13.** 1.7 **15.** 12.6 **17.** 15.0 **19.** 571.4 **21.** 65.8 **23.** 2.6 **25.** 22.47 **27.** 25.00 **29.** 32.50 **31.** 238.46 **33.** 53.33 **35.** 1266.67

Written Exercises, page 353 **1.** 0.125 **3.** 0.6 **5.** 0.375 **7.** 0.285714285714 . . . **9.** 0.571428571428 . . . **11.** 0.11 . . . **13.** 0.5454 . . . **15.** 0.4 **17.** 0.720 **19.** 0.654

Written Exercises, page 355 **1.** 50 **3.** 4 **5.** 8.6363 . . . **7.** 2.6 **9.** 5 **11.** 13.33 . . . **13.** 2 **15.** 6600 **17.** 86.25 **19.** 71.6 **21.** 7.5 **23.** 5 **25.** 5 **27.** 2 **29.** 2.4 **31.** 2 **33.** 2

Written Exercises, page 357 **1.** 2.4 and 3.3 **3.** 43 cm and 90.3 cm **5.** 11.35 km

Calculator Activities, page 357 **1.** 2.5 **3.** 0.75 **5.** 10.73

Written Exercises, pages 361-362
1. 40% **3.** 65% **5.** 50% **7.** 7% **9.** 9% **11.** 4.5 % **13.** 21.4% **15.** 12.5% **17.** 20% **19.** 75% **21.** 0.64 **23.** 0.95 **25.** 0.20 **27.** 0.67 **29.** 0.47 **31.** 0.75 **33.** 0.08 **35.** 0.09 **37.** 0.065 **39.** 0.055 **41.** $\frac{1}{3}$
43. $\frac{1}{4}$ **45.** $\frac{1}{2}$ **47.** $\frac{1}{10}$ **49.** $\frac{3}{5}$ **51.** about 10% **53.** about 40% **55.** 55%; 45%; 35% **57.** 10%; 15%; 20%; 25% **59.** 32% **61.** 4 **63.** See graph.

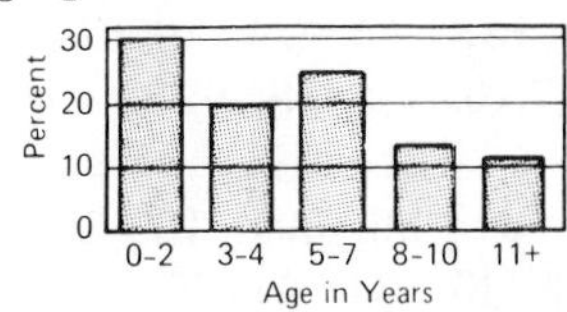

Written Exercises, page 364 **1.** 36 **3.** 4.5 **5.** 8.4 **7.** 28 **9.** 90 **11.** 37.5 **13.** 18 **15.** 288 **17.** $5.40 **19.** $5.00 **21.** $4.76 **23.** $50

Written Exercises, pages 367–368 **1.** 22 **3.** 40% **5.** 2.65 **7.** 55 **9.** 950 **11.** $.15 **13.** 31.8 kg **15.** 816 **17.** 20% **19.** $2.32 **21.** 24%

Calculator Activities, page 368 **1.** 4.5 **3.** 75% **5.** 10

Written Exercises, page 370 **1.** $40 **3.** $50 **5.** about 16.7 years **7.** $37.50

Written Exercises, page 372 **1.** $1000 at 8.5% **3.** $80,000 at 7% and $40,000 at 9% **5.** Stocks, $5500; bonds, $3500

Written Exercises, pages 374–376 **1.** 40 kg at $.70 per kg and 60 kg at $.95 per kg. **3.** 380 children's tickets **5.** 20 pounds at $3.20 per lb, 40 pounds at $4.00 per lb **7.** 200 kg **9.** 5 dimes, 14 quarters

Calculator Activities, page 377 **1.** 0.625 **3.** 0.533 . . . **5.** 0.1 **7.** 0.33 . . .

Problem Solving Strategies, page 379
1. A. Number of boys + Number of girls = 23 B. 5 = Number of girls − Number of boys **3.** Distance Marla travels + Distance Danny travels = 21; $1\frac{1}{2}$ hours **5.** apple, 40¢; pear, 80¢

Skills Review, page 380 **1.** increase: $.52 **3.** decrease: $.75 **5.** increase: $2.24 **7.** 50% **9.** 50% **11.** 20% **13.** 25% **15.** 40% **17.** 50% **19.** 20% **21.** 30% **23.** 25%

Chapter Review Exercises, pages 381–382 **1.** 5.70 **2.** 17.155 **3.** 51.15 **4.** 67.12 **5.** 4.9 **6.** 11.75 **7.** 78.08 **8.** 2.345 **9.** 12.24 **10.** 0.45 **11.** 0.812 **12.** 0.015 **13.** 52.4 **14.** 1000.0 **15.** 2142.5 **16.** 1.7 **17.** 0.375 **18.** 0.125 **19.** 0.833 . . . **20.** 0.0909 . . . **21.** 0.44 . . . **22.** 4 **23.** 110 **24.** 7 **25.** 30 **26.** 300 **27.** 16.5 **28.** 2 **29.** 10 **30.** 15 **31.** $44.25 **32.** 1.9 cm **33.** 4% **34.** 32% **35.** 13.6% **36.** 2.5% **37.** 90% **38.** 30% **39.** 25% **40.** 20% **41.** 12.5% **42.** $33\frac{1}{3}$% **43.** 0.36 **44.** 0.85 **45.** 0.025 **46.** 0.0315 **47.** 0.075 **48.** $\frac{1}{4}$ **49.** $\frac{1}{10}$ **50.** $\frac{2}{3}$ **51.** $\frac{1}{2}$ **52.** $\frac{3}{4}$ **53.** 20 **54.** 5 **55.** 50% **56.** 12.5% **57.** 100 **58.** 150 **59.** $1300 at 6%, $700 at 9% **60.** 400 people at the $15 dinner, 150 people at the $100 dinner **61.** 5 liters

Cumulative Review, page 384 **1.** $\frac{7xy}{3}$ **3.** $\frac{3}{4}$ **5.** 150 cm **7.** $-\frac{1}{2(r+3)}$ **9.** $\frac{15x+7}{3x^2}$ **11.** 1 **13.** 8.38 **15.** 1.125 **17.** 34.32 **19.** 25 **21. a.** 0.88 **b.** 0.88 . . . **c.** 0.03 **23. a.** $\frac{17}{20}$ **b.** $\frac{6}{25}$ **c.** $\frac{3}{10}$ **25.** 70% **27.** 480 L

CHAPTER 11

Written Exercises, page 389 **1.** 2 **3.** 5 **5.** 4 **7.** −2 **9.** 1 **11.** −7 **13.** −6 **15.** 8 **17.** −8 **19.** 6 **21.** 8 **23.** 3 **25.** −10 **27.** 20 cm **29.** 16 ft **31.** 40 m

Puzzle Problems, page 389
Answers may vary.
$5 = (\sqrt{4} \times \sqrt{4}) + (4 \div 4)$;
$6 = \sqrt{4} + (4 \times 4) \div 4$;
$7 = 4 + 4 - (4 \div 4)$;
$8 = 4 + (4 \times 4) \div 4$;
$9 = 4 + 4 + (4 \div 4)$;
$10 = 4 + 4 + 4 - \sqrt{4}$;
$11 = 44 \div (\sqrt{4} + \sqrt{4})$;
$12 = \sqrt{4} \times (4 + 4) - 4$;
$13 = (44 \div 4) + \sqrt{4}$;
$14 = 4 + 4 + 4 + \sqrt{4}$;
$15 = (4 \times 4) - (4 \div 4)$;
$16 = 4 + 4 + 4 + 4$

Written Exercises, page 391 **1.** 2.828 **3.** 6.325 **5.** 4.243 **7.** 4.472 **9.** 8.888

11. 9.434 **13.** 5.916 **15.** 8.660 **17.** −6.164 **19.** −7.071 **21.** 7.550 **23.** ±4 **25.** ±8.185 **27.** 3.7 **29.** 7.6 **31.** 7.1 **33.** 4.4 **35.** 6.1 **37.** −4.2 **39.** −8.4 **41.** 5.7 **43.** −8.2 **45.** 9.1 **47.** 7.5 cm **49.** 5.9 cm **51.** 20.4 cm **53.** 32.2 cm **55.** 31.0 cm **57.** 33.9 cm

Written Exercises, page 393 **1.** rational **3.** rational **5.** rational **7.** rational **9.** irrational **11.** irrational **13.** rational **15.** rational **17.** rational **19.** rational **21.** rational **23.** rational **25.** rational **27.** rational **29.** irrational **31.** 153.86 cm^2 **33.** 38.465 cm^2 **35.** perfect square **37.** $\frac{1}{2}$ **39.** $-\frac{11}{18}$ **41.** Answers may vary. $\sqrt{2} - \sqrt{2} = 0$, which is rational.

Written Exercises, page 395 **1.** 12 **3.** 15 **5.** $10\sqrt{10}$ **7.** $20\sqrt{5}$ **9.** $8\sqrt{2}$ **11.** $6\sqrt{5}$ **13.** $6\sqrt{3}$ **15.** $8\sqrt{3}$ **17.** $2y\sqrt{3}$ **19.** $xy\sqrt{15}$ **21.** $6y\sqrt{x}$ **23.** $ab\sqrt{3}$

Written Exercises, page 397 **1.** ±8 **3.** ±4 **5.** ±3 **7.** ±11 **9.** ±3 **11.** ±6 **13.** ±6 **15.** ±5 **17.** $\pm 2\sqrt{3}$ **19.** $\pm\sqrt{6}$ **21.** $\pm\sqrt{6}$ **23.** $\pm\sqrt{7}$ **25.** $\pm 3\sqrt{2}$ **27.** ±4 **29.** ±2 **31.** ±6 **33.** ±6.3 **35.** ±6.2 **37.** ±8.7 **39.** ±6.7

Written Exercises, pages 400–401 **1.** $2\sqrt{5}$ **3.** $2\sqrt{10}$ **5.** $2\sqrt{15}$ **7.** 5.9 m **9.** 1.4 km **11.** 48.5 cm **13.** $d = e\sqrt{3}$

Written Exercises, page 403 **1.** $\frac{2}{5}$ **3.** $\frac{11}{2}$ **5.** $\frac{1}{4}$ **7.** $\frac{3}{4}$ **9.** $\frac{3}{14}$ **11.** $\frac{1}{12}$ **13.** $\frac{x}{6}$ **15.** $\frac{2y}{3}$ **17.** $\frac{x\sqrt{3}}{2}$ **19.** $\frac{a\sqrt{7}}{8}$ **21.** 0.87 **23.** 1.15 **25.** 0.89 **27.** 1.53

Puzzle Problems, page 403 If you examine the figure closely, you'll see that the longer side of the large rectangle is equal to the length of the diagonal of the square. They are both $x\sqrt{2}$, because the ratio of the length to the width of the large rectangle must be $\sqrt{2}$.

Written Exercises, page 405 **1.** 3 **3.** 7 **5.** $3\sqrt{2}$ **7.** 4 **9.** $4\sqrt{3}$ **11.** 12 **13.** $15\sqrt{2}$ **15.** 6 **17.** 6 **19.** $5\sqrt{3}$ **21.** x **23.** $8a$ **25.** $8x$ **27.** $\sqrt{5}$ **29.** $\sqrt{7}$ **31.** 2 **33.** $10\sqrt{2}$ **35.** 8 **37.** a **39.** $\sqrt{2n}$ **41.** $\frac{3a}{b}$ **43.** $\frac{1}{6}$ **45.** 1 **47.** $\sqrt{5}$ **49.** $6x^2$ **51.** $-12x^2$ **53.** $\frac{x}{2}$ **55.** $3x\sqrt{2}$ **57.** $10a^4\sqrt{15}$ **59.** $-x^2$

Written Exercises, page 407 **1.** $\frac{\sqrt{2}}{2}$ **3.** $\frac{\sqrt{3}}{2}$ **5.** $\frac{\sqrt{3}}{3}$ **7.** $\frac{\sqrt{10}}{4}$ **9.** $\frac{\sqrt{70}}{10}$ **11.** $\frac{2\sqrt{7}}{7}$ **13.** $\frac{3x\sqrt{5}}{5}$ **15.** $\frac{a\sqrt{2}}{2}$ **17.** $\frac{-x^2\sqrt{y}}{y}$ **19.** 4 **21.** 3.54 **23.** 3.79 **25.** 0.33 **27.** 0.61

Written Exercises, page 409 **1.** $8\sqrt{2}$ **3.** $13\sqrt{5}$ **5.** $4\sqrt{5}$ **7.** $\sqrt{3}$ **9.** $4\sqrt{2}$ **11.** $3\sqrt{2}$ **13.** $-6\sqrt{3}$ **15.** $4\sqrt{5} - \sqrt{3}$ **17.** $10\sqrt{5} - 2\sqrt{2}$ **19.** $-2\sqrt{3} + \sqrt{6}$ **21.** $5\sqrt{5} - 4\sqrt{7}$ **23.** $(3 + 3b)\sqrt{b}$ **25.** 0 **27.** $-7\sqrt{6}$ **29.** $\frac{7\sqrt{5}}{5} + 5\sqrt{3}$

Written Exercises, page 411 **1.** −1 **3.** −13 **5.** −5 **7.** 2 **9.** 7 **11.** 44 **13.** $\frac{\sqrt{3} - 1}{2}$ **15.** $\frac{2\sqrt{2} - 6}{-7}$ **17.** $\frac{\sqrt{6} + 6}{-5}$ **19.** $\frac{3\sqrt{10} - 2\sqrt{15}}{3}$ **21.** $\frac{7 + \sqrt{7}}{3}$ **23.** $-12\sqrt{2} - 6\sqrt{10}$ **25.** $2 - \sqrt{3}$ **27.** $\frac{1 - 2\sqrt{7}}{-9}$ **29.** $7 - 4\sqrt{3}$ **31.** $6 - 2\sqrt{5}$ **33.** $54 + 14\sqrt{5}$ **35.** $26 - 8\sqrt{10}$ **37.** $57 - 12\sqrt{15}$ **39.** $92 - 24\sqrt{10}$ **41.** $61 + 28\sqrt{3}$

Calculator Activities, page 411 **1.** 10.954 **3.** −20 **5.** ±0.5 **7.** 92.326

Skills Review, page 412 **1.** $7x + 12y$ **3.** −11 **5.** $6ab^2 - 6a + 3$ **7.** $y^2 + 9y + 14$ **9.** $b^2 + 13b + 36$ **11.** $t^2 - 11t + 24$ **13.** $c^2 - c - 72$ **15.** $k^2 - 13k - 30$ **17.** $g^2 - 100z^2$ **19.** $6(t + 2)$ **21.** $3(x^2 + 5y)$ **23.** $(t + 4)(t + 5)$ **25.** $(z - 4)(z - 8)$ **27.** $(x - 2)(x - 10)$

29. $(y + 5z)^2$ **31.** $(a - 7)(a + 3)$ **33.** $(t + 12)(t - 5)$ **35.** $(1 + 6c)(1 - 6c)$ **37.** $2(x + 2)(x + 3)$ **39.** $5(t - 6)(t + 3)$ **41.** $2d(d + 5)(d - 5)$ **43.** $(4 + x)(4 - x)$ **45.** $-(m - 8)^2$

Chapter Review Exercises, pages 413–414 **1.** 5 **2.** −6 **3.** 9 **4.** −10 **5.** 3 **6.** 8 **7.** −7 **8.** 8 **9.** 12 **10.** −15 **11.** 4.4 **12.** 5.4 **13.** 7.7 **14.** 6.5 **15.** −2.6 **16.** rational **17.** rational **18.** irrational **19.** rational **20.** rational **21.** rational **22.** rational **23.** rational **24.** rational **25.** irrational **26.** $3\sqrt{2}$ **27.** $4\sqrt{2}$ **28.** $2\sqrt{7}$ **29.** $5\sqrt{6}$ **30.** $3\sqrt{5}$ **31.** $5\sqrt{7}$ **32.** $2\sqrt{35}$ **33.** $10\sqrt{2}$ **34.** ± 3 **35.** ± 9 **36.** $\pm 4\sqrt{3}$ **37.** ± 7 **38.** $\sqrt{74}$ **39.** $3\sqrt{5}$ **40.** $2\sqrt{33}$ **41.** $\frac{1}{6}$ **42.** $\frac{11}{8}$ **43.** $\frac{\sqrt{3}}{9}$ **44.** $\frac{x\sqrt{14}}{8}$ **45.** $\frac{xy}{3}$ **46.** 10 **47.** 18 **48.** $10\sqrt{5}$ **49.** $18n$ **50.** $\sqrt{5}$ **51.** $\sqrt{3}$ **52.** $\sqrt{2r}$ **53.** 2 **54.** $\frac{\sqrt{5}}{5}$ **55.** $\frac{\sqrt{14}}{4}$ **56.** $\frac{a\sqrt{5}}{5}$ **57.** $\frac{\sqrt{10x}}{5}$ **58.** $\frac{4\sqrt{2\pi}}{\pi}$ **59.** $\frac{n\sqrt{2}}{2}$ **60.** $2\sqrt{3}$ **61.** $3\sqrt{2} - 2\sqrt{3}$ **62.** $5\sqrt{5} + 5\sqrt{3}$ **63.** $9\sqrt{3} - 2$ **64.** $6\sqrt{6} + 2\sqrt{2} - 12\sqrt{5}$ **65.** $\sqrt{7} - 3\sqrt{21} + 7$ **66.** −4 **67.** 17 **68.** $28 - 16\sqrt{3}$

Mixed Review, pages 416–417 **1.** 4.5 cm, 7.2 cm, 7.2 cm **3.** $2x\sqrt{3}$ **5.** $-\frac{1}{x - 5}$ **7.** 2 **9.** 2 **11.** 0.583 **13.** $(y + 3)(y + 16)$ **15.** $2xy(x^2 + 2xy + 7y^2)$ **17.** $\frac{2r^3}{3}$ **19.** $3m^3 - 14m^2 + 3m + 20$ **21.** $x = 30$ **23.** 12.5 years **25.** 62.5% **27.** $3\frac{3}{5}$ hours **29.** $y = 7$ **31.** $4x^2 - 4x - 24$ **33.** $\frac{n - 8}{3n}$ **35.** (−1, −9), (2, 0) **37.** $2\frac{1}{2}$ hours **39.** $x = -5$

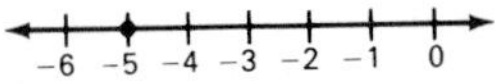

CHAPTER 12

Written Exercises, page 421 **1.** 0, −3 **3.** 0, −9 **5.** 0, 1 **7.** 0, 9 **9,** 3, 1 **11.** −4, −3 **13.** −3, −8 **15.** 14, 2 **17.** −5, −6 **19.** 8, −8 **21.** −12, 12 **23.** −100, 100 **25.** 12, −5 **27.** 3, −7 **29.** 6, −4 **31.** 0, 1, −2 **33.** 4, 6

Written Exercises, page 423 **1.** $x(x - 2)$ **3.** $2x(x - 2)$ **5.** $(x + 3)(x - 1)$ **7.** $(x + 2)(x + 4)$ **9.** $(x - 6)(x + 3)$ **11.** 0 or 3 **13.** 0 or −3 **15.** 0 or 4 **17.** 8 or 1 **19.** 7 or 1 **21.** 3 or 7 **23.** −7 or 3 **25.** 4 or −2 **27.** 7 or −2 **29.** −5 **31.** 7 or 4 **33.** −3 or 3 **35.** 11 or 2 **37.** 11 or −5 **39.** 5 or 8 **41.** 4 or −10 **43.** −1 or 7 **45.** 5 or −16

Written Exercises, page 425 **1.** 0 or 3 **3.** 0 or −4 **5.** −8 or 4 **7.** 4 or −2 **9.** 3 **11.** −10 or −2 **13.** −7 or 6 **15.** 9 or −2 **17.** 7 or 5 **19.** −6 or 2 **21.** 7 or −7 **23.** −7 or 4 **25.** −4 or 3 **27.** 4 or −3 **29.** −7 or 7 **31.** $-\frac{3}{2}$ or 1 **33.** $\frac{5}{2}$ or −2 **35.** $-\frac{9}{2}$ or 3

Written Exercises, page 427 **1.** ± 2 **3.** ± 4 **5.** ± 8 **7.** ± 11 **9.** ± 20 **11.** $\pm\sqrt{5}$ **13.** $\pm 2\sqrt{2}$ **15.** ± 1 **17.** $\pm 2\sqrt{3}$ **19.** ± 2 **21.** No solution **23.** ± 2 **25.** ± 7 **27.** $\pm 2\sqrt{6}$ **29.** No solution **31.** $\pm\sqrt{15}$ **33.** $\pm 2\sqrt{15}$ **35.** $\pm 3\sqrt{3}$ **37.** ± 3 **39.** $\pm\sqrt{10}$ **41.** $\pm\frac{1}{3}$ **43.** ± 6

45.

time	1	2	4	5	10
distance	5	20	80	125	500

Written Exercises, page 429 **1.** 9, −3 **3.** 9, −1 **5.** 2, −18 **7.** 11, −5 **9.** 18, −4 **11.** 2, −1 **13.** 5, −2 **15.** 1, −4 **17.** 3, −6 **19.** 2 **21.** 2, 0 **23.** $2 + \sqrt{7}, 2 - \sqrt{7}$ **25.** $-1 + \sqrt{2}, -1 - \sqrt{2}$ **27.** $5 + 2\sqrt{2}, 5 - 2\sqrt{2}$ **29.** 10, 4 **31.** 1, −5 **33.** −1, −5 **35.** 1, −3 **37.** 1, −5

Written Exercises, page 432 **1.** $-\frac{1}{2}, -1$ **3.** $\frac{1}{3}, 2$ **5.** $\frac{1}{4}, -2$ **7.** $-4, \frac{1}{2}$ **9.** $\frac{1}{3}, 2$ **11.** $-\frac{2}{3}, 1$ **13.** $1, -1$ **15.** $\frac{1}{2}, -\frac{1}{2}$ **17.** $\frac{-1 + \sqrt{17}}{4}, \frac{-1 - \sqrt{17}}{4}$ **19.** $-\frac{2}{5}, -1$ **21.** $1 + \sqrt{6}, 1 - \sqrt{6}$ **23.** $\frac{-1 + \sqrt{57}}{4}, \frac{-1 - \sqrt{57}}{4}$ **25.** $\frac{1 + \sqrt{41}}{4}, \frac{1 - \sqrt{41}}{4}$ **27.** $3 + \sqrt{6}, 3 - \sqrt{6}$ **29.** $\frac{7 + \sqrt{33}}{4}, \frac{7 - \sqrt{33}}{4}$ **31.** 0, 5 **33.** −1, 5 **35.** 2, 5 **37.** 2, 4 **39.** −3, 4 **41.** 2, −5 **43.** 1, −6 **45.** −2, 4 **47.** 5, −6

Mixed Practice Exercises, pages 432–433 **1.** 0, 2 **3.** 0, 4 **5.** 0, 2 **7.** −2, −1 **9.** −4, 2 **11.** 6, −5 **13.** ±3 **15.** $\pm\sqrt{35}$ **17.** $\pm\sqrt{19}$ **19.** $\pm 3\sqrt{2}$ **21.** 5, −3 **23.** $3 + \sqrt{17}, 3 - \sqrt{17}$ **25.** −1, 3 **27.** −4, 2 **29.** $-1, -\frac{1}{3}$ **31.** $-1, \frac{5}{3}$ **33.** 1, 2 **35.** $\frac{1 + \sqrt{33}}{4}, \frac{1 - \sqrt{33}}{4}$

Written Exercises, page 435 **1.** Length is 8 cm; width is 5 cm. **3.** −5 and −7 or 5 and 7 **5.** 2 **7.** 5 **9.** 12 **11.** Length is 12 cm; width is 4 cm. **13.** Al bikes 9 km/h; Vince bikes 12 km/h.

Written Exercises, page 438

1. a.

$y = x^2$	
x	y
−2	4
−1	1
0	0
1	1
2	4

b.

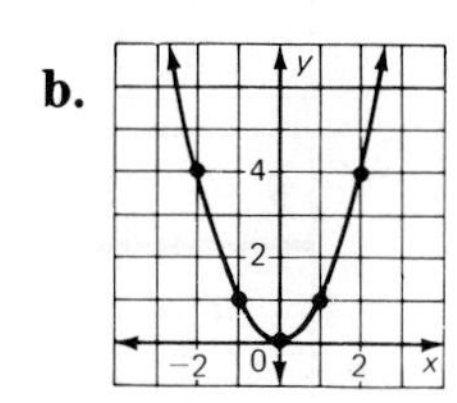

c. (0, 0)

3. a.

$y = x^2 - 4$	
x	y
−2	0
−1	−3
0	−4
1	−3
2	0

b.

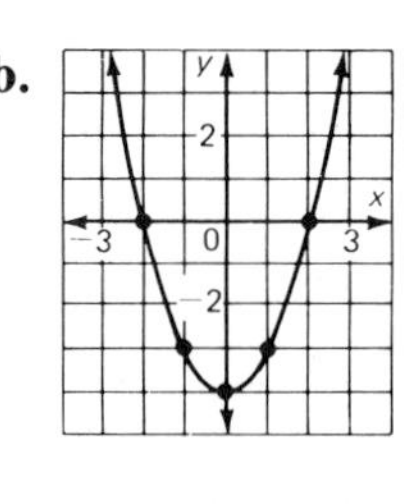

c. (0, −4)

5. a.

$y = -2x^2$	
x	y
−2	−8
−1	−2
0	0
1	−2
2	−8

b.

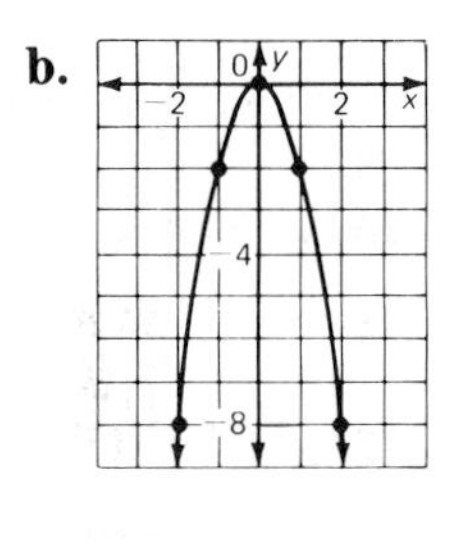

c. (0, 0)

7. a.

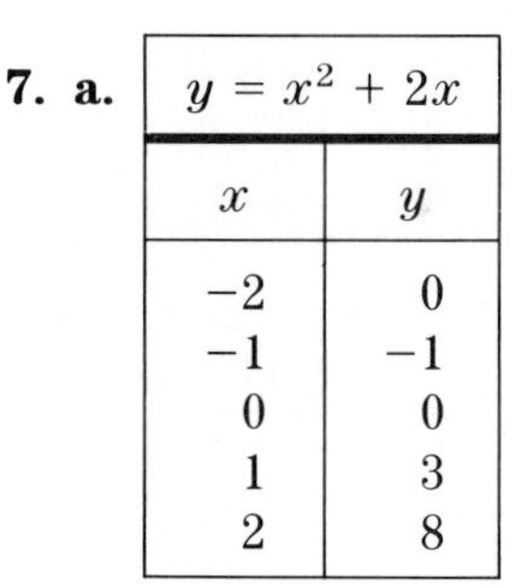

$y = x^2 + 2x$	
x	y
−2	0
−1	−1
0	0
1	3
2	8

b.

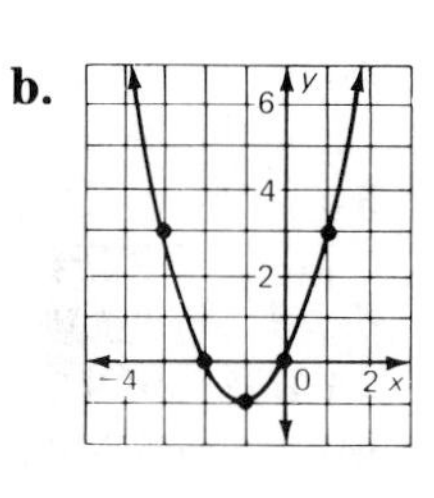

c. (−1, −1)

9. a.

$y = 2x - x^2$	
x	y
−1	−3
0	0
1	1
2	0
3	−3

b.

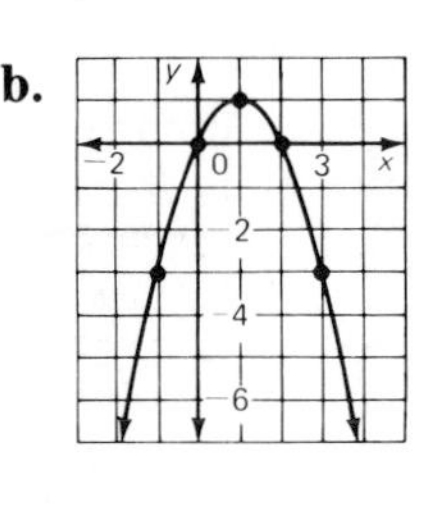

c. (1, 1)

11. a.

$y = x^2 - 4x + 3$	
x	y
0	3
1	0
2	-1
3	0
4	3

b.

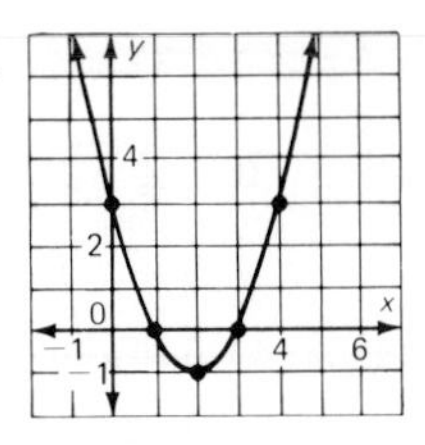

c. $(2, -1)$

13. a.

$y = x^2 - 2x - 2$	
x	y
-1	1
0	-2
1	-3
2	-2
3	1

b.

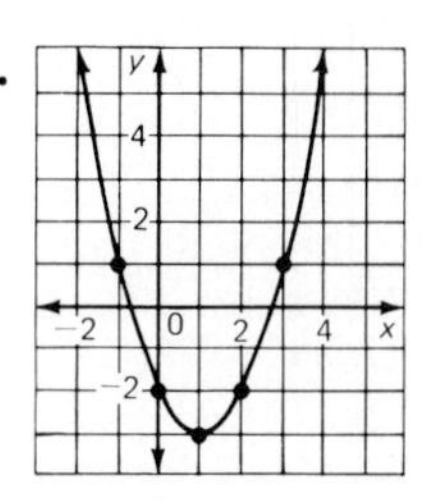

c. $(1, -3)$

15. a.

$y = x^2 - 6x + 9$	
x	y
1	4
2	1
3	0
4	1
5	4

b.

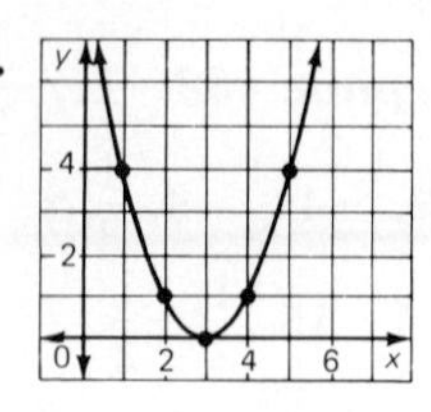

c. $(3, 0)$

17. a.

$y = (3 - x)(3 + x)$	
x	y
-3	0
-2	5
-1	8
0	9
1	8
2	5
3	0

b.

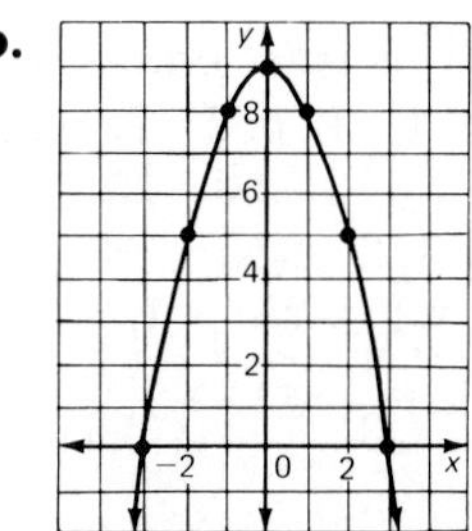

c. $(0, 9)$

19.

$y = x^2 + 2$	
x	y
-2	6
-1	3
0	2
1	3
2	6

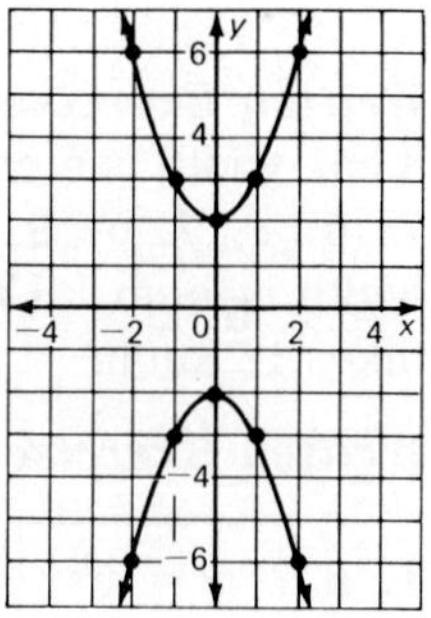

$y = -(x^2 + 2)$	
x	y
-2	-6
-1	-3
0	-2
1	-3
2	-6

21. 1 and 3 **23.** two

25.

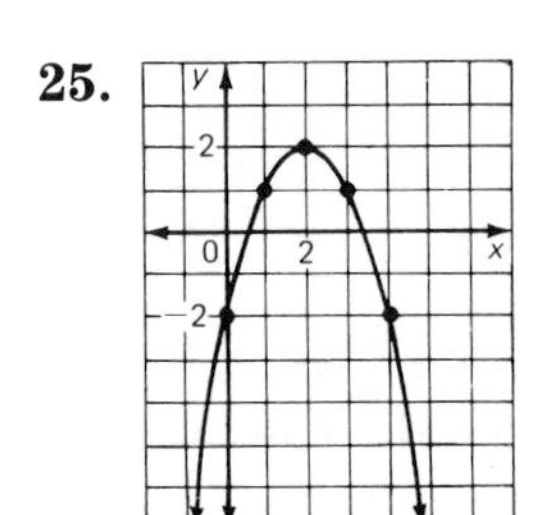

27. of $y = x^2 - 2x + 4$, (1, 3); of $y = x^2 - 2x + 2$, (1, 1); of $y = x^2 - 2x$, (1, −1)
29. 2 times **31.** 1 time

Problem Solving Strategies, page 441
1. No solution **3.** Not enough information
5. Not enough information **7.** 18 min
9. No solution

Computer Activities, page 443
1. a. 0.625 **b.** $0.\overline{5}$ **c.** $0.\overline{692307}$
d. $0.\overline{571428}$ **e.** 0.65 **f.** 0.64
g. $0.\overline{5294117647058823}$ **h.** $0.6\overline{1}$
i. $0.\overline{6956521739130434782608}$ **j.** 0.384
k. $0.11\overline{6}$ **l.** $0.\overline{358974}$
3. Answers may vary. Example:
$\frac{12}{29} = 0.\overline{4137931034482758620689655172}$
5. $\frac{13}{48}$

Skills Review, page 444 **1.** 30 cm
3. $26x$ **5.** 20 **7.** 25π **9.** 4π **11.** $28x^3$

Chapter Review Exercises, pages 445–446 **1.** 0, 2 **2.** 0, −4 **3.** 0, 7
4. 0, 9 **5.** 0, −6 **6.** 0, −8 **7.** −1, −5
8. 3, −6 **9.** −5, 3 **10.** 0, 3 **11.** 0, −4
12. 0, −5 **13.** 0, 2 **14.** 3, 5 **15.** 2, 4
16. −4, 5 **17.** 3, −6 **18.** 5, −7 **19.** 0, 2
20. 0, −3 **21.** 0, −1 **22.** 0, −4 **23.** −2, −5 **24.** −3 **25.** −5, −6 **26.** 2, 7
27. −6, 7 **28.** ±3 **29.** ±9 **30.** $\pm 2\sqrt{5}$
31. $\pm 3\sqrt{3}$ **32.** ±5 **33.** $\pm\sqrt{6}$ **34.** ±4
35. ±6 **36.** ±7 **37.** 5, −3 **38.** 6, −12
39. 4, −12 **40.** 8, −4 **41.** 2, −12 **42.** 17, −3 **43.** 0, −2 **44.** 5, 1 **45.** 5, 1 **46.** 2, −5 **47.** $\frac{1}{2}$, −3 **48.** $-\frac{2}{3}$, −1 **49.** 1, $-\frac{1}{2}$
50. $\frac{3}{2}$, −1 **51.** −1, $-\frac{3}{2}$ **52.** $\frac{-1 \pm \sqrt{11}}{5}$
53. $\frac{-7 \pm \sqrt{17}}{8}$ **54.** $\frac{1 \pm \sqrt{13}}{3}$ **55.** width, 4 cm; length, 9 cm **56.** 3 and 9, or −9 and −3

57.

$y = x^2 + 2$	
x	y
−3	11
−2	6
−1	3
0	2
1	3
2	6
3	11

If $y = 6$, $x = -2$ or $x = 2$.

58.

$y = x^2 + 3x - 5$	
x	y
−3	−5
−2	−7
−1	−7
0	−5
1	−1
2	5
3	13

If $y = -7$, $x = -2$ or $x = -1$.

59.

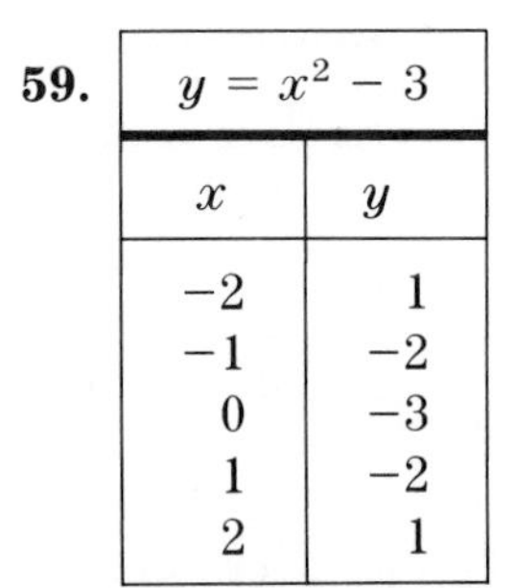

$y = x^2 - 3$	
x	y
−2	1
−1	−2
0	−3
1	−2
2	1

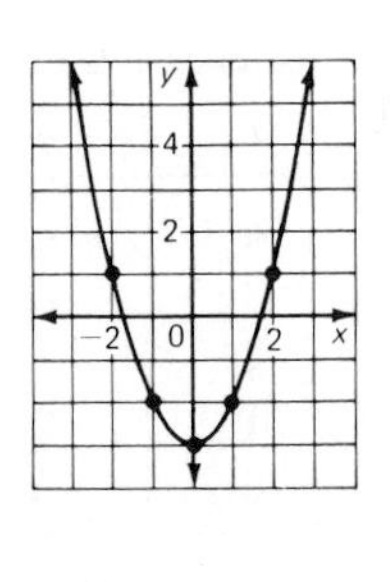

60.

$y = 2x^2 + 1$	
x	y
-3	19
-2	9
-1	3
0	1
1	3
2	9
3	19

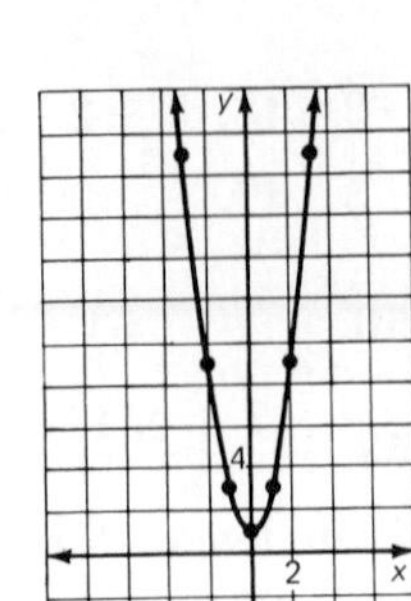

61.

$y = x^2 + 2x + 1$	
x	y
-3	4
-2	1
-1	0
0	1
1	4

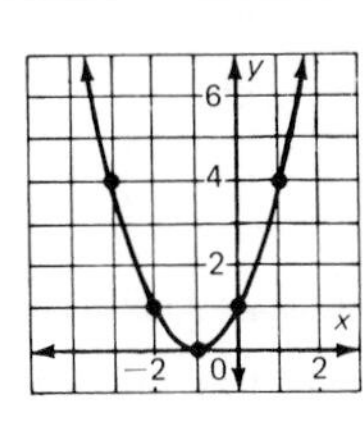

Cumulative Review, page 448 **1.** 20 **3.** 9 **5.** -23 **7.** -4.123 **9.** -4.690 **11.** $3\sqrt{6}$ **13.** $12\sqrt{3}$ **15.** $6x\sqrt{2y}$ **17.** 4, -4 **19.** 7, -7 **21.** $\frac{3}{4}$ **23.** $6a^2$ **25.** $\frac{\sqrt{6}}{4}$ **27.** $2\sqrt{3}$ **29.** 0, -4 **31.** 2, -4 **33.** 1, 7 **35.** 0, -5 **37.** 4 **39.** $\pm 2\sqrt{5}$ **41.** 5, -3 **43.** 12, -4 **45.** $\frac{1}{2}, -\frac{3}{2}$

47.

$y = x^2 - 4x + 5$	
x	y
4	5
3	2
2	1
1	2
0	5

(2, 1)

49.

$y = x^2 - 2x$	
x	y
3	3
2	0
1	-1
0	0
-1	3

(1, -1)

EXTRA PRACTICE: CHAPTER 1

Page 451
For pages 4–9 **1.** 42 **3.** 63 **5.** 10 **7.** 15 **9.** 10 **11.** 2 **13.** 27 **15.** 6 **17.** 15 **19.** 25 **21.** 24 **23.** 4
For pages 10–13 **1.** $2n$ **3.** $3n$ **5.** $2x$ **7.** $10m$ **9.** $9a$ **11.** $3a$ **13.** $10x$ **15.** $3a - b$ **17.** $5x - 5$ **19.** $11m + 1$ **21.** $11y + 1$ **23.** $16x + 2y$ **25.** $n + 2r$

Page 452
For pages 14–17 **1.** 9 **3.** 4 **5.** 4 **7.** 27 **9.** y^3 **11.** a^3 **13.** rs^2 **15.** xy^3 **17.** $3a^2$ **19.** $10n^2$ **21.** $8r$ **23.** $16n$ **25.** $8r^2$ **27.** $56y^2$ **29.** $8ab$ **31.** $10a^2$
For pages 18–19 **1.** $3x + 9$ **3.** $5a - 5$ **5.** $8n + 32$ **7.** $4x + 12$ **9.** $x^2 - 2x$ **11.** $n^2 + 2n$ **13.** $12n - 48$ **15.** $6a - 3$ **17.** $6a - 9b$ **19.** $6a^2 + ab$ **21.** $2n + 7$ **23.** $3n + 11$ **25.** 8 **27.** $a + 13$ **29.** $y + 14$ **31.** $10a + 14$

Page 453
For pages 20–23 **1.** 0 **3.** 0 **5.** 0 **7.** 6 **9.** 0 **11.** impossible **13.** 0 **15.** 3 **17.** 1 **19.** 1 **21.** $2x$ **23.** $6a$ **25.** n **27.** x
For pages 25–28 **1.** 1 **3.** 3 **5.** 2 **7.** 1, 2 **9.** 15, 16 **11.** 3, 4, 10

EXTRA PRACTICE: CHAPTER 2

Page 454
For pages 38–41 **1.** 6 **3.** 6 **5.** 11 **7.** 13 **9.** 4 **11.** 1 **13.** 15 **15.** 6 **17.** 25 **19.** 13 **21.** 8 **23.** 7 **25.** 6 **27.** 24 **29.** 7 **31.** 24
For pages 42–45 **1.** 6 **3.** 3 **5.** 5 **7.** 4 **9.** 12 **11.** 12 **13.** 6 **15.** 24 **17.** 9 **19.** 6 **21.** 35 **23.** 64 **25.** 10 **27.** 28

Page 455
For pages 46–49 **1.** 1 **3.** 6 **5.** 2 **7.** 10 **9.** 10 **11.** 7 **13.** 1 **15.** 12 **17.** 7 **19.** 9
For pages 52–53 **1.** 3 **3.** 2 **5.** 4 **7.** 5 **9.** 2 **11.** 3 **13.** 8 **15.** 8 **17.** 6
For pages 58–61 **1.** 8 **3.** \$10

Page 456

For pages 62–63 **1.** 10 **3.** 1 **5.** 10 **7.** 4 **9.** 2 **11.** 3 **13.** 1 **15.** 2
For pages 64–65 **1.** 12 **3.** 2 **5.** 2 **7.** 6 **9.** 1 **11.** 7 **13.** 9 **15.** 7
For pages 66–67 **1.** Ed, \$5; Bob, \$15; Al, \$45 **3.** 21 years old

EXTRA PRACTICE: CHAPTER 3

Page 457

For pages 78–82 **1.** > **3.** > **5.** > **7.** >

9. $n = 3$

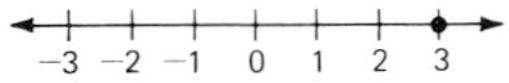

11. $x = 2$

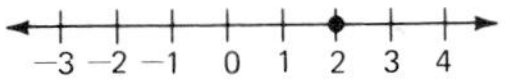

13.

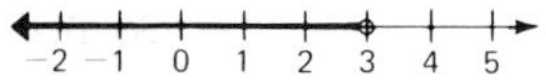

15.

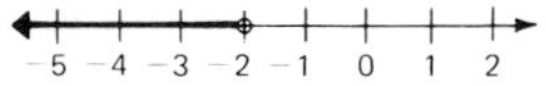

For pages 83–87 **1.** −4 **3.** 0 **5.** −5 **7.** −6 **9.** −2 **11.** −4 **13.** −10 **15.** 12 **17.** 13 **19.** 3 **21.** 9
For pages 89–90 **1.** $-3n$ **3.** $-10b$ **5.** $-6y$ **7.** $a + 1$ **9.** 3 **11.** $x + 2$ **13.** $3x^2 - y$ **15.** $5n - n^2$

Page 458

For pages 91–94 **1.** 48 **3.** 56 **5.** 25 **7.** −25 **9.** $-21n$ **11.** $4a - 2b$ **13.** $-m + 4$ **15.** −2
For pages 95–96 **1.** −9 **3.** 7 **5.** −7 **7.** −3 **9.** −8 **11.** −4
For pages 97–98 **1.** 3 **3.** −3 **5.** 0 **7.** 5 **9.** −4 **11.** −4 **13.** 1 **15.** −2 **17.** 6 **19.** −6 **21.** −4 **23.** 7 **25.** 40 **27.** −3

EXTRA PRACTICE: CHAPTER 4

Page 459

For pages 112–114 **1.** 28 cm **3.** 20 m **5.** $4a$ **7.** $20x$ **9.** $x = 4$
For pages 115–118 **1.** 28 **3.** $4x^2$ **5.** \$47.40

Page 460

For pages 119–122 **1.** $V = Bh$ **3.** 216 cm^3
For pages 124–133 **1.** hK **3.** 80 km
For pages 134–137 **1.** $n - y$ **3.** $4y - 12$ **5.** $\frac{y}{n}$ **7.** $15 - 10y$ **9.** $\frac{y}{3}$ **11.** $k + y$

EXTRA PRACTICE: CHAPTER 5

Page 461

For pages 152–155 **1.** $4n + 4$ **3.** $2a^2$ **5.** $5z$ **7.** $-3x - 2y$ **9.** $13a + 3b$ **11.** $3x^2$ **13.** a **15.** 8 **17.** $-2n^2$ **19.** $-4a - 8$ **21.** $x + 5$ **23.** $5a + 3b$ **25.** $(5n + 5)$ km

Pages 461–462

For pages 156–159 **1.** y^4 **3.** $2n^3$ **5.** $-x^2y$ **7.** n^7 **9.** n^{12} **11.** y^{10} **13.** a^2b^2 **15.** m^4n^4 **17.** $-x^3y^3$ **19.** $-27x^6$ **21.** $-a^3b^4$ **23.** $512r^3s^6$ **25.** $8x^8y^8$

Page 462

For pages 160–163 **1.** $3n + 15$ **3.** $6a^2 + 6b^2$ **5.** $ax - ay$ **7.** $-6a - 12b - 18c$ **9.** $4x^3y + 4x^2y^2 + 12xy^3$ **11.** $-3x^4 + 5x^3 - x^2 + x$ **13.** $20 - 9a + a^2$ **15.** $x^2 + 12x + 32$ **17.** $25x^2 - 30x + 8$ **19.** $a^3 + a^2 - a - 1$ **21.** $m^3 - mn^2 - m^2n + n^3$ **23.** $a^4 - 1$

Pages 462–463

For pages 164–169 **1.** $n^2 + 7n + 12$ **3.** $m^2 - 7m + 10$ **5.** $25y^2 - 1$ **7.** $8x^2 + 2xy - y^2$ **9.** $15x^2 - 8xy + y^2$ **11.** $n^2 + 4n + 4$ **13.** $n^2 + 14n + 49$ **15.** $x^2 + 6xy + 9y^2$ **17.** $a^2 - 20ab + 100b^2$ **19.** $-n^2 - 4n - 4$ **21.** $16r^2 + 16rs + 4s^2$ **23.** $16n^2 + 2n$

Page 463

For pages 170–171, 174–175 **1.** 3 **3.** x **5.** $3x^2$ **7.** $-9n^2$ **9.** $2x - 4y$ **11.** $xy - 3$ **13.** $4n^2 - 2n - 1$ **15.** $3a^2b^2 - 2ab + 1$ **17.** $-1 + 6a^2 - 8a^4$ **19.** y **21.** $2x^2y^2$ **23.** $8a^3b^2 + 6a^2b^3 - ab^4 + 3b^5$

EXTRA PRACTICE: CHAPTER 6

Page 464

For pages 186–189 **1.** $2 \cdot 2 \cdot 5$
3. $2 \cdot 5 \cdot 5 \cdot 5$ **5.** $2 \cdot 2 \cdot 5 \cdot 7$ **7.** $2 \cdot 3 \cdot 5 \cdot 5$
9. $2 \cdot 2 \cdot 3 \cdot 3 \cdot 5$ **11.** 5 **13.** 13 **15.** 15
17. 35 **19.** $2(7 - x)$ **21.** $12(1 - 3x)$
23. $y(y - 3)$ **25.** $y(10x - y)$
27. $6ab(b + 4a)$ **29.** $4(x^2 + 3xy + 6y^2)$
31. $5(1 - 5n + 15n^2 + 25n^3)$
33. $ab(a^3 - a^2b + ab^2 - b^3)$
35. $25xy(4x^3y^3 - 3x^2y^2 + 2xy + 1)$

Pages 464–465

For pages 192–197 **1.** $(y + 3)(y + 4)$
3. $(n + 6)(n + 10)$ **5.** $(x + 4)(x + 9)$
7. $(b + 9)(b + 10)$ **9.** $(n + 12)(n + 1)$
11. $(n + 17)(n + 2)$ **13.** $(x + 13)(x + 1)$
15. $(x + 14)(x + 2)$ **17.** $(m - 4)(m - 5)$
19. $(y - 9)(y - 5)$ **21.** $(y - 12)(y - 5)$
23. $(n - 10)(n - 8)$ **25.** $(x + 40)(x + 2)$
27. $(x - 100)(x - 1)$ **29.** $(m - 30)(m - 2)$
31. $(x + 16)(x + 4)$ **33.** $(n + 4)(n + 17)$

Page 465

For pages 198–203 **1.** $(y - 3)^2$
3. $(n - 7)^2$ **5.** $(n + 10)^2$
7. $(n + 4)(n + 1)$ **9.** $(y - 11)^2$
11. $(mn - 2)^2$ **13.** $(y - 8)(y + 1)$
15. $(y - 4)(y + 10)$ **17.** $(m - 10)(m + 5)$
19. $(n - 9)(n + 1)$ **21.** $(n - 4)(n + 1)$

Pages 465–466

For pages 204–207 **1.** $x^2 - 16$
3. $1 - 25x^2$ **5.** $4x^2 - y^2$ **7.** $25x^2 - 4y^2$
9. $16n^2 - 9$ **11.** $81 - 4b^2$
13. $(x - 5)(x + 5)$ **15.** $(x - 10)(x + 10)$
17. $9(x - 2y)(x + 2y)$
19. $(4a - 1)(4a + 1)$
21. $4(2s - t)(2s + t)$
23. $(11 - 3x)(11 + 3x)$ **25.** $n^4 - 1$
27. $y^4 - 64$ **29.** $(4y - 3)(4y + 3)$
31. $(a - 8)(a + 8)$
33. $(9a - 4b)(9a + 4b)$
35. $(n - 25)(n + 25)$
37. $(20 + r)(20 - r)$
39. $(mn + 5p)(mn - 5p)$

Page 466

For pages 208–209 **1.** $2(x - 5)(x + 5)$
3. $4(n + 4)(n + 1)$ **5.** $4(n + 4)(n - 1)$
7. $10(x - 2)(x + 2)$ **9.** $3(n - 4)(n + 1)$
11. $12(x + 1)(x + 2)$ **13.** $3(n + 2)^2$
15. $5(a - b)(a + b)$ **17.** $4(x - 3)^2$
19. $4(n - 6)(n + 4)$ **21.** $x(1 - 3x)$
23. $y^2(x^2 - y)$ **25.** $10(a + 2b)(a - 3b)$
27. $(3a - 2b)(3a + 2b)$
29. $4(r - 2s)(r - 4s)$
31. $5(x - 15)(x + 4)$
33. $(x^2 + y^2)(x - y)(x + y)$

EXTRA PRACTICE: CHAPTER 7

Page 467

For pages 224–227

1.

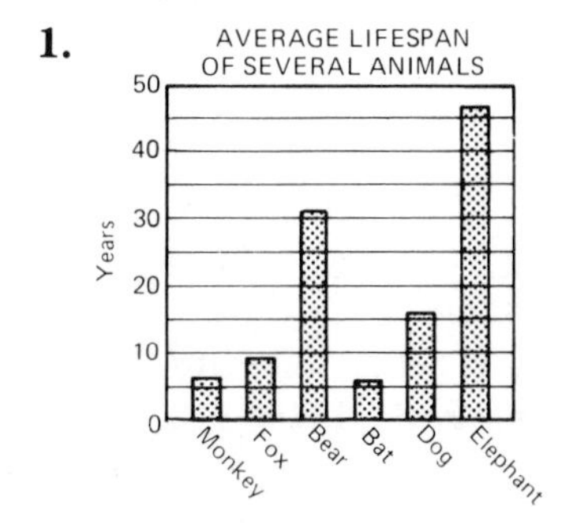

For pages 233–236

1–4.

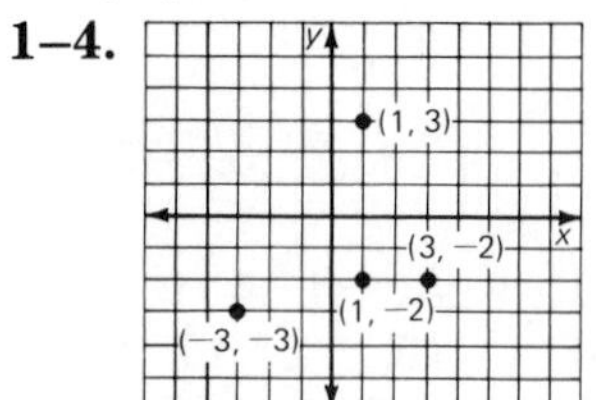

5.

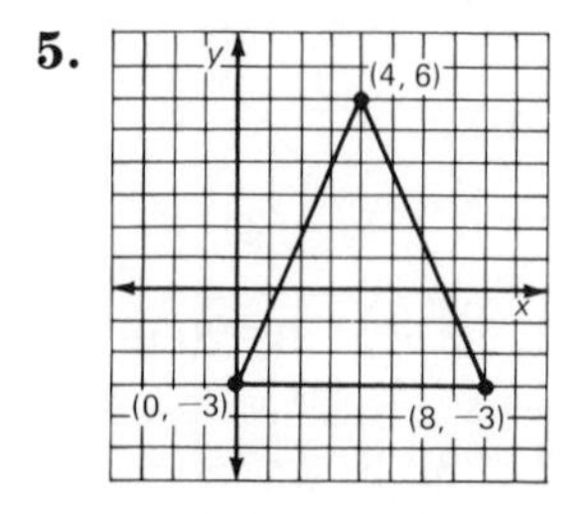

Pages 467–468

For pages 237–240 **1.** 4, 5, 3, 6 **3.** 2, 5, 11, 14 **5.** $7 - x$ **7.** $x + 3$ **9.** $x - 5$ **11.** $-2 - 10x$
13–19. Answers may vary.
13. $(-1, -4), (0, -3), (1, -2)$ **15.** $(-1, 3), (0, 1), (1, -1)$ **17.** $(-1, 2), (0, 0), (1, -2)$ **19.** $(-1, -2), (0, 0), (1, 2)$ **21.** $y = x - 5$ **23.** $y = 3x - 1$

Page 468

For pages 241–244

1.

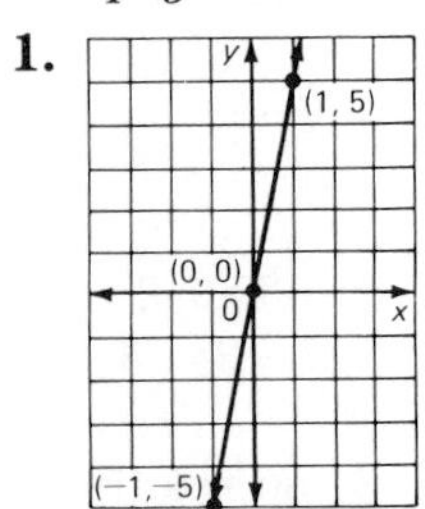

3.

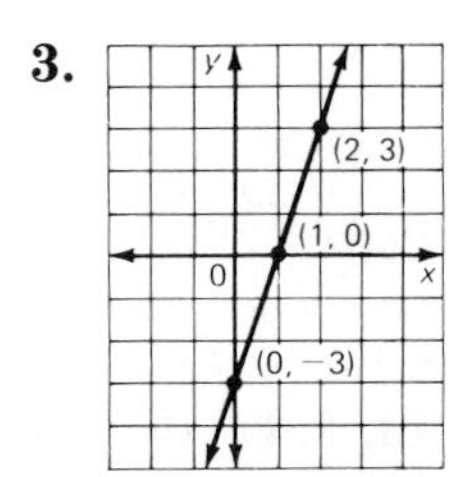

5.

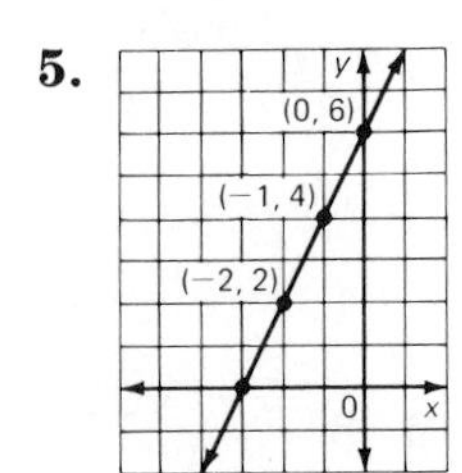

7.

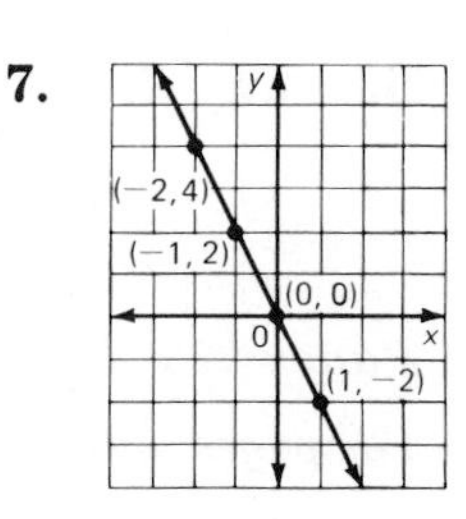

9.

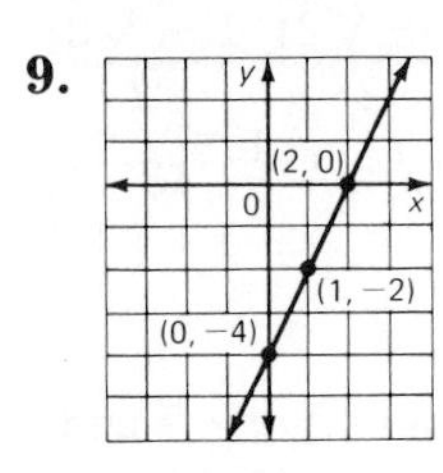

11. -3 **13.** $\frac{1}{2}$

For pages 245–248 **1.** $y = 3x$ **3.** 90; 120

EXTRA PRACTICE: CHAPTER 8

Page 469

For pages 262–265 **1.** $(-1, 1)$ **3.** $(2, 1)$
5. about $\left(-2\frac{1}{2}, -5\frac{1}{2}\right)$
7. about $\left(\frac{3}{4}, -1\frac{3}{4}\right)$

For pages 266–269 **1.** 3 **3.** -3 **5.** no **7.** all

Pages 469–470

For pages 270–275 **1.** $(3, 9)$ **3.** $(0, 2)$ **5.** $(5, -1)$ **7.** $(-2, -3)$ **9.** $(-9, -34)$
11. $(1, 1)$ **13.** $(45, 9)$ **15.** $\left(\frac{3}{4}, \frac{-5}{4}\right)$

Page 470

For pages 276–283 **1.** $(-7, 6)$ **3.** $(-2, -1)$ **5.** $(2, 0)$ **7.** $(1, -1)$ **9.** 5, 10 **11.** \$12 **13.** Al, 11; Maya, 5 **15.** $(18, -12)$ **17.** $(0, 0)$ **19.** $(-8, -5)$ **21.** 24 pencils, 1 pen

EXTRA PRACTICE: CHAPTER 9

Page 471

For pages 298–303 **1.** $\frac{1}{4}$ **3.** $\frac{1}{4}$ **5.** $\frac{1}{3}$
7. $-\frac{1}{7}$ **9.** $-\frac{1}{4xy}$ **11.** $-\frac{y^2}{3x}$ **13.** $-\frac{8}{ns}$
15. $2xy$ **17.** $\frac{3}{10}$ **19.** $\frac{1}{3}$ **21.** $-1(-x)$
23. $-1(n^2 - z^2)$ **25.** -1 **27.** $-x - y$
29. $-\frac{1}{3 + x}$ **31.** $-\frac{1}{3n + 3}$ **33.** $\frac{x - 3}{x + 2}$

Pages 471–472

For pages 304–313 **1.** $\frac{3}{2}$ **3.** $\frac{1}{4}$
5. 12 boys, 9 girls **7.** 12 **9.** 24 **11.** 3
13. -5 **15.** 6 km **17.** $\frac{1}{5}$ **19.** $\frac{1}{12}$

Page 472

For pages 314–324 **1.** $\frac{1}{16}$ **3.** $\frac{2a^2}{b^2}$ **5.** 6
7. $\frac{x - 3}{3}$ **9.** $-\frac{4n}{3}$ **11.** $\frac{3n - 1}{n - 1}$
13. $\frac{5}{4}$, or $1\frac{1}{4}$ **15.** $\frac{3x}{10}$ **17.** $\frac{y + 3x}{xy}$
19. $\frac{9a - b}{a^2b^2}$ **21.** $\frac{x - y}{x + 8}$ **23.** $3a^2 - 9a$

Page 473

For pages 327–333 **1.** -100 **3.** 2 **5.** 2 **7.** 6 **9.** -12 **11.** $2\frac{2}{5}$ hours **13.** -30 **15.** $7\frac{1}{2}$ hours

EXTRA PRACTICE: CHAPTER 10

Page 474

For pages 348–353 **1.** 0.61 **3.** 5.30 **5.** 0.505 **7.** 1.64 **9.** 47.5 **11.** 62.5 **13.** 3.5577 **15.** 4.275 **17.** 1.2 **19.** 6.9 **21.** 0.5 **23.** 52.8 **25.** 0.4 **27.** 0.142857142857 . . . **29.** 0.714285714285 . . . **31.** 0.0833 . . . **33.** 0.66 . . .
For pages 354–357 **1.** 10 **3.** 6000 **5.** -355.4 **7.** 20 and 30 **9.** 2000

Page 475

For pages 359–364 **1.** 0.05; $\frac{1}{20}$ **3.** 80%; 0.80 **5.** 0.09; $\frac{9}{100}$ **7.** 6%; $\frac{3}{50}$ **9.** \$.63 **11.** about \$3.07 **13.** \$3.10 **15.** \$18 **17.** \$240

Pages 475–476

For pages 365–368 **1.** 20% **3.** 350 **5.** 60 **7.** \$.98 **9.** 2.5%

Page 476

For pages 369–372 **1.** \$390 **3.** \$6000 **5.** \$150 at 9% and \$50 at 18%
For pages 373–376 **1.** 500 student tickets

EXTRA PRACTICE: CHAPTER 11

Page 477

For pages 388–393 **1.** 7 **3.** 5 **5.** 1 **7.** 0 **9.** 6 **11.** -3.464 **13.** -1.732 **15.** -8.718 **17.** 9.9 **19.** 7.9 **21.** rational **23.** irrational **25.** irrational
For pages 394–395 **1.** $10\sqrt{5}$ **3.** 80 **5.** $7\sqrt{3}$ **7.** 27 **9.** $9x$ **11.** $x^2y\sqrt{5}$
For pages 396–397 **1.** ± 7 **3.** ± 6 **5.** $\pm 2\sqrt{7}$ **7.** ± 4.2 **9.** ± 4.5

Page 478

For pages 398–401 **1.** 7.2 cm **3.** 7.5 cm
For pages 402–407 **1.** $\frac{1}{3}$ **3.** $\frac{3}{10}$ **5.** $\frac{x}{5}$ **7.** 6 **9.** 12 **11.** $90x$ **13.** 15 **15.** $2x^2$ **17.** 0.79 **19.** 0.26
For pages 408–409 **1.** $-6\sqrt{3}$ **3.** $3\sqrt{2}$ **5.** $17\sqrt{2}$ **7.** $-2\sqrt{3n}$ **9.** $-\sqrt{2}$

EXTRA PRACTICE: CHAPTER 12

Page 479

For pages 420–425 **1.** 0, -5 **3.** 0, 2 **5.** 5, 2 **7.** 0, -1 **9.** 0, 9 **11.** -3, -8 **13.** 2, 1 **15.** -5, 2 **17.** 6, 1
For pages 426–429 **1.** 0 **3.** no solution **5.** -4, 8 **7.** $4 + \sqrt{2}$, $4 - \sqrt{2}$ **9.** $\frac{2\sqrt{2} - 1}{3}$, $\frac{-2\sqrt{2} - 1}{3}$ **11.** $\pm 2\sqrt{2}$

Pages 479–480

For pages 430–433 **1.** 3, 2 **3.** $\frac{5 + \sqrt{21}}{2}$, $\frac{5 - \sqrt{21}}{2}$ **5.** 4, -5 **7.** $\frac{1 + \sqrt{21}}{2}$, $\frac{1 - \sqrt{21}}{2}$ **9.** $\frac{1 + \sqrt{33}}{4}$, $\frac{1 - \sqrt{33}}{4}$ **11.** 1, $-\frac{1}{4}$ **13.** -2, 7 **15.** 4 **17.** 1, 2

Page 480

For pages 434–435 **1.** 7 and 2 or -7 and -2 **3.** 14 and 15 or -14 and -15 **5.** 2, 3, and 4 or -2, -3, and -4
For pages 436–439

1. a.

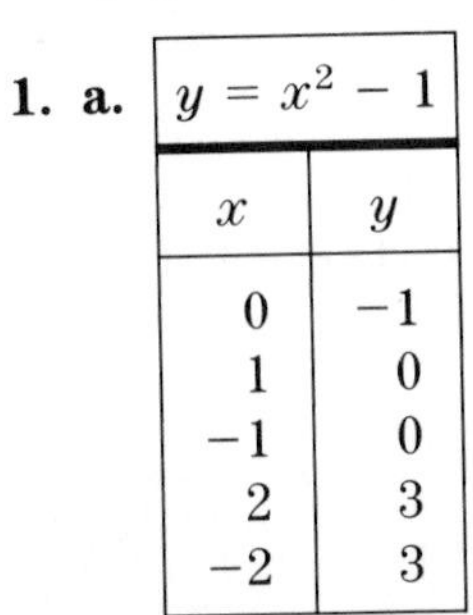

$y = x^2 - 1$	
x	y
0	-1
1	0
-1	0
2	3
-2	3

b.

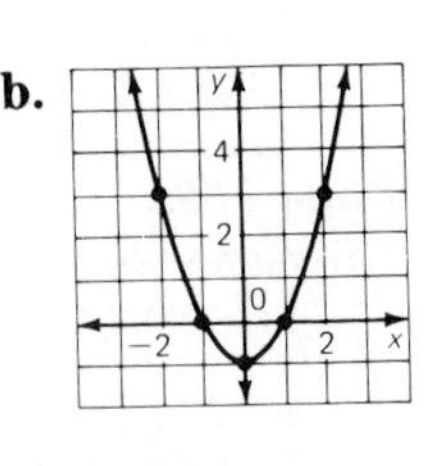

c. (0, -1)

3. a.

$y = -x^2$	
x	y
-2	-4
-1	-1
0	0
1	-1
2	-4

b.

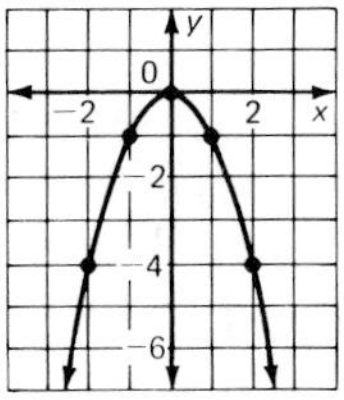

c. (0, 0)

5. a.

$y = x^2 - 2x + 1$	
x	y
1	0
0	1
-1	4
2	1
3	4

b.

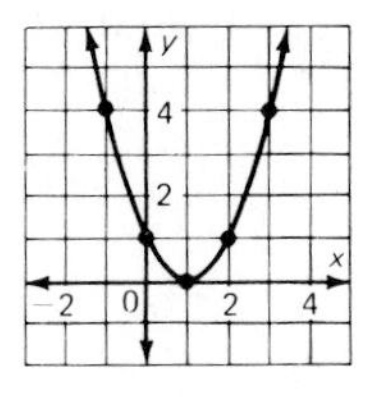

c. (1, 0)

7. a.

$y = x^2 + 4x + 2$	
x	y
-2	-2
-1	-1
0	2
-4	2
-3	-1

b.

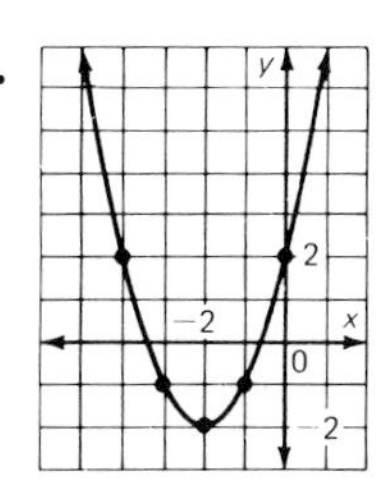

c. $(-2, -2)$

9. a.

$y = x^2 + 4x - 3$	
x	y
-2	-7
-1	-6
0	-3
-4	-3
-3	-6

b.

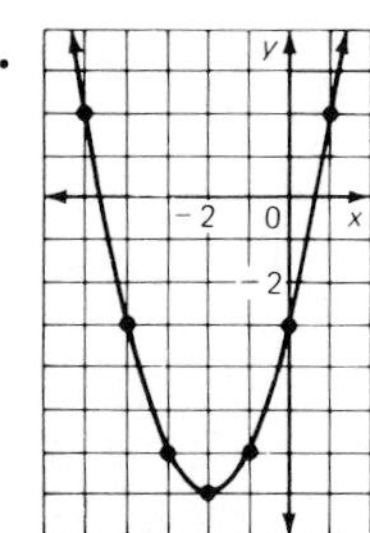

c. $(-2, -7)$

CUMULATIVE REVIEW: CHAPTERS 1–3

Basic Skills, page 481 **1.** 64 **3.** 12 **5.** 5 **7.** 0 **9.** 1 **11.** 5 **13.** -5 **15.** 3 **17.** -16 **19.** 3 **21.** -2 **23.** 8 **25.** -4 **27.** -6 **29.** $9b$ **31.** $9x - 2$ **33.** 4 **35.** $3a^2b^2$ **37.** x **39.** $-a + 2$ **41.** $3x + 3$

Equations and Inequalities, pages 481–482 **1.** no **3. a.** 4 **b.** yes

c.

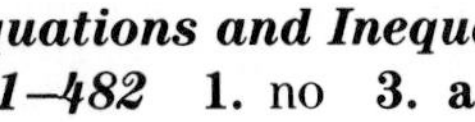

5. -1 **7.** 6 **9.** -6 **11.** 7 **13.** 9 **15.** -8 **17.** 8 **19.** 1 **21.** 9 **23.** $<$ **25.** 0; $\frac{6}{0}$ has no value.

Problems, page 482 **1.** 38 **3.** Mary has 70, Don has 35. **5.** 10

CUMULATIVE REVIEW: CHAPTERS 1–6

Basic Skills, page 483 **1.** 0 **3.** $-10a$ **5.** $9x^2$ **7.** $4x$ **9.** $4a^4$ **11.** $-27a^3b^3$ **13.** $x^2 + 5x - 6$ **15.** $y^2 - y - 6$ **17.** $x^2 + 8x + 16$ **19.** $9x^2 - 12xy + 4y^2$ **21.** $4n^2$ **23.** $-9x$ **25.** $5a - a^2 - a^3$ **27.** $3xy(1 - 2xy)$ **29.** $(y + 1)(y + 6)$ **31.** $(n + 11)(n - 2)$

Equations and Formulas, pages 483–484 **1.** 25 **3.** 3 **5.** 8 **7.** 4 **9.** 4 **11.** 1 **13.** $\frac{y}{6}$ **15.** 3 **17.** $4x$ **19.** $\frac{D}{t}$

Problems, page 484 **1.** 16 cm **3.** 16 cm^2 **5.** 12, 13

CUMULATIVE REVIEW: CHAPTERS 1–9

Basic Skills, page 485 **1.** impossible **3.** impossible **5.** 0 **7.** $-9x^2y^2$ **9.** $4a^2b^4$ **11.** $270y^6$ **13.** $-x^2$ **15.** $x^2 + 6x + 8$ **17.** $8x^2 + 2x - 3$ **19.** $x^3 + x^2 - 6x$ **21.** $2(x^2 - 2y^2)$ **23.** $(n - 1)(n - 2)$ **25.** $(x + 4)(x + 4)$ **27.** $(2y + 3)(2y - 3)$ **29.** $(x + 3)(x - 2)$

Equations and Graphs, pages 485–486 **1.** 3 **3.** -2 **5.** 4 **7.** $x - 3$ **9.** $5x - 10$ **11.** $8 + 2x$

13.

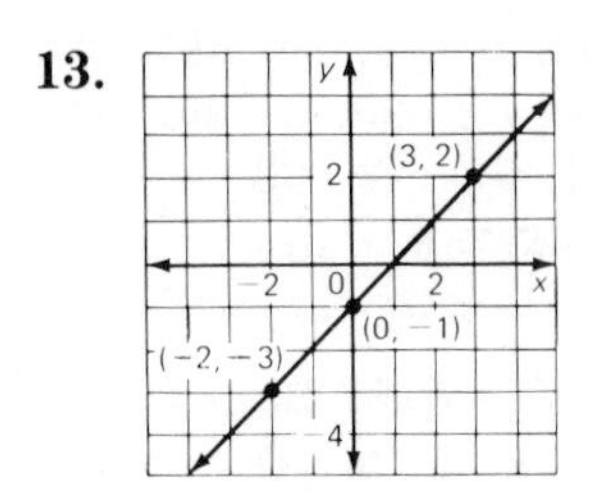

15.

17. $\frac{1}{2}$ **19.** (3, 3) **21.** (−2, 1) **23.** (6, 3) **25.** (2, 9) **27.** (1, 1)

Problems, page 486 **1.** $1\frac{1}{3}$ hours

CUMULATIVE REVIEW: CHAPTERS 1–12

Basic Skills, page 487 **1.** $-2x - 10$ **3.** $5a + 1$ **5.** $-x$ **7.** $b - 2$ **9.** $24x - 24$ **11.** $n^2 - 16$ **13.** $10n^2 + 21n - 10$ **15.** $-2x^2 + 7x + 49$ **17.** $3(n + 1)(n + 2)$ **19.** $(x - 1)(x - 6)$ **21.** $(x - 7)(x + 2)$ **23.** $\frac{2y}{3}$ **25.** $\frac{1}{x}$ **27.** $6x$ **29.** $\frac{\sqrt{10}}{5}$ **31.** $-\sqrt{3}$ **33.** $\sqrt{2}$

Equations and Inequalities, pages 487–488 **1.** 8 **3.** 8 **5.** 4 or −4 **7.** −8 **9.** $\pm 5\sqrt{2}$ **11.** 12 **13.** 50 **15.** 0 or 4 **17. a.** 8 **b.** $\frac{24}{a}$ **c.** $\frac{c}{a}$ **19.** $x = 1$; $y = 1$ **21.** 1 or 6 **23.** 4 or −4 **25.** $\frac{1 + \sqrt{10}}{3}$ or $\frac{1 - \sqrt{10}}{3}$

Problems, page 488 **1.** 40 kg of the 65¢ kind, 60 kg of the other

CUMULATIVE REVIEW: WORD PROBLEMS

Pages 489–491 **1.** 10 **3.** 16 years old **5.** 24 erasers **7.** 12 **9.** 7 L **11.** 12 km **13.** $1\frac{1}{5}$ hours **15.** 66 **17.** 240 **19.** 31, 42 **21.** width, 6 cm; length, 17 cm **23.** shirt, \$9; pants, \$26 **25.** width, 10 cm; length, 15 cm **27.** Brazil nuts, 20 kg; cashews, 10 kg **29.** $7\frac{1}{2}$ hours **31.** \$40.32 **33.** length, 16 cm; width, 4 cm **35.** 12, 14